$\|\|\mathbf{v}\|\|$	the length of the vector **v**, pp. 183, 210
$\|\|\mathbf{u} - \mathbf{v}\|\|$	the distance between vectors **u** and **v**, pp. 184, 213
$\cos \theta$	the cosine of the angle between two nonzero vectors **u** and **v**, pp. 185, 211
$\mathbf{u} \cdot \mathbf{v}$	the standard inner product on R^2 or R^3, p. 186
$(\mathbf{u}, \mathbf{v})$	an inner product, pp. 186, 203, 204
$\mathbf{u} \times \mathbf{v}$	the cross product operation, p. 193
$\text{proj}_W \mathbf{u}$	the orthogonal projection of the vector **u** on the subspace W, p. 226
L, L_1, L_2	linear transformations, p. 245
$p'(t)$	the derivative of $p(t)$ with respect to t, p. 253 (Exercise 3)
$\ker L$	the kernel of the linear transformation L, p. 256
range L	the range of the linear transformation L, p. 258
$j_1 j_2 \cdots j_n$	a permutation of $S = \{1, 2, \ldots, n\}$, p. 301
$\det(A)$, $\|A\|$	the determinant of the matrix A, p. 302
$\|M_{ij}\|$	the minor of a_{ij}, p. 317
A_{ij}	the cofactor of a_{ij}, p. 317
adj A	the adjoint of the matrix A, p. 323
D	a diagonal matrix, p. 338
λ_j	an eigenvalue of the linear transformation L or matrix A, pp. 339, 344, 451
$\mathbf{x}_j$	an eigenvector of L or A associated with the eigenvalue λ_j, pp. 339, 344, 451
$\|\lambda I_n - A\| = f(\lambda)$	the characteristic polynomial of A, pp. 349, 452
$\|\lambda I_n - A\| = f(\lambda) = 0$	the characteristic equation of A, p. 349
$g(\mathbf{x}) = \mathbf{x}^T A \mathbf{x}$	a real quadratic form in n variables, p. 375
$\mathbf{x}(t) = \begin{bmatrix} x_1(t) \\ x_2(t) \\ \vdots \\ x_n(t) \end{bmatrix}$	an $n \times 1$ matrix whose entries are functions of t, p. 410
$\mathbf{x}'(t) = \begin{bmatrix} x_1'(t) \\ x_2'(t) \\ \vdots \\ x_n'(t) \end{bmatrix}$	p. 410
$A(t)$	a matrix function, p. 427 (Exercise 1)
e^{At}	the matrix exponential function, p. 428
$c = a + bi$	a complex number, p. 435
$\overline{c} = a - bi$	the conjugate of $c = a + bi$, p. 436
$\|c\| = \|a + bi\| = \sqrt{a^2 + b^2}$	the absolute value or modulus of the complex number $c = a + bi$, p. 439
$\overline{A} = [\overline{a}_{ij}]$	the conjugate of the matrix $A = [a_{ij}]$, p. 441

Elementary Linear Algebra

BERNARD KOLMAN
Department of Mathematics, Drexel University

Elementary Linear Algebra

FIFTH EDITION

Macmillan Publishing Company
NEW YORK
Collier Macmillan, Inc.
TORONTO
Maxwell Macmillan International
NEW YORK OXFORD SINGAPORE SYDNEY

Editor: Robert W. Pirtle
Production Supervisor: Elaine W. Wetterau
Production Manager: Pamela Kennedy
Text Designer: Robert Freese
Cover Designer: Robert Freese
Cover illustration: *Cint Choe,* by Vasarely, copyright © 1991 ARS N.Y./SPADEM

This book was set in Elan Book, Times Roman, and Helvetica types by York Graphic Services, Inc.; printed and bound by R. R. Donnelley & Sons Company.
The cover was printed by Lehigh Press.

Macmillan Publishing Company
866 Third Avenue, New York, New York 10022

Collier Macmillan Canada, Inc.
1200 Eglinton Avenue, E.
Suite 200
Don Mills, Ontario, M3C 3N1

LIBRARY OF CONGRESS CATALOGING-IN-PUBLICATION DATA
Kolman, Bernard
Elementary linear algebra/Bernard Kolman.—5th ed.
p. cm.
Includes index.
ISBN 0-02-366045-7
1. Algebras, Linear. I. Title.
QA184.K668 1991
512′.5—dc20 90-6042
CIP

Printing: 1 2 3 4 5 6 7 8 Year: 1 2 3 4 5 6 7 8 9 0

To Lisa, Stephen
and to the memory of Lillie

Preface

Linear algebra has now become a standard part of the undergraduate mathematical training of a diverse number of students for at least two reasons. First, few subjects can claim to have such widespread applications in other areas of mathematics—multivariable calculus, differential equations, and probability theory, for example—as well as in physics, biology, chemistry, economics, psychology, sociology, and all fields of engineering. Second, the subject provides the student with his or her first introduction to postulational or axiomatic mathematics. Engineering, the sciences, and the social sciences today are becoming more analytically oriented; that is, more mathematical in flavor, and the mere ability to manipulate matrices is no longer adequate. Linear algebra affords, at the sophomore level, an excellent opportunity to develop a capability for handling abstract concepts.

Both the author and the publisher have been very pleased by the widespread acceptance of the first four editions of this book. Although many changes have been made in this edition, my objective has remained the same as in the first four editions: *to present the basic ideas of linear algebra in a manner that the student will find understandable*. To achieve this objective, author and publisher have made extensive use of faculty and student suggestions to develop the following helpful features.

Features

Presentation

I have learned from experience that at the sophomore level, abstract ideas must be introduced quite gradually and must be based on some firm foundations. Thus we begin the study of linear algebra with the treatment of matrices as mere arrays of numbers that arise naturally in the solution of systems of linear equations, a problem already familiar to the student. Considerable attention has been devoted from one edition to the next to refining and improving the pedagogical aspects of the exposition. The abstract ideas are carefully balanced by the considerable emphasis on the geometrical and computational material of the subject.

When teaching linear algebra. I have often found that eigenvalues and eigenvectors had to be covered much too hastily, because of time limitations, to do this material justice. This book has been written so that this very important topic can be comfortably covered in a one-quarter or one-semester course.

In using the first four editions of this book, for a one-quarter linear algebra course meeting four times a week, no difficulty has been encountered in covering the first six chapters, except for the optional material and Sections 6.3, 6.4, and 6.5. A suggested pace for covering the basic material, based on 21 years of experience with the first four editions is given on page x.

Material Covered

Chapter 1 deals with matrices and their properties. Methods for solving systems of linear equations are covered in this chapter. Section 1.8, new to this edition, provides an introduction to LU-factorization. In **Chapter 2,** we come to a more abstract notion, that of a vector space. Starting first with vectors in the plane and in 3-space, we then restrict our attention to finite-dimensional, real vector spaces. Here we tap some of the many geometric ideas that arise naturally. Thus we prove that an n-dimensional, real vector space is isomorphic to R^n, the vector space of all ordered n-tuples of real numbers, or the vector space of all $n \times 1$ matrices with real entries. Since R^n is but a slight generalization of R^2 and R^3, two- and three-dimensional space discussed at the beginning of the chapter, this shows that the notion of a finite-dimensional real vector space is not as remote as it may have seemed when first introduced. **Chapter 3** covers inner product spaces and has a strong geometric orientation. Section 3.2, new to this edition, presents cross products in R^3 and some of its applications. Section 3.5, also new to this edition, presents an application of this material to least squares. **Chapter 4** deals with matrices and linear transformations; in it we consider the dimension theorems and also applications to the solution of systems of linear equations. **Chapter 5** introduces the basic properties of determinants and some of their applications. **Chapter 6** considers eigenvalues and eigenvectors, and real quadratic forms. In this chapter we completely solve the diagonalization problem for symmetric matrices. Section 6.4, Conic Sections, and Section 6.5, Quadric Surfaces, are new to this edition. **Chapter 7** provides an introduction to the important application of linear algebra in the solution of differential equations. It is possible to go from Section 6.1 directly to Chapter 7, providing an immediate application of the material in Section 6.1. Moreover, Sections 6.3, 6.4, and 6.5 provide an application of the material in Section 6.2. **Appendix A** reviews some very basic material dealing with sets and functions. It can be consulted at any time, as needed. **Appendix B,** on complex numbers, introduces in a brief but thorough manner complex numbers and their use in linear algebra.

Exercises

The exercises form an integral part of the text. Many of them are numerical in nature, whereas others are of a theoretical type. This edition contains 314 new exercises as well as 58 exercises, indicated by the symbol **C**, requiring the use of a computer, allowing for discovery and exploration. The answers to all odd-numbered numerical exercises appear in the back of the book. An **Answer Manual,** containing answers to all even-numbered exercises and solutions to all theoretical exercises, is available (to instructors only) at no cost from the publisher.

Student Solutions Manual and Study Guide

The Student Solutions Manual and Study Guide, prepared by David R. Hill, Temple University, contains solutions to all odd-numbered exercises, both numerical and theoretical. It also includes an introductory chapter on logic and methods of proof in mathematics.

Suggested Pace for Basic Material

Chapters 1 to 6, omitting all optional material and Sections 6.3, 6.4, and 6.5.

Chapter 1	8 lectures	
Chapter 2	9 lectures	
Chapter 3	4 lectures	(For most students Section 3.1 is a review of known material.)
Chapter 4	7 lectures	
Chapter 5	5 lectures	
Chapter 6	5 lectures	
	38 lectures	

Using Computers in Linear Algebra

This book does not require the use of a computer. However, if the instructor or student wishes to support the instruction or learning of linear algebra with computing, we provide an opportunity for such activities. Since more students are entering a linear algebra course with computing backgrounds, linear algebra can be used to introduce mathematical computing. The increased availability of quality software and the accessibility of microcomputers make such activities quite feasible. A number of software packages are available that use a highly interactive command structure and hence require no programming by the student or instructor. It is with such software in mind that we have included 58 optional computer-oriented exercises, which are designated by the symbol **C**. These exercises have been designed so that they can be solved by a number of available software packages, thereby giving instructors the opportunity to select the package that will best suit their course. Naturally, there are other exercises throughout the book that can more easily be solved by the use of such software.

Among the many software packages that can be used for solving linear algebra problems, we mention the following which possess the attributes mentioned above and with which we have had some experience:

MATLAB, MATRIXPAD, MAX, Derive, Mathematica, and Macsyma

The table on page xi lists the commands used by these software packages to carry out some common linear algebra tasks. The designation (NA) indicates that such a command is not available.

Software Commands

	MATLAB	*MATRIXPAD*	*MAX*	*Derive*	*Mathematica*	*Macsyma*
Transpose	A′	Press F1	tra	A`	transpose[A]	transpose(A)
Reduced row echelon form	rref(A)	Press F6	(NA) but can compute row echelon form	row_reduce(A)	row reduce[A]	echelon(A)
Solve a system Ax = b	x = A\b	Enter [A ¦ b], press F6, solve for x	solve	row_reduce (A ¦ b) and solve for x	linear solve [A,b]	echelon (A ¦ b) and solve for x
Inverse	inv(A)	Press F7	inv	A^{-1}	inverse[A]	invert(A)
Rank	rank(A)	Press F6 and count number of nonzero rows	rank	row_reduce (A) and count number of nonzero rows	row reduce[A] and count number of nonzero rows	rank(A)
Length or norm of a vector	norm(x)	(NA)	(NA)	(NA)	(NA)	(NA)
Determinant	det(A)	Press F5	det	det(A)	det[A]	determinant(A)
Eigenvalues and eigenvectors	eig(A) or [p,d] = eig(A)	Indirectly available	eig	eigenvalues(A,x)	eigenvalues[A]	eigenvalues(A) eigenvectors(A)
Handles complex arithmetic	yes	no	for eigenproblems	yes	yes	yes
Graphics	yes	no	no	yes	yes	yes
Handles symbolic computations	no	no	no	yes	yes	yes
Addresses	The Mathworks Inc. 21 Eliot St. South Natick, MA 01760 (508)653-1415	D. C. Heath Inc. 125 Spring St. Lexington, MA 02173 (800)235-3565	Brooks Cole Wadsworth Inc. Pacific Grove, CA 93950-5098 (408)373-0728	Soft Warehouse Inc. 3615 Harding Ave. Suite 505 Honolulu, Hawaii 96816-3735 (808)734-5801	Wolfram Research Inc. P.O. Box 6059 Champaign, IL 61821 (217)398-0700	Symbolics Inc. 8 New England Executive Park E Burlington, MA 01803 (800)MACSYMA

A Brief Guide to the Software Packages

For computational work in linear algebra the most powerful package available is MATLAB, which uses the most up-to-date numerical techniques. Although it lacks symbolic computational capabilities, it has graphics and handles a wide variety of numerical analysis tasks. Moreover, it is easy to learn and use. Mathematica and Macsyma both carry out symbolic computations and can solve many mathematical problems in numerous areas. The graphics capabilities in Mathematica are outstanding, but this package requires a computer with lots of memory. Derive is an inexpensive, easy-to-use package that has graphics and symbolic computing, as well as the ability to deal with problems in a number of different areas. MATRIXPAD and MAX are simpler packages devoted solely to linear algebra, with no graphics or symbolic computing. They are inexpensive, easy to learn and use.

Acknowledgments

I am pleased to express my thanks to the following reviewers of the first four editions: Edward Norman, University of Central Florida; the late Charles S. Duris and Herbert J. Nichol, both at Drexel University; Stephen D. Kerr, Weber State College; Norman Lee, Ball State University; William Briggs, University of Colorado; Richard Roth, University of Colorado; David Stanford, College of William and Mary; and David R. Hill, Temple University; and of the fifth edition: David L. Abrahamson, Rhode Island College; Ruth Berger, Memphis State University; Michael A. Geraghty, University of Iowa; You-Feng Lin, University of South Florida; Lothar Redlin, Pennsylvania State University, Abington; Richard Sot, University of Nevada, Reno; Raymond Southworth, Professor Emeritus, College of William and Mary; J. Barry Turett, Oakland University; and David R. Hill, Temple University. The numerous suggestions, comments, and criticisms of these people greatly improved the manuscript.

I should like to express my deep gratitude to David R. Hill, Temple University, for developing the material in Sections 1.8, 3.5, 6.5, and Appendix B. He also created many of the new exercises and all the computer exercises. Moreover, he prepared the Answer Manual and the Student Solutions Manual and critically read galley and page proofs. It was a pleasure working with him.

I also thank John Edenhofner Jr., Merck Inc., for solving the new 314 exercises, for reading galley and page proofs, and for preparing the index; Nina Edelman, Spring Garden College, for reading galley and page proofs; Robert Davis, University of Nevada, Reno, and Heinz Gonska, University of Duisburg for making several valuable suggestions; Jerrold Grossman, Oakland University, for pointing out a number of errors in the previous edition; the students at Drexel University who used earlier editions of the book; and the instructors from numerous institutions who communicated to me their classroom experiences with the book.

Finally, a sincere expression of thanks goes to Robert W. Pirtle, Executive Editor; Elaine Wetterau, Production Supervisor; Pamela Kennedy, Production Manager; Robert Freese, Designer; Charles Healy, Senior Marketing Manager; and the entire staff of the Macmillan Publishing Company for their interest and unfailing cooperation during the conception, design, production, and marketing phases of this book.

B. K.

New Features in the Fifth Edition

- New sections have been added on
  - LU-Factorization
  - Vectors in the Plane and in 3-Space
  - Cross Product in R^3
  - Least Squares
  - Conic Sections
  - Quadric Surfaces
- Many more new exercises (314 of them) at all levels have been added.
- Exercises (58 of them) requiring the use of a computer have been added, allowing for exploration and discovery.
- More material on computational techniques has been incorporated.
- More material on geometry has been included.
- Additional material on isometries has been added.
- Many new figures (58 of them) have been added to enhance the geometric flavor of the text.
- The notation for the transition matrix from one basis to another has been reversed to the more conventional usage.
- Vectors are now denoted by Roman letters and not by Greek letters.
- Answers in the back of the book are now given to all odd-numbered exercises.
- A Student Solutions Manual and Study Guide, including an introductory chapter on logic and methods of proof in mathematics, has been prepared.
- The Answer Manual contains answers to the even-numbered exercises and solutions to all theoretical exercises. It is available, to instructors only, at no cost from the publisher.
- Additional concrete material and pedagogy has been incorporated into the book.
- A second color has been used to improve pedagogy.
- The clarity of writing has been improved throughout the book.

Contents

3 Inner Product Spaces 181

4 Linear Transformations and Matrices 245

5 Determinants 301

6 Eigenvalues and Eigenvectors 337

Elementary Linear Algebra

1 Linear Equations and Matrices

1.1. Systems of Linear Equations

One of the most frequently recurring practical problems in almost all fields of study—such as mathematics, physics, biology, chemistry, economics, all phases of engineering, operations research, the social sciences, and so forth—is that of solving a system of linear equations. The equation

$$b = a_1x_1 + a_2x_2 + \cdots + a_nx_n, \tag{1}$$

which expresses b in terms of the variables $x_1, x_2, \ldots, x_n$ and the constants $a_1, a_2, \ldots, a_n$, is called a **linear equation**. In many applications we are given b and must find numbers $x_1, x_2, \ldots, x_n$ satisfying (1).

A **solution** to a linear equation (1) is a sequence of n numbers $s_1, s_2, \ldots, s_n$ such that (1) is satisfied when $x_1 = s_1, x_2 = s_2, \ldots, x_n = s_n$ are substituted in (1). Thus $x_1 = 2$, $x_2 = 3$, and $x_3 = -4$ is a solution to the linear equation

$$6x_1 - 3x_2 + 4x_3 = -13,$$

because

$$6(2) - 3(3) + 4(-4) = -13.$$

Note: The appendix reviews some very basic material dealing with sets and functions. It can be consulted at any time, as needed.

More generally, a **system of m linear equations in n unknowns**, or a **linear system**, is a set of m linear equations each in n unknowns. A linear system can be conveniently written as

$$\begin{aligned} a_{11}x_1 + a_{12}x_2 + \cdots + a_{1n}x_n &= b_1 \\ a_{21}x_1 + a_{22}x_2 + \cdots + a_{2n}x_n &= b_2 \\ \vdots \qquad \vdots \qquad\quad \vdots \quad &\quad \vdots \\ a_{m1}x_1 + a_{m2}x_2 + \cdots + a_{mn}x_n &= b_m. \end{aligned} \tag{2}$$

Thus the ith equation is

$$a_{i1}x_1 + a_{i2}x_2 + \cdots + a_{in}x_n = b_i.$$

In (2) the a_{ij} are known constants. Given values of $b_1, b_2, \ldots, b_n$, we want to find values of $x_1, x_2, \ldots, x_n$ that will satisfy each equation in (2).

A **solution** to a linear system (2) is a sequence of n numbers $s_1, s_2, \ldots, s_n$, which have the property that each equation in (2) is satisfied when $x_1 = s_1$, $x_2 = s_2, \ldots, x_n = s_n$ are substituted.

If the linear system (2) has no solution, it is said to be **inconsistent**; if it has a solution, it is called **consistent**. If $b_1 = b_2 = \cdots = b_m = 0$, then (2) is called a **homogeneous system**. The solution $x_1 = x_2 = \cdots = x_n = 0$ to a homogeneous system is called the **trivial solution**. A solution to a homogeneous system in which not all of $x_1, x_2, \ldots, x_n$ are zero is called a **nontrivial solution**.

Consider another system of r linear equations in n unknowns:

$$\begin{aligned} c_{11}x_1 + c_{12}x_2 + \cdots + c_{1n}x_n &= d_1 \\ c_{21}x_1 + c_{22}x_2 + \cdots + c_{2n}x_n &= d_2 \\ \vdots \qquad \vdots \qquad\quad \vdots \quad &\quad \vdots \\ c_{r1}x_1 + c_{r2}x_2 + \cdots + c_{rn}x_n &= d_r. \end{aligned} \tag{3}$$

We say that (2) and (3) are **equivalent** if they both have exactly the same solutions.

Example 1. The linear system

$$\begin{aligned} x_1 - 3x_2 &= -7 \\ 2x_1 + x_2 &= 7 \end{aligned} \tag{4}$$

has only the solution $x_1 = 2$ and $x_2 = 3$. The linear system

$$\begin{aligned} 8x_1 - 3x_2 &= 7 \\ 3x_1 - 2x_2 &= 0 \\ 10x_1 - 2x_2 &= 14 \end{aligned} \tag{5}$$

also has only the solution $x_1 = 2$ and $x_2 = 3$. Thus (4) and (5) are equivalent. ■

To find solutions to a linear system, we shall use a technique called the **method of elimination**; that is, we eliminate some variables by adding a multiple of one equation to another equation. Elimination merely amounts to the development of a new linear system which is equivalent to the original system but is much simpler to solve. Readers have probably confined their earlier work in this area to linear systems in which $m = n$, that is, linear systems having as many equations as unknowns. In this course we shall broaden our outlook by dealing with systems in which we have $m = n$, $m < n$, and $m > n$. Indeed, there are numerous applications in which $m \neq n$.

Example 2. Consider the linear system

$$\begin{aligned} x_1 - 3x_2 &= -3 \\ 2x_1 + x_2 &= 8. \end{aligned} \tag{6}$$

To eliminate x_1, we add (-2) times the first equation to the second one, obtaining

$$7x_2 = 14,$$

an equation having no x_1 term. Thus we have eliminated the unknown x_1. Then solving for x_2, we have

$$x_2 = 2,$$

and substituting into the first equation of (6), we obtain

$$x_1 = 3.$$

Then $x_1 = 3$, $x_2 = 2$ is the only solution to the given linear system. ■

Example 3. Consider the linear system

$$\begin{aligned} x_1 - 3x_2 &= -7 \\ 2x_1 - 6x_2 &= 7. \end{aligned} \tag{7}$$

Again, we decide to eliminate x_1. We add (-2) times the first equation to the second one, obtaining

$$0 = 21,$$

which makes no sense. This means that (7) has no solution; it is inconsistent. We could have come to the same conclusion from observing that in (7) the left side of the second equation is twice the left side of the first equation, but the right side of the second equation is not twice the right side of the first equation. ■

Example 4. Consider the linear system

$$\begin{aligned} x_1 + 2x_2 + 3x_3 &= 6 \\ 2x_1 - 3x_2 + 2x_3 &= 14 \\ 3x_1 + x_2 - x_3 &= -2. \end{aligned} \tag{8}$$

To eliminate x_1, we add (-2) times the first equation to the second one and (-3) times the first equation to the third one, obtaining

$$\begin{aligned} -7x_2 - 4x_3 &= 2 \\ -5x_2 - 10x_3 &= -20. \end{aligned} \tag{9}$$

This is a system of two equations in the unknowns x_2 and x_3. We multiply the second equation of (9) by $-\frac{1}{5}$, obtaining

$$\begin{aligned} -7x_2 - 4x_3 &= 2 \\ x_2 + 2x_3 &= 4, \end{aligned}$$

which we write, by interchanging equations, as

$$\begin{aligned} x_2 + 2x_3 &= 4 \\ -7x_2 - 4x_3 &= 2. \end{aligned} \tag{10}$$

We now eliminate x_2 in (10) by adding 7 times the first equation to the second one, to obtain

$$10x_3 = 30,$$

or

$$x_3 = 3. \tag{11}$$

Substituting this value of x_3 into the first equation of (10), we find that $x_2 = -2$. Substituting these values of x_2 and x_3 into the first equation of (8), we find that $x_1 = 1$. We might observe further that our elimination procedure has actually produced the linear system

$$\begin{aligned} x_1 + 2x_2 + 3x_3 &= 6 \\ x_2 + 2x_3 &= 4 \\ x_3 &= 3, \end{aligned} \tag{12}$$

obtained by using the first equations of (8) and (10) as well as (11). The importance of the procedure is that although the linear systems (8) and (12) are equivalent, (12) has the advantage that it is easier to solve. ■

Example 5. Consider the linear system

$$\begin{aligned} x_1 + 2x_2 - 3x_3 &= -4 \\ 2x_1 + x_2 - 3x_3 &= 4. \end{aligned} \tag{13}$$

Eliminating x_1, we add (-2) times the first equation to the second equation, to obtain

$$-3x_2 + 3x_3 = 12. \tag{14}$$

We must now solve (14). A solution is

$$x_2 = x_3 - 4,$$

where x_3 can be any real number. Then from the first equation of (13),

$$\begin{aligned} x_1 &= -4 - 2x_2 + 3x_3 \\ &= -4 - 2(x_3 - 4) + 3x_3 \\ &= x_3 + 4. \end{aligned}$$

Thus a solution to the linear system (13) is

$$\begin{aligned} x_1 &= x_3 + 4 \\ x_2 &= x_3 - 4 \\ x_3 &= \text{any real number.} \end{aligned}$$

This means that the linear system (13) has infinitely many solutions. Every time we assign a value to x_3 we obtain another solution to (13). Thus, if $x_3 = 1$, then

$$x_1 = 5, \qquad x_2 = -3, \qquad \text{and} \qquad x_3 = 1$$

is a solution, while if $x_3 = -2$, then

$$x_1 = 2, \qquad x_2 = -6, \qquad \text{and} \qquad x_3 = -2$$

is another solution. ■

These examples suggest that a linear system may have a unique solution, no solution, or infinitely many solutions.

Consider next a linear system of two equations in the unknowns x_1 and x_2:

$$\begin{aligned} a_1x_1 + a_2x_2 &= c_1 \\ b_1x_1 + b_2x_2 &= c_2. \end{aligned} \tag{15}$$

The graph of each of these equations is a straight line, which we denote by ℓ_1 and ℓ_2, respectively. If $x_1 = s_1$, $x_2 = s_2$ is a solution to the linear system (15), then the point (s_1, s_2) lies on both lines ℓ_1 and ℓ_2. Conversely, if the point (s_1, s_2) lies on both lines ℓ_1 and ℓ_2, then $x_1 = s_1$, $x_2 = s_2$ is a solution to the linear system (15). Thus we are led geometrically to the same three possibilities mentioned above. See Figure 1.1.

Figure 1.1

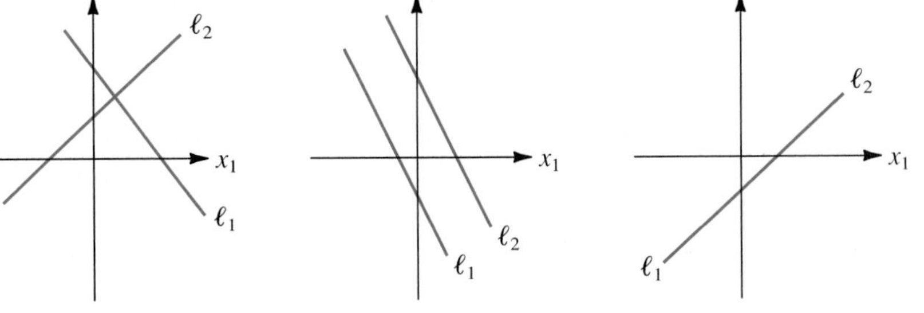

(a) A unique solution. (b) No solution. (c) Infinitely many solutions.

If we examine the method of elimination more closely, we find that it involves three manipulations that can be performed on a linear system to convert it into an equivalent system. These manipulations are as follows:

1. Interchange the ith and jth equations.
2. Multiply an equation by a nonzero constant.

3. Replace the ith equation by c times the jth equation plus the ith equation, $i \neq j$. That is, replace

$$a_{i1}x_1 + a_{i2}x_2 + \cdots + a_{in}x_n = b_i$$

by

$$(a_{i1} + ca_{j1})x_1 + (a_{i2} + ca_{j2})x_2 + \cdots + (a_{in} + ca_{jn})x_n = b_i + cb_j.$$

It is not difficult to prove that performing these manipulations on a linear system leads to an equivalent system.

Example 6. Suppose that the ith equation of the linear system (2) is multiplied by the nonzero constant c, obtaining the linear system

$$\begin{aligned} a_{11}x_1 &+ a_{12}x_2 + \cdots + a_{1n}x_n = b_1 \\ a_{21}x_1 &+ a_{22}x_2 + \cdots + a_{2n}x_n = b_2 \\ &\vdots \\ ca_{i1}x_1 &+ ca_{i2}x_2 + \cdots + ca_{in}x_n = cb_i \\ &\vdots \\ a_{m1}x_1 &+ a_{m2}x_2 + \cdots + a_{mn}x_n = b_m. \end{aligned} \tag{16}$$

If $x_1 = s_1, x_2 = s_2, \ldots, x_n = s_n$ is a solution to (2), then it is a solution to all the equations in (16) except possibly for the ith equation. For the ith equation we have

$$c(a_{i1}s_1 + a_{i2}s_2 + \cdots + a_{in}s_n) = cb_i$$

or

$$ca_{i1}s_1 + ca_{i2}s_2 + \cdots + ca_{in}s_n = cb_i.$$

Thus the ith equation of (16) is also satisfied. Hence every solution to (2) is also a solution to (16). Conversely, every solution to (16) also satisfies (2). Hence (2) and (16) are equivalent systems. ■

Before continuing the study of solving linear systems, we now introduce the notion of a matrix, which will greatly simplify our notational problems, and develop tools to solve many important applied problems.

1.1 EXERCISES

In Exercises 1 through 14, solve the given linear system by the method of elimination.

1. $$\begin{aligned} x_1 + 2x_2 &= 8 \\ 3x_1 - 4x_2 &= 4. \end{aligned}$$

2. $$\begin{aligned} 2x_1 - 3x_2 + 4x_3 &= -12 \\ x_1 - 2x_2 + x_3 &= -5 \\ 3x_1 + x_2 + 2x_3 &= 1. \end{aligned}$$

3. $$\begin{aligned} 3x_1 + 2x_2 + x_3 &= 2 \\ 4x_1 + 2x_2 + 2x_3 &= 8 \\ x_1 - x_2 + x_3 &= 4. \end{aligned}$$

4. $$\begin{aligned} x_1 + x_2 &= 5 \\ 3x_1 + 3x_2 &= 10. \end{aligned}$$

5. $$\begin{aligned} 2x_1 + 4x_2 + 6x_3 &= -12 \\ 2x_1 - 3x_2 - 4x_3 &= 15 \\ 3x_1 + 4x_2 + 5x_3 &= -8. \end{aligned}$$

6. $$\begin{aligned} x_1 + x_2 - 2x_3 &= 5 \\ 2x_1 + 3x_2 + 4x_3 &= 2. \end{aligned}$$

7. $$\begin{aligned} x_1 + 4x_2 - x_3 &= 12 \\ 3x_1 + 8x_2 - 2x_3 &= 4. \end{aligned}$$

8. $$\begin{aligned} 3x_1 + 4x_2 - x_3 &= 8 \\ 6x_1 + 8x_2 - 2x_3 &= 3. \end{aligned}$$

9. $$\begin{aligned} x_1 + x_2 + 3x_3 &= 12 \\ 2x_1 + 2x_2 + 6x_3 &= 6. \end{aligned}$$

10. $$\begin{aligned} x_1 + x_2 &= 1 \\ 2x_1 - x_2 &= 5 \\ 3x_1 + 4x_2 &= 2. \end{aligned}$$

11. $$\begin{aligned} 2x_1 + 3x_2 &= 13 \\ x_1 - 2x_2 &= 3 \\ 5x_1 + 2x_2 &= 27. \end{aligned}$$

12. $$\begin{aligned} x_1 - 5x_2 &= 6 \\ 3x_1 + 2x_2 &= 1 \\ 5x_1 + 2x_2 &= 1. \end{aligned}$$

13. $$\begin{aligned} x_1 + 3x_2 &= -4 \\ 2x_1 + 5x_2 &= -8 \\ x_1 + 3x_2 &= -5. \end{aligned}$$

14. $$\begin{aligned} 2x_1 + 3x_2 - x_3 &= 6 \\ 2x_1 - x_2 + 2x_3 &= -8 \\ 3x_1 - x_2 + x_3 &= -7. \end{aligned}$$

15. Given the linear system

$$\begin{aligned} 2x_1 - x_2 &= 5 \\ 4x_1 - 2x_2 &= t. \end{aligned}$$

(a) Determine a value of t so that the system is consistent.

(b) Determine a value of t so that the system is inconsistent.

(c) How many different values of t can be selected in part (b)?

16. Is every homogeneous linear system always consistent?

17. Given the linear system

$$\begin{aligned} 2x_1 + 3x_2 - x_3 &= 0 \\ x_1 - 4x_2 + 5x_3 &= 0. \end{aligned}$$

(a) Verify that $x_1 = 1$, $x_2 = -1$, $x_3 = -1$ is a solution.

(b) Verify that $x_1 = -2$, $x_2 = 2$, $x_3 = 2$ is a solution.

(c) Adding the corresponding x-values of the solutions in parts (a) and (b) gives $x_1 = -1$, $x_2 = 1$, $x_3 = 1$. Is this a solution to the linear system?

(d) Multiply each of the x-values in part (a) by 3. Are the resulting values a solution to the linear system?

18. Without using the method of elimination solve the linear system

$$\begin{aligned} 2x_1 + x_2 - 2x_3 &= -5 \\ 3x_2 + x_3 &= 7 \\ x_3 &= 4. \end{aligned}$$

19. Without using the method of elimination solve the linear system

$$\begin{aligned} 4x_1 \qquad\qquad &= 8 \\ -2x_1 + 3x_2 \qquad &= -1 \\ 3x_1 + 5x_2 - 2x_3 &= 11. \end{aligned}$$

20. Is there a value of r so that $x_1 = 1, x_2 = 2, x_3 = r$ is a solution to the following linear system? If there is, find it.

$$\begin{aligned} 2x_1 + 3x_2 - x_3 &= 11 \\ x_1 - x_2 + 2x_3 &= -7 \\ 4x_1 + x_2 - 2x_3 &= 12 \end{aligned}$$

21. Is there a value of r so that $x_1 = r, x_2 = 2, x_3 = 1$ is a solution to the following linear system? If there is, find it.

$$\begin{aligned} 3x_1 \qquad - 2x_3 &= 4 \\ x_1 - 4x_2 + x_3 &= -5 \\ -2x_1 + 3x_2 + 2x_3 &= 9 \end{aligned}$$

22. Show that the linear system obtained by interchanging two equations in (2) is equivalent to (2).

23. Show that the linear system obtained by adding a multiple of an equation in (2) to another equation is equivalent to (2).

C 24. For the software you are using, determine the command that "automatically" solves a linear system of equations.

C 25. Use the command from Exercise 24 to solve Exercises 3 and 4 and compare the output with the results you obtained by the method of elimination.

C 26. Solve the linear system

$$\begin{aligned} x_1 + \tfrac{1}{2}x_2 + \tfrac{1}{3}x_3 &= 1 \\ \tfrac{1}{2}x_1 + \tfrac{1}{3}x_2 + \tfrac{1}{4}x_3 &= \tfrac{11}{18} \\ \tfrac{1}{3}x_1 + \tfrac{1}{4}x_2 + \tfrac{1}{5}x_3 &= \tfrac{9}{20} \end{aligned}$$

using your software. Compare the computed solution with the exact solution $x_1 = \frac{1}{2}, x_2 = \frac{1}{3}, x_3 = 1$.

1.2. *Matrices; Matrix Operations*

If we examine the method of elimination described in Section 1.1, we make the following observation: Only the numbers in front of the unknowns $x_1, x_2, \ldots, x_n$ are being changed as we perform the steps in the method of elimination. Thus we might think of looking for a way of writing a linear system without having to carry along the unknowns. Matrices enable us to do this—that is, to write linear systems in a compact form that makes it easier to automate the elimination method on an electronic computer in order to obtain a fast and efficient procedure for finding solutions. Their use is not, however, merely that of a convenient notation. We now develop operations on matrices and will work with matrices according to the rules they obey; this will enable us to solve systems of linear equations and to do other computational problems in a fast and efficient manner. Of course, as any good definition should do, the notion of a matrix provides not only a new way of looking at old problems but also gives rise to a great many new questions, some of which we study in this book.

Definition 1.1. A **matrix** is a rectangular array of numbers denoted by

$$A = \begin{bmatrix} a_{11} & a_{12} & \cdots & a_{1n} \\ a_{21} & a_{22} & \cdots & a_{2n} \\ \vdots & \vdots & & \vdots \\ a_{m1} & a_{m2} & \cdots & a_{mn} \end{bmatrix}.$$ ■

Unless stated otherwise, we assume that all our matrices are composed entirely of real numbers. The ***i*th row** of A is

$$[a_{i1} \quad a_{i2} \quad \cdots \quad a_{in}] \qquad (1 \le i \le m)$$

while the ***j*th column** of A is

$$\begin{bmatrix} a_{1j} \\ a_{2j} \\ \vdots \\ a_{mj} \end{bmatrix} \qquad (1 \le j \le n).$$

If a matrix A has m rows and n columns, we say that A is an ***m* by *n* matrix** (written $m \times n$). If $m = n$, we say that A is a **square matrix of order *n*** and that the elements $a_{11}, a_{22}, \ldots, a_{nn}$ are on the **main diagonal** of A. We refer to a_{ij} as the **(*i*, *j*) entry** (entry in ith row and jth column) or **(*i*, *j*)th element** and we often write

$$A = [a_{ij}].$$

We shall also write ${}_mA_n$ to indicate that A has m rows and n columns. If A is $n \times n$, we merely write A_n.

Example 1. The following are matrices:

$$A = \begin{bmatrix} 1 & 2 & 3 \\ 2 & -1 & 4 \\ 0 & -3 & 2 \end{bmatrix}, \qquad B = [1 \quad 3 \quad -7],$$

$$C = \begin{bmatrix} 2 \\ -1 \\ 3 \\ 4 \end{bmatrix}, \quad \text{and} \quad D = \begin{bmatrix} 0 & 3 \\ -1 & -2 \end{bmatrix}.$$

In A, $a_{32} = -3$; in C, $c_{41} = 4$. Here A is 3×3, B is 1×3, C is 4×1, and D is 2×2. In A, the elements $a_{11} = 1$, $a_{22} = -1$, and $a_{33} = 2$ are on the main diagonal. ■

Whenever a new object is introduced in mathematics, one must determine when two such objects are equal. For example, in the set of all rational numbers the numbers $\frac{2}{3}$ and $\frac{4}{6}$ are called equal, although they are not represented in the same manner. What we have in mind is the definition that a/b equals c/d when $ad = bc$. Accordingly, we now have the following definition.

Definition 1.2. Two $m \times n$ matrices $A = [a_{ij}]$ and $B = [b_{ij}]$ are **equal** if they agree entry by entry, that is, if $a_{ij} = b_{ij}$ for $i = 1, 2, \ldots, m$ and $j = 1, 2, \ldots, n$. ■

Example 2. The matrices

$$A = \begin{bmatrix} 1 & 2 & -1 \\ 2 & -3 & 4 \\ 0 & -4 & 5 \end{bmatrix} \quad \text{and} \quad B = \begin{bmatrix} 1 & 2 & w \\ 2 & x & 4 \\ y & -4 & z \end{bmatrix}$$

are equal if and only if $w = -1$, $x = -3$, $y = 0$, and $z = 5$. ■

We next define a number of operations that will produce new matrices out of given matrices; this will enable us to compute with the matrices and not deal with the equations from which they arise. These operations are also useful in the applications of matrices.

Matrix Addition

Definition 1.3. If $A = [a_{ij}]$ and $B = [b_{ij}]$ are both $m \times n$ matrices, then the **sum** $A + B$ is an $m \times n$ matrix $C = [c_{ij}]$ defined by $c_{ij} = a_{ij} + b_{ij}$, $i = 1, 2, \ldots, m$; $j = 1, 2, \ldots, n$. Thus, to obtain the sum of A and B, we merely add corresponding entries. ■

Example 3. Let

$$A = \begin{bmatrix} 1 & -2 & 3 \\ 2 & -1 & 4 \end{bmatrix} \quad \text{and} \quad B = \begin{bmatrix} 0 & 2 & 1 \\ 1 & 3 & -4 \end{bmatrix}.$$

Then

$$A + B = \begin{bmatrix} 1+0 & -2+2 & 3+1 \\ 2+1 & -1+3 & 4+-(4) \end{bmatrix} = \begin{bmatrix} 1 & 0 & 4 \\ 3 & 2 & 0 \end{bmatrix}. \quad \blacksquare$$

It should be noted that the sum of the matrices A and B is defined only when A and B have the same number of rows and the same number of columns, that is, only when A and B are of the same size. We now establish the convention that when $A + B$ is formed, both A and B are of the same size. The basic properties of matrix addition are considered in the following section and are similar to those satisfied by the real numbers.

Scalar Multiplication

Definition 1.4. If $A = [a_{ij}]$ is an $m \times n$ matrix and c is a real number, then the **scalar multiple** of A by r, rA, is the $m \times n$ matrix $C = [c_{ij}]$, where $c_{ij} = ra_{ij}$, $i = 1, 2, \ldots, m$, $j = 1, 2, \ldots, n$; that is, the matrix C is obtained by multiplying each entry of A by r. ■

Example 4. We have

$$-2\begin{bmatrix} 4 & -2 & -3 \\ 7 & -3 & 2 \end{bmatrix} = \begin{bmatrix} (-2)(4) & (-2)(-2) & (-2)(-3) \\ (-2)(7) & (-2)(-3) & (-2)(\ 2) \end{bmatrix} = \begin{bmatrix} -8 & 4 & 6 \\ -14 & 6 & -4 \end{bmatrix}$$

■

If A and B are $m \times n$ matrices, we write $A + (-1)B$ as $A - B$ and call this the **difference between A and B**. We also note that $-A$ is $(-1)A$.

Example 5. Let

$$A = \begin{bmatrix} 2 & 3 & -5 \\ 4 & 2 & 1 \end{bmatrix} \quad \text{and} \quad B = \begin{bmatrix} 2 & -1 & 3 \\ 3 & 5 & -2 \end{bmatrix}.$$

Then

$$A - B = \begin{bmatrix} 2-2 & 3+1 & -5-3 \\ 4-3 & 2-5 & 1+2 \end{bmatrix} = \begin{bmatrix} 0 & 4 & -8 \\ 1 & -3 & 3 \end{bmatrix}. \quad \blacksquare$$

We shall sometimes use the **summation notation**, and we now review this useful and compact notation.

By $\sum_{i=1}^{n} r_i a_i$ we mean $r_1 a_1 + r_2 a_2 + \cdots + r_n a_n$. The letter i is called the **index of summation**; it is a dummy variable that can be replaced by another letter. Hence we can write

$$\sum_{i=1}^{n} r_i a_i = \sum_{j=1}^{n} r_j a_j = \sum_{k=1}^{n} r_k a_k.$$

Thus

$$\sum_{i=1}^{4} r_i a_i = r_1 a_1 + r_2 a_2 + r_3 a_3 + r_4 a_4.$$

The summation notation satisfies the following properties:

1. $\sum_{i=1}^{n} (r_i + s_i) a_i = \sum_{i=1}^{n} r_i a_i + \sum_{i=1}^{n} s_i a_i.$

2. $\sum_{i=1}^{n} c(r_i a_i) = c \sum_{i=1}^{n} r_i a_i.$

3. $\sum_{j=1}^{m} \sum_{i=1}^{n} a_{ij} = \sum_{i=1}^{n} \sum_{j=1}^{m} a_{ij}.$

Property 3 can be interpreted as follows. The left side is obtained by adding all the entries in each column and then adding all the resulting numbers. The right side is obtained by adding all the entries in each row and then adding all the resulting numbers.

Matrix Multiplication

Definition 1.5. If $A = [a_{ij}]$ is an $m \times n$ matrix and $B = [b_{ij}]$ is an $n \times p$ matrix, then the **product** of A and B, $AB = C = [c_{ij}]$, is an $m \times p$ matrix defined by

$$c_{ij} = \sum_{k=1}^{n} a_{ik}b_{kj} = a_{i1}b_{1j} + a_{i2}b_{2j} + \cdots + a_{in}b_{nj} \qquad \begin{aligned} i &= 1, 2, \ldots, m \\ j &= 1, 2, \ldots, p. \end{aligned}$$ ■

Note that AB is defined only when the number of columns of A is the same as the number of rows of B. We also observe that the (i, j) entry in C is obtained by using the ith row of A and the jth column of B. Thus

$$\begin{bmatrix} a_{11} & a_{12} & \cdots & a_{1n} \\ a_{21} & a_{22} & \cdots & a_{2n} \\ \vdots & \vdots & & \vdots \\ a_{i1} & a_{i2} & \cdots & a_{in} \\ \vdots & \vdots & & \vdots \\ a_{m1} & a_{m2} & \cdots & a_{mn} \end{bmatrix} \begin{bmatrix} b_{11} & b_{12} & \cdots & b_{1j} & \cdots & b_{1p} \\ b_{21} & b_{22} & \cdots & b_{2j} & \cdots & b_{2p} \\ \vdots & \vdots & & \vdots & & \vdots \\ b_{n1} & b_{n2} & \cdots & b_{nj} & \cdots & b_{np} \end{bmatrix}$$

$$= \begin{bmatrix} c_{11} & c_{12} & \cdots & c_{1p} \\ c_{21} & c_{22} & \cdots & c_{2p} \\ \vdots & \vdots & c_{ij} & \vdots \\ c_{m1} & c_{m2} & \cdots & c_{mp} \end{bmatrix}.$$

Example 6. Let

$$A = \begin{bmatrix} 1 & 2 & -1 \\ 3 & 1 & 4 \end{bmatrix} \quad \text{and} \quad B = \begin{bmatrix} -2 & 5 \\ 4 & -3 \\ 2 & 1 \end{bmatrix}.$$

Then

$$\begin{aligned} AB &= \begin{bmatrix} (1)(-2) + (2)(4) + (-1)(2) & (1)(5) + (2)(-3) + (-1)(1) \\ (3)(-2) + (1)(4) + (4)(2) & (3)(5) + (1)(-3) + (4)(1) \end{bmatrix} \\ &= \begin{bmatrix} 4 & -2 \\ 6 & 16 \end{bmatrix}. \end{aligned}$$ ■

The basic properties of matrix multiplication are considered in the following section. However, we note here that multiplication of matrices requires much more care than their addition, since the algebraic properties of matrix multiplication differ from those satisfied by the real numbers. Part of the problem is due to the fact that AB is defined only when the number of columns of A is the same as the number of rows of B. Thus, if A is an $m \times n$ matrix and B is an $n \times p$ matrix, then AB is an $m \times p$ matrix. What about BA? Three different situations may occur:

1. BA may not be defined. This will take place if $p \neq m$.
2. If BA is defined, BA will be $n \times n$ and AB will be $m \times m$, and if $m \neq n$, AB and BA are of different sizes.
3. If BA and AB are of the same size, they may be unequal.

As in the case of addition, we establish the convention that when AB is written, it is defined.

Example 7. Let A be a 2×3 matrix and let B be a 3×4 matrix. Then AB is 2×4 and BA is not defined. ■

Example 8. Let A be 2×3 and let B be 3×2. Then AB is 2×2 and BA is 3×3. ■

Example 9. Let

$$A = \begin{bmatrix} 1 & 2 \\ -1 & 3 \end{bmatrix} \quad \text{and} \quad B = \begin{bmatrix} 2 & 1 \\ 0 & 1 \end{bmatrix}.$$

Then

$$AB = \begin{bmatrix} 2 & 3 \\ -2 & 2 \end{bmatrix} \quad \text{while} \quad BA = \begin{bmatrix} 1 & 7 \\ -1 & 3 \end{bmatrix}.$$

Thus $AB \neq BA$. ■

One might ask why matrix equality and matrix addition are defined in such a natural way while matrix multiplication appears to be much more complicated. Only a thorough understanding of the composition of functions and the relationship that exists between matrices and what are called linear transformations would show that the definition of multiplication given above is the natural one. These topics will be covered later in the book.

It is sometimes useful to be able to find a column in the matrix product AB without having to multiply the two matrices. It is not difficult to show (Exercise 26) that the jth column of the matrix product AB is equal to the matrix product AB_j, where B_j is the jth column of B.

Example 10. Let

$$A = \begin{bmatrix} 1 & 2 \\ 3 & 4 \\ -1 & 5 \end{bmatrix} \quad \text{and} \quad B = \begin{bmatrix} -2 & 3 & 4 \\ 3 & 2 & 1 \end{bmatrix}.$$

Then the second column of AB is

$$AB_2 = \begin{bmatrix} 1 & 2 \\ 3 & 4 \\ -1 & 5 \end{bmatrix} \begin{bmatrix} 3 \\ 2 \end{bmatrix} = \begin{bmatrix} 7 \\ 17 \\ 7 \end{bmatrix}.$$

■

Linear Systems

We now return to the linear system (2) in Section 1.1:

$$\begin{aligned} a_{11}x_1 + a_{12}x_2 + \cdots + a_{1n}x_n &= b_1 \\ a_{21}x_1 + a_{22}x_2 + \cdots + a_{2n}x_n &= b_2 \\ \vdots \qquad\quad \vdots \qquad\qquad\quad \vdots \quad &\quad \vdots \\ a_{m1}x_1 + a_{m2}x_2 + \cdots + a_{mn}x_n &= b_m, \end{aligned}$$

and define the following matrices:

$$A = \begin{bmatrix} a_{11} & a_{12} & \cdots & a_{1n} \\ a_{21} & a_{22} & \cdots & a_{2n} \\ \vdots & \vdots & & \vdots \\ a_{m1} & a_{m2} & \cdots & a_{mn} \end{bmatrix}, \quad X = \begin{bmatrix} x_1 \\ x_2 \\ \vdots \\ x_n \end{bmatrix}, \quad B = \begin{bmatrix} b_1 \\ b_2 \\ \vdots \\ b_m \end{bmatrix}.$$

We can then write this linear system as $AX = B$. The matrix A is called the **coefficient matrix** of the system and the matrix

$$\left[\begin{array}{cccc|c} a_{11} & a_{12} & \cdots & a_{1n} & b_1 \\ a_{21} & a_{22} & \cdots & a_{2n} & b_2 \\ \vdots & \vdots & & \vdots & \vdots \\ a_{m1} & a_{m2} & \cdots & a_{mn} & b_m \end{array}\right]$$

is called the **augmented matrix** of the system. The coefficient and augmented matrices of a linear system will play key roles in our methods of solving linear systems.

Example 11. Consider the following linear system:

$$\begin{aligned} 2x_1 + 3x_2 - 4x_3 + x_4 &= 5 \\ -2x_1 + x_3 &= 7 \\ 3x_1 + 2x_2 - 4x_4 &= 3. \end{aligned}$$

We can write this in matrix form as

$$\begin{bmatrix} 2 & 3 & -4 & 1 \\ -2 & 0 & 1 & 0 \\ 3 & 2 & 0 & -4 \end{bmatrix} \begin{bmatrix} x_1 \\ x_2 \\ x_3 \\ x_4 \end{bmatrix} = \begin{bmatrix} 5 \\ 7 \\ 3 \end{bmatrix}.$$

The coefficient matrix of this system is

$$\begin{bmatrix} 2 & 3 & -4 & 1 \\ -2 & 0 & 1 & 0 \\ 3 & 2 & 0 & -4 \end{bmatrix},$$

and the augmented matrix is

$$\left[\begin{array}{cccc|c} 2 & 3 & -4 & 1 & 5 \\ -2 & 0 & 1 & 0 & 7 \\ 3 & 2 & 0 & -4 & 3 \end{array}\right].$$

■

Example 12. The matrix

$$\left[\begin{array}{ccc|c} 2 & -1 & 3 & 4 \\ 3 & 0 & 2 & 5 \end{array}\right]$$

is the augmented matrix of the linear system

$$\begin{aligned} 2x_1 - x_2 + 3x_3 &= 4 \\ 3x_1 + 2x_3 &= 5. \end{aligned}$$

■

Definition 1.6. If $A = [a_{ij}]$ is an $m \times n$ matrix, then the **transpose** of A, $A^T = [a_{ij}^T]$, is the $n \times m$ matrix defined by $a_{ij}^T = a_{ji}$. Thus the transpose of A is obtained from A by interchanging the rows and columns of A. ■

Example 13. If

$$A = \begin{bmatrix} 1 & 2 & -1 \\ -3 & 2 & 7 \end{bmatrix}, \quad \text{then} \quad A^T = \begin{bmatrix} 1 & -3 \\ 2 & 2 \\ -1 & 7 \end{bmatrix}.$$ ■

1.2. EXERCISES

Consider the following matrices for Exercises 1 through 5.

$$A = \begin{bmatrix} 1 & 2 & 3 \\ 2 & 1 & 4 \end{bmatrix}, \quad B = \begin{bmatrix} 1 & 0 \\ 2 & 1 \\ 3 & 2 \end{bmatrix},$$

$$C = \begin{bmatrix} 3 & -1 & 3 \\ 4 & 1 & 5 \\ 2 & 1 & 3 \end{bmatrix},$$

$$D = \begin{bmatrix} 3 & -2 \\ 2 & 5 \end{bmatrix}, \quad E = \begin{bmatrix} 2 & -4 & 5 \\ 0 & 1 & 4 \\ 3 & 2 & 1 \end{bmatrix},$$

and $$F = \begin{bmatrix} -1 & 2 \\ 0 & 4 \\ 3 & 5 \end{bmatrix}$$

1. If possible, compute:
(a) $C + E$. (b) AB and BA.
(c) $2C - 3E$. (d) $CB + D$.
(e) $AB + D^2$, where $D^2 = DD$.

2. If possible, compute:
(a) $DA + B$. (b) EC.
(c) CE. (d) $EB + F$.
(e) $FC + D$.

3. If possible, compute:
(a) $FD - 3B$.
(b) $AB - 2D$.
(c) $3(2A)$ and $6A$.
(d) $2F - 3(AE)$.
(e) $BD + AE$.

4. If possible, compute:
(a) $A(BD)$. (b) $(AB)D$. (c) $A(C + E)$.
(d) $AC + AE$. (e) $3A + 2A$ and $5A$.
(f) $A(C - 3E)$.

5. If possible, compute:
(a) A^T. (b) $(A^T)^T$. (c) $(AB)^T$.
(d) B^TA^T. (e) $(C + E)^T$ and $C^T + E^T$.
(f) $A(2B)$ and $2(AB)$.

6. Let $A = [1 \;\; 2 \;\; -3]$, $B = [-1 \;\; 4 \;\; 2]$, and $C = [-3 \;\; 0 \;\; 1]$. If possible, compute:
(a) AB^T. (b) CA^T. (c) $(BA^T)C$. (d) A^TB.
(e) CC^T. (f) C^TC. (g) B^TCAA^T.

7. Let $A = \begin{bmatrix} 2 & -3 & 1 \\ 1 & 2 & 4 \end{bmatrix}$ and $B = \begin{bmatrix} 3 \\ 5 \\ 2 \end{bmatrix}$.

(a) Verify that $AB = 3A_1 + 5A_2 + 2A_3$, where A_j is the jth column of A for $j = 1, 2, 3$.

(b) Verify that $AB = \begin{bmatrix} p_1 \\ p_2 \end{bmatrix}$, where $p_j = (\text{row}_j\,(A))B$, $j = 1, 2$.

8. Find a value of r so that $AB^T = 0$, where $A = [r \quad 1 \quad -2]$ and $B = [1 \quad 3 \quad -1]$.

9. Find a value of r and a value of s so that $AB^T = 0$, where $A = [1 \quad r \quad 1]$ and $B = [-2 \quad 2 \quad s]$.

10. (a) Let A be an $m \times n$ matrix with a row consisting entirely of zeros. Prove that if B is an $n \times p$ matrix, then AB has a row of zeros.
(b) Let A be an $m \times n$ matrix with a column consisting entirely of zeros and let B be $p \times m$. Prove that BA has a column of zeros.

11. Let $A = \begin{bmatrix} -3 & 2 & 1 \\ 4 & 5 & 0 \end{bmatrix}$ with c_j = the jth column of A, $j = 1, 2$. Verify that

$$A^TA = \begin{bmatrix} c_1^Tc_1 & c_1^Tc_2 & c_1^Tc_3 \\ c_2^Tc_1 & c_2^Tc_2 & c_2^Tc_3 \\ c_3^Tc_1 & c_3^Tc_2 & c_3^Tc_3 \end{bmatrix}.$$

12. True or False:

(a) $\sum_{i=1}^{n} (a_i + 1) = \sum_{i=1}^{n} a_i + n.$

(b) $\sum_{i=1}^{n} \sum_{j=1}^{m} 1 = mn.$

(c) $\sum_{j=1}^{m} \sum_{i=1}^{n} a_ib_j = \left[\sum_{i=1}^{n} a_i\right]\left[\sum_{j=1}^{m} b_j\right].$

13. Let $A = \begin{bmatrix} 1 & 2 \\ 3 & 2 \end{bmatrix}$ and $B = \begin{bmatrix} 2 & -1 \\ -3 & 4 \end{bmatrix}$. Show that $AB \neq BA$.

14. Consider the following linear system:

$$\begin{aligned} 2x_1 + 3x_2 - 3x_3 + x_4 + x_5 &= 7 \\ 3x_1 + 2x_3 + 3x_5 &= -2 \\ 2x_1 + 3x_2 - 4x_4 &= 3 \\ x_3 + x_4 + x_5 &= 5. \end{aligned}$$

(a) Find the coefficient matrix.
(b) Write the linear system in matrix form.
(c) Find the augmented matrix.

15. Write the linear system whose augmented matrix is

$$\left[\begin{array}{rrrr|r} -2 & -1 & 0 & 4 & 5 \\ -3 & 2 & 7 & 8 & 3 \\ 1 & 0 & 0 & 2 & 4 \\ 3 & 0 & 1 & 3 & 6 \end{array}\right].$$

16. If $\begin{bmatrix} a+b & c+d \\ c-d & a-b \end{bmatrix} = \begin{bmatrix} 4 & 6 \\ 10 & 2 \end{bmatrix}$, find a, b, c, and d.

17. Write the following linear system in matrix form.

$$\begin{aligned} 2x_1 + 3x_2 &= 0 \\ 3x_2 + x_3 &= 0 \\ 2x_1 - x_3 &= 0. \end{aligned}$$

18. Write the linear system whose augmented matrix is:

(a) $\left[\begin{array}{rrrr|r} 2 & 1 & 3 & 4 & 0 \\ 3 & -1 & 2 & 0 & 3 \\ -2 & 1 & -4 & 3 & 2 \end{array}\right].$

(b) $\left[\begin{array}{rrrr|r} 2 & 1 & 3 & 4 & 0 \\ 3 & -1 & 2 & 0 & 3 \\ -2 & 1 & -4 & 3 & 2 \\ 0 & 0 & 0 & 0 & 0 \end{array}\right].$

19. How are the linear systems obtained in Exercise 18 related?

20. Write each of the following as a linear system in matrix form.

(a) $x_1\begin{bmatrix} 1 \\ 2 \end{bmatrix} + x_2\begin{bmatrix} 2 \\ 5 \end{bmatrix} + x_3\begin{bmatrix} 0 \\ 3 \end{bmatrix} = \begin{bmatrix} 1 \\ 1 \end{bmatrix}.$

(b) $x_1\begin{bmatrix} 1 \\ 1 \\ 2 \end{bmatrix} + x_2\begin{bmatrix} 2 \\ 1 \\ 0 \end{bmatrix} + x_3\begin{bmatrix} 1 \\ 2 \\ 2 \end{bmatrix} = \begin{bmatrix} 0 \\ 0 \\ 0 \end{bmatrix}.$

21. Write each of the following as a linear system in matrix form.
(a) $x_1[1 \quad 2 \quad 1] + x_2[3 \quad 0 \quad -1] = [3 \quad 1 \quad 4]$.
(b) $x_1[2 \quad 1 \quad 0 \quad 1] + x_2[3 \quad -1 \quad 2 \quad 2] + x_3[0 \quad 1 \quad -1 \quad 3] = [0 \quad 0 \quad 0 \quad 0]$.

22. Let X_1 and X_2 be solutions to the homogeneous linear system $AX = O$.
(a) Show that $X_1 + X_2$ is a solution.
(b) Show that $X_1 - X_2$ is a solution.
(c) For any scalar r, show that rX_1 is a solution.
(d) For any scalars r and s, show that $rX_1 + sX_2$ is a solution.

23. If $A = [a_{ij}]$ is an $n \times n$ matrix, then the **trace** of A, $\text{Tr}(A)$, is defined as the sum of all elements on the main diagonal of A, $\text{Tr}(A) = \sum_{i=1}^{n} a_{ii}$. Prove:
(a) $\text{Tr}(cA) = c\,\text{Tr}(A)$, where c is a real number.
(b) $\text{Tr}(A + B) = \text{Tr}(A) + \text{Tr}(B)$.
(c) $\text{Tr}(AB) = \text{Tr}(BA)$.
(d) $\text{Tr}(A^T) = \text{Tr}(A)$.
(e) $\text{Tr}(A^TA) \geq 0$.

24. Compute the trace (see Exercise 23) of each of the following matrices.

(a) $\begin{bmatrix} 1 & 0 \\ 2 & 3 \end{bmatrix}$. (b) $\begin{bmatrix} 2 & 2 & 3 \\ 2 & 4 & 4 \\ 3 & -2 & -5 \end{bmatrix}$.

(c) $\begin{bmatrix} 1 & 0 & 0 \\ 0 & 1 & 0 \\ 0 & 0 & 1 \end{bmatrix}$.

25. Show that there are no 2×2 matrices A and B such that $AB - BA = \begin{bmatrix} 1 & 0 \\ 0 & 1 \end{bmatrix}$.

26. Show that the jth column of the matrix product AB is equal to the matrix product AB_j, where B_j is the jth column of B. It follows that the product AB can be written in terms of columns as

$$AB = [AB_1 \quad AB_2 \quad \cdots \quad AB_n].$$

27. Show that if $AX = B$ has more than one solution, then it has infinitely many solutions. (*Hint:* If X_1 and X_2 are solutions, consider $X_3 = rX_1 + sX_2$, where $r + s = 1$.)

28. Show that if X_1 and X_2 are solutions to the linear system $AX = B$, then $X_1 - X_2$ is a solution to the associated homogeneous system $AX = O$.

29. For the software you are using, determine the commands for matrix addition, scalar multiplication, matrix multiplication, and the transpose.

30. Use the matrices A and C in Exercise 1 and the matrix multiplication command to compute AC and CA.

31. For the software you are using, determine the command for obtaining the powers $A^2, A^3, \ldots$ of a square matrix A. Then for

$$A = \begin{bmatrix} 0 & 1 & 0 & 0 & 0 \\ 0 & 0 & 1 & 0 & 0 \\ 0 & 0 & 0 & 1 & 0 \\ 0 & 0 & 0 & 0 & 1 \\ 0 & 0 & 0 & 0 & 0 \end{bmatrix}$$

compute the matrix sequence A^k, $k = 2, 3, 4, 5, 6$. Describe the behavior of A^k as $k \to \infty$.

32. Experiment to determine the behavior of the matrix sequence A^k as $k \to \infty$ for each of the following matrices.

(a) $A = \begin{bmatrix} \frac{1}{2} & \frac{1}{3} \\ \frac{1}{4} & \frac{1}{5} \end{bmatrix}$. (b) $A = \begin{bmatrix} 1 & -1 & 0 \\ 0 & 1 & -1 \\ -1 & 0 & 1 \end{bmatrix}$.

33. Compute B^TB and BB^T for

$$B = [1 \quad \tfrac{1}{2} \quad \tfrac{1}{3} \quad \tfrac{1}{4} \quad \tfrac{1}{5} \quad \tfrac{1}{6}].$$

1.3. *Algebraic Properties of Matrix Operations*

In this section we consider the algebraic properties of the matrix operations just defined. Many of these properties are similar to the familiar properties holding for the real numbers. However, there will be striking differences between the set of real numbers and the set of matrices in their algebraic behavior under certain operations, for example, under multiplication (as seen in Section 1.2). The proofs of most of the properties will be left as exercises.

Proof for Group matrix

Theorem 1.1 (*Properties of Matrix Addition*). *Let A, B, and C be $m \times n$ matrices.*

(a) $A + B = B + A$.

(b) $A + (B + C) = (A + B) + C$.

(c) *There is a unique $m \times n$ matrix ${}_mO_n$ such that*

$$A + O = A \tag{1}$$

for any $m \times n$ matrix A. The matrix ${}_mO_n$ is called the $m \times n$ **zero matrix.** *When $m = n$, we write O_n. When m and n are understood, we shall write ${}_mO_n$ simply as O.*

(d) *For each $m \times n$ matrix A, there is a unique $m \times n$ matrix D such that*

$$A + D = O. \tag{2}$$

We shall write D as $-A$ so that (2) can be written as

$$A + (-A) = O.$$

The matrix $-A$ is called the **negative** *of A.*

Proof: (a) Let

$$A = [a_{ij}],\ B = [b_{ij}],\quad A + B = C = [c_{ij}],\quad \text{and}\quad B + A = D = [d_{ij}].$$

We must show that $c_{ij} = d_{ij}$ for all i, j. Now $c_{ij} = a_{ij} + b_{ij}$ and $d_{ij} = b_{ij} + a_{ij}$ for all i, j. Since a_{ij} and b_{ij} are real numbers, we have $a_{ij} + b_{ij} = b_{ij} + a_{ij}$, which implies that $c_{ij} = d_{ij}$ for all i, j.

(c) Let $U = [u_{ij}]$. Then $A + U = A$ if and only if[†] $a_{ij} + u_{ij} = a_{ij}$, which holds if and only if $u_{ij} = 0$. Thus U is the $m \times n$ matrix all of whose entries are zero; U is denoted by ${}_mO_n$, or simply by O. ■

[†] The connector "if and only if" means that both statements are true or both statements are false. Thus (i) if $A + U = A$, then $a_{ij} + u_{ij} = a_{ij}$ and (ii) if $a_{ij} + u_{ij} = a_{ij}$, then $A + U = A$.

Example 1. The 2×2 zero matrix is

$$O_2 = \begin{bmatrix} 0 & 0 \\ 0 & 0 \end{bmatrix}.$$

If

$$A = \begin{bmatrix} 4 & -1 \\ 2 & 3 \end{bmatrix},$$

then

$$\begin{bmatrix} 4 & -1 \\ 2 & 3 \end{bmatrix} + \begin{bmatrix} 0 & 0 \\ 0 & 0 \end{bmatrix} = \begin{bmatrix} 4+0 & -1+0 \\ 2+0 & 3+0 \end{bmatrix} = \begin{bmatrix} 4 & -1 \\ 2 & 3 \end{bmatrix}.$$

The 2×3 zero matrix is

$$_2O_3 = \begin{bmatrix} 0 & 0 & 0 \\ 0 & 0 & 0 \end{bmatrix}.$$

■

Example 2. If $A = \begin{bmatrix} 1 & 3 & -2 \\ -2 & 4 & 3 \end{bmatrix}$, then $-A = \begin{bmatrix} -1 & -3 & 2 \\ 2 & -4 & -3 \end{bmatrix}$. ■

Theorem 1.2 *(Properties of Matrix Multiplication)*
(a) *If A, B, and C are of the appropriate sizes, then*

$$A(BC) = (AB)C.$$

(b) *If A, B, and C are of the appropriate sizes, then*

$$(A + B)C = AC + BC.$$

(c) *If A, B, and C are of the appropriate sizes, then*

$$C(A + B) = CA + CB.$$

Proof: (a) Suppose that A is $m \times n$, B is $n \times p$, and C is $p \times q$. We shall prove the result for $m = 2$, $n = 3$, $p = 4$, and $q = 3$. The general proof is completely analogous.

Let $A = [a_{ij}]$, $B = [b_{ij}]$, $C = [c_{ij}]$, $AB = D = [d_{ij}]$, $BC = E = [e_{ij}]$,

$(AB)C = F = [f_{ij}]$, and $A(BC) = G = [g_{ij}]$. We must show that $f_{ij} = g_{ij}$ for all i, j. Now

$$f_{ij} = \sum_{k=1}^{4} d_{ik}c_{kj} = \sum_{k=1}^{4} \left(\sum_{r=1}^{3} a_{ir}b_{rk} \right) c_{kj}$$

and

$$g_{ij} = \sum_{r=1}^{3} a_{ir}e_{rj} = \sum_{r=1}^{3} a_{ir} \left(\sum_{k=1}^{4} b_{rk}c_{kj} \right).$$

Then

$$\begin{aligned} f_{ij} &= \sum_{k=1}^{4} (a_{i1}b_{1k} + a_{i2}b_{2k} + a_{i3}b_{3k})c_{kj} \\ &= a_{i1} \sum_{k=1}^{4} b_{1k}c_{kj} + a_{i2} \sum_{k=1}^{4} b_{2k}c_{kj} + a_{i3} \sum_{k=1}^{4} b_{3k}c_{kj} \\ &= \sum_{r=1}^{3} a_{ir} \left(\sum_{k=1}^{4} b_{rk}c_{kj} \right) = g_{ij}. \end{aligned}$$

(b) and (c) are left as exercises. ■

Example 3. Let

$$A = \begin{bmatrix} 5 & 2 & 3 \\ 2 & -3 & 4 \end{bmatrix}, \quad B = \begin{bmatrix} 2 & -1 & 1 & 0 \\ 0 & 2 & 2 & 2 \\ 3 & 0 & -1 & 3 \end{bmatrix},$$

and

$$C = \begin{bmatrix} 1 & 0 & 2 \\ 2 & -3 & 0 \\ 0 & 0 & 3 \\ 2 & 1 & 0 \end{bmatrix}.$$

Then

$$A(BC) = \begin{bmatrix} 5 & 2 & 3 \\ 2 & -3 & 4 \end{bmatrix} \begin{bmatrix} 0 & 3 & 7 \\ 8 & -4 & 6 \\ 9 & 3 & 3 \end{bmatrix} = \begin{bmatrix} 43 & 16 & 56 \\ 12 & 30 & 8 \end{bmatrix}$$

and

$$(AB)C = \begin{bmatrix} 19 & -1 & 6 & 13 \\ 16 & -8 & -8 & 6 \end{bmatrix} \begin{bmatrix} 1 & 0 & 2 \\ 2 & -3 & 0 \\ 0 & 0 & 3 \\ 2 & 1 & 0 \end{bmatrix} = \begin{bmatrix} 43 & 16 & 56 \\ 12 & 30 & 8 \end{bmatrix}.$$

■

Example 4. Let

$$A = \begin{bmatrix} 2 & 2 & 3 \\ 3 & -1 & 2 \end{bmatrix}, \quad B = \begin{bmatrix} 0 & 0 & 1 \\ 2 & 3 & -1 \end{bmatrix}, \quad \text{and} \quad C = \begin{bmatrix} 1 & 0 \\ 2 & 2 \\ 3 & -1 \end{bmatrix}.$$

Then

$$(A + B)C = \begin{bmatrix} 2 & 2 & 4 \\ 5 & 2 & 1 \end{bmatrix} \begin{bmatrix} 1 & 0 \\ 2 & 2 \\ 3 & -1 \end{bmatrix} = \begin{bmatrix} 18 & 0 \\ 12 & 3 \end{bmatrix}$$

and (verify)

$$AC + BC = \begin{bmatrix} 15 & 1 \\ 7 & -4 \end{bmatrix} + \begin{bmatrix} 3 & -1 \\ 5 & 7 \end{bmatrix} = \begin{bmatrix} 18 & 0 \\ 12 & 3 \end{bmatrix}.$$

■

Recall Example 9 in Section 1.2, which shows that AB need not always equal BA. This is the first significant difference between multiplication of matrices and multiplication of real numbers.

Theorem 1.3 (Properties of Scalar Multiplication)

If r and s are real numbers and A and B are matrices, then
(a) $r(sA) = (rs)A$.
(b) $(r + s)A = rA + sA$.
(c) $r(A + B) = rA + rB$.
(d) $A(rB) = r(AB) = (rA)B$.

Proof: Exercises. ■

Example 5. Let

$$A = \begin{bmatrix} 4 & 2 & 3 \\ 2 & -3 & 4 \end{bmatrix} \quad \text{and} \quad B = \begin{bmatrix} 3 & -2 & 1 \\ 2 & 0 & -1 \\ 0 & 1 & 2 \end{bmatrix}.$$

Then

$$2(3A) = 2\begin{bmatrix} 12 & 6 & 9 \\ 6 & -9 & 12 \end{bmatrix} = \begin{bmatrix} 24 & 12 & 18 \\ 12 & -18 & 24 \end{bmatrix} = 6A.$$

We also have

$$A(2B) = \begin{bmatrix} 4 & 2 & 3 \\ 2 & -3 & 4 \end{bmatrix}\begin{bmatrix} 6 & -4 & 2 \\ 4 & 0 & -2 \\ 0 & 2 & 4 \end{bmatrix} = \begin{bmatrix} 32 & -10 & 16 \\ 0 & 0 & 26 \end{bmatrix} = 2(AB).$$ ■

So far we have seen that multiplication and addition of matrices have much in common with multiplication and addition of real numbers. We now look at some properties of the transpose.

Theorem 1.4 (Properties of Transpose)

If r is a scalar and A and B are matrices, then
(a) $(A^T)^T = A$.
(b) $(A + B)^T = A^T + B^T$.
(c) $(AB)^T = B^TA^T$.
(d) $(rA)^T = rA^T$.

Proof: We leave the proofs of (a), (b), and (d) as exercises.

(c) Let $A = [a_{ij}]$ and $B = [b_{ij}]$; let $AB = C = [c_{ij}]$. We must prove that c_{ij}^T is the (i, j) entry in $B^T A^T$. Now

$$c_{ij}^T = c_{ji} = \sum_{k=1}^{n} a_{jk} b_{ki} = \sum_{k=1}^{n} a_{kj}^T b_{ik}^T = \sum_{k=1}^{n} b_{ik}^T a_{kj}^T = \text{the } (i, j) \text{ entry in } B^T A^T. \quad \blacksquare$$

Example 6. Let

$$A = \begin{bmatrix} 1 & 2 & 3 \\ -2 & 0 & 1 \end{bmatrix} \quad \text{and} \quad B = \begin{bmatrix} 3 & -1 & 2 \\ 3 & 2 & -1 \end{bmatrix}.$$

Then

$$A^T = \begin{bmatrix} 1 & -2 \\ 2 & 0 \\ 3 & 1 \end{bmatrix} \quad \text{and} \quad B^T = \begin{bmatrix} 3 & 3 \\ -1 & 2 \\ 2 & -1 \end{bmatrix}.$$

Also,

$$A + B = \begin{bmatrix} 4 & 1 & 5 \\ 1 & 2 & 0 \end{bmatrix} \quad \text{and} \quad (A + B)^T = \begin{bmatrix} 4 & 1 \\ 1 & 2 \\ 5 & 0 \end{bmatrix}.$$

Now

$$A^T + B^T = \begin{bmatrix} 4 & 1 \\ 1 & 2 \\ 5 & 0 \end{bmatrix} = (A + B)^T.$$

$\blacksquare$

Example 7. Let

$$A = \begin{bmatrix} 1 & 3 & 2 \\ 2 & -1 & 3 \end{bmatrix} \quad \text{and} \quad B = \begin{bmatrix} 0 & 1 \\ 2 & 2 \\ 3 & -1 \end{bmatrix}.$$

Then

$$AB = \begin{bmatrix} 12 & 5 \\ 7 & -3 \end{bmatrix} \quad \text{and} \quad (AB)^T = \begin{bmatrix} 12 & 7 \\ 5 & -3 \end{bmatrix}.$$

On the other hand,

$$A^T = \begin{bmatrix} 1 & 2 \\ 3 & -1 \\ 2 & 3 \end{bmatrix} \quad \text{and} \quad B^T = \begin{bmatrix} 0 & 2 & 3 \\ 1 & 2 & -1 \end{bmatrix},$$

and then

$$B^TA^T = \begin{bmatrix} 12 & 7 \\ 5 & -3 \end{bmatrix} = (AB)^T.$$

■

We also note two other peculiarities of matrix multiplication. If a and b are real numbers, then $ab = 0$ can hold only if a or b is zero. However, this is not true for matrices.

Example 8. If $A = \begin{bmatrix} 1 & 2 \\ 2 & 4 \end{bmatrix}$ and $B = \begin{bmatrix} 4 & -6 \\ -2 & 3 \end{bmatrix}$, then neither A nor B is the zero matrix, but $AB = \begin{bmatrix} 0 & 0 \\ 0 & 0 \end{bmatrix}$. ■

If a, b, and c are real numbers for which $ab = ac$ and $a \neq 0$, it follows that $b = c$. That is, we can cancel out the nonzero factor a. However, the cancellation law does not hold for matrices, as the following example shows.

Example 9. If

$$A = \begin{bmatrix} 1 & 2 \\ 2 & 4 \end{bmatrix}, \quad B = \begin{bmatrix} 2 & 1 \\ 3 & 2 \end{bmatrix}, \quad \text{and} \quad C = \begin{bmatrix} -2 & 7 \\ 5 & -1 \end{bmatrix},$$

then

$$AB = AC = \begin{bmatrix} 8 & 5 \\ 16 & 10 \end{bmatrix},$$

but $B \neq C$. ■

We summarize the differences between matrix multiplication and the multiplication of real numbers as follows:

1. AB need not equal BA.
2. AB may be the zero matrix with $A \neq O$ and $B \neq O$.
3. AB may equal AC with $B \neq C$.

In this section we have developed a number of properties about matrices and their transposes. If a future problem either asks a question about these ideas or involves these concepts, refer to these properties to help answer the question. These results can be used to develop many more results.

1.3. EXERCISES

1. Prove Theorem 1.1(b).

2. Prove Theorem 1.1(d).

3. Verify Theorem 1.2(a) for the following matrices:

$$A = \begin{bmatrix} 1 & 3 \\ 2 & -1 \end{bmatrix}, \quad B = \begin{bmatrix} -1 & 3 & 2 \\ 1 & -3 & 4 \end{bmatrix},$$

$$\text{and} \quad C = \begin{bmatrix} 1 & 0 \\ 3 & -1 \\ 1 & 2 \end{bmatrix}.$$

4. Prove Theorem 1.2(b) and (c).

5. Verify Theorem 1.2(c) for the following matrices:

$$A = \begin{bmatrix} 2 & -3 & 2 \\ 3 & -1 & -2 \end{bmatrix}, \quad B = \begin{bmatrix} 0 & 1 & 2 \\ 1 & 3 & -2 \end{bmatrix},$$

$$\text{and} \quad C = \begin{bmatrix} 1 & -3 \\ -3 & 4 \end{bmatrix}.$$

6. Let A be an $m \times n$ matrix and $B = \begin{bmatrix} b_1 \\ b_2 \\ \vdots \\ b_n \end{bmatrix}$, an $n \times 1$ matrix. Prove that

$$AB = \sum_{j=1}^{n} b_j A_j,$$

where A_j is the jth column of A.

7. Let A be an $m \times n$ matrix and $C = [c_1 \; c_2 \; \cdots \; c_m]$ a $1 \times m$ matrix. Prove that

$$CA = \sum_{j=1}^{m} c_j A_j,$$

where A_j is the jth row of A.

8. Let $A = \begin{bmatrix} \cos\theta & \sin\theta \\ -\sin\theta & \cos\theta \end{bmatrix}$.

(a) Determine a simple expression for A^2.
(b) Determine a simple expression for A^3.
(c) Conjecture the form of a simple expression for A^k, k a positive integer.
(d) Prove or disprove your conjecture in part (c).

9. Find a pair of unequal 2×2 matrices A and B, other than those given in Example 8, such that $AB = O_2$.

10. Find two different 2×2 matrices A such that

$$A^2 = AA = \begin{bmatrix} 1 & 0 \\ 0 & 1 \end{bmatrix}.$$

11. Find two unequal 2×2 matrices A and B such that $AB = \begin{bmatrix} 1 & 0 \\ 0 & 1 \end{bmatrix}$.

12. Find two different 2×2 matrices A such that $A^2 = AA = O_2$.

13. Prove Theorem 1.3(a).

14. Prove Theorem 1.3(b).

15. Verify Theorem 1.3(b) for $r = 4$, $s = -2$, and

$$A = \begin{bmatrix} 2 & -3 \\ 4 & 2 \end{bmatrix}.$$

16. Prove Theorem 1.3(c).

17. Verify Theorem 1.3(c) for $r = -3$ and

$$A = \begin{bmatrix} 4 & 2 \\ 1 & -3 \\ 3 & 2 \end{bmatrix} \quad \text{and} \quad B = \begin{bmatrix} 0 & 2 \\ 4 & 3 \\ -2 & 1 \end{bmatrix}.$$

18. Prove Theorem 1.3(d).

19. Verify Theorem 1.3(d) for the following matrices:

$$A = \begin{bmatrix} 1 & 3 \\ 2 & -1 \end{bmatrix}, \quad B = \begin{bmatrix} -1 & 3 & 2 \\ 1 & -3 & 4 \end{bmatrix},$$

and $r = -3$.

20. Determine a scalar r such that $AX = rX$, where

$$A = \begin{bmatrix} 2 & 1 \\ 1 & 2 \end{bmatrix} \quad \text{and} \quad X = \begin{bmatrix} 1 \\ 1 \end{bmatrix}.$$

21. Determine a scalar r such that $AX = rX$, where

$$A = \begin{bmatrix} 1 & 2 & -1 \\ 1 & 0 & 1 \\ 4 & -4 & 5 \end{bmatrix} \quad \text{and} \quad X = \begin{bmatrix} -\frac{1}{2} \\ \frac{1}{4} \\ 1 \end{bmatrix}.$$

22. Prove that if $AX = rX$ for $n \times n$ matrix A, $n \times 1$ matrix X, and scalar r, then $AY = rY$, where $Y = sX$ for any scalar s.

23. Determine a scalar s such that $A^2X = sX$ when $AX = rX$.

24. Prove Theorem 1.4(a).

25. Prove Theorem 1.4(b) and (d).

26. Verify Theorem 1.4(a), (b), and (d).

$$A = \begin{bmatrix} 1 & 3 & 2 \\ 2 & 1 & -3 \end{bmatrix}, \quad B = \begin{bmatrix} 4 & 2 & -1 \\ -2 & 1 & 5 \end{bmatrix},$$

and $r = -4$.

27. Verify Theorem 1.4(c) for $A = \begin{bmatrix} 1 & 3 & 2 \\ 2 & 1 & -3 \end{bmatrix}$

and $B = \begin{bmatrix} 3 & -1 \\ 2 & 4 \\ 1 & 2 \end{bmatrix}$.

28. Let $A = \begin{bmatrix} 2 \\ -1 \\ 3 \end{bmatrix}$, $B = \begin{bmatrix} 3 \\ -2 \\ -4 \end{bmatrix}$, and $C = \begin{bmatrix} -1 \\ 5 \\ 1 \end{bmatrix}$.

(a) Compute $(AB^T)C$.
(b) Compute B^TC and multiply the result times A.
(c) Explain why $(AB^T)C = (B^TC)A$.

29. Determine a constant k such that $(kA)^T(kA) = 1$, where $A = \begin{bmatrix} -2 \\ 1 \\ -1 \end{bmatrix}$. Is there more than one value of k that could be used?

30. Find three 2×2 matrices, A, B, and C, such that $AB = AC$ with $B \neq C$ and $A \neq O_2$.

31. Let A be an $m \times n$ matrix and c a real number. Show that if $cA = O_n$, then $c = 0$ or $A = O_n$.

32. Determine all 2×2 matrices A such that $AB = BA$ for any 2×2 matrix B.

33. Show that $(A - B)^T = A^T - B^T$.

34. Let $A = \begin{bmatrix} 6 & -1 & 1 \\ 0 & 13 & -16 \\ 0 & 8 & -11 \end{bmatrix}$ and $X = \begin{bmatrix} 10.5 \\ 21 \\ 10.5 \end{bmatrix}$.

(a) Determine a scalar r such that $AX = rX$.

(b) Is it true that $A^T X = rX$, for the value r determined in part (a)?

35. Repeat Exercise 34 with

$$A = \begin{bmatrix} -3.35 & -3.00 & 3.60 \\ 1.20 & 2.05 & -6.20 \\ -3.60 & -2.40 & 3.85 \end{bmatrix} \quad \text{and}$$

$$X = \begin{bmatrix} 12.5 \\ -12.5 \\ 6.25 \end{bmatrix}.$$

36. Let $A = \begin{bmatrix} 0.1 & 0.01 \\ 0.001 & 0.0001 \end{bmatrix}$. In your software, set the display format to show as many decimal places as possible, then compute

$$B = 10 * A,$$

$$C = \underbrace{A + A + A + A + A + A + A + A + A + A}_{10 \text{ summands}},$$

$$\text{and} \quad D = B - C.$$

If D is not O_2, then you have verified that scalar multiplication by a positive integer and successive addition are not the same in your computing environment. (It is not unusual that $D \neq O_2$ since many computing environments use only a "model" of exact arithmetic, called floating point arithmetic.)

37. Let $A = \begin{bmatrix} 1 & 1 \\ 1 & 1 \end{bmatrix}$. In your software, set the display to show as many decimal places as possible. Experiment to find a positive integer k such that $A + 10^{-k} * A$ is equal to A. If you find such an integer k, you have verified that there is more than one matrix in your computational environment that plays the role of O_2.

1.4. *Special Types of Matrices and Partitioned Matrices*

We have already introduced one special type of matrix, ${}_mO_n$, the matrix all of whose entries are zero. We now consider several other types of matrices whose structure is rather specialized and for which it will be convenient to have special names.

An $n \times n$ matrix $A = [a_{ij}]$ is called a **diagonal matrix** if $a_{ij} = 0$ for $i \neq j$. Thus, for a diagonal matrix, the terms *off* the main diagonal are all zero. Note that O_n is a diagonal matrix. A **scalar matrix** is a diagonal matrix whose diagonal elements are equal. The scalar matrix $I_n = [a_{ij}]$, where $a_{ii} = 1$ and $a_{ij} = 0$ for $i \neq j$, is called the $n \times n$ **identity matrix**.

Example 1. Let

$$A = \begin{bmatrix} 1 & 0 & 0 \\ 0 & 2 & 0 \\ 0 & 0 & 3 \end{bmatrix}, \quad B = \begin{bmatrix} 2 & 0 & 0 \\ 0 & 2 & 0 \\ 0 & 0 & 2 \end{bmatrix}, \quad \text{and} \quad I_3 = \begin{bmatrix} 1 & 0 & 0 \\ 0 & 1 & 0 \\ 0 & 0 & 1 \end{bmatrix}.$$

Then A, B, and I_3 are diagonal matrices; B and I_3 are scalar matrices; and I_3 is the 3×3 identity matrix. ■

It is easy to show that if A is any $m \times n$ matrix, then

$$AI_n = A \qquad \text{and} \qquad I_m A = A.$$

Also, if A is a scalar matrix, then $A = rI_n$ for some scalar r.

Suppose that A is a square matrix. If p is a positive integer, then we define

$$A^p = \underbrace{A \cdot A \cdot \cdots \cdot A}_{p \text{ factors}}.$$

If A is $n \times n$, we also define

$$A^0 = I_n.$$

For nonnegative integers p and q, the familiar laws of exponents for the real numbers can also be proved for matrix multiplication of a square matrix A (Exercise 6):

$$A^p A^q = A^{p+q}$$

and

$$(A^p)^q = A^{pq}.$$

It should be noted that the rule

$$(AB)^p = A^p B^p$$

does not hold for square matrices unless $AB = BA$ (Exercise 7).

An $n \times n$ matrix $A = [a_{ij}]$ is called **upper triangular** if $a_{ij} = 0$ for $i > j$. It is called **lower triangular** if $a_{ij} = 0$ for $i < j$. A diagonal matrix is both upper triangular and lower triangular.

Example 2. The matrix

$$A = \begin{bmatrix} 1 & 3 & 3 \\ 0 & 3 & 5 \\ 0 & 0 & 2 \end{bmatrix}$$

is upper triangular and

$$B = \begin{bmatrix} 1 & 0 & 0 \\ 2 & 3 & 0 \\ 3 & 5 & 2 \end{bmatrix}$$

is lower triangular. ■

Definition 1.7. A matrix A is called **symmetric** if $A^T = A$. ■

Definition 1.8. A matrix A is called **skew symmetric** if $A^T = -A$. ■

Example 3. $A = \begin{bmatrix} 1 & 2 & 3 \\ 2 & 4 & 5 \\ 3 & 5 & 6 \end{bmatrix}$ is a symmetric matrix. ■

Example 4. $B = \begin{bmatrix} 0 & 2 & 3 \\ -2 & 0 & -4 \\ -3 & 4 & 0 \end{bmatrix}$ is a skew symmetric matrix. ■

We can make a few observations about symmetric and skew symmetric matrices; the proofs of most of these statements will be left as exercises.

It follows from the definitions above that if A is symmetric or skew symmetric, then A is a square matrix. If A is a symmetric matrix, then the entries of A are symmetric with respect to the main diagonal of A. Also, A is symmetric if and only if $a_{ij} = a_{ji}$ and A is skew symmetric if and only if $a_{ij} = -a_{ji}$. Moreover, if A is skew symmetric, then the entries on the main diagonal of A are all zero. An important property of symmetric and skew symmetric matrices is the following. If A is an $n \times n$ matrix, then we can show that $A = S + K$, where S is symmetric and K is skew symmetric. Moreover, this decomposition is unique.

Partitioned Matrices (Optional)

If we start out with an $m \times n$ matrix $A = [a_{ij}]$ and cross out some, but not all, of its rows or columns, we obtain a **submatrix** of A.

Example 5. Let $A = \begin{bmatrix} 1 & 2 & 3 & 4 \\ -2 & 4 & -3 & 5 \\ 3 & 0 & 5 & -3 \end{bmatrix}$. If we cross out the second row and third column, we obtain the submatrix $\begin{bmatrix} 1 & 2 & 4 \\ 3 & 0 & -3 \end{bmatrix}$. ■

We may now consider a matrix A as partitioned into submatrices. Of course, the partitioning may be done in many different ways.

Example 6. The matrix

$$A = \left[\begin{array}{ccc|cc} a_{11} & a_{12} & a_{13} & a_{14} & a_{15} \\ a_{21} & a_{22} & a_{23} & a_{24} & a_{25} \\ \hline a_{31} & a_{32} & a_{33} & a_{34} & a_{35} \\ a_{41} & a_{42} & a_{43} & a_{44} & a_{45} \end{array}\right]$$

is partitioned as

$$A = \begin{bmatrix} A_{11} & A_{12} \\ A_{21} & A_{22} \end{bmatrix}.$$

We could also write

$$A = \left[\begin{array}{cc|cc|c} a_{11} & a_{12} & a_{13} & a_{14} & a_{15} \\ a_{21} & a_{22} & a_{23} & a_{24} & a_{25} \\ \hline a_{31} & a_{32} & a_{33} & a_{34} & a_{35} \\ a_{41} & a_{42} & a_{43} & a_{44} & a_{45} \end{array}\right] = \begin{bmatrix} \hat{A}_{11} & \hat{A}_{12} & \hat{A}_{13} \\ \hat{A}_{21} & \hat{A}_{22} & \hat{A}_{23} \end{bmatrix}, \tag{1}$$

which gives another partitioning of A. We thus speak of **partitioned matrices**. ■

Example 7. The augmented matrix (defined in Section 1.2) of a linear system is a partitioned matrix. Thus, if $AX = B$, we can write the augmented matrix of this system as $[A \mid B]$. ■

If A is partitioned as shown in (1) and

$$B = \left[\begin{array}{cc:cc} b_{11} & b_{12} & b_{13} & b_{14} \\ b_{21} & b_{22} & b_{23} & b_{24} \\ \hdashline b_{31} & b_{32} & b_{33} & b_{34} \\ b_{41} & b_{42} & b_{43} & b_{44} \\ \hdashline b_{51} & b_{52} & b_{53} & b_{54} \end{array}\right] = \begin{bmatrix} B_{11} & B_{12} \\ B_{21} & B_{22} \\ B_{31} & B_{32} \end{bmatrix},$$

then by straightforward computation we can show that

$$AB = \left[\begin{array}{c:c} (A_{11}B_{11} + A_{12}B_{21} + A_{13}B_{31}) & (A_{11}B_{12} + A_{12}B_{22} + A_{13}B_{32}) \\ \hdashline (A_{21}B_{11} + A_{22}B_{21} + A_{23}B_{31}) & (A_{21}B_{12} + A_{22}B_{22} + A_{23}B_{32}) \end{array}\right]. \quad \blacksquare$$

Example 8. Let

$$A = \left[\begin{array}{cc:cc} 1 & 0 & 1 & 0 \\ 0 & 2 & 3 & -1 \\ \hdashline 2 & 0 & -4 & 0 \\ 0 & 1 & 0 & 3 \end{array}\right] = \begin{bmatrix} A_{11} & A_{12} \\ A_{21} & A_{22} \end{bmatrix}$$

and let

$$B = \left[\begin{array}{ccc:ccc} 2 & 0 & 0 & 1 & 1 & -1 \\ 0 & 1 & 1 & -1 & 2 & 2 \\ \hdashline 1 & 3 & 0 & 0 & 1 & 0 \\ -3 & -1 & 2 & 1 & 0 & -1 \end{array}\right] = \begin{bmatrix} B_{11} & B_{12} \\ B_{21} & B_{22} \end{bmatrix}.$$

Then

$$AB = C = \left[\begin{array}{ccc:ccc} 3 & 3 & 0 & 1 & 2 & -1 \\ 6 & 12 & 0 & -3 & 7 & 5 \\ \hdashline 0 & -12 & 0 & 2 & -2 & -2 \\ -9 & -2 & 7 & 2 & 2 & -1 \end{array}\right] = \begin{bmatrix} C_{11} & C_{12} \\ C_{21} & C_{22} \end{bmatrix},$$

where C_{11} should be $A_{11}B_{11} + A_{12}B_{21}$. We verify that C_{11} is this expression as follows

$$A_{11}B_{11} + A_{12}B_{21} = \begin{bmatrix} 1 & 0 \\ 0 & 2 \end{bmatrix}\begin{bmatrix} 2 & 0 & 0 \\ 0 & 1 & 1 \end{bmatrix} + \begin{bmatrix} 1 & 0 \\ 3 & -1 \end{bmatrix}\begin{bmatrix} 1 & 3 & 0 \\ -3 & -1 & 2 \end{bmatrix}$$

$$= \begin{bmatrix} 2 & 0 & 0 \\ 0 & 2 & 2 \end{bmatrix} + \begin{bmatrix} 1 & 3 & 0 \\ 6 & 10 & -2 \end{bmatrix}$$

$$= \begin{bmatrix} 3 & 3 & 0 \\ 6 & 12 & 0 \end{bmatrix} = C_{11}.$$

■

Partitioned matrices can be used to great advantage in dealing with matrices that exceed the memory capacity of a computer. Thus, in multiplying two partitioned matrices, one can keep the matrices on disk and only bring into memory the submatrices required to form the submatrix products. The latter, of course, can be put out on disk as they are formed. The partitioning must be done so that the products of corresponding submatrices are defined. The addition of partitioned matrices is performed in the obvious manner.

Nonsingular Matrices

We now come to a special type of square matrix and formulate the notion corresponding to the reciprocal of a nonzero real number.

Definition 1.9. An $n \times n$ matrix A is called **nonsingular**, or **invertible**, if there exists an $n \times n$ matrix B such that $AB = BA = I_n$; B is called an **inverse** of A. Otherwise, A is called **singular** or **noninvertible.** ■

Example 9. Let $A = \begin{bmatrix} 2 & 3 \\ 2 & 2 \end{bmatrix}$ and let $B = \begin{bmatrix} -1 & \frac{3}{2} \\ 1 & -1 \end{bmatrix}$. Since $AB = BA = I_2$, we conclude that B is an inverse of A. ■

Theorem 1.5. *The inverse of a matrix, if it exists, is unique.*

Proof: Let B and C be inverses of A. Then

$$AB = BA = I_n \qquad \text{and} \qquad AC = CA = I_n.$$

We then have $B = BI_n = B(AC) = (BA)C = I_nC = C$, which proves that the inverse of a matrix, if it exists, is unique. ■

We now write the inverse of a nonsingular matrix A, as A^{-1}. Thus

$$AA^{-1} = A^{-1}A = I_n.$$

Example 10. Let

$$A = \begin{bmatrix} 1 & 2 \\ 3 & 4 \end{bmatrix}.$$

To find A^{-1}, we let

$$A^{-1} = \begin{bmatrix} a & b \\ c & d \end{bmatrix}.$$

Then we must have

$$AA^{-1} = \begin{bmatrix} 1 & 2 \\ 3 & 4 \end{bmatrix}\begin{bmatrix} a & b \\ c & d \end{bmatrix} = I_2 = \begin{bmatrix} 1 & 0 \\ 0 & 1 \end{bmatrix}$$

so that

$$\begin{bmatrix} a + 2c & b + 2d \\ 3a + 4c & 3b + 4d \end{bmatrix} = \begin{bmatrix} 1 & 0 \\ 0 & 1 \end{bmatrix}.$$

Equating corresponding entries of these two matrices, we obtain the linear systems

$$\begin{aligned} a + 2c &= 1 \\ 3a + 4c &= 0 \end{aligned} \quad \text{and} \quad \begin{aligned} b + 2d &= 0 \\ 3b + 4d &= 1. \end{aligned}$$

The solutions are (verify) $a = -2$, $c = \frac{3}{2}$, $b = 1$, and $d = -\frac{1}{2}$. Moreover, since the matrix

$$\begin{bmatrix} a & b \\ c & d \end{bmatrix} = \begin{bmatrix} -2 & 1 \\ \frac{3}{2} & -\frac{1}{2} \end{bmatrix}$$

also satisfies the property that

$$\begin{bmatrix} -2 & 1 \\ \frac{3}{2} & -\frac{1}{2} \end{bmatrix} \begin{bmatrix} 1 & 2 \\ 3 & 4 \end{bmatrix} = \begin{bmatrix} 1 & 0 \\ 0 & 1 \end{bmatrix},$$

we conclude that A is nonsingular and that

$$A^{-1} = \begin{bmatrix} -2 & 1 \\ \frac{3}{2} & -\frac{1}{2} \end{bmatrix}.$$

■

Example 11. Let

$$A = \begin{bmatrix} 1 & 2 \\ 2 & 4 \end{bmatrix}.$$

To find A^{-1}, we let

$$A^{-1} = \begin{bmatrix} a & b \\ c & d \end{bmatrix}.$$

Then we must have

$$AA^{-1} = \begin{bmatrix} 1 & 2 \\ 2 & 4 \end{bmatrix} \begin{bmatrix} a & b \\ c & d \end{bmatrix} = I_2 = \begin{bmatrix} 1 & 0 \\ 0 & 1 \end{bmatrix}$$

so that

$$\begin{bmatrix} a + 2c & b + 2d \\ 2a + 4c & 2b + 4d \end{bmatrix} = \begin{bmatrix} 1 & 0 \\ 0 & 1 \end{bmatrix}.$$

Equating corresponding entries of these two matrices, we obtain the linear systems

$$\begin{aligned} a + 2c &= 1 \\ 2a + 4c &= 0 \end{aligned} \quad \text{and} \quad \begin{aligned} b + 2d &= 0 \\ 2b + 4d &= 1. \end{aligned}$$

These linear systems have no solutions, so A has no inverse. ■

We next establish several properties of inverses of matrices.

Theorem 1.6. *If A and B are both nonsingular $n \times n$ matrices, then AB is nonsingular and $(AB)^{-1} = B^{-1}A^{-1}$.*

Proof: We have $(AB)(B^{-1}A^{-1}) = A(BB^{-1})A^{-1} = (AI_n)A^{-1} = AA^{-1} = I_n$. Similarly, $(B^{-1}A^{-1})(AB) = I_n$. Therefore, AB is nonsingular. Since the inverse of a matrix is unique, we conclude that $(AB)^{-1} = B^{-1}A^{-1}$. ■

Corollary 1.1. *If $A_1, A_2, \ldots, A_r$ are $n \times n$ nonsingular matrices, then $A_1A_2 \cdots A_r$ is nonsingular and $(A_1A_2 \cdots A_r)^{-1} = A_r^{-1}A_{r-1}^{-1} \cdots A_1^{-1}$.*

Proof: Exercise. ■

Theorem 1.7. *If A is a nonsingular matrix, then A^{-1} is nonsingular and $(A^{-1})^{-1} = A$.*

Proof: Exercise. ■

Theorem 1.8. *If A is a nonsingular matrix, then A^T is nonsingular and $(A^T)^{-1} = (A^{-1})^T$.*

Proof: We have $AA^{-1} = I_n$. Taking transposes of both sides, we obtain

$$(A^{-1})^T A^T = I_n^T = I_n.$$

Taking transposes of both sides of the equation $A^{-1}A = I_n$, we find, similarly, that

$$(A^T)(A^{-1})^T = I_n.$$

These equations imply that $(A^{-1})^T = (A^T)^{-1}$. ■

Example 12. If

$$A = \begin{bmatrix} 1 & 2 \\ 3 & 4 \end{bmatrix},$$

then from Example 10

$$A^{-1} = \begin{bmatrix} -2 & 1 \\ \frac{3}{2} & -\frac{1}{2} \end{bmatrix} \quad \text{and} \quad (A^{-1})^T = \begin{bmatrix} -2 & \frac{3}{2} \\ 1 & -\frac{1}{2} \end{bmatrix}.$$

Also (verify)

$$A^T = \begin{bmatrix} 1 & 3 \\ 2 & 4 \end{bmatrix} \quad \text{and} \quad (A^T)^{-1} = \begin{bmatrix} -2 & \frac{3}{2} \\ 1 & -\frac{1}{2} \end{bmatrix}. \qquad \blacksquare$$

It follows from Theorem 1.8 that if A is a symmetric nonsingular matrix, then A^{-1} is symmetric (see Exercise 36).

Suppose that A is nonsingular. Then $AB = AC$ implies that $B = C$ (Exercise 32) and $AB = O_n$ implies that $B = O_n$ (Exercise 33).

Linear Systems and Inverses

If A is an $n \times n$ matrix, then the linear system $AX = B$ is a system of n equations in n unknowns. Suppose that A is nonsingular. Then A^{-1} exists and we can multiply $AX = B$ by A^{-1} on both sides, obtaining

$$A^{-1}(AX) = A^{-1}B,$$

or

$$I_n X = X = A^{-1}B. \tag{2}$$

Moreover, $X = A^{-1}B$ is clearly a solution to the given linear system. Thus, if A is nonsingular, we have a unique solution.

If A is a nonsingular $n \times n$ matrix, Equation (2) implies that if the linear system $AX = B$ needs to be solved repeatedly for different B's, we need only compute A^{-1} once; then whenever we change B, we find the corresponding solution X by forming $A^{-1}B$. Although this is certainly a valid approach, its value is of a more theoretical rather than practical sense, since a more efficient procedure for solving such problems will be presented in Section 1.8.

Example 13. Suppose that A is the matrix of Example 10 so that

$$A^{-1} = \begin{bmatrix} -2 & 1 \\ \frac{3}{2} & -\frac{1}{2} \end{bmatrix}.$$

If

$$B = \begin{bmatrix} 8 \\ 6 \end{bmatrix},$$

then the solution to the linear system $AX = B$ is

$$X = A^{-1}B = \begin{bmatrix} -2 & 1 \\ \frac{3}{2} & -\frac{1}{2} \end{bmatrix} \begin{bmatrix} 8 \\ 6 \end{bmatrix} = \begin{bmatrix} -10 \\ 9 \end{bmatrix}.$$

On the other hand, if

$$B = \begin{bmatrix} 10 \\ 20 \end{bmatrix},$$

then

$$X = A^{-1} \begin{bmatrix} 10 \\ 20 \end{bmatrix} = \begin{bmatrix} 0 \\ 5 \end{bmatrix}.$$

■

1.4. EXERCISES

1. (a) Show that if A is any $m \times n$ matrix, then $I_m A = A$ and $AI_n = A$.
(b) Show that if A is an $n \times n$ scalar matrix, then $A = rI_n$ for some real number r.

2. Prove that the sum, product, and scalar multiple of diagonal, scalar, and upper (lower) triangular matrices is diagonal, scalar, and upper (lower) triangular, respectively.

3. Prove: If A and B are $n \times n$ diagonal matrices, then $AB = BA$.

4. Let

$$A = \begin{bmatrix} 3 & 2 & -1 \\ 0 & -4 & 3 \\ 0 & 0 & 0 \end{bmatrix} \quad \text{and}$$

$$B = \begin{bmatrix} 6 & -3 & 2 \\ 0 & 2 & 4 \\ 0 & 0 & 3 \end{bmatrix}.$$

Verify that $A + B$ and AB are upper triangular.

5. Describe all matrices that are both upper and lower triangular.

6. Let p and q be nonnegative integers and let A be a square matrix. Show that

$$A^p A^q = A^{p+q} \quad \text{and} \quad (A^p)^q = A^{pq}.$$

7. If $AB = BA$, and p is a nonnegative integer, show that $(AB)^p = A^p B^p$.

8. If p is a nonnegative integer and c is a scalar, show that $(cA)^p = c^p A^p$.

9. Find a 2×2 matrix $B \neq O_2$ and $B \neq I_2$ such that $AB = BA$, where $A = \begin{bmatrix} 1 & 2 \\ 2 & 1 \end{bmatrix}$. How many such matrices B are there?

10. Find a 2×2 matrix $B \neq O_2$ and $B \neq I_2$ such that $AB = BA$, where $A = \begin{bmatrix} 1 & 2 \\ 0 & 1 \end{bmatrix}$. How many such matrices B are there?

11. Prove or disprove: For any $n \times n$ matrix A, $A^TA = AA^T$.

12. (a) Show that A is symmetric if and only if $a_{ij} = a_{ji}$ for all i, j.
(b) Show that A is skew symmetric if and only if $a_{ij} = -a_{ji}$ for all i, j.
(c) Show that if A is skew symmetric, then the elements on the main diagonal of A are all zero.

13. Show that if A is a symmetric matrix, then A^T is symmetric.

14. Describe all skew symmetric scalar matrices.

15. Show that if A is any $n \times n$ matrix, then:
(a) AA^T and A^TA are symmetric.
(b) $A + A^T$ is symmetric.
(c) $A - A^T$ is skew symmetric.

16. Let A and B be symmetric matrices.
(a) Show that $A + B$ is symmetric.
(b) Show that AB is symmetric if and only if $AB = BA$.

17. Show that if A is an $n \times n$ matrix, then $A = S + K$, where S is symmetric and K is skew symmetric. Also show that this decomposition is unique.

18. Let

$$A = \begin{bmatrix} 1 & 3 & -2 \\ 4 & 6 & 2 \\ 5 & 1 & 3 \end{bmatrix}.$$

Find the matrices S and K described in Exercise 17.

19. Show that the matrix $A = \begin{bmatrix} 2 & 3 \\ 4 & 6 \end{bmatrix}$ is singular.

20. If

$$D = \begin{bmatrix} 4 & 0 & 0 \\ 0 & -2 & 0 \\ 0 & 0 & 3 \end{bmatrix},$$

find D^{-1}.

21. Find the inverse of each of the following matrices.

(a) $A = \begin{bmatrix} 1 & 3 \\ 5 & 2 \end{bmatrix}$. (b) $A = \begin{bmatrix} 1 & 2 \\ 2 & 1 \end{bmatrix}$.

22. If A is a nonsingular matrix whose inverse is $\begin{bmatrix} 2 & 1 \\ 4 & 1 \end{bmatrix}$, find A.

23. If

$$A^{-1} = \begin{bmatrix} 3 & 2 \\ 1 & 3 \end{bmatrix} \quad \text{and} \quad B^{-1} = \begin{bmatrix} 2 & 5 \\ 3 & -2 \end{bmatrix},$$

find $(AB)^{-1}$.

24. Suppose that

$$A^{-1} = \begin{bmatrix} 1 & 2 \\ 1 & 3 \end{bmatrix}.$$

Solve the linear system $AX = B$ for each of the following matrices B.

(a) $\begin{bmatrix} 4 \\ 6 \end{bmatrix}$. (b) $\begin{bmatrix} 8 \\ 15 \end{bmatrix}$.

25. Consider the linear system $AX = B$, where A is the matrix defined in Exercise 21(a).

(a) Find a solution if $B = \begin{bmatrix} 3 \\ 4 \end{bmatrix}$.

(b) Find a solution if $B = \begin{bmatrix} 5 \\ 6 \end{bmatrix}$.

26. Find two 2×2 singular matrices whose sum is nonsingular.

27. Find two 2×2 nonsingular matrices whose sum is singular.

28. Prove Corollary 1.1.

29. Prove Theorem 1.7.

30. Prove that if one row (column) of the $n \times n$ matrix A consists entirely of zeros, then A is singular. [*Hint:* Assume that A is nonsingular; that is, there exists an $n \times n$ matrix B such that $AB = BA = I_n$. Establish a contradiction.]

31. Prove: If A is a diagonal matrix with nonzero diagonal entries $a_{11}, a_{22}, \ldots, a_{nn}$, then A is nonsingular and A^{-1} is a diagonal matrix with diagonal entries $1/a_{11}, 1/a_{22}, \ldots, 1/a_{nn}$.

32. Show that if $AB = AC$ and A is nonsingular, then $B = C$.

33. Show that if A is nonsingular and $AB = O_n$ for an $n \times n$ matrix B, then $B = O_n$.

34. Let $A = \begin{bmatrix} a & b \\ c & d \end{bmatrix}$. Show that A is nonsingular if and only if $ad - bc \neq 0$.

35. Consider the homogeneous system $AX = O$, where A is $n \times n$. If A is nonsingular, show that the only solution is the trivial one, $X = O$.

36. Prove that if A is symmetric and nonsingular, then A^{-1} is symmetric.

37. **(Optional)** Formulate the method for adding partitioned matrices and verify your method by partitioning the matrices

$$A = \begin{bmatrix} 1 & 3 & -1 \\ 2 & 1 & 0 \\ 2 & -3 & 1 \end{bmatrix} \quad \text{and}$$

$$B = \begin{bmatrix} 3 & 2 & 1 \\ -2 & 3 & 1 \\ 4 & 1 & 5 \end{bmatrix}$$

in two different ways and finding their sum.

38. **(Optional)** Let A and B be the following matrices:

$$A = \begin{bmatrix} 2 & 1 & 3 & 4 & 2 \\ 1 & 2 & 3 & -1 & 4 \\ 2 & 3 & 2 & 1 & 4 \\ 5 & -1 & 3 & 2 & 6 \\ 3 & 1 & 2 & 4 & 6 \\ 2 & -1 & 3 & 5 & 7 \end{bmatrix} \quad \text{and}$$

$$B = \begin{bmatrix} 1 & 2 & 3 & 4 & 1 \\ 2 & 1 & 3 & 2 & -1 \\ 1 & 5 & 4 & 2 & 3 \\ 2 & 1 & 3 & 5 & 7 \\ 3 & 2 & 4 & 6 & 1 \end{bmatrix}.$$

Find AB by partitioning A and B in two different ways.

39. For the software you are using, determine the command(s) or procedures required to do the following.

(a) Adjoin a row or column to an existing matrix.

(b) Construct the partitioned matrix

$$\begin{bmatrix} A & O \\ O & B \end{bmatrix}$$

from existing matrices A and B using appropriate size zero matrices.

(c) Extract a submatrix from an existing matrix.

40. Most software for linear algebra has specific commands for extracting the diagonal, upper triangular part, and lower triangular part of a matrix. Determine the corresponding commands for the software that you are using and experiment with them.

41. Determine the command for computing the inverse of a matrix in the software you use. Usually, if such a command is applied to a singular matrix, a warning message is displayed. Experiment with your inverse command to determine which of the following matrices are singular.

(a) $\begin{bmatrix} 1 & 2 & 3 \\ 4 & 5 & 6 \\ 7 & 8 & 0 \end{bmatrix}$. (b) $\begin{bmatrix} 1 & 2 & 3 \\ 4 & 5 & 6 \\ 7 & 8 & 9 \end{bmatrix}$.

(c) $\begin{bmatrix} 1 & 2 & 4 \\ -1 & 1 & -1 \\ 2 & -1 & 3 \end{bmatrix}$.

42. If B is the inverse of $n \times n$ matrix A, then Theorem 1.9 guarantees that $AB = BA = I_n$. The unstated assumption is that exact arithmetic is used. If computer arithmetic is used to compute AB, then AB need not equal I_n and in fact BA need not equal AB. However, both AB and BA should be close to I_n. In your software use the inverse command (see Exercise 41) and form the products AB and BA for the following matrices.

(a) $A = \begin{bmatrix} 1 & \frac{1}{3} \\ 0 & \frac{1}{3} \end{bmatrix}$. (b) $A = \begin{bmatrix} \frac{1}{2} & \frac{1}{4} \\ \frac{1}{4} & \frac{1}{2} \end{bmatrix}$.

(c) $A = \begin{bmatrix} 1 & \frac{1}{2} & \frac{1}{3} \\ \frac{1}{2} & \frac{1}{3} & \frac{1}{4} \\ \frac{1}{3} & \frac{1}{4} & \frac{1}{5} \end{bmatrix}$.

43. In Section 1.1 we studied the method of elimination for solving linear systems $AX = B$. In Equation (2) of this section we showed that the solution is given by $X = A^{-1}B$, provided that A is nonsingular. Using your software's command for automatically solving linear systems and its inverse command, compare these two solution techniques on the following linear systems.

(a) $A = \begin{bmatrix} \frac{1}{3} & \frac{2}{3} & \frac{4}{3} \\ 0 & \frac{2}{3} & \frac{4}{3} \\ 0 & 0 & \frac{5}{3} \end{bmatrix}$, $B = \begin{bmatrix} 2 \\ 2 \\ \frac{10}{3} \end{bmatrix}$.

(b) $A = [a_{ij}]$, where $a_{ij} = \dfrac{1}{i+j-1}$, $i, j = 1, 2, \ldots, 10$, and $B =$ the first column of I_{10}.

1.5. *Echelon Form of a Matrix*

In this section we take the elimination method for solving linear systems learned in high school, and systematize it by introducing the language of matrices. This will result in two methods for solving a system of m linear equations in n unknowns. These methods take the augmented matrix of the linear system, perform certain operations on it, and obtain a new matrix that represents an equivalent linear system (i.e., has the same solutions as the original linear system). The important point here is that the latter linear system can be solved very easily.

For example, if

$$\left[\begin{array}{ccc|c} 1 & 2 & 0 & 3 \\ 0 & 1 & 1 & 2 \\ 0 & 0 & 1 & -1 \end{array}\right]$$

represents the augmented matrix of a linear system, then the solution is easily found from the corresponding equations

$$\begin{aligned} x_1 + 2x_2 &= 3 \\ x_2 + x_3 &= 2 \\ x_3 &= -1. \end{aligned}$$

The task of this section is to manipulate the augmented matrix representing a given linear system into a form from which the solution can easily be found.

Definition 1.10. An $m \times n$ matrix A is said to be in **reduced row echelon form** if it satisfies the following properties:

(a) All rows consisting entirely of zeros, if any, are at the bottom of the matrix.
(b) The first nonzero entry in each row that does not consist entirely of zeros is a 1, called the **leading entry** of its row.
(c) If rows i and $i + 1$ are two successive rows that do not consist entirely of zeros, then the leading entry of row $i + 1$ is to the right of the leading entry of row i.
(d) If a column contains a leading entry of some row, then all other entries in that column are zero. ■

If A satisfies properties (a), (b), and (c), it is said to be in **row echelon form**. In Definition 1.10, there may be no rows that consist entirely of zeros.

A similar definition can be formulated in the obvious manner for **reduced column echelon form** and **column echelon form**.

Example 1. The following are matrices in row echelon form:

$$A = \begin{bmatrix} 1 & 5 & 0 & 2 & -2 & 4 \\ 0 & 1 & 0 & 3 & 4 & 8 \\ 0 & 0 & 0 & 1 & 7 & -2 \\ 0 & 0 & 0 & 0 & 0 & 0 \\ 0 & 0 & 0 & 0 & 0 & 0 \end{bmatrix}, \qquad B = \begin{bmatrix} 1 & 0 & 0 & 0 \\ 0 & 1 & 0 & 0 \\ 0 & 0 & 1 & 0 \\ 0 & 0 & 0 & 1 \end{bmatrix},$$

and

$$C = \begin{bmatrix} 0 & 0 & 1 & 3 & 5 & 7 & 9 \\ 0 & 0 & 0 & 0 & 1 & -2 & 3 \\ 0 & 0 & 0 & 0 & 0 & 1 & 2 \\ 0 & 0 & 0 & 0 & 0 & 0 & 1 \\ 0 & 0 & 0 & 0 & 0 & 0 & 0 \end{bmatrix}.$$

■

Example 2. The following are matrices in reduced row echelon form:

$$B = \begin{bmatrix} 1 & 0 & 0 & 0 \\ 0 & 1 & 0 & 0 \\ 0 & 0 & 1 & 0 \\ 0 & 0 & 0 & 1 \end{bmatrix}, \qquad D = \begin{bmatrix} 1 & 0 & 0 & 0 & -2 & 4 \\ 0 & 1 & 0 & 0 & 4 & 8 \\ 0 & 0 & 0 & 1 & 7 & -2 \\ 0 & 0 & 0 & 0 & 0 & 0 \\ 0 & 0 & 0 & 0 & 0 & 0 \end{bmatrix},$$

and

$$E = \begin{bmatrix} 1 & 2 & 0 & 0 & 1 \\ 0 & 0 & 1 & 2 & 3 \\ 0 & 0 & 0 & 0 & 0 \end{bmatrix}.$$

The following matrices are not in reduced row echelon form. (Why not?)

$$F = \begin{bmatrix} 1 & 2 & 0 & 4 \\ 0 & 0 & 0 & 0 \\ 0 & 0 & 1 & -3 \end{bmatrix}, \qquad G = \begin{bmatrix} 1 & 0 & 3 & 4 \\ 0 & 2 & -2 & 5 \\ 0 & 0 & 1 & 2 \end{bmatrix},$$

$$H = \begin{bmatrix} 1 & 0 & 3 & 4 \\ 0 & 1 & -2 & 5 \\ 0 & 1 & 2 & 2 \\ 0 & 0 & 0 & 0 \end{bmatrix}, \qquad J = \begin{bmatrix} 1 & 2 & 3 & 4 \\ 0 & 1 & -2 & 5 \\ 0 & 0 & 1 & 2 \\ 0 & 0 & 0 & 0 \end{bmatrix}.$$

■

We shall now show that every matrix can be put into row (column) echelon form, or into reduced row (column) echelon form, by means of certain row (column) operations.

Definition 1.11. An **elementary row (column) operation** on a matrix A is any one of the following operations:

(a) Interchange rows (columns) i and j of A. This is called a **type I operation.**
(b) Multiply row (column) i of A by $c \neq 0$. This is called a **type II operation.**
(c) Add c times row (column) i of A to row (column) j of A, $i \neq j$. This is called a **type III operation.** ■

Observe that when a matrix is viewed as the augmented matrix of a linear system, the elementary row operations are equivalent, respectively, to interchanging two equations, multiplying an equation by a nonzero constant, and adding a multiple of one equation to another equation.

Example 3. Let

$$A = \begin{bmatrix} 0 & 0 & 1 & 2 \\ 2 & 3 & 0 & -2 \\ 3 & 3 & 6 & -9 \end{bmatrix}.$$

Interchanging rows 1 and 3 of A, we obtain

$$B = \begin{bmatrix} 3 & 3 & 6 & -9 \\ 2 & 3 & 0 & -2 \\ 0 & 0 & 1 & 2 \end{bmatrix}.$$

Multiplying the third row of A by $\frac{1}{3}$, we obtain

$$C = \begin{bmatrix} 0 & 0 & 1 & 2 \\ 2 & 3 & 0 & -2 \\ 1 & 1 & 2 & -3 \end{bmatrix}.$$

Adding -2 times row 2 of A to row 3 of A, we obtain

$$D = \begin{bmatrix} 0 & 0 & 1 & 2 \\ 2 & 3 & 0 & -2 \\ -1 & -3 & 6 & -5 \end{bmatrix}.$$

■

Definition 1.12. An $m \times n$ matrix A is said to be **row (column) equivalent** to an $m \times n$ matrix B if B can be obtained by applying a finite sequence of elementary row (column) operations to A. ■

Example 4. The matrix

$$A = \begin{bmatrix} 1 & 2 & 4 & 3 \\ 2 & 1 & 3 & 2 \\ 1 & -1 & 2 & 3 \end{bmatrix}$$

is row equivalent to

$$D = \begin{bmatrix} 2 & 4 & 8 & 6 \\ 1 & -1 & 2 & 3 \\ 4 & -1 & 7 & 8 \end{bmatrix},$$

because if we add twice row 3 of A to its second row, we obtain

$$B = \begin{bmatrix} 1 & 2 & 4 & 3 \\ 4 & -1 & 7 & 8 \\ 1 & -1 & 2 & 3 \end{bmatrix}.$$

Interchanging rows 2 and 3 of B, we obtain

$$C = \begin{bmatrix} 1 & 2 & 4 & 3 \\ 1 & -1 & 2 & 3 \\ 4 & -1 & 7 & 8 \end{bmatrix}.$$

Multiplying row 1 of C by 2, we obtain D. ■

We can easily show (see Exercise 4) that (a) every matrix is row equivalent to itself; (b) if A is row equivalent to B, then B is row equivalent to A; and (c) if A is row equivalent to B and B is row equivalent to C, then A is row equivalent to C. In view of (b), both statements "A is row equivalent to B" and "B is row equivalent to A" can be replaced by "A and B are row equivalent." A similar statement holds for column equivalence.

Theorem 1.9. Every nonzero $m \times n$ matrix $A = [a_{ij}]$ is row (column) equivalent to a matrix in row (column) echelon form.

Proof: We shall prove that A is row equivalent to a matrix in row echelon form, that is, that by using only elementary row operations we can transform A into a matrix in row echelon form. A completely analogous proof using elementary column operations establishes the result for column equivalence. We look in matrix A for the first column with a nonzero entry; say this is column j and say that this first nonzero entry in column j occurs in row i. Now interchange, if necessary, rows 1 and i, getting matrix $B = [b_{ij}]$. Thus $b_{1j} \neq 0$. Multiply all entries in row 1 of B by $1/b_{1j}$, obtaining $C = [c_{ij}]$. Note that $c_{1j} = 1$. Now if c_{hj}, $2 \leq h \leq m$, is not zero, then to row h of C we add $-c_{hj}$ times row 1; all elements in column j, rows 2, 3, . . . , m are zero. Denote the resulting matrix by D. Note that we have used only elementary row operations. Next, consider the $(m - 1) \times n$ submatrix A_1 of D obtained by deleting the first row of D. We now repeat the procedure above with matrix A_1 instead of matrix A. Continuing this way, we obtain a matrix in row echelon form which is row equivalent to A. ■

Example 5. Let

$$A = \begin{bmatrix} 0 & 2 & 3 & -4 & 1 \\ 0 & 0 & 2 & 3 & 4 \\ 2 & 2 & -5 & 2 & 4 \\ 2 & 0 & -6 & 9 & 7 \end{bmatrix}.$$

Column 1 is the first (counting from left to right) column in A with a nonzero entry. The first (counting from top to bottom) nonzero entry in the first column occurs in the third row. We interchange the first and third rows of A, obtaining

$$B = \begin{bmatrix} 2 & 2 & -5 & 2 & 4 \\ 0 & 0 & 2 & 3 & 4 \\ 0 & 2 & 3 & -4 & 1 \\ 2 & 0 & -6 & 9 & 7 \end{bmatrix}.$$

Multiply the first row of B by $\dfrac{1}{b_{11}} = \dfrac{1}{2}$, to obtain

$$C = \begin{bmatrix} 1 & 1 & -\frac{5}{2} & 1 & 2 \\ 0 & 0 & 2 & 3 & 4 \\ 0 & 2 & 3 & -4 & 1 \\ 2 & 0 & -6 & 9 & 7 \end{bmatrix}.$$

Add -2 times the first row of C to the fourth row of C to produce a matrix D in which the only nonzero entry in the first column is $d_{11} = 1$:

$$D = \begin{bmatrix} 1 & 1 & -\frac{5}{2} & 1 & 2 \\ 0 & 0 & 2 & 3 & 4 \\ 0 & 2 & 3 & -4 & 1 \\ 0 & -2 & -1 & 7 & 3 \end{bmatrix}.$$

Identify A_1 as the submatrix of D obtained by deleting the first row of D: do not erase the first row of D. Repeat the steps above with A_1 instead of A.

$$\begin{matrix} & 1 & 1 & -\frac{5}{2} & 1 & 2 \end{matrix}$$
$$A_1 = \begin{bmatrix} 0 & 0 & 2 & 3 & 4 \\ 0 & 2 & 3 & -4 & 1 \\ 0 & -2 & -1 & 7 & 3 \end{bmatrix}$$

Interchange the first and second rows of A_1 to obtain

$$\begin{array}{ccccc} 1 & 1 & -\frac{5}{2} & 1 & 2 \end{array}$$
$$B_1 = \begin{bmatrix} 0 & 2 & 3 & -4 & 1 \\ 0 & 0 & 2 & 3 & 4 \\ 0 & -2 & -1 & 7 & 3 \end{bmatrix}$$

Multiply the first row of B_1 by $\frac{1}{2}$ to obtain

$$\begin{array}{ccccc} 1 & 1 & -\frac{5}{2} & 1 & 2 \end{array}$$
$$C_1 = \begin{bmatrix} 0 & 1 & \frac{3}{2} & -2 & \frac{1}{2} \\ 0 & 0 & 2 & 3 & 4 \\ 0 & -2 & -1 & 7 & 3 \end{bmatrix}$$

Add two times the first row of C_1 to its third row to obtain

$$\begin{array}{ccccc} 1 & 1 & -\frac{5}{2} & 1 & 2 \end{array}$$
$$D_1 = \begin{bmatrix} 0 & 1 & \frac{3}{2} & -2 & \frac{1}{2} \\ 0 & 0 & 2 & 3 & 4 \\ 0 & 0 & 2 & 3 & 4 \end{bmatrix}.$$

Deleting the first row of D_1 yields the matrix A_2. We repeat the procedure above with A_2 instead of A. No rows of A_2 have to be interchanged.

$$\begin{array}{ccccc} 1 & 1 & -\frac{5}{2} & 1 & 2 \\ 0 & 1 & \frac{3}{2} & -2 & \frac{1}{2} \end{array}$$
$$A_2 = \begin{bmatrix} 0 & 0 & 2 & 3 & 4 \\ 0 & 0 & 2 & 3 & 4 \end{bmatrix} = B_2$$

Multiply the first row of B_2 by $\frac{1}{2}$ to obtain

$$\begin{array}{ccccc} 1 & 1 & -\frac{5}{2} & 1 & 2 \\ 0 & 1 & \frac{3}{2} & -2 & \frac{1}{2} \end{array}$$
$$C_2 = \begin{bmatrix} 0 & 0 & 1 & \frac{3}{2} & 2 \\ 0 & 0 & 2 & 3 & 4 \end{bmatrix}$$

Finally, add -2 times the first row of C_2 to its second row to obtain

$$\begin{array}{ccccc} 1 & 1 & -\frac{5}{2} & 1 & 2 \\ 0 & 1 & \frac{3}{2} & -2 & \frac{1}{2} \end{array}$$
$$D_2 = \begin{bmatrix} 0 & 0 & 1 & \frac{3}{2} & 2 \\ 0 & 0 & 0 & 0 & 0 \end{bmatrix}.$$

The matrix

$$H = \begin{bmatrix} 1 & 1 & -\frac{5}{2} & 1 & 2 \\ 0 & 1 & \frac{3}{2} & -2 & \frac{1}{2} \\ 0 & 0 & 1 & \frac{3}{2} & 2 \\ 0 & 0 & 0 & 0 & 0 \end{bmatrix}$$

is in row echelon form and is row equivalent to A. ■

When doing hand computations, it is sometimes possible to avoid messy fractions by suitably modifying the steps in the procedure.

Theorem 1.10. *Every nonzero $m \times n$ matrix $A = [a_{ij}]$ is row (column) equivalent to a matrix in reduced row (column) echelon form.*

Proof: We proceed as in Theorem 1.9, obtaining a matrix H in row echelon form that is row equivalent to A. In H, if row i contains a nonzero element, then its first (counting from left to right) nonzero element is 1. Suppose that it occurs in column c_i. Then $c_1 < c_2 < \cdots < c_r$, where $r (1 \le r \le m)$ is the number of nonzero rows in H. Now subtract suitable multiples of row i of H to make all elements in column c_i and rows $i-1, i-2, \ldots, 1$ of H equal to zero. The result is a matrix K in reduced row echelon form which has been obtained from H by elementary row operations, and is thus row equivalent to H. Since A is row equivalent to H and H is row equivalent to K, then A is row equivalent to K. An analogous proof can be given to show that A is column equivalent to a matrix in reduced column echelon form. ■

It can be shown, with some difficulty, that there is only one matrix in reduced row echelon form that is row equivalent to a given matrix. We omit the proof.

Example 6. Suppose that we wish to find a matrix in reduced row echelon form that is row equivalent to the matrix A of Example 5. Starting with the matrix H obtained there, we add -1 times the second row to the first row, obtaining

$$\begin{bmatrix} 1 & 0 & -4 & 3 & \frac{3}{2} \\ 0 & 1 & \frac{3}{2} & -2 & \frac{1}{2} \\ 0 & 0 & 1 & \frac{3}{2} & 2 \\ 0 & 0 & 0 & 0 & 0 \end{bmatrix}.$$

In this matrix we add $-\frac{3}{2}$ times the third row to its second row and 4 times the third row to its first row. This yields

$$K = \begin{bmatrix} 1 & 0 & 0 & 9 & \frac{19}{2} \\ 0 & 1 & 0 & -\frac{17}{4} & -\frac{5}{2} \\ 0 & 0 & 1 & \frac{3}{2} & 2 \\ 0 & 0 & 0 & 0 & 0 \end{bmatrix},$$

which is in reduced row echelon form and is row equivalent to A. ■

We now apply these results to the solution of linear systems.

Theorem 1.11. *Let $AX = B$ and $CX = D$ be two linear systems each of m equations in n unknowns. If the augmented matrices $[A \mid B]$ and $[C \mid D]$ are row equivalent, then the linear systems are equivalent; that is, they have exactly the same solutions.*

Proof: This follows from the definition of row equivalence and from the fact that the three elementary row operations on the augmented matrix are the three manipulations on linear systems, discussed in Section 1.1, which yield equivalent linear systems. We also note that if one system has no solution, then the other system has no solution. ■

Corollary 1.2. *If A and B are row equivalent $m \times n$ matrices, then the homogeneous systems $AX = O$ and $BX = O$ are equivalent.*

Proof: Exercise. ■

We now pause to observe that we have developed the essential features of two very straightforward methods for solving linear systems. The idea consists of starting with the linear system $AX = B$, then obtaining a partitioned matrix $[C \mid D]$ in either row echelon form or reduced row echelon form that is row equivalent to the augmented matrix $[A \mid B]$. Now $[C \mid D]$ represents the linear system $CX = D$, which is quite simple to solve because of the structure of $[C \mid D]$, and the set of solutions to this system gives precisely the set of solutions to $AX = B$. The method where $[C \mid D]$ is in row echelon form is called **Gaussian elimination**; the method where $[C \mid D]$ is in reduced row echelon form is called **Gauss†–Jordan‡ reduction**. These methods are used often and computer codes of their implementations are widely available.

† Carl Friedrich Gauss (1777–1855) was born into a poor working-class family in Brunswick, Germany, and died in Göttingen, Germany, the most famous mathematician in the world. He was a child prodigy with a genius that did not impress his father, who called him a "star-gazer." However, his teachers were impressed enough to arrange for the Duke of Brunswick to provide a scholarship for Gauss at the local secondary school. As a teenager there, he made original discoveries in number theory and began to speculate about non-Euclidean geometry. His scientific publications include important contributions in number theory, mathematical astronomy, mathematical geography, statistics, differential geometry, and magnetism. His diaries and private notes contain many other discoveries that he never published.

An austere, conservative man who had few friends and whose private life was generally unhappy, he was very concerned that proper credit be given for scientific discoveries. When he relied on the results of others, he was careful to acknowledge them, and when others independently discovered results in his private notes, he was quick to claim priority.

Gaussian elimination consists of two steps:

STEP 1. The transformation of the augmented matrix $[A \mid B]$ to the matrix $[C \mid D]$ in row echelon form using elementary row operations.

STEP 2. Solution of the linear system corresponding to the augmented matrix $[C \mid D]$ using **back substitution**.

For the case in which A is $n \times n$, and the linear system $AX = B$ has a unique solution, the matrix $[C \mid D]$ has the following form:

$$\left[\begin{array}{cccccc|c} 1 & c_{12} & c_{13} & \cdots & & c_{1n} & d_1 \\ 0 & 1 & c_{23} & \cdots & & c_{2n} & d_2 \\ \vdots & \vdots & & & & \vdots & \vdots \\ 0 & 0 & 0 & \cdots & 1 & c_{n-1\,n} & d_{n-1} \\ 0 & 0 & 0 & \cdots & 0 & 1 & d_n \end{array}\right].$$

(The remaining cases are treated below.) This augmented matrix represents the linear system

$$\begin{aligned} x_1 + c_{12}x_2 + c_{13}x_3 + \cdots \quad + \quad c_{1n}x_n &= d_1 \\ x_2 + c_{23}x_3 + \cdots \quad + \quad c_{2n}x_n &= d_2 \\ \vdots \qquad\qquad \vdots \qquad &\quad \vdots \\ x_{n-1} + c_{n-1\,n}x_n &= d_{n-1} \\ x_n &= d_n. \end{aligned}$$

In his research he used a method of calculation that later generations generalized to row reduction of matrices and named in his honor, although the method was used in China almost 2000 years earlier.

‡ Wilhelm Jordan (1842–1899) was born in southern Germany. He attended college in Stuttgart and in 1868 became full professor of geodesy at the technical college in Karlsruhe, Germany. He participated in surveying several regions of Germany. Jordan was a prolific writer whose major work, *Handbuch der Vermessungskunde (Handbook of Geodesy),* was translated into French, Italian, and Russian. He was considered a superb writer and an excellent teacher. Unfortunately, the Gauss–Jordan reduction method has been widely attributed to Camille Jordan (1838–1922), a well-known French mathematician. Moreover, it seems that the method was also discovered independently at the same time by B. I. Clasen, a priest who lived in Luxembourg. This biographical sketch is based on an excellent article: S. C. Althoen and R. McLaughlin, "Gauss–Jordan reduction: a brief history," *MAA Monthly,* 94 (1987), 130–142.

Back substitution proceeds from the nth equation upward, solving for one variable from each equation:

$$\begin{aligned} x_n &= d_n \\ x_{n-1} &= d_{n-1} - c_{n-1n}x_n \\ &\vdots \\ x_2 &= d_2 - c_{23}x_3 - c_{24}x_4 - \cdots - c_{2n}x_n \\ x_1 &= d_1 - c_{12}x_2 - c_{13}x_3 - \cdots - c_{1n}x_n. \end{aligned}$$

Example 7. The linear system

$$\begin{aligned} x_1 + 2x_2 + 3x_3 &= 9 \\ 2x_1 - x_2 + x_3 &= 8 \\ 3x_1 - x_3 &= 3 \end{aligned}$$

has the augmented matrix

$$[A \mid B] = \left[\begin{array}{ccc|c} 1 & 2 & 3 & 9 \\ 2 & -1 & 1 & 8 \\ 3 & 0 & -1 & 3 \end{array}\right].$$

Transforming this matrix to row echelon form, we obtain (verify)

$$[C \mid D] = \left[\begin{array}{ccc|c} 1 & 2 & 3 & 9 \\ 0 & 1 & 1 & 2 \\ 0 & 0 & 1 & 3 \end{array}\right].$$

Using back substitution, we now have

$$\begin{aligned} x_3 &= 3 \\ x_2 &= 2 - 1x_3 = 2 - 3 = -1 \\ x_1 &= 9 - 2x_2 - 3x_3 = 9 + 2 - 9 = 2; \end{aligned}$$

thus the solution is $x_1 = 2$, $x_2 = -1$, $x_3 = 3$. ■

The general case in which A is $m \times n$ is handled in a similar fashion, but we need to elaborate upon several situations that can occur. We thus consider $CX = D$, where C is $m \times n$, and $[C \mid D]$ is in row echelon form. Then, for example, $[C \mid D]$ is of the following form:

$$\left[\begin{array}{cccccccc|c} 1 & c_{12} & c_{13} & \cdots & & & & c_{1n} & d_1 \\ 0 & 0 & 1 & c_{24} & & \cdots & & c_{2n} & d_2 \\ \vdots & \vdots & \vdots & \vdots & & & & \vdots & \vdots \\ 0 & 0 & \cdots & & 0 & 1 & & c_{k-1n} & d_{k-1} \\ 0 & \cdots & & & \vdots & 0 & & 1 & d_k \\ 0 & \cdots & & & & \vdots & & 0 & d_{k+1} \\ \vdots & & & & & & & \vdots & \vdots \\ 0 & \cdots & & & & & & 0 & d_m \end{array}\right].$$

This augmented matrix represents the linear system

$$\begin{array}{rcrcl} x_1 + c_{12}x_2 + c_{13}x_3 + \cdots & & + c_{1n} & x_n & = d_1 \\ x_3 + c_{24}x_4 + \cdots & & + c_{2n} & x_n & = d_2 \\ \vdots & & & & \\ & & x_{n-1} + c_{k-1n}x_n & & = d_{k-1} \\ & & & x_n & = d_k \\ 0x_1 + \cdots & & + & 0x_n & = d_{k+1} \\ \vdots & \vdots & & \vdots & \\ 0x_1 + \cdots & & + & 0x_n & = d_m. \end{array}$$

First, if $d_{k+1} = 1$, then $CX = D$ has no solution, since at least one equation is not satisfied. If $d_{k+1} = 0$, which implies that $d_{k+2} = \cdots = d_m = 0$, we then obtain $x_n = d_k$, $x_{n-1} = d_{k-1} - c_{k-1n}x_n = d_{k-1} - c_{k-1n}d_k$, and continue using back substitution to find the remaining unknowns corresponding to the leading entry in each row. Of course, in the solution some of the unknowns may be expressed in terms of others that can take on any values whatever. This merely indicates that $CX = D$ has infinitely many solutions. On the other hand, every unknown may have a determined value, indicating that the solution is unique.

Example 8. Let

$$[C \mid D] = \left[\begin{array}{ccccc|c} 1 & 2 & 3 & 4 & 5 & 6 \\ 0 & 1 & 2 & 3 & -1 & 7 \\ 0 & 0 & 1 & 2 & 3 & 7 \\ 0 & 0 & 0 & 1 & 2 & 9 \end{array}\right].$$

Then

$$\begin{aligned} x_4 &= 9 - 2x_5 \\ x_3 &= 7 - 2x_4 - 3x_5 = 7 - 2(9 - 2x_5) - 3x_5 = -11 + x_5 \\ x_2 &= 7 - 2x_3 - 3x_4 + x_5 = 2 + 5x_5 \\ x_1 &= 6 - 2x_2 - 3x_3 - 4x_4 - 5x_5 = -1 - 10x_5 \\ x_5 &= \text{any real number.} \end{aligned}$$

Thus all solutions are of the form

$$\begin{aligned} x_1 &= -1 - 10r \\ x_2 &= 2 + 5r \\ x_3 &= -11 + r \\ x_4 &= 9 - 2r \\ x_5 &= r, \text{ any real number.} \end{aligned}$$

Since r can be assigned any real number, the given linear system has infinitely many solutions. ■

Example 9. If

$$[C \mid D] = \left[\begin{array}{cccc|c} 1 & 2 & 3 & 4 & 5 \\ 0 & 1 & 2 & 3 & 6 \\ 0 & 0 & 0 & 0 & 1 \end{array}\right],$$

then $CX = D$ has no solution, for the last equation is

$$0x_1 + 0x_2 + 0x_3 + 0x_4 = 1,$$

which can never be satisfied. ■

If $[C \mid D]$ is in reduced row echelon form, then we can solve $CX = D$ without back substitution, but, of course, it takes more effort to put a matrix in reduced row echelon form than in row echelon form.

Example 10. If

$$[C \mid D] = \left[\begin{array}{cccc|c} 1 & 0 & 0 & 0 & 5 \\ 0 & 1 & 0 & 0 & 6 \\ 0 & 0 & 1 & 0 & 7 \\ 0 & 0 & 0 & 1 & 8 \end{array}\right],$$

then

$$\begin{aligned} x_1 &= 5 \\ x_2 &= 6 \\ x_3 &= 7 \\ x_4 &= 8. \end{aligned}$$ ■

Example 11. If

$$[C \mid D] = \left[\begin{array}{ccccc|c} 1 & 1 & 2 & 0 & -\frac{5}{2} & \frac{2}{3} \\ 0 & 0 & 0 & 1 & \frac{1}{2} & \frac{1}{2} \\ 0 & 0 & 0 & 0 & 0 & 0 \end{array}\right],$$

then

$$\begin{aligned} x_4 &= \tfrac{1}{2} - \tfrac{1}{2}x_5 \\ x_1 &= \tfrac{2}{3} - x_2 - 2x_3 + \tfrac{5}{2}x_5, \end{aligned}$$

where x_2, x_3, and x_5 can take on any real numbers. Thus a solution is of the form

$$\begin{aligned} x_1 &= \tfrac{2}{3} - r - 2s + \tfrac{5}{2}t \\ x_2 &= r \\ x_3 &= s \\ x_4 &= \tfrac{1}{2} - \tfrac{1}{2}t \\ x_5 &= t, \end{aligned}$$

where r, s, and t are any real numbers. ■

We now solve a linear system both by Gaussian elimination and by Gauss–Jordan reduction.

Example 12. Consider the linear system

$$\begin{aligned} x_1 + 2x_2 + 3x_3 &= 6 \\ 2x_1 - 3x_2 + 2x_3 &= 14 \\ 3x_1 + x_2 - x_3 &= -2. \end{aligned}$$

We form the augmented matrix

$$\left[\begin{array}{ccc|c} 1 & 2 & 3 & 6 \\ 2 & -3 & 2 & 14 \\ 3 & 1 & -1 & -2 \end{array}\right].$$

Add (-2) times the first row to the second row:

$$\left[\begin{array}{ccc|c} 1 & 2 & 3 & 6 \\ 0 & -7 & -4 & 2 \\ 3 & 1 & -1 & -2 \end{array}\right].$$

Add (-3) times the first row to the third row:

$$\left[\begin{array}{ccc|c} 1 & 2 & 3 & 6 \\ 0 & -7 & -4 & 2 \\ 0 & -5 & -10 & -20 \end{array}\right].$$

Multiply the third row by $-\frac{1}{5}$ and interchange the second and third rows:

$$\left[\begin{array}{ccc|c} 1 & 2 & 3 & 6 \\ 0 & 1 & 2 & 4 \\ 0 & -7 & -4 & 2 \end{array}\right].$$

Add seven times the second row to the third row:

$$\left[\begin{array}{ccc|c} 1 & 2 & 3 & 6 \\ 0 & 1 & 2 & 4 \\ 0 & 0 & 10 & 30 \end{array}\right].$$

Multiply the third row by $\frac{1}{10}$:

$$\left[\begin{array}{ccc|c} 1 & 2 & 3 & 6 \\ 0 & 1 & 2 & 4 \\ 0 & 0 & 1 & 3 \end{array}\right].$$

This matrix is in row echelon form. This means that $x_3 = 3$ and from the second row

$$x_2 + 2x_3 = 4$$

so that

$$x_3 = 4 - 2(3) = -2.$$

From the first row

$$x_1 + 2x_2 + 3x_3 = 6,$$

which implies that

$$x_1 = 6 - 2x_2 - 3x_3 = 6 - 2(-2) - 3(3) = 1.$$

Thus $x_1 = 1$, $x_2 = -2$, and $x_3 = 3$ is the solution. This gives the solution by Gaussian elimination.

If, instead, we wish to use Gauss–Jordan reduction, we would transform the last matrix into reduced row echelon form by the following steps:

Add (-2) times the second row to the first row:

$$\left[\begin{array}{ccc|c} 1 & 0 & -1 & -2 \\ 0 & 1 & 2 & 4 \\ 0 & 0 & 1 & 3 \end{array}\right].$$

Add (-2) times the third row to the second row:

$$\left[\begin{array}{ccc|c} 1 & 0 & -1 & -2 \\ 0 & 1 & 0 & -2 \\ 0 & 0 & 1 & 3 \end{array}\right].$$

Add the third row to the first row:

$$\left[\begin{array}{ccc|c} 1 & 0 & 0 & 1 \\ 0 & 1 & 0 & -2 \\ 0 & 0 & 1 & 3 \end{array}\right].$$

The solution is $x_1 = 1$, $x_2 = -2$, and $x_3 = 3$, as before. ■

REMARK. In both Gaussian elimination and Gauss–Jordan reduction, we can only use row operations. Do not try to use any column operations.

Homogeneous Systems

Now we study a homogeneous system $AX = O$ of m linear equations in n unknowns.

Example 13. Consider the homogeneous system whose augmented matrix is

$$\left[\begin{array}{ccccc|c} 1 & 0 & 0 & 0 & 2 & 0 \\ 0 & 0 & 1 & 0 & 3 & 0 \\ 0 & 0 & 0 & 1 & 4 & 0 \\ 0 & 0 & 0 & 0 & 0 & 0 \end{array}\right].$$

Since the augmented matrix is in reduced row echelon form, the solution is easily seen to be

$$x_1 = -2r$$
$$x_2 = s$$
$$x_3 = -3r$$
$$x_4 = -4r$$
$$x_5 = r,$$

where r and s are any real numbers. ■

In Example 13 we solved a homogeneous system of $m(=4)$ linear equations in $n(=5)$ unknowns, where $m < n$ and the augmented matrix A was in reduced row echelon form. We can ignore any row of the augmented matrix that consists entirely of zeros. Thus let rows $1, 2, \ldots, r$ of A be the nonzero rows, and let the 1 in row i occur in column c_i. We are then solving a homogeneous system of r equations in n unknowns, $r < n$, and in this special case (A is in reduced row echelon form) we can solve for $x_{c_1}, x_{c_2}, \ldots, x_{c_r}$, in terms of the remaining $n - r$ unknowns. Since the latter can take on any real values, there are infinitely many solutions to the system $AX = O$; in particular, there is a nontrivial solution. We now show that this situation holds whenever we have $m < n$; A does not have to be in reduced row echelon form.

Theorem 1.12. *A homogeneous system of m linear equations in n unknowns always has a nontrivial solution if $m < n$, that is, if the number of unknowns exceeds the number of equations.*

Proof: Let B be a matrix in reduced row echelon form that is row equivalent to A. Then the homogeneous systems $AX = O$ and $BX = O$ are equivalent. If we let r be the number of nonzero rows of B, then $r \leq m$. If $m < n$, we conclude that $r < n$. We are then solving r equations in n unknowns and can solve for r unknowns in terms of the remaining $n - r$ unknowns, the latter being free to take on any values we please. Thus $BX = O$, and hence $AX = O$ has a nontrivial solution. ■

We shall soon use this result in the following equivalent form: If A is $m \times n$ and $AX = O$ has only the trivial solution, then $m \geq n$.

Example 14. Consider the homogeneous system

$$\begin{aligned} x_1 + x_2 + x_3 + x_4 &= 0 \\ x_1 \qquad\qquad + x_4 &= 0 \\ x_1 + 2x_2 + x_3 \qquad &= 0. \end{aligned}$$

The augmented matrix

$$A = \left[\begin{array}{cccc|c} 1 & 1 & 1 & 1 & 0 \\ 1 & 0 & 0 & 1 & 0 \\ 1 & 2 & 1 & 0 & 0 \end{array}\right]$$

is row equivalent to (verify)

$$\left[\begin{array}{cccc|c} 1 & 0 & 0 & 1 & 0 \\ 0 & 1 & 0 & -1 & 0 \\ 0 & 0 & 1 & 1 & 0 \end{array}\right].$$

Hence the solution is

$$\begin{aligned} x_1 &= -r \\ x_2 &= r \\ x_3 &= -r \\ x_4 &= r, \text{ any real number.} \end{aligned}$$

■

A useful property of matrices in reduced row echelon form (see Exercise 3) is that if A is an $n \times n$ matrix in reduced row echelon form $\neq I_n$, then A has a row consisting entirely of zeros.

1.5. EXERCISES

1. Let

$$A = \begin{bmatrix} 1 & 2 & -3 & 1 \\ -1 & 0 & 3 & 4 \\ 0 & 1 & 2 & -1 \\ 2 & 3 & 0 & -3 \end{bmatrix}.$$

(a) Find a matrix B in row echelon form that is row equivalent to A.

(b) Find a matrix C in reduced row echelon form that is row equivalent to A.

2. Let

$$A = \begin{bmatrix} 1 & -2 & 0 & 2 \\ 2 & -3 & -1 & 5 \\ 1 & 3 & 2 & 5 \\ 1 & 1 & 0 & 2 \end{bmatrix}.$$

(a) Find a matrix B in row echelon form that is row equivalent to A.
(b) Find a matrix C in reduced row echelon form that is row equivalent to A.

3. Let A be an $n \times n$ matrix in reduced row echelon form. Prove that if $A \neq I_n$, then A has a row consisting entirely of zeros.

4. Prove:
(a) Every matrix is row equivalent to itself.
(b) If A is row equivalent to B, then B is row equivalent to A.
(c) If A is row equivalent to B and B is row equivalent to C, then A is row equivalent to C.

5. Consider the linear system

$$\begin{aligned} x_1 + x_2 + 2x_3 &= -1 \\ x_1 - 2x_2 + x_3 &= -5 \\ 3x_1 + x_2 + x_3 &= 3. \end{aligned}$$

(a) Find all solutions, if any exist, by using the Gaussian elimination method.
(b) Find all solutions, if any exist, by using the Gauss–Jordan reduction method.

6. Repeat Exercise 5 for each of the following linear systems.

(a) $$\begin{aligned} x_1 + x_2 + 2x_3 + 3x_4 &= 13 \\ x_1 - 2x_2 + x_3 + x_4 &= 8 \\ 3x_1 + x_2 + x_3 - x_4 &= 1. \end{aligned}$$

(b) $$\begin{aligned} x_1 + x_2 + x_3 &= 1 \\ x_1 + x_2 - 2x_3 &= 3 \\ 2x_1 + x_2 + x_3 &= 2. \end{aligned}$$

(c) $$\begin{aligned} 2x_1 + x_2 + x_3 - 2x_4 &= 1 \\ 3x_1 - 2x_2 + x_3 - 6x_4 &= -2 \\ x_1 + x_2 - x_3 - x_4 &= -1 \\ 6x_1 + x_3 - 9x_4 &= -2 \\ 5x_1 - x_2 + 2x_3 - 8x_4 &= 3. \end{aligned}$$

In Exercises 7, 8, and 9, solve the linear system, if it is consistent, with given *augmented* matrix.

7. (a) $\left[\begin{array}{ccc|c} 1 & 1 & 1 & 0 \\ 1 & 1 & 0 & 3 \\ 0 & 1 & 1 & 1 \end{array}\right]$. (b) $\left[\begin{array}{ccc|c} 1 & 2 & 3 & 0 \\ 1 & 1 & 1 & 0 \\ 1 & 1 & 2 & 0 \end{array}\right]$.

(c) $\left[\begin{array}{ccc|c} 1 & 2 & 3 & 0 \\ 1 & 1 & 1 & 0 \\ 5 & 7 & 9 & 0 \end{array}\right]$. (d) $\left[\begin{array}{ccc|c} 1 & 2 & 3 & 0 \\ 1 & 2 & 1 & 0 \end{array}\right]$.

8. (a) $\left[\begin{array}{cccc|c} 1 & 2 & 3 & 1 & 8 \\ 1 & 3 & 0 & 1 & 7 \\ 1 & 0 & 2 & 1 & 3 \end{array}\right]$.

(b) $\left[\begin{array}{cccc|c} 1 & 1 & 3 & -3 & 0 \\ 0 & 2 & 1 & -3 & 3 \\ 1 & 0 & 2 & -1 & -1 \end{array}\right]$.

9. (a) $\left[\begin{array}{ccc|c} 1 & 2 & 1 & 7 \\ 2 & 0 & 1 & 4 \\ 1 & 0 & 2 & 5 \\ 1 & 2 & 3 & 11 \\ 2 & 1 & 4 & 12 \end{array}\right]$. (b) $\left[\begin{array}{ccc|c} 1 & 2 & 1 & 0 \\ 2 & 3 & 0 & 0 \\ 0 & 1 & 2 & 0 \\ 2 & 1 & 4 & 0 \end{array}\right]$.

10. Find a 2×1 matrix X with entries not all zero such that $AX = 4X$, where $A = \begin{bmatrix} 4 & 1 \\ 0 & 2 \end{bmatrix}$. [*Hint:* Rewrite the matrix equation $AX = 4X$ as $AX - 4X = (A - 4I_2)X = O$ and solve the homogeneous linear system.]

11. Find a 2×1 matrix X with entries not all zero such that $AX = 3X$, where $A = \begin{bmatrix} 2 & 1 \\ 1 & 2 \end{bmatrix}$.

12. Find a 3×1 matrix X with entries not all zero such that

$$AX = 3X, \text{ where } A = \begin{bmatrix} 1 & 2 & -1 \\ 1 & 0 & 1 \\ 4 & -4 & 5 \end{bmatrix}.$$

13. Find a 3×1 matrix X with entries not all zero such that

$$AX = 1X, \text{ where } A = \begin{bmatrix} 1 & 2 & -1 \\ 1 & 0 & 1 \\ 4 & -4 & 5 \end{bmatrix}.$$

14. In the following linear system, determine all values of a for which the resulting linear system has:
(a) No solution.
(b) A unique solution.
(c) Infinitely many solutions.

$$\begin{aligned} x_1 + x_2 \quad\quad - x_3 &= 2 \\ x_1 + 2x_2 \quad\quad + x_3 &= 3 \\ x_1 + x_2 + (a^2 - 5)x_3 &= a. \end{aligned}$$

15. Repeat Exercise 14 for the linear system

$$\begin{aligned} x_1 + x_2 \quad\quad + x_3 &= 2 \\ 2x_1 + 3x_2 \quad\quad + 2x_3 &= 5 \\ 2x_1 + 3x_2 + (a^2 - 1)x_3 &= a + 1. \end{aligned}$$

16. Repeat Exercise 14 for the linear system

$$\begin{aligned} x_1 + x_2 \quad\quad + x_3 &= 2 \\ x_1 + 2x_2 \quad\quad + x_3 &= 3 \\ x_1 + x_2 + (a^2 - 5)x_3 &= a. \end{aligned}$$

17. Repeat Exercise 14 for the linear system

$$\begin{aligned} x_1 \quad\quad + x_2 &= 3 \\ x_1 + (a^2 - 8)x_2 &= a. \end{aligned}$$

18. Let $A = \begin{bmatrix} a & b \\ c & d \end{bmatrix}$ and $X = \begin{bmatrix} x_1 \\ x_2 \end{bmatrix}$. Show that the linear system $AX = O$ has only the trivial solution if and only if $ad - bc \neq 0$.

19. Show that $A = \begin{bmatrix} a & b \\ c & d \end{bmatrix}$ is row equivalent to I_2 if and only if $ad - bc \neq 0$.

Exercises 20 through 23 are optional.

20. (a) Formulate the definitions of column echelon form and reduced column echelon form of a matrix.
(b) Prove that every $m \times n$ matrix is column equivalent to a matrix in column echelon form.

21. Prove that every $m \times n$ matrix is column equivalent to a matrix in reduced column echelon form.

22. Let A be the matrix in Exercise 1.
(a) Find a matrix in column echelon form that is column equivalent to A.
(b) Find a matrix in reduced column echelon form that is column equivalent to A.

23. Repeat Exercise 22 for the matrix

$$\begin{bmatrix} 1 & 2 & 3 & 4 & 5 \\ 2 & 1 & 3 & -1 & 2 \\ 3 & 1 & 2 & 4 & 1 \end{bmatrix}.$$

24. Determine the reduced row echelon form of

$$A = \begin{bmatrix} \cos\theta & \sin\theta \\ -\sin\theta & \cos\theta \end{bmatrix}.$$

25. Show that if the homogeneous system

$$\begin{aligned} (a - r)x_1 \quad\quad + dx_2 &= 0 \\ cx_1 + (b - r)x_2 &= 0 \end{aligned}$$

has a nontrivial solution, then r satisfies the equation $(a - r)(b - r) - cd = 0$.

26. Let $AX = B$, $B \neq O$ be a consistent linear system.
(a) Show that if X_1 is a solution to the linear system $AX = B$ and Y_1 is a solution to the associated homogeneous system $AX = O$, then $X_1 + Y_1$ is a solution to the system $AX = B$.
(b) Show that every solution X to $AX = B$ can be written as $X_1 + Y_1$, where X_1 is a particular solution to $AX = B$ and Y_1 is a solution to $AX = O$. [*Hint:* Let $X = X_1 + (X - X_1)$.]

C 27. Determine if the software you are using has a command for computing the reduced row echelon form of a matrix. If it does, experiment with this command on some of the previous exercises.

1.6. Elementary Matrices; Finding A^{-1}

In this section we develop a method for finding the inverse of a matrix if it exists. This method is such that we do not have to first find out whether A^{-1} exists. We start to find A^{-1}; if in the course of the computation we hit a certain situation, then we know that A^{-1} does not exist. Otherwise, we proceed to the end and obtain A^{-1}. This method requires that elementary row operations of types I, II, and III be performed on A. We clarify these notions by starting with the following definition.

Definition 1.13. An **$n \times n$ elementary matrix of type I, type II, or type III** is a matrix obtained from the identity matrix I_n by performing a single elementary row or elementary column operation of type I, type II, or type III, respectively. ■

Example 1. The following are elementary matrices:

$$E_1 = \begin{bmatrix} 0 & 0 & 1 \\ 0 & 1 & 0 \\ 1 & 0 & 0 \end{bmatrix}, \qquad E_2 = \begin{bmatrix} 1 & 0 & 0 \\ 0 & -2 & 0 \\ 0 & 0 & 1 \end{bmatrix},$$

$$E_3 = \begin{bmatrix} 1 & 2 & 0 \\ 0 & 1 & 0 \\ 0 & 0 & 1 \end{bmatrix}, \qquad \text{and} \qquad E_4 = \begin{bmatrix} 1 & 0 & 3 \\ 0 & 1 & 0 \\ 0 & 0 & 1 \end{bmatrix}.$$

Matrix E_1 is of type I—we interchanged the first and third rows of I_3; E_2 is of type II—we multiplied the second row of I_3 by -2; E_3 is of type III—we added twice the second row of I_3 to the first row of I_3; and E_4 is of type III—we added three times the first column of I_3 to the third column of I_3. ■

Theorem 1.13. *Let A be an $m \times n$ matrix, and let an elementary row (column) operation of type* I, *type* II, *or type* III *be performed on A to yield matrix B. Let E be the elementary matrix obtained from I_m (I_n) by performing the same elementary row (column) operation as was performed on A. Then $B = EA$ ($B = AE$).*

Proof: Exercise. ■

Theorem 1.13 says that an elementary row operation on A can be achieved by premultiplying A (multiplying A on the left) by the corresponding elementary matrix E; an elementary column operation on A can be obtained by postmultiplying A (multiplying A on the right) by the corresponding elementary matrix.

Example 2. Let $A = \begin{bmatrix} 1 & 3 & 2 & 1 \\ -1 & 2 & 3 & 4 \\ 3 & 0 & 1 & 2 \end{bmatrix}$ and let B result from A by adding -2 times the third row of A to the first row of A. Thus $B = \begin{bmatrix} -5 & 3 & 0 & -3 \\ -1 & 2 & 3 & 4 \\ 3 & 0 & 1 & 2 \end{bmatrix}$.
Now let E be the matrix that is obtained from I_3 by adding -2 times the third row of I_3 to the first row of I_3. Thus $E = \begin{bmatrix} 1 & 0 & -2 \\ 0 & 1 & 0 \\ 0 & 0 & 1 \end{bmatrix}$. It is easy to verify that $B = EA$. ■

Theorem 1.14. *If A and B are $m \times n$ matrices, then A is row (column) equivalent to B if and only if $B = E_kE_{k-1} \cdots E_2E_1A$ ($B = AE_1E_2 \cdots E_{k-1}E_k$), where $E_1, E_2, \ldots, E_{k-1}, E_k$ are elementary matrices.*

Proof: We prove only the theorem for row equivalence. If A is row equivalent to B, then B results from A by a sequence of elementary row operations. This implies that there exist elementary matrices $E_1, E_2, \ldots, E_k$ such that $B = E_kE_{k-1} \cdots E_2E_1A$.

Conversely, if $B = E_kE_{k-1} \cdots E_2E_1A$, where the E_i are elementary matrices, then B results from A by a sequence of elementary row operations, which implies that A is row equivalent to B. ■

Theorem 1.15. *An elementary matrix E is nonsingular and its inverse is an elementary matrix of the same type.*

Proof: Exercise. ■

Thus an elementary row operation can be "undone" by another elementary row operation of the same type.

We now obtain an algorithm for finding A^{-1} if it exists; first, we prove the following lemma.

Lemma 1.1† *Let A be an $n \times n$ matrix and let the homogeneous system $AX = O$ have only the trivial solution $X = O$. Then A is row equivalent to I_n.*

† A lemma is a theorem that is established for the purpose of proving another theorem.

Proof: Let B be a matrix in reduced row echelon form that is row equivalent to A. Then the homogeneous systems $AX = O$ and $BX = O$ are equivalent, and thus $BX = O$ also has only the trivial solution. It is clear that if r is the number of nonzero rows of B, then the homogeneous system $BX = O$ is equivalent to the homogeneous system whose coefficient matrix consists of the nonzero rows of B and is therefore $r \times n$. Since this last homogeneous system only has the trivial solution, we conclude from Theorem 1.12 that $r \geq n$. Since B is $n \times n$, $r \leq n$. Hence $r = n$, which means that B has no zero rows. Thus $B = I_n$. ∎

Theorem 1.16. *A is nonsingular if and only if A is a product of elementary matrices.*

Proof: If A is a product of elementary matrices $E_1, E_2, \ldots, E_k$, then $A = E_1E_2 \cdots E_k$. Now each elementary matrix is nonsingular, and the product of nonsingular matrices is nonsingular; therefore, A is nonsingular.

Conversely, if A is nonsingular, then $AX = O$ implies that $A^{-1}(AX) = A^{-1}O = O$, so $I_nX = O$ or $X = O$. Thus $AX = O$ has only the trivial solution. Lemma 1.1 then implies that A is row equivalent to I_n. This means that there exist elementary matrices $E_1, E_2, \ldots, E_k$ such that

$$I_n = E_kE_{k-1} \cdots E_2E_1A.$$

It then follows that $A = (E_kE_{k-1} \cdots E_2E_1)^{-1} = E_1^{-1}E_2^{-1} \cdots E_{k-1}^{-1}E_k^{-1}$. Since the inverse of an elementary matrix is an elementary matrix, we have established the result. ∎

Corollary 1.3. *A is nonsingular if and only if A is row equivalent to I_n.*

Proof: If A is row equivalent to I_n, then $I_n = E_kE_{k-1} \cdots E_2E_1A$, where $E_1, E_2, \ldots, E_k$ are elementary matrices. Therefore, it follows that $A = E_1^{-1}E_2^{-1} \cdots E_k^{-1}$. Now the inverse of an elementary matrix is an elementary matrix, and so by Theorem 1.16, A is nonsingular.

Conversely, if A is nonsingular, then A is a product of elementary matrices, $A = E_kE_{k-1} \cdots E_2E_1$. Now $A = AI_n = E_kE_{k-1} \cdots E_2E_1I_n$, which implies that A is row equivalent to I_n. ∎

We can see that Lemma 1.1 and Corollary 1.3 imply that if the homogeneous system $AX = O$, where A is $n \times n$, has only the trivial solution $X = O$, then A is nonsingular. Conversely, consider $AX = O$, where A is $n \times n$, and let A be nonsingular. Then A^{-1} exists and we have $A^{-1}(AX) = A^{-1}O = O$. Thus $X = O$, which means that the homogeneous system has only the trivial solution. We have thus proved the following important theorem.

Theorem 1.17. The homogeneous system of n linear equations in n unknowns $AX = O$ has a nontrivial solution if and only if A is singular. ■

Example 3. Let $A = \begin{bmatrix} 1 & 2 \\ 2 & 4 \end{bmatrix}$ be the matrix defined in Example 11 of Section 1.4, which is singular. Consider the homogeneous system $AX = O$; that is,

$$\begin{bmatrix} 1 & 2 \\ 2 & 4 \end{bmatrix}\begin{bmatrix} x_1 \\ x_2 \end{bmatrix} = \begin{bmatrix} 0 \\ 0 \end{bmatrix}.$$

The reduced row echelon form of the augmented matrix is $\left[\begin{array}{cc|c} 1 & 2 & 0 \\ 0 & 0 & 0 \end{array}\right]$ (verify), so a solution is

$$x_1 = -2r$$
$$x_2 = r,$$

where r is any real number. Thus the homogeneous system has a nontrivial solution. ■

> Note that at this point we have shown that the following statements are equivalent for an $n \times n$ matrix A.
>
> 1. A is nonsingular.
> 2. $AX = O$ has only the trivial solution.
> 3. A is row (column) equivalent to I_n.
> 4. The system $AX = B$ has a unique solution for every $n \times 1$ matrix B.
> 5. A is a product of elementary matrices.

At the end of the proof of Theorem 1.16, A was nonsingular and

$$A = E_1^{-1}E_2^{-1} \cdots E_{k-1}^{-1}E_k^{-1},$$

from which it follows that

$$A^{-1} = (E_1^{-1}E_2^{-1} \cdots E_{k-1}^{-1}E_k^{-1})^{-1} = E_kE_{k-1} \cdots E_2E_1.$$

This now provides an algorithm for finding A^{-1}. Thus we perform elementary row operations on A until we get I_n; the product of the elementary matrices $E_kE_{k-1} \cdots E_2E_1$ then gives A^{-1}. A convenient way of organizing the computing process is to write down the partitioned matrix $[A \mid I_n]$. Then

$$(E_kE_{k-1}\cdots E_2E_1)[A \mid I_n] = [E_kE_{k-1}\cdots E_2E_1A \mid E_kE_{k-1}\cdots E_2E_1] = [I_n \mid A^{-1}].$$

That is, we transform the partitioned matrix $[A \mid I_n]$ to reduced row echelon form, obtaining $[I_n \mid A^{-1}]$.

Example 4. Let $A = \begin{bmatrix} 1 & 1 & 1 \\ 0 & 2 & 3 \\ 5 & 5 & 1 \end{bmatrix}$. Assuming that A is nonsingular, we form

$$[A \mid I_3] = \left[\begin{array}{ccc|ccc} 1 & 1 & 1 & 1 & 0 & 0 \\ 0 & 2 & 3 & 0 & 1 & 0 \\ 5 & 5 & 1 & 0 & 0 & 1 \end{array}\right].$$

We now perform elementary row operations that transform $[A \mid I_3]$ to $[I_3 \mid A^{-1}]$; we consider $[A \mid I_3]$ as a 3×6 matrix, and whatever we do to a row of A we also do to the corresponding row of I_3. Thus we arrange our computations as follows:

	A			I_3		
1	1	1	1	0	0	Add (-5) times the first row to the third row to obtain
0	2	3	0	1	0	
5	5	1	0	0	1	
1	1	1	1	0	0	Multiply the second row by $\frac{1}{2}$ to obtain
0	2	3	0	1	0	
0	0	-4	-5	0	1	
1	1	1	1	0	0	Multiply the third row by $-\frac{1}{4}$ to obtain
0	1	$\frac{3}{2}$	0	$\frac{1}{2}$	0	
0	0	-4	-5	0	1	
1	1	1	1	0	0	Add -1 times the second row to the first row to obtain
0	1	$\frac{3}{2}$	0	$\frac{1}{2}$	0	
0	0	1	$\frac{5}{4}$	0	$-\frac{1}{4}$	
1	0	$-\frac{1}{2}$	1	$-\frac{1}{2}$	0	Add $-\frac{3}{2}$ times the third row to the second row and add $\frac{1}{2}$ times the third row to the first row to obtain
0	1	$\frac{3}{2}$	0	$\frac{1}{2}$	0	
0	0	1	$\frac{5}{4}$	0	$-\frac{1}{4}$	
1	0	0	$\frac{13}{8}$	$-\frac{1}{2}$	$-\frac{1}{8}$	
0	1	0	$-\frac{15}{8}$	$\frac{1}{2}$	$\frac{3}{8}$	
0	0	1	$\frac{5}{4}$	0	$-\frac{1}{4}$	

Hence

$$A^{-1} = \begin{bmatrix} \frac{13}{8} & -\frac{1}{2} & -\frac{1}{8} \\ -\frac{15}{8} & \frac{1}{2} & \frac{3}{8} \\ \frac{5}{4} & 0 & -\frac{1}{4} \end{bmatrix}.$$

It is easy to verify that $AA^{-1} = A^{-1}A = I_3$. ■

The question that arises at this point is how to tell when A is singular. The answer is that A is singular if and only if A is row equivalent to a matrix B having at least one row that consists entirely of zeros. We now prove this result.

Theorem 1.18. *An $n \times n$ matrix A is singular if and only if A is row equivalent to a matrix B that has a row of zeros.*

Proof: First, let A be row equivalent to a matrix B that has a row consisting entirely of zeros. From Exercise 30 of Section 1.4 it follows that B is singular. Now $B = E_kE_{k-1} \cdots E_1A$, where $E_1, E_2, \ldots, E_k$ are elementary matrices. If A is nonsingular, then B is nonsingular, a contradiction. Thus A is singular.

Conversely, if A is singular, then A is not row equivalent to I_n, by Corollary 1.3. Thus A is row equivalent to a matrix $B \neq I_n$, which is in reduced row echelon form. From Exercise 3 of Section 1.5 it follows that B must have a row of zeros. ■

This means that in order to find A^{-1}, we do not have to determine, in advance, whether it exists. We merely start to calculate A^{-1}; if at any point in the computation we find a matrix B that is row equivalent to A and has a row of zeros, then A^{-1} does not exist. That is, we transform the partitioned matrix $[A \mid I_n]$ to reduced row echelon form, obtaining $[C \mid D]$. If $C = I_n$, then $D = A^{-1}$. If $C \neq I_n$, then C has a row of zeros and we conclude that A is singular.

Example 5. Let $A = \begin{bmatrix} 1 & 2 & -3 \\ 1 & -2 & 1 \\ 5 & -2 & -3 \end{bmatrix}$. To find A^{-1}, we proceed as follows:

$$\begin{array}{rrr|rrr} & A & & & I_3 & \\ 1 & 2 & -3 & 1 & 0 & 0 \\ 1 & -2 & 1 & 0 & 1 & 0 \\ 5 & -2 & -3 & 0 & 0 & 1 \end{array}$$

Add (-1) times the first row to the second row to obtain

$$\begin{array}{rrr|rrr} 1 & 2 & -3 & 1 & 0 & 0 \\ 0 & -4 & 4 & -1 & 1 & 0 \\ 5 & -2 & -3 & 0 & 0 & 1 \end{array}$$

Add (-5) times the first row to the third row to obtain

$$\begin{array}{rrr|rrr} 1 & 2 & -3 & 1 & 0 & 0 \\ 0 & -4 & 4 & -1 & 1 & 0 \\ 0 & -12 & 12 & -5 & 0 & 1 \\ \hline 1 & 2 & -3 & 1 & 0 & 0 \\ 0 & -4 & 4 & -1 & 1 & 0 \\ 0 & 0 & 0 & -2 & -3 & 1 \end{array}$$

Add (-3) times the second row to the third row to obtain

At this point A is row equivalent to

$$B = \begin{bmatrix} 1 & 2 & -3 \\ 0 & -4 & 4 \\ 0 & 0 & 0 \end{bmatrix},$$

the last matrix under A. Since B has a row of zeros, we stop and conclude that A is a singular matrix. ■

In Section 1.4 we defined an $n \times n$ matrix B to be the inverse of the $n \times n$ matrix A if $AB = I_n$ and $BA = I_n$. We now show that one of these equations follows from the other.

Theorem 1.19. *If A and B are $n \times n$ matrices such that $AB = I_n$, then $BA = I_n$. Thus $B = A^{-1}$.*

Proof: We first show that if $AB = I_n$, then A is nonsingular. Suppose that A is singular. Then A is row equivalent to a matrix C with a row of zeros. Now $C = E_kE_{k-1} \cdots E_1A$, where $E_1, E_2, \ldots, E_k$ are elementary matrices. Then $CB = E_kE_{k-1} \cdots E_1AB$, so AB is row equivalent to CB. Since CB has a row of zeros, we conclude from Theorem 1.18 that AB is singular. Then $AB = I_n$ is impossible, because I_n is nonsingular. This contradiction shows that A is nonsingular, and so A^{-1} exists. Multiplying both sides of the equation $AB = I_n$ by A^{-1} on the left, we then obtain (verify) $B = A^{-1}$. ■

1.6. EXERCISES

1. Prove Theorem 1.13.

2. Let A be a 4×3 matrix. Find the elementary matrix E, which as a premultiplier of A, that is, as EA, performs the following elementary row operations on A.

(a) Multiplies the second row of A by -2.

(b) Adds 3 times the third row of A to the fourth row of A.

(c) Interchanges the first and third rows of A.

3. Let A be a 3×4 matrix. Find the elementary matrix F, which as a postmultiplier of A, that is, as AF, performs the following elementary column operations on A.

(a) Adds -4 times the first column of A to the second column of A.

(b) Interchanges the second and third columns of A.

(c) Multiplies the third column of A by 4.

4. Prove Theorem 1.15 (*Hint:* Find the inverse of the elementary matrix of type I, type II, and type III.)

5. Find the inverse of $A = \begin{bmatrix} 1 & 3 \\ 2 & 4 \end{bmatrix}$.

6. Find the inverse of $A = \begin{bmatrix} 1 & 2 & 3 \\ 0 & 2 & 3 \\ 1 & 2 & 4 \end{bmatrix}$.

7. Which of the following matrices are singular? For the nonsingular ones find the inverse.

(a) $\begin{bmatrix} 1 & 3 \\ 2 & 6 \end{bmatrix}$. (b) $\begin{bmatrix} 1 & 3 \\ -2 & 6 \end{bmatrix}$.

(c) $\begin{bmatrix} 1 & 2 & 3 \\ 1 & 1 & 2 \\ 0 & 1 & 2 \end{bmatrix}$. (d) $\begin{bmatrix} 1 & 2 & 3 \\ 1 & 1 & 2 \\ 0 & 1 & 1 \end{bmatrix}$.

8. Invert the following matrices, if possible.

(a) $\begin{bmatrix} 1 & 2 & -3 & 1 \\ -1 & 3 & -3 & -2 \\ 2 & 0 & 1 & 5 \\ 3 & 1 & -2 & 5 \end{bmatrix}$. (b) $\begin{bmatrix} 3 & 1 & 2 \\ 2 & 1 & 2 \\ 1 & 2 & 2 \end{bmatrix}$.

(c) $\begin{bmatrix} 1 & 2 & 3 \\ 1 & 1 & 2 \\ 1 & 1 & 0 \end{bmatrix}$. (d) $\begin{bmatrix} 2 & 1 & 3 \\ 0 & 1 & 2 \\ 1 & 0 & 3 \end{bmatrix}$.

9. Find the inverse, if it exists, of:

(a) $\begin{bmatrix} 1 & 1 & 1 \\ 1 & 2 & 3 \\ 0 & 1 & 1 \end{bmatrix}$. (b) $\begin{bmatrix} 1 & 1 & 1 & 1 \\ 1 & 2 & -1 & 2 \\ 1 & -1 & 2 & 1 \\ 1 & 3 & 3 & 2 \end{bmatrix}$.

(c) $\begin{bmatrix} 1 & 1 & 1 & 1 \\ 1 & 3 & 1 & 2 \\ 1 & 2 & -1 & 1 \\ 5 & 9 & 1 & 6 \end{bmatrix}$. (d) $\begin{bmatrix} 1 & 2 & 1 \\ 1 & 3 & 2 \\ 1 & 0 & 1 \end{bmatrix}$.

(e) $\begin{bmatrix} 1 & 2 & 2 \\ 1 & 3 & 1 \\ 1 & 1 & 3 \end{bmatrix}$.

10. Find the inverse, if it exists, of:

(a) $A = \begin{bmatrix} 1 & 1 & 2 & 1 \\ 0 & -2 & 0 & 0 \\ 1 & 2 & 1 & -2 \\ 0 & 3 & 2 & 1 \end{bmatrix}$.

(b) $A = \begin{bmatrix} 1 & 1 & 1 & 1 \\ 1 & 3 & 1 & 2 \\ 1 & 2 & -1 & 1 \\ 5 & 9 & 1 & 6 \end{bmatrix}$.

In Exercises 11 and 12, prove that the given matrix A is nonsingular and write it as a product of elementary matrices. (*Hint:* First, write the inverse as a product of elementary matrices, then use Theorem 1.15.)

11. $A = \begin{bmatrix} 1 & 2 \\ 3 & 4 \end{bmatrix}$.

12. $A = \begin{bmatrix} 1 & 2 & 3 \\ 0 & 1 & 2 \\ 1 & 0 & 3 \end{bmatrix}$.

13. If A is a nonsingular matrix whose inverse is $\begin{bmatrix} 4 & 2 \\ 1 & 1 \end{bmatrix}$, find A.

14. If

$$A^{-1} = \begin{bmatrix} 1 & 1 & 1 \\ 1 & 1 & 2 \\ 1 & -1 & 1 \end{bmatrix},$$

find A.

15. Which of the following homogeneous systems have a nontrivial solution?

(a) $$\begin{aligned} x_1 + 2x_2 + 3x_3 &= 0 \\ 2x_2 + 2x_3 &= 0 \\ x_1 + 2x_2 + 3x_3 &= 0. \end{aligned}$$

(b) $$\begin{aligned} 2x_1 + x_2 - x_3 &= 0 \\ x_1 - 2x_2 - 3x_3 &= 0 \\ -3x_1 - x_2 + 2x_3 &= 0. \end{aligned}$$

(c) $$\begin{aligned} 3x_1 + x_2 + 3x_3 &= 0 \\ -2x_1 + 2x_2 - 4x_3 &= 0 \\ 2x_1 - 3x_2 + 5x_3 &= 0. \end{aligned}$$

16. Which of the following homogeneous systems have a nontrivial solution?

(a) $$\begin{aligned} x_1 + x_2 + 2x_3 &= 0 \\ 2x_1 + x_2 + x_3 &= 0 \\ 3x_1 - x_2 + x_3 &= 0. \end{aligned}$$

(b) $$\begin{aligned} x_1 - x_2 + x_3 &= 0 \\ 2x_1 + x_2 \phantom{{}+ x_3} &= 0 \\ 2x_1 - 2x_2 + 2x_3 &= 0. \end{aligned}$$

(c) $$\begin{aligned} 2x_1 - x_2 + 5x_3 &= 0 \\ 3x_1 + 2x_2 - 3x_3 &= 0 \\ x_1 - x_2 + 4x_3 &= 0. \end{aligned}$$

17. Find all value(s) of a for which the inverse of

$$A = \begin{bmatrix} 1 & 1 & 0 \\ 1 & 0 & 0 \\ 1 & 2 & a \end{bmatrix}$$

exists. What is A^{-1}?

18. For what values of a does the homogeneous system

$$\begin{aligned} (a-1)x_1 \qquad + 2x_2 &= 0 \\ 2x_1 + (a-1)x_2 &= 0 \end{aligned}$$

have a nontrivial solution?

19. Prove that $A = \begin{bmatrix} a & b \\ c & d \end{bmatrix}$ is nonsingular if and only if $ad - bc \neq 0$. If this condition holds, show that

$$A^{-1} = \begin{bmatrix} \dfrac{d}{ad-bc} & \dfrac{-b}{ad-bc} \\ \dfrac{-c}{ad-bc} & \dfrac{a}{ad-bc} \end{bmatrix}.$$

20. Let $A = \begin{bmatrix} 2 & 3 & -1 \\ 1 & 0 & 3 \\ 0 & 2 & -3 \\ -2 & 1 & 3 \end{bmatrix}$. Find the elementary matrix that as a postmultiplier of A performs the following elementary column operations on A.

(a) Multiplies the third column of A by -3.
(b) Interchanges the second and third columns of A.
(c) Adds -5 times the first column of A to the third column of A.

21. Prove that two $m \times n$ matrices A and B are row equivalent if and only if there exists a nonsingular matrix P such that $B = PA$. (*Hint:* Use Theorems 1.14 and 1.16.)

22. Let A and B be row equivalent $n \times n$ matrices. Prove that A is nonsingular if and only if B is nonsingular.

23. Let A and B be $n \times n$ matrices. Show that if AB is nonsingular, then A and B must be nonsingular. (*Hint:* Use Theorem 1.17.)

24. Let A be an $m \times n$ matrix. Show that A is row equivalent to ${}_mO_n$ if and only if $A = {}_mO_n$.

25. Let A and B be $m \times n$ matrices. Show that A is row equivalent to B if and only if A^T is column equivalent to B^T.

26. Show that a square matrix which has a row or a column consisting entirely of zeros must be singular.

27. (a) Is $(A + B)^{-1} = A^{-1} + B^{-1}$?

(b) Is $(cA)^{-1} = \dfrac{1}{c}A^{-1}$?

28. If A is an $n \times n$ matrix, prove that A is nonsingular if and only if the linear system $AX = B$ is consistent for every $n \times 1$ matrix B.

29. Prove that the inverse of a nonsingular upper (lower) triangular matrix is upper (lower) triangular.

30. If the software you use has a command for computing reduced row echelon form, use it to determine if the matrices A in Exercises 7, 8, and 9 have an inverse by operating on the matrix $[A \mid I_n]$ (see Example 4).

31. Repeat Exercise 30 on the matrices given in Exercise 41 of Section 1.4.

1.7. *Equivalent Matrices*

We have so far considered A to be row (column) equivalent to B if B results from A by a finite sequence of elementary row (column) operations. A natural extension of this idea is that of considering B to arise from A by a finite sequence of elementary row *or* elementary column operations. This leads to the notion of equivalence of matrices. The material discussed in this section will also be used in Section 2.6.

Definition 1.14. If A and B are two $m \times n$ matrices, then A is **equivalent** to B if we obtain B from A by a finite sequence of elementary row or elementary column operations. ■

As we have seen in the case of row equivalence, we can easily show (see Exercise 1) that (a) every matrix is equivalent to itself; (b) if A is equivalent to B, then B is equivalent to A; (c) if A is equivalent to B and B is equivalent to C, then A is equivalent to C. In view of (b) both statements "A is equivalent to B" and "B is equivalent to A" can be replaced by "A and B are equivalent." It is also easy to show that if two matrices are row equivalent, then they are equivalent (see Exercise 4).

Theorem 1.20. *If A is any nonzero $m \times n$ matrix, then A is equivalent to a partitioned matrix of the form*

$$\begin{bmatrix} I_r & {}_rO_{n-r} \\ {}_{m-r}O_r & {}_{m-r}O_{n-r} \end{bmatrix}.$$

Proof: By Theorem 1.10, A is row equivalent to a matrix B that is in reduced row echelon form. Using elementary column operations of type I, we get B to be equivalent to a matrix C of the form

$$\begin{bmatrix} I_r & {}_rU_{n-r} \\ {}_{m-r}O_r & {}_{m-r}O_{n-r} \end{bmatrix},$$

where r is the number of nonzero rows in B. By elementary column operations of type III, C is equivalent to a matrix D of the form

$$\begin{bmatrix} I_r & {}_rO_{n-r} \\ {}_{m-r}O_r & {}_{m-r}O_{n-r} \end{bmatrix}.$$

From Exercise 1 it then follows that A is equivalent to D. ■

Of course, in Theorem 1.20, r may equal m, in which case there will not be any zero rows at the bottom of the matrix. (What happens if $r = n$? If $r = m = n$?)

Example 1. Let $A = \begin{bmatrix} 1 & 1 & 2 & -1 \\ 1 & 2 & 1 & 0 \\ -1 & -4 & 1 & -2 \\ 1 & -2 & 5 & -4 \end{bmatrix}$. To find a matrix of the form described in Theorem 1.20, which is equivalent to A, we proceed as follows. Add (-1) times the first row of A to the second row of A to obtain

$$\begin{bmatrix} 1 & 1 & 2 & -1 \\ 0 & 1 & -1 & 1 \\ -1 & -4 & 1 & -2 \\ 1 & -2 & 5 & -4 \end{bmatrix}$$

Add the first row to the third row to obtain

$$\begin{bmatrix} 1 & 1 & 2 & -1 \\ 0 & 1 & -1 & 1 \\ 0 & -3 & 3 & -3 \\ 1 & -2 & 5 & -4 \end{bmatrix}$$

Add (-1) times the first row to the fourth row to obtain

$$\begin{bmatrix} 1 & 1 & 2 & -1 \\ 0 & 1 & -1 & 1 \\ 0 & -3 & 3 & -3 \\ 0 & -3 & 3 & -3 \end{bmatrix}$$

Add (-1) times the third row to the fourth row to obtain

$$\begin{bmatrix} 1 & 1 & 2 & -1 \\ 0 & 1 & -1 & 1 \\ 0 & -3 & 3 & -3 \\ 0 & 0 & 0 & 0 \end{bmatrix}$$

Multiply the third row by $-\frac{1}{3}$ to obtain

$$\begin{bmatrix} 1 & 1 & 2 & -1 \\ 0 & 1 & -1 & 1 \\ 0 & 1 & -1 & 1 \\ 0 & 0 & 0 & 0 \end{bmatrix}$$ Add (-1) times the second row to the third row to obtain

$$\begin{bmatrix} 1 & 1 & 2 & -1 \\ 0 & 1 & -1 & 1 \\ 0 & 0 & 0 & 0 \\ 0 & 0 & 0 & 0 \end{bmatrix}$$ Add (-1) times the second row to the first row to obtain

$$\begin{bmatrix} 1 & 0 & 3 & -2 \\ 0 & 1 & -1 & 1 \\ 0 & 0 & 0 & 0 \\ 0 & 0 & 0 & 0 \end{bmatrix}$$ Add (-3) times the first column to the third column to obtain

$$\begin{bmatrix} 1 & 0 & 0 & -2 \\ 0 & 1 & -1 & 1 \\ 0 & 0 & 0 & 0 \\ 0 & 0 & 0 & 0 \end{bmatrix}$$ Add 2 times the first column to the fourth column to obtain

$$\begin{bmatrix} 1 & 0 & 0 & 0 \\ 0 & 1 & -1 & 1 \\ 0 & 0 & 0 & 0 \\ 0 & 0 & 0 & 0 \end{bmatrix}$$ Add 1 times the second column to the third column to obtain

$$\begin{bmatrix} 1 & 0 & 0 & 0 \\ 0 & 1 & 0 & 1 \\ 0 & 0 & 0 & 0 \\ 0 & 0 & 0 & 0 \end{bmatrix}$$ Add (-1) times the second column to the fourth column to obtain

$$\begin{bmatrix} 1 & 0 & 0 & 0 \\ 0 & 1 & 0 & 0 \\ 0 & 0 & 0 & 0 \\ 0 & 0 & 0 & 0 \end{bmatrix}.$$

This is the desired matrix. ■

The following theorem gives another useful way to look at the equivalence of matrices.

Theorem 1.21. Two $m \times n$ matrices A and B are equivalent if and only if $B = PAQ$ for some nonsingular matrices P and Q.

Proof: Exercise. ■

We next prove a theorem that is analogous to Corollary 1.3.

Theorem 1.22. An $n \times n$ matrix A is nonsingular if and only if A is equivalent to I_n.

Proof: If A is equivalent to I_n, then I_n arises from A by a sequence of elementary row or elementary column operations. Thus there exist elementary matrices $E_1, E_2, \ldots, E_r, F_1, F_2, \ldots, F_s$ such that

$$I_n = E_r E_{r-1} \cdots E_2 E_1 A F_1 F_2 \cdots F_s.$$

Let $E_r E_{r-1} \cdots E_2 E_1 = P$, and $F_1 F_2 \cdots F_s = Q$. Then $I_n = PAQ$, where P and Q are nonsingular. It then follows that $A = P^{-1}Q^{-1}$, and since P^{-1} and Q^{-1} are nonsingular, A is nonsingular.

Conversely, if A is nonsingular. then from Corollary 1.3 it follows that A is row equivalent to I_n. Hence A is equivalent to I_n. ■

1.7. EXERCISES

1. (a) Prove that every matrix A is equivalent to itself.

(b) Prove that if A is equivalent to B, then B is equivalent to A.

(c) Prove that if A is equivalent to B and B is equivalent to C, then A is equivalent to C.

2. For each of the following matrices, find a matrix of the form described in Theorem 1.20, equivalent to the given matrix.

(a) $\begin{bmatrix} 1 & 2 & -1 & 4 \\ 5 & 1 & 2 & -3 \\ 2 & 1 & 4 & 3 \\ 2 & 0 & 1 & 2 \\ 5 & 1 & 2 & 3 \end{bmatrix}$. (b) $\begin{bmatrix} 1 & 2 & 1 \\ 2 & 3 & 1 \\ 2 & 1 & 3 \end{bmatrix}$.

(c) $\begin{bmatrix} 1 & -2 & 1 \\ 2 & 3 & 2 \\ 3 & 1 & 3 \end{bmatrix}$. (d) $\begin{bmatrix} 1 & 3 & -1 & 2 \\ 2 & -4 & -2 & 1 \\ 3 & 1 & 2 & -3 \\ 7 & 3 & -2 & 5 \end{bmatrix}$.

3. Repeat Exercise 2 for the following matrices.

(a) $\begin{bmatrix} 1 & 2 & 3 & -1 \\ 1 & 0 & 2 & 3 \\ 3 & 4 & 8 & 1 \end{bmatrix}$. (b) $\begin{bmatrix} 3 & 4 & 1 \\ 1 & 2 & -2 \\ 5 & 6 & 4 \\ 5 & 8 & -1 \end{bmatrix}$.

(c) $\begin{bmatrix} 2 & 3 & 4 & -1 \\ 1 & 2 & 1 & -1 \\ 2 & -1 & 1 & 1 \\ 4 & 2 & 5 & 0 \\ 4 & 3 & 3 & -1 \end{bmatrix}$. (d) $\begin{bmatrix} 1 & 2 & 3 \\ 1 & -1 & 0 \\ 0 & 1 & 2 \end{bmatrix}$.

4. Show that if A and B are row equivalent, then they are equivalent.

5. Prove Theorem 1.21.

6. Let $A = \begin{bmatrix} 1 & 2 & 3 \\ 1 & 1 & 2 \\ 0 & 1 & 1 \end{bmatrix}$. Find a matrix B of the form described in Theorem 1.20 which is equivalent to A. Also, find nonsingular matrices P and Q such that $B = PAQ$.

7. Repeat Exercise 6 for

$$A = \begin{bmatrix} 1 & -1 & 2 & 3 \\ 2 & -1 & 3 & 1 \\ 4 & -3 & 7 & 7 \\ 0 & -1 & 1 & 5 \end{bmatrix}.$$

8. Let A be an $m \times n$ matrix. Show that A is equivalent to ${}_mO_n$ if and only if $A = {}_mO_n$.

9. Let A and B be $m \times n$ matrices. Show that A is equivalent to B if and only if A^T is equivalent to B^T.

10. For each of the following matrices A, find a matrix $B \neq A$ that is equivalent to A.

(a) $A = \begin{bmatrix} 1 & -2 & 3 & 1 \\ 0 & -1 & 4 & 3 \\ 1 & 0 & -2 & -1 \end{bmatrix}$.

(b) $A = \begin{bmatrix} 1 & 3 \\ 2 & 6 \end{bmatrix}$.

(c) $A = \begin{bmatrix} 1 & 2 & 3 & 4 & 3 \\ 0 & 1 & -2 & 0 & 2 \\ -1 & 3 & 2 & 0 & 1 \end{bmatrix}$.

11. Let A and B be equivalent square matrices. Prove that A is nonsingular if and only if B is nonsingular.

1.8. *LU-Factorization (Optional)*

Gaussian elimination is more efficient than Gauss–Jordan elimination. In the special case that A is *square, upper triangular,* and has *all its diagonal entries different from zero,* the linear system $AX = B$ can be solved directly. There is no need to obtain the reduced row echelon form or even the row echelon form of the augmented matrix $[A \mid B]$. The augmented matrix of such a system is given by

$$\left[\begin{array}{cccccc|c} a_{11} & a_{12} & a_{13} & \cdots & a_{1n} & & b_1 \\ 0 & a_{22} & a_{23} & \cdots & a_{2n} & & b_2 \\ 0 & 0 & a_{33} & \cdots & a_{3n} & & b_3 \\ \vdots & \vdots & \vdots & \cdots & \vdots & & \vdots \\ 0 & 0 & 0 & \cdots & a_{nn} & & b_n \end{array}\right].$$

The solution is obtained by the following algorithm:

$$x_n = \frac{b_n}{a_{nn}}$$

$$x_{n-1} = \frac{b_{n-1} - a_{n-1\,n}x_n}{a_{n-1\,n-1}}$$

$$\vdots$$

$$x_j = \frac{b_j - \sum_{k=n}^{j-1} a_{jk}x_k}{a_{jj}} \qquad j = n, n-1, \ldots, 1.$$

This procedure is **back substitution**, which we used in conjunction with Gaussian elimination in Section 1.5, where it was required that the diagonal entries be one.

In a similar manner, if A is *square, lower triangular,* and has *all its diagonal entries different from zero,* the system $AX = B$ can be solved by **forward substitution**, which is the algorithm we present next. The augmented matrix $[A \mid B]$ has the form

$$\left[\begin{array}{cccccc|c} a_{11} & 0 & 0 & \cdots & 0 & & b_1 \\ a_{21} & a_{22} & 0 & \cdots & 0 & & b_2 \\ a_{31} & a_{32} & a_{33} & \cdots & 0 & & b_3 \\ \vdots & \vdots & \vdots & \cdots & \vdots & & \vdots \\ a_{n1} & a_{n2} & a_{n3} & \cdots & a_{nn} & & b_n \end{array}\right]$$

and the solution is given by

$$x_1 = \frac{b_1}{a_{11}}$$

$$x_2 = \frac{b_2 - a_{21}x_1}{a_{22}}$$

$$\vdots$$

$$x_j = \frac{b_j - \sum_{k=1}^{j-1} a_{jk}x_k}{a_{jj}} \qquad j = 1, 2, \ldots, n.$$

We illustrate forward substitution in the following example.

Example 1. To solve the linear system

$$\begin{aligned} 5x_1 \phantom{{}+3x_2+4x_3} &= 10 \\ 4x_1 - 2x_2 \phantom{{}+4x_3} &= 28 \\ 2x_1 + 3x_2 + 4x_3 &= 26 \end{aligned}$$

we use forward substitution. Hence we obtain from the previous algorithm

$$x_1 = \frac{10}{5} = 2$$

$$x_2 = \frac{29 - 4x_1}{-2} = -10$$

$$x_3 = \frac{26 - 2x_1 - 3x_2}{4} = 13,$$

which implies that the solution to the given lower triangular system of equations is

$$X = \begin{bmatrix} 2 \\ -10 \\ 13 \end{bmatrix}.$$

■

The ease with which systems of equations with upper or lower triangular coefficient matrices can be solved is quite attractive. The forward substitution and back substitution algorithms are fast and simple to use. In order to introduce another important numerical procedure for solving linear systems of equations, we begin with the following definition.

Definition 1.15. An $n \times n$ matrix A is said to be in **triangular form** when it satisfies the following properties:
(a) A is either an upper or a lower triangular matrix.
(b) All the diagonal entries of A are different from zero. ■

We use the terms *upper triangular form* and *lower triangular form* to describe particular matrices in triangular form. We also say that a linear system $AX = B$ is in **triangular form** if its coefficient matrix A is in triangular form. Any linear system of equations that is in triangular form can be solved directly by either forward substitution or back substitution.

Suppose that a nonsingular matrix A can be written as a product of a matrix L in lower triangular form and a matrix U in upper triangular form; that is,

$$A = LU.$$

In this case we say that A has an **LU-factorization** or an **LU-decomposition**. The LU-factorization of a matrix A can be used to efficiently solve a linear system $AX = B$. It follows that $AX = B$ is equivalent to $(LU)X = B$. Letting $UX = Z$, we have

$$LZ = B.$$

Since L is in lower triangular form, we solve directly for Z by forward substitution. Next, since U is in upper triangular form we solve $UX = Z$ for X by back substitution. In summary, *if a nonsingular matrix A has an LU-factorization, then the solution of $AX = B$ can be determined by a forward substitution followed by a back substitution.* We illustrate this procedure in the next example.

Example 2. Consider the linear system.

$$\begin{aligned} 6x_1 - 2x_2 - 4x_3 + 4x_4 &= 2 \\ 3x_1 - 3x_2 - 6x_3 + x_4 &= -4 \\ -12x_1 + 8x_2 + 21x_3 - 8x_4 &= 8 \\ -6x_1 \qquad - 10x_3 + 7x_4 &= -43 \end{aligned}$$

whose coefficient matrix

$$A = \begin{bmatrix} 6 & -2 & -4 & 4 \\ 3 & -3 & -6 & 1 \\ 12 & 8 & 21 & -8 \\ -6 & 0 & -10 & 7 \end{bmatrix}$$

has an LU-factorization where

$$L = \begin{bmatrix} 2 & 0 & 0 & 0 \\ 1 & -1 & 0 & 0 \\ -4 & 2 & 1 & 0 \\ -2 & -1 & -2 & 2 \end{bmatrix} \quad \text{and} \quad U = \begin{bmatrix} 3 & -1 & -2 & 2 \\ 0 & 2 & 4 & 1 \\ 0 & 0 & 5 & -2 \\ 0 & 0 & 0 & 4 \end{bmatrix}$$

(verify). To solve the given system using this LU-factorization, we proceed as follows. Let

$$B = \begin{bmatrix} 2 \\ -4 \\ 8 \\ -43 \end{bmatrix}.$$

Then we solve $AX = B$ by writing it as $LUX = B$. First, let $UX = Z$ and now solve $LZ = B$:

$$\begin{bmatrix} 2 & 0 & 0 & 0 \\ 1 & -1 & 0 & 0 \\ -4 & 2 & 1 & 0 \\ -2 & -1 & -2 & 2 \end{bmatrix} \begin{bmatrix} z_1 \\ z_2 \\ z_3 \\ z_4 \end{bmatrix} = \begin{bmatrix} 2 \\ -4 \\ 8 \\ -43 \end{bmatrix}$$

by forward substitution. We obtain

$$z_1 = \frac{2}{2} = 1$$

$$z_2 = \frac{-4 - z_1}{-1} = 5$$

$$z_3 = \frac{8 + 4z_1 - 2z_2}{1} = 2$$

$$z_4 = \frac{-43 + 2z_1 + z_2 + 2z_3}{2} = -16.$$

Next we solve $UX = Z$,

$$\begin{bmatrix} 3 & -1 & -2 & 2 \\ 0 & 2 & 4 & 1 \\ 0 & 0 & 5 & -2 \\ 0 & 0 & 0 & 4 \end{bmatrix} \begin{bmatrix} x_1 \\ x_2 \\ x_3 \\ x_4 \end{bmatrix} = \begin{bmatrix} 1 \\ 5 \\ 2 \\ -16 \end{bmatrix},$$

by back substitution. We obtain

$$x_4 = \frac{-16}{4} = -4$$

$$x_3 = \frac{2 + 2x_4}{5} = -1.2$$

$$x_2 = \frac{5 - 4x_3 - x_4}{2} = 6.9$$

$$x_1 = \frac{1 + x_2 + 2x_3 - 2x_4}{3} = 4.5.$$

Thus the solution to the linear system is

$$X = \begin{bmatrix} 4.5 \\ 6.9 \\ -1.2 \\ -4 \end{bmatrix}.$$

■

Next we show how to obtain an LU-factorization of a matrix by modifying the Gaussian elimination procedure from Section 1.5. No row interchanges will be permitted and we do not require that the diagonal entries have value 1. The following LU-factorization procedure may fail when row interchanges are not permitted. At the end of this section we discuss how to enhance the LU-factorization scheme presented to prevent such failures. We observe that the only elementary row operation permitted is the one that adds a multiple of one row to a different row.

To describe the LU-factorization, we present a step-by-step procedure in the next example.

Example 3. Let A be the coefficient matrix of the linear system of Example 2.

$$A = \begin{bmatrix} 6 & -2 & -4 & 4 \\ 3 & -3 & -6 & 1 \\ -12 & 8 & 21 & -8 \\ -6 & 0 & -10 & 7 \end{bmatrix}.$$

We proceed to "zero out" entries below the diagonal entries using only the row operation that adds a multiple of one row to a different row.

PROCEDURE

STEP 1. "Zero out" below the first diagonal entry of A. Add $-\frac{1}{2}$ times the first row of A to the second row of A. Add 2 times the first row of A to the third row of A. Add 1 times the first row of A to the fourth row of A. Call the new resulting matrix U_1.

MATRICES USED

$$U_1 = \begin{bmatrix} 6 & -2 & -4 & 4 \\ 0 & -2 & -4 & -1 \\ 0 & 4 & 13 & 0 \\ 0 & -2 & -14 & 11 \end{bmatrix}$$

We begin building a lower triangular matrix, L_1, with 1's on the main diagonal, to record the row operations. Enter the *negatives of the multipliers* used in the row operations in the first column of L_1, below the first diagonal entry of L_1.

$$L_1 = \begin{bmatrix} 1 & 0 & 0 & 0 \\ \frac{1}{2} & 1 & 0 & 0 \\ -2 & * & 1 & 0 \\ -1 & * & * & 1 \end{bmatrix}$$

STEP 2. "Zero out" below the second diagonal entry of U_1. Add 2 times the second row of U_1 to the third row of U_1. Add -1 times the second row of U_1 to the fourth row of U_1. Call the new resulting matrix U_2.

$$U_2 = \begin{bmatrix} 6 & -2 & -4 & 4 \\ 0 & -2 & -4 & -1 \\ 0 & 0 & 5 & -2 \\ 0 & 0 & -10 & 12 \end{bmatrix}$$

Enter the negatives of the multipliers from the row operations below the second diagonal entry of L_1. Call the new matrix L_2.

$$L_2 = \begin{bmatrix} 1 & 0 & 0 & 0 \\ \frac{1}{2} & 1 & 0 & 0 \\ -2 & -2 & 1 & 0 \\ -1 & 1 & * & 1 \end{bmatrix}$$

STEP 3. "Zero out" below the third diagonal entry of U_2. Add 2 times the third row of U_2 to the fourth row of U_2. Call the new resulting matrix U_3.

$$U_3 = \begin{bmatrix} 6 & -2 & -4 & 4 \\ 0 & -2 & -4 & -1 \\ 0 & 0 & 5 & -2 \\ 0 & 0 & 0 & 8 \end{bmatrix}$$

Enter the negative of the multiplier below the third diagonal entry of L_2. Call the new matrix L_3.

$$L_3 = \begin{bmatrix} 1 & 0 & 0 & 0 \\ \frac{1}{2} & 1 & 0 & 0 \\ -2 & -2 & 1 & 0 \\ -1 & 1 & -2 & 1 \end{bmatrix}$$

Let $L = L_3$ and $U = U_3$. Then the product LU gives the original matrix A (verify). The linear system of equations in Example 2 can be solved using the LU-factorization just obtained. ■

In the preceding discussions we have asserted that Gaussian elimination and LU-factorization are more efficient than Gauss–Jordan reduction. A way to measure efficiency of techniques for solving systems of equations is to count the number of arithmetic operations (additions, subtractions, multiplications, and divisions) required to compute the solution. The usual procedure is to count only the multiplications and divisions involved. This custom arose because the first digital computers required significantly more time to perform multiplications and divisions than additions and subtractions. Also, the number of additions and subtractions is approximately the same as the number of multiplications for methods based on row operations. We call the number of multiplications and divisions the **operation count**.

We consider the operation count for back substitution. To compute x_n, one division is required. To compute x_{n-1}, one division and one multiplication are required. In general, to compute x_j, one division and $n - j$ multiplications are required. Thus the operation count for back substitution is

$$1 + 2 + \cdots + (n - j) + 1 + \cdots + (n - 1) + 1.$$

Using summation notation, we have that the operation count for back substitution is

$$\sum_{k=1}^{n} k,$$

which simplifies to $n(n + 1)/2$. Forward substitution has the same operation count.

Next consider the operation count for a row operation that adds a multiple of one row to a different row. In the elimination-type methods used to solve linear systems, this type of row operation is specifically used to "zero out" entries. Consider a pair of rows from an augmented matrix that has undergone a number of elimination steps with $q_1 \neq 0$:

$$\begin{array}{cccccccc|c} 0 & \cdots & 0 & q_1 & q_2 & \cdots & q_k & & q_{k+1} \\ 0 & \cdots & 0 & s_1 & s_2 & \cdots & s_k & & s_{k+1} \end{array}.$$

Gaussian elimination as described in Section 1.5 used $k + 1$ divisions to get to the form

$$\begin{array}{cccccccc|c} 0 & \cdots & 0 & 1 & Q_2 & \cdots & Q_k & & Q_{k+1} \\ 0 & \cdots & 0 & s_1 & s_2 & \cdots & s_k & & s_{k+1} \end{array},$$

where $Q_j = q_j/q_1$ for $j = 2, \ldots, k + 1$. Then k multiplications are required to obtain the form

$$\begin{array}{cccccccc|c} 0 & \cdots & 0 & 1 & Q_2 & \cdots & Q_k & & Q_{k+1} \\ 0 & \cdots & 0 & 0 & S_2 & \cdots & S_k & & S_{k+1} \end{array},$$

where $S_j = s_j - s_1 Q_j$ for $j = 2, \ldots, k + 1$. The entry that is "zeroed out" is set to zero, so no multiplication is required. Thus one such step of Gaussian elimination has an operation count of $2k + 1$, except the last step, which requires only two divisions. Carefully totaling the operations involved, we find that the operation count to obtain upper triangular form is

$$\frac{2n^3 + 3n^2 + 7n}{6}.$$

Since the diagonal entries are 1, back substitution for an upper triangular matrix requires no divisions; hence the operation count is $(n(n + 1)/2) - n$. Thus Gaussian elimination with back substitution has an operation count of

$$\frac{2n^3 + 3n^2 + 7n}{6} + \frac{n(n + 1)}{2} - n = \frac{2n^3 + 6n^2 + 4n}{6}.$$

A similar development shows that Gauss–Jordan has an operation count of

$$\frac{n^3 + 2n^2 - n}{2}.$$

It can be shown that the method of solving a linear system by LU-factorization using storage of multipliers followed by a forward and back substitution has an operation count of

$$\frac{n^3 + 3n^2 - n}{3}.$$

Table 1.1 gives a sample of comparisons for the number of multiplications and divisions required for the three solution methods of Gaussian elimination, Gauss–Jordan reduction and LU-factorization. These results clearly imply that Gauss–Jordan requires more operations. Hence Gauss–Jordan is seldom used in computer algorithms.

Table 1.1 Operation Count Sample

n	*Gaussian Elimination*	*Gauss–Jordan Reduction*	*LU-Factorization*
3	20	21	17
4	40	46	36
5	70	85	65
10	440	595	430
15	1,360	1,905	1,345
20	3,080	4,390	3,060
50	44,200	64,975	44,150

Note that a matrix A can have more than one LU-factorization. This is illustrated by Examples 2 and 3. There are many methods for obtaining an LU-factorization besides the scheme for **storage of multipliers** described in Example 3. It is important to note that if $a_{11} = 0$, then the procedure in Example 3 fails. Also, if the second diagonal entry of U_1 is zero or if the third diagonal entry of U_2 is zero, the procedure fails. In such cases we can try rearranging the equations of the system and beginning again, or using one of the other methods for LU-factorization. Most computer programs for LU-factorization incorporate row interchanges into the storage of multipliers scheme and use additional strategies to help control roundoff error. In this case the product of L and U is not necessarily A. The resulting matrix consists of a permutation of the rows of A. The book *Experiments in Computational Matrix Algebra,* by David R. Hill (New York: Random House, Inc., 1988) explores such a modification of the procedure for LU-factorization.

1.8. EXERCISES

In Exercises 1 through 4, solve the linear system $AX = B$ with the given LU-factorization of the coefficient matrix A. Solve the linear system using a forward substitution followed by a back substitution.

1. $A = \begin{bmatrix} 2 & 8 & 0 \\ 2 & 2 & -3 \\ 1 & 2 & 7 \end{bmatrix}, \quad B = \begin{bmatrix} 18 \\ 3 \\ 12 \end{bmatrix},$

$L = \begin{bmatrix} 2 & 0 & 2 \\ 2 & -3 & 0 \\ 1 & -1 & 4 \end{bmatrix}, \quad U = \begin{bmatrix} 1 & 4 & 0 \\ 0 & 2 & 1 \\ 0 & 0 & 2 \end{bmatrix}.$

2. $A = \begin{bmatrix} 8 & 12 & -4 \\ 6 & 5 & 7 \\ 2 & 1 & 6 \end{bmatrix}, \quad B = \begin{bmatrix} -36 \\ 11 \\ 16 \end{bmatrix},$

$L = \begin{bmatrix} 4 & 0 & 0 \\ 3 & 2 & 0 \\ 1 & 1 & 1 \end{bmatrix}, \quad U = \begin{bmatrix} 2 & 3 & -1 \\ 0 & -2 & 5 \\ 0 & 0 & 2 \end{bmatrix}.$

3. $A = \begin{bmatrix} 2 & 3 & 0 & 1 \\ 4 & 5 & 3 & 3 \\ -2 & -6 & 7 & 7 \\ 8 & 9 & 5 & 21 \end{bmatrix}, \quad B = \begin{bmatrix} -2 \\ -2 \\ -16 \\ -66 \end{bmatrix},$

$L = \begin{bmatrix} 1 & 0 & 0 & 0 \\ 2 & 1 & 0 & 0 \\ -1 & 3 & 1 & 0 \\ 4 & 3 & 2 & 1 \end{bmatrix},$

$U = \begin{bmatrix} 2 & 3 & 0 & 1 \\ 0 & -1 & 3 & 1 \\ 0 & 0 & -2 & 5 \\ 0 & 0 & 0 & 4 \end{bmatrix}.$

4. $A = \begin{bmatrix} 4 & 2 & 1 & 0 \\ -4 & -6 & 1 & 3 \\ 8 & 16 & -3 & -4 \\ 20 & 10 & 4 & -3 \end{bmatrix}, \quad B = \begin{bmatrix} 6 \\ 13 \\ -20 \\ 15 \end{bmatrix},$

$L = \begin{bmatrix} 1 & 0 & 0 & 0 \\ -1 & 1 & 0 & 0 \\ 2 & -3 & 1 & 0 \\ 5 & 0 & -1 & 1 \end{bmatrix},$

$U = \begin{bmatrix} 4 & 2 & 1 & 0 \\ 0 & -4 & 2 & 3 \\ 0 & 0 & 1 & 5 \\ 0 & 0 & 0 & 2 \end{bmatrix}.$

In Exercises 5 through 10, find an LU-factorization of the coefficient matrix of the given linear system $AX = B$. Solve the linear system using a forward substitution followed by a back substitution.

5. $A = \begin{bmatrix} 2 & 3 & 4 \\ 4 & 5 & 10 \\ 4 & 8 & 2 \end{bmatrix}$, $B = \begin{bmatrix} 6 \\ 16 \\ 2 \end{bmatrix}$.

6. $A = \begin{bmatrix} -3 & 1 & -2 \\ -12 & 10 & -6 \\ 15 & 13 & 12 \end{bmatrix}$, $B = \begin{bmatrix} 15 \\ 82 \\ -5 \end{bmatrix}$.

7. $A = \begin{bmatrix} 4 & 2 & 3 \\ 2 & 0 & 5 \\ 1 & 2 & 1 \end{bmatrix}$, $B = \begin{bmatrix} 1 \\ -1 \\ -3 \end{bmatrix}$.

8. $A = \begin{bmatrix} -5 & 4 & 0 & 1 \\ -30 & 27 & 2 & 7 \\ 5 & 2 & 0 & 2 \\ 10 & 1 & -2 & 1 \end{bmatrix}$, $B = \begin{bmatrix} -17 \\ -102 \\ -7 \\ -6 \end{bmatrix}$.

9. $A = \begin{bmatrix} 2 & 1 & 0 & -4 \\ 1 & 0 & 0.25 & -1 \\ -2 & -1.1 & 0.25 & 6.2 \\ 4 & 2.2 & 0.3 & -2.4 \end{bmatrix}$,

$B = \begin{bmatrix} -3 \\ -1.5 \\ 5.6 \\ 2.2 \end{bmatrix}$.

10. $A = \begin{bmatrix} 4 & 1 & 0.25 & -0.5 \\ 0.8 & 0.6 & 1.25 & -2.6 \\ -1.6 & -0.08 & 0.01 & 0.2 \\ 8 & 1.52 & -0.6 & -1.3 \end{bmatrix}$,

$B = \begin{bmatrix} -0.15 \\ 9.77 \\ 1.69 \\ -4.576 \end{bmatrix}$.

C 11. In the software you are using investigate to see if there is a command for obtaining an LU-factorization of a matrix. If there is, use it to find the LU-factorization of matrix A in Example 2. The result obtained in your software need not be that given in Examples 2 or 3 because there are many ways to compute such a factorization. Also, some software does not explicitly display L and U, but gives a matrix from which L and U can be ''decoded.'' See the documentation on your software for more details.

C 12. In the software you are using investigate to see if there are commands for doing forward substitution or back substitution. Experiment with the use of such commands on the linear systems using L and U from Examples 2 and 3.

Supplementary Exercises

1. Determine the number of entries on or above the main diagonal of a $k \times k$ matrix when (a) $k = 2$; (b) $k = 3$; (c) $k = 4$; (d) $k = n$.

2. Let

$$A = \begin{bmatrix} 0 & 2 \\ 0 & 5 \end{bmatrix}.$$

(a) Find a $2 \times k$ matrix $B \neq O$ such that $AB = O$ for $k = 1, 2, 3, 4$.

(b) Are your answers to part (a) unique? Explain.

3. Find all 2×2 matrices with real entries of the form

$$A = \begin{bmatrix} a & b \\ 0 & c \end{bmatrix}$$

such that $A^2 = I_2$.

4. An $n \times n$ matrix A (with real entries) is called a **square root** of the $n \times n$ matrix B (with real entries) if $A^2 = B$.

(a) Find a square root of $B = \begin{bmatrix} 1 & 1 \\ 0 & 1 \end{bmatrix}$.

(b) Find a square root of $B = \begin{bmatrix} 1 & 0 & 0 \\ 0 & 0 & 0 \\ 0 & 0 & 0 \end{bmatrix}$.

(c) Find a square root of $B = I_4$.
(d) Show that there is no square root of

$$B = \begin{bmatrix} 0 & 1 \\ 0 & 0 \end{bmatrix}.$$

5. Let A be an $m \times n$ matrix.
(a) Describe the diagonal entries of A^TA in terms of the columns of A.
(b) Prove that the diagonal entries of A^TA are nonnegative.
(c) When is $A^TA = O_n$?

6. If A is an $n \times n$ matrix, show that $(A^k)^T = (A^T)^k$ for any positive integer k.

7. Prove that every symmetric upper (or lower) triangular matrix is diagonal.

8. An $n \times n$ matrix A is called **block diagonal** if it can be partitioned in such a way that all the nonzero entries are contained in square blocks A_{ii}.
(a) Partition the following matrix into a block diagonal matrix.

$$\begin{bmatrix} 1 & 2 & 0 & 0 \\ 3 & 4 & 0 & 0 \\ 0 & 0 & 1 & 0 \\ 0 & 0 & 2 & 3 \end{bmatrix}.$$

(b) If A is block diagonal, then the linear system $AX = B$ is said to be **uncoupled** because it can be solved by considering the linear systems with coefficient matrices A_{ii} and right-hand sides an appropriate portion of B. Solve $AX = B$ by "uncoupling" the linear system when A is the 4×4 matrix of part (a) and

$$B = \begin{bmatrix} 1 \\ 1 \\ 0 \\ 3 \end{bmatrix}.$$

9. Let A be an upper triangular matrix. Show that A is nonsingular if and only if all the entries on the main diagonal of A are nonzero.

10. Prove that if A is skew symmetric and nonsingular, then A^{-1} is skew symmetric.

11. If A is an $n \times n$ matrix, then A is called **idempotent** if $A^2 = A$.
(a) Verify that I_n and O_n are idempotent.
(b) Find an idempotent matrix that is not I_n or O_n.
(c) Prove that the only $n \times n$ nonsingular idempotent matrix is I_n.

12. If A is an $n \times n$ matrix, then A is called **nilpotent** if $A^k = O_n$ for some positive integer k.
(a) Prove that every nilpotent matrix is singular.
(b) Verify that $A = \begin{bmatrix} 0 & 1 & 1 \\ 0 & 0 & 1 \\ 0 & 0 & 0 \end{bmatrix}$ is nilpotent.
(c) If A is nilpotent, prove that $I_n - A$ is nonsingular. [*Hint:* Find $(I_n - A)^{-1}$ in the cases $A^k = O_n$, $k = 1, 2, \ldots$, and look for a pattern.]

13. Find all values of a for which the following linear systems have solutions.

(a)
$$\begin{aligned} x + 2y + z &= a^2 \\ x + y + 3z &= a \\ 3x + 4y + 7z &= 8. \end{aligned}$$

(b)
$$\begin{aligned} x + 2y + z &= a^2 \\ x + y + 3z &= a \\ 3x + 4y + 8z &= 8. \end{aligned}$$

14. Find all values of a for which the following homogeneous system has nontrivial solutions.

$$\begin{aligned} (1 - a)x + z &= 0 \\ -ay + z &= 0 \\ y - az &= 0. \end{aligned}$$

15. For an $n \times n$ matrix A, the main counter diagonal elements are $a_{1n}, a_{2n-1}, \ldots, a_{n1}$. (Note that a_{ij} is a main counter diagonal element provided that $i + j = n + 1$.) The sum of the main counter diagonal elements is denoted $\text{Mcd}(A)$ and we have

$$\text{Mcd}(A) = \sum_{i+j=n+1} a_{ij},$$

meaning the sum of all the entries of A whose subscripts add to $n + 1$.

(a) Prove: $\text{Mcd}(cA) = c\,\text{Mcd}(A)$, where c is a real number.
(b) Prove: $\text{Mcd}(A + B) = \text{Mcd}(A) + \text{Mcd}(B)$.
(c) Prove: $\text{Mcd}(A^T) = \text{Mcd}(A)$.
(d) Show by example that $\text{Mcd}(AB)$ need not be equal to $\text{Mcd}(BA)$.

16. Show that the product of two 2×2 skew symmetric matrices is diagonal. Is this true for $n \times n$ skew symmetric matrices with $n > 2$?

17. Prove that if $\text{Tr}(A^TA) = 0$, then $A = O$.

18. Let X and Y be column matrices with n elements. The outer product of X and Y is the matrix product XY^T which gives the $n \times n$ matrix

$$\begin{bmatrix} x_1y_1 & x_1y_2 & \cdots & x_1y_n \\ x_2y_1 & x_2y_2 & \cdots & x_2y_n \\ \vdots & \vdots & \cdots & \vdots \\ x_ny_1 & x_ny_2 & \cdots & x_ny_n \end{bmatrix}.$$

(a) Form the outer product of X and Y, where

$$X = \begin{bmatrix} 1 \\ 2 \\ 3 \end{bmatrix} \quad \text{and} \quad Y = \begin{bmatrix} 4 \\ 5 \\ 0 \end{bmatrix}.$$

(b) Form the outer product of X and Y, where

$$X = \begin{bmatrix} 1 \\ 2 \\ 1 \\ 2 \end{bmatrix} \quad \text{and} \quad Y = \begin{bmatrix} -1 \\ 0 \\ 3 \\ 5 \end{bmatrix}.$$

19. Prove or disprove: The outer product of X and Y equals the outer product of Y and X.

20. Prove that $\text{Tr}(XY^T) = X^TY$.

21. Prove that the outer product of X and Y is row equivalent to either O_n or to a matrix with $n - 1$ rows of zeros.

22. Let $A = \begin{bmatrix} 1 & 7 \\ 3 & 9 \\ 5 & 11 \end{bmatrix}$ and $B = \begin{bmatrix} 2 & 4 \\ 6 & 8 \end{bmatrix}$. Verify that

$$AB = \sum_{i=1}^{2} \text{outer product of } \text{col}_i(A) \text{ with } \text{row}_i(B).$$

23. Let W be an $n \times 1$ matrix such that $W^TW = 1$. The $n \times n$ matrix

$$H = I_n - 2W\,W^T$$

is called a **Householder matrix**. (Note that a Householder matrix is the identity matrix plus a scalar multiple of an outer product.)
(a) Show that H is symmetric.
(b) Show that $H^{-1} = H^T$.

24. A **circulant** of order n is the $n \times n$ matrix defined by

$$C = \text{circ}(c_1, c_2, \ldots, c_n) = \begin{bmatrix} c_1 & c_2 & c_3 & \cdots & c_n \\ c_n & c_1 & c_2 & \cdots & c_{n-1} \\ c_{n-1} & c_n & c_1 & \cdots & c_{n-2} \\ \vdots & \vdots & \vdots & \cdots & \vdots \\ c_2 & c_3 & c_4 & \cdots & c_1 \end{bmatrix}$$

The elements of each row of C are the same as those in the previous rows, but shifted one position to the right and wrapped around.
(a) Form the circulant $C = \text{circ}(1, 2, 3)$.
(b) Form the circulant $C = \text{circ}(1, 2, 5, -1)$.
(c) Form the circulant $C = \text{circ}(1, 0, 0, 0, 0)$.
(d) Form the circulant $C = \text{circ}(1, 2, 1, 0, 0)$.

25. Let $C = \text{circ}(c_1, c_2, c_3)$. Under what conditions is C symmetric?

26. Let $C = \text{circ}(c_1, c_2, \ldots, c_n)$ and let X be the $n \times 1$ matrix of all ones. Determine a simple expression for CX.

27. Verify that for $C = \text{circ}(c_1, c_2, c_3)$, $C^TC = CC^T$.

28. For $n \times n$ matrices A and B, when does $(A + B)(A - B) = A^2 - B^2$?

29. Develop a simple expression for the entries of A^n, where n is a positive integer and $A = \begin{bmatrix} 1 & \frac{1}{2} \\ 0 & \frac{1}{2} \end{bmatrix}$.

2 Real Vector Spaces

2.1. Vectors in the Plane and in 3-Space

In many applications we deal with measurable quantities, such as pressure, mass, and speed, which can be completely described by giving their magnitude. They are called **scalars** and will be denoted by lowercase italic letters such as c, d, r, s, and t. There are many other measurable quantities, such as velocity, force, and acceleration, which require for their description not only magnitude, but also a sense of direction. These are called **vectors** and their study comprises this chapter. Vectors will be denoted by lowercase boldface letters, such as **u**, **v**, **x**, **y**, and **z**. The reader has already encountered vectors in elementary physics and in calculus.

Vectors in the Plane

We draw a pair of perpendicular lines intersecting at a point O, called the **origin**. One of the lines, the ***x*-axis** is usually taken in a horizontal position. The other line, the ***y*-axis**, is then taken in a vertical position. The x- and y-axes together are called **coordinate axes** (Figure 2.1) and they form a **rectangular coordinate system** or a **Cartesian** (after René Descartes[†]) **coordinate system**. We now

[†] René Descartes (1596–1650) was one of the best known scientists and philosophers of his day; he was considered by some to be the founder of modern philosophy. After completing a university degree in law, he turned to the private study of mathematics, simultaneously pursuing interests in Parisian night life and in the military, volunteering for brief periods in the Dutch, Bavarian, and French armies. The most productive period of his life was 1628–1648, when he lived in Holland. In

choose a point on the x-axis to the right of O and a point on the y-axis above O to fix the units of length and positive directions on the x- and y-axes. Frequently, but not always, these points are chosen so that they are both equidistant from O, that is, so that the same unit of length is used for both axes.

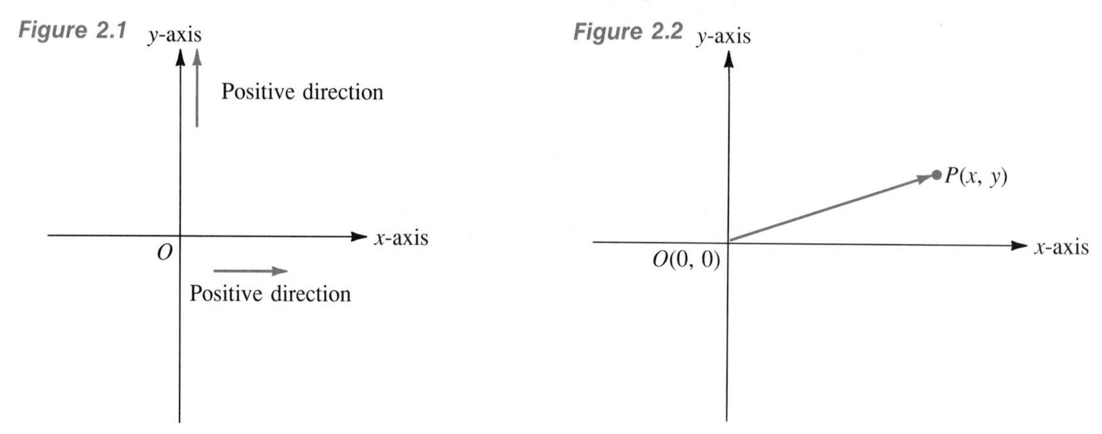

Figure 2.1

Figure 2.2

With each point P in the plane we associate an ordered pair (x, y) of real numbers, its **coordinates**. Conversely, we can associate a point in the plane with each ordered pair of real numbers. The point P with coordinates (x, y) is denoted by $P(x, y)$, or simply by (x, y). The set of all points in the plane is denoted by R^2; it is called **2-space**.

Consider the 2×1 matrix

$$X = \begin{bmatrix} x \\ y \end{bmatrix},$$

where x and y are real numbers. With X we associate the directed line segment with the initial point at the origin and terminal point at $P(x, y)$. The directed line segment from O to P is denoted by $\overrightarrow{OP}$; O is called its **tail** and P its **head**. We distinguish tail and head by placing an arrow at the head (Figure 2.2). A directed

1649 he accepted an invitation from Queen Christina of Sweden to be her private tutor and to establish an Academy of Sciences there. Unfortunately, he had no time for this project, for he died of pneumonia in 1650.

In 1619, Descartes had a dream in which he realized that the method of mathematics is the best way for obtaining truth. However, his only mathematical publication was *La Géométrie,* which appeared as an appendix to his major philosophical work *Discours de la méthode pour bien conduir sa raison, et chercher la vérité dans les sciences.* In *La Géométrie* he proposes the radical idea of doing geometry algebraically. To express a curve algebraically, one chooses any convenient line of reference and, on the line, a point of reference. If y represents the distance from any point of the curve to the reference line and x represents the distance along the line to the reference point, there is an equation relating x and y that represents the curve. The systematic use of "Cartesian" coordinates described above was introduced later in the seventeenth century by authors following up on Descartes' work.

line segment has a **direction**, indicated by the arrow at its head. The **magnitude** of a directed line segment is its length. Thus a directed line segment can be used to describe force, velocity, or acceleration. Conversely, with the directed line segment $\overrightarrow{OP}$ with tail $O(0, 0)$ and head $P(x, y)$ we can associate the matrix

$$\begin{bmatrix} x \\ y \end{bmatrix}.$$

Definition 2.1. A **vector in the plane** is a 2×1 matrix $\mathbf{x} = \begin{bmatrix} x \\ y \end{bmatrix}$, where x and y are real numbers, called the **components** of $\mathbf{x}$. We refer to a vector in the plane merely as a **vector**. ■

Thus, with every vector, we can associate a directed line segment and conversely, with every directed line segment we can associate a vector. Frequently, the notions of directed line segment and vector are used interchangeably, and a directed line segment is called a **vector**.

Since a vector is a matrix, the vectors

$$\mathbf{u} = \begin{bmatrix} x_1 \\ y_1 \end{bmatrix} \quad \text{and} \quad \mathbf{v} = \begin{bmatrix} x_2 \\ y_2 \end{bmatrix}$$

are said to be **equal** if $x_1 = x_2$ and $y_1 = y_2$. That is, two vectors are equal if their respective components are equal.

Example 1. The vectors

$$\begin{bmatrix} a + b \\ 2 \end{bmatrix} \quad \text{and} \quad \begin{bmatrix} 3 \\ a - b \end{bmatrix}$$

are equal if

$$\begin{aligned} a + b &= 3 \\ a - b &= 2, \end{aligned}$$

which means (verify) that $a = \frac{5}{2}$ and $b = \frac{1}{2}$. ■

Frequently, in physical applications it is necessary to deal with a directed line segment $\overrightarrow{PQ}$, from the point $P(x, y)$ (not the origin) to the point $Q(x', y')$, as shown in Figure 2.3(a). Such a directed line segment will also be called a **vector in the plane**, or simply a **vector** with **tail** $P(x, y)$ and **head** $Q(x', y')$. The **components** of such a vector are $x' - x$ and $y' - y$. Thus the vector $\overrightarrow{PQ}$ in Figure 2.3(a) can also be represented by the vector

$$\begin{bmatrix} x' - x \\ y' - y \end{bmatrix}$$

with tail O and head $P''(x' - x, y' - y)$. Two such vectors in the plane will be called **equal** if their respective components are equal. Consider the vectors $\overrightarrow{P_1Q_1}$, $\overrightarrow{P_2Q_2}$, and $\overrightarrow{P_3Q_3}$ joining the points $P_1(3, 2)$ and $Q_1(5, 5)$, $P_2(0, 0)$ and $Q_2(2, 3)$, $P_3(-3, 1)$ and $Q_3(-1, 4)$, respectively, as shown in Figure 2.3(b). Since they all have the same components, they are equal. Moreover, the head $Q_4(x_4', y_4')$ of the vector $\overrightarrow{P_4Q_4} = \begin{bmatrix} 2 \\ 3 \end{bmatrix} = \overrightarrow{P_2Q_2}$, with tail $P_4(-5, 2)$, can be determined as follows. We must have $x_4' - (-5) = 2$ and $y_4' - 2 = 3$ so that $x_4' = 2 - 5 = -3$ and $y_4' = 3 + 2 = 5$. Similarly, the tail $P_5(x_5, y_5)$ of the vector $\overrightarrow{P_5Q_5} = \begin{bmatrix} 2 \\ 3 \end{bmatrix}$ with head $Q_5(8, 6)$ is determined as follows. We must have $8 - x_5 = 2$ and $6 - y_5 = 3$ so that $x_5 = 8 - 2 = 6$ and $y_5 = 6 - 3 = 3$.

Figure 2.3

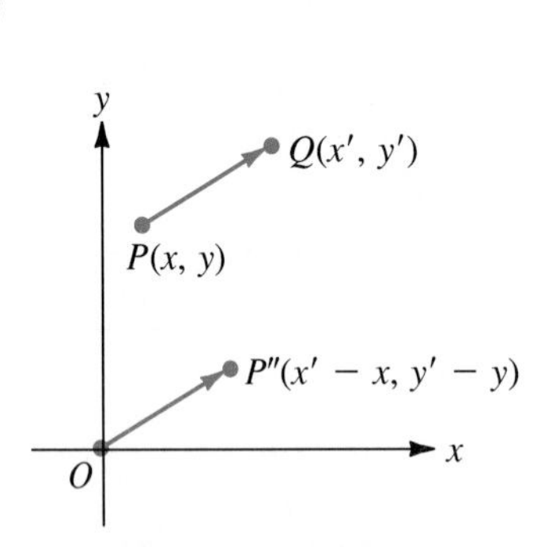

(**a**) Different directed line segments representing the same vector.

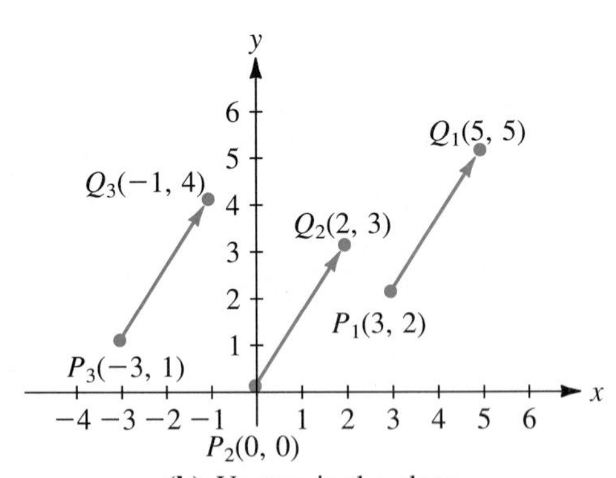

(**b**) Vectors in the plane.

With each vector $\mathbf{x} = \begin{bmatrix} x \\ y \end{bmatrix}$ we can also associate the unique point $P(x, y)$; conversely, with each point $P(x, y)$ we associate the unique vector $\begin{bmatrix} x \\ y \end{bmatrix}$. Hence we also write the vector $\mathbf{x}$ as (x, y). Of course, this association is carried out by means of the directed line segment $\overrightarrow{OP}$, where O is the origin and P is the point with coordinates (x, y) (Figure 2.2).

Thus the plane may be viewed both as the set of all points or as the set of all vectors. For this reason, and depending upon the context, we sometimes take R^2 as the set of all ordered pairs (x, y) and sometimes as the set of all 2×1 matrices $\begin{bmatrix} x \\ y \end{bmatrix}$ (or directed line segments).

Definition 2.2. Let

$$\mathbf{u} = \begin{bmatrix} u_1 \\ u_2 \end{bmatrix} \quad \text{and} \quad \mathbf{v} = \begin{bmatrix} v_1 \\ v_2 \end{bmatrix}$$

be two vectors in the plane. The **sum** of the vectors $\mathbf{u}$ and $\mathbf{v}$ is the vector

$$\mathbf{u} + \mathbf{v} = \begin{bmatrix} u_1 + v_1 \\ u_2 + v_2 \end{bmatrix}. \quad \blacksquare$$

Example 2. Let $\mathbf{u} = \begin{bmatrix} 2 \\ 3 \end{bmatrix}$ and $\mathbf{v} = \begin{bmatrix} 3 \\ -4 \end{bmatrix}$. Then

$$\mathbf{u} + \mathbf{v} = \begin{bmatrix} 2 + 3 \\ 3 + (-4) \end{bmatrix} = \begin{bmatrix} 5 \\ -1 \end{bmatrix}.$$

See Figure 2.4. ■

Figure 2.4

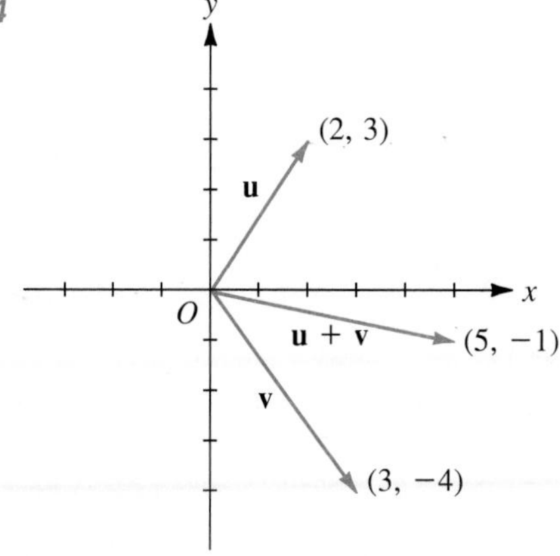

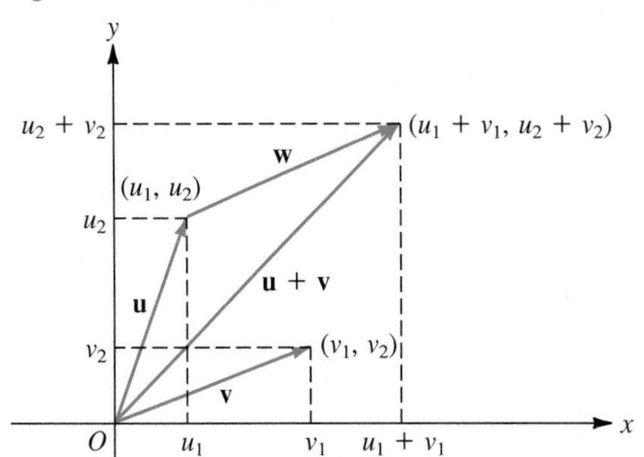

Figure 2.5 Vector addition.

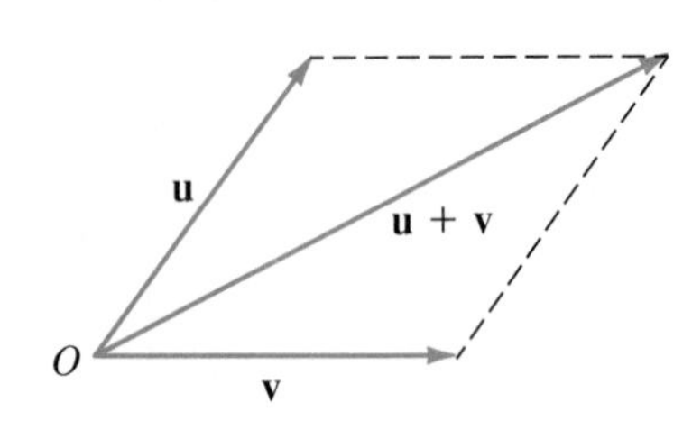

Figure 2.6 Vector addition.

We can interpret vector addition geometrically, as follows. In Figure 2.5 the directed line segment **w** is parallel to **v**, it has the same length as **v**, and its tail is the head (u_1, u_2) of **u**, so its head is $(u_1 + v_1, u_2 + v_2)$. Thus the vector with tail O and head $(u_1 + v_1, u_2 + v_2)$ is $\mathbf{u} + \mathbf{v}$. We can also describe $\mathbf{u} + \mathbf{v}$ as the diagonal of the parallelogram defined by **u** and **v**, as shown in Figure 2.6.

Definition 2.3. If $\mathbf{u} = \begin{bmatrix} u_1 \\ u_2 \end{bmatrix}$ is a vector and c is a scalar (a real number), then the **scalar multiple** $c\mathbf{u}$ of **u** by c is the vector $\begin{bmatrix} cu_1 \\ cu_2 \end{bmatrix}$. Thus the scalar multiple $c\mathbf{u}$ is obtained by multiplying each component of **u** by c. If $c > 0$, then $c\mathbf{u}$ is in the same direction as **u**, whereas if $d < 0$, then $d\mathbf{u}$ is in the opposite direction (Figure 2.7). ■

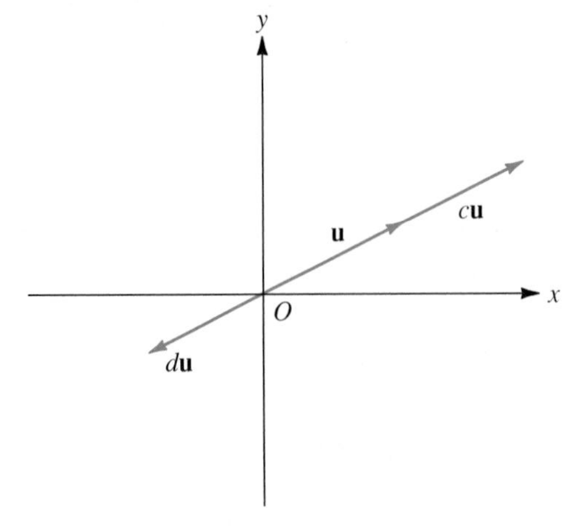

Figure 2.7 Scalar multiplication.

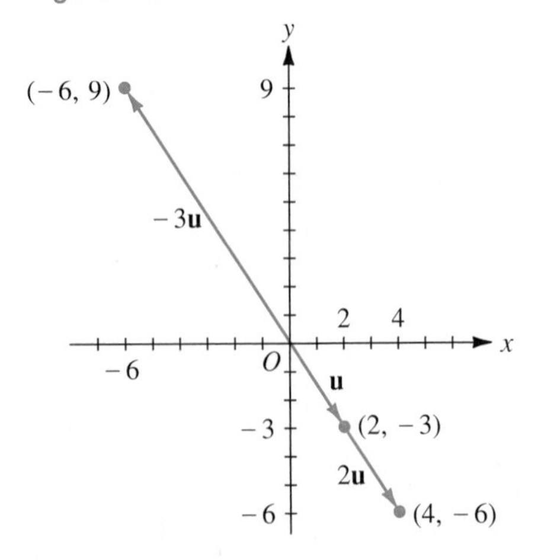

Figure 2.8

Example 3. If $c = 2$, $d = -3$, and $\mathbf{u} = \begin{bmatrix} 2 \\ -3 \end{bmatrix}$, then

$$c\mathbf{u} = 2\begin{bmatrix} 2 \\ -3 \end{bmatrix} = \begin{bmatrix} 2(2) \\ 2(-3) \end{bmatrix} = \begin{bmatrix} 4 \\ -6 \end{bmatrix}$$

and

$$d\mathbf{u} = -3\begin{bmatrix} 2 \\ -3 \end{bmatrix} = \begin{bmatrix} (-3)(2) \\ (-3)(-3) \end{bmatrix} = \begin{bmatrix} -6 \\ 9 \end{bmatrix},$$

which are shown in Figure 2.8. ■

The vector

$$\begin{bmatrix} 0 \\ 0 \end{bmatrix}$$

is called the **zero vector** and is denoted by $\mathbf{0}$. If $\mathbf{u}$ is any vector, it follows that (Exercise 21)

$$\mathbf{u} + \mathbf{0} = \mathbf{u}.$$

We can also show (Exercise 22) that

$$\mathbf{u} + (-1)\mathbf{u} = \mathbf{0},$$

and we write $(-1)\mathbf{u}$ as $-\mathbf{u}$ and call it the **negative** of $\mathbf{u}$. Moreover, we write $\mathbf{u} + (-1)\mathbf{v}$ as $\mathbf{u} - \mathbf{v}$ and call it the **difference between u and v**. It is shown in Figure 2.9(a). Observe that while vector addition gives one diagonal of a parallelogram, vector subtraction gives the other diagonal [see Figure 2.9(b)].

Figure 2.9

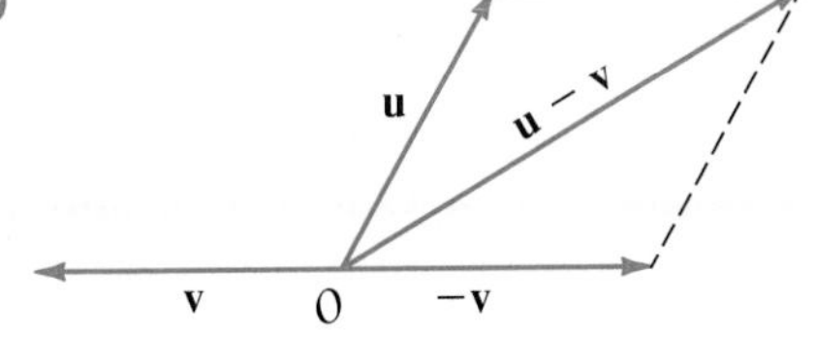

(**a**) Difference between vectors.

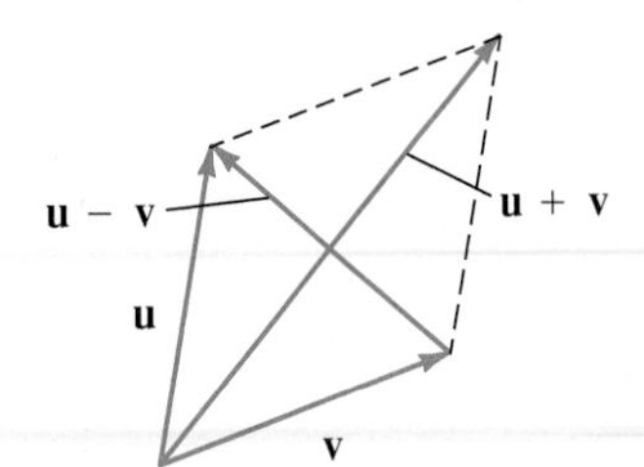

(**b**) Vector sum and vector difference.

Vectors in Space

The foregoing discussion of vectors in the plane can be generalized to vectors in space as follows. We first fix a **coordinate system** by choosing a point, called the **origin**, and three lines, called the **coordinate axes**, each passing through the origin, so that each line is perpendicular to the other two. These lines are individually called the ***x*-, *y*-, and *z*-axes**. On each of these axes we choose a point fixing the units of length and positive directions on the coordinate axes. Frequently, but not always, the same unit of length is used for all the coordinate axes. In Figure 2.10 we show two of the many possible coordinate systems.

Figure 2.10

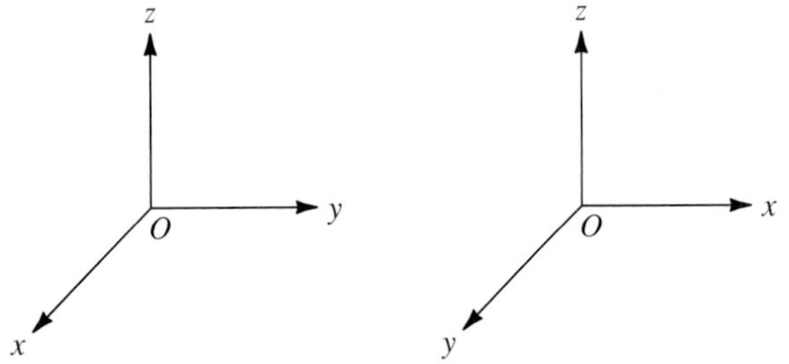

(**a**) Right-handed coordinate system.

(**b**) Left-handed coordinate system.

Figure 2.11

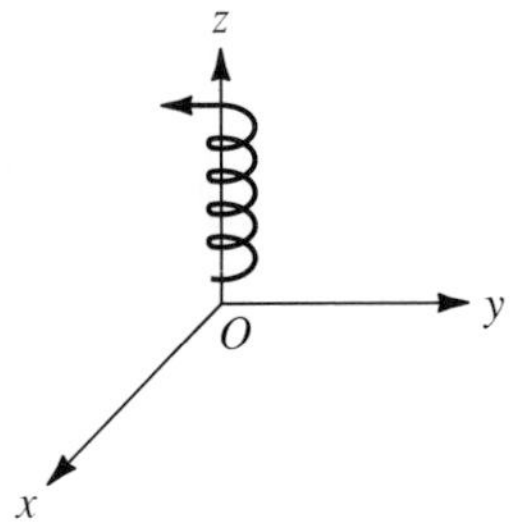

The coordinate system shown in Figure 2.10(a) is called a **right-handed coordinate system**; the one shown in Figure 2.10(b) is called **left-handed**. A right-handed system is characterized by the following property. If we rotate the x-axis counterclockwise toward the y-axis, then a right-hand screw will move in the positive z direction (Figure 2.11).

With each point P in space we associate an ordered triple (x, y, z) of real numbers, its coordinates. Conversely, we can associate a point in space with each ordered triple of real numbers. The point P with coordinates x, y, and z is denoted by $P(x, y, z)$ or simply by (x, y, z). The set of all points in space is called **3-space** and it is denoted by R^3.

A **vector in space**, or simply a **vector**, is a 3×1 matrix

$$\mathbf{x} = \begin{bmatrix} x \\ y \\ z \end{bmatrix},$$

where x, y, and z are real numbers, called the **components** of vector $\mathbf{x}$. Two vectors in space are said to be **equal** if their respective components are equal.

As in the plane, with the vector $\mathbf{x} = \begin{bmatrix} x \\ y \\ z \end{bmatrix}$

we associate the directed line segment $\overrightarrow{OP}$, whose tail is $O(0, 0, 0)$ and whose head is $P(x, y, z)$; conversely, with each such directed line segment we associate the vector $\mathbf{x}$ [see Figure 2.12(a)]. Thus we can also write the vector $\mathbf{x}$ as (x, y, z). Again as in the plane, in physical applications we often deal with a directed line segment $\overrightarrow{PQ}$, from the point $P(x, y, z)$ (not the origin) to the point $Q(x', y', z')$, as shown in Figure 2.12(b). Such a directed line segment will also be called a **vector in R^3**, or simply a **vector** with tail at $P(x, y, z)$ and head at $Q(x', y', z')$. The components of such a vector are $x' - x$, $y' - y$, and $z' - z$. Two such vectors in R^3 will be called **equal** if their respective components are equal. Thus the vector $\overrightarrow{PQ}$ in Figure 2.12(b) can also be represented by the vector $\begin{bmatrix} x - x' \\ y - y' \\ z - z' \end{bmatrix}$ with tail at O and head at $P''(x' - x, y' - y, z' - z)$.

Figure 2.12

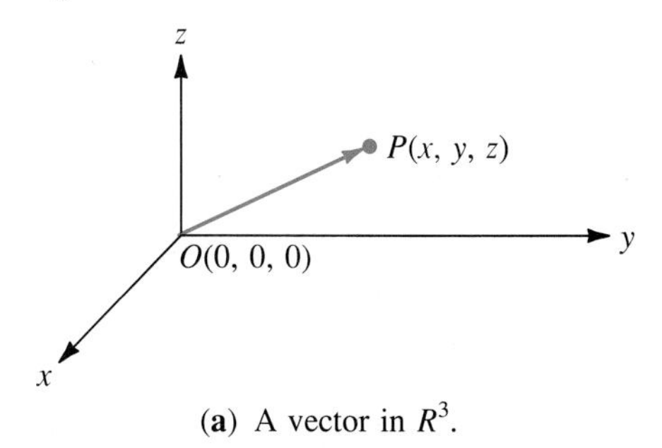

(a) A vector in R^3.

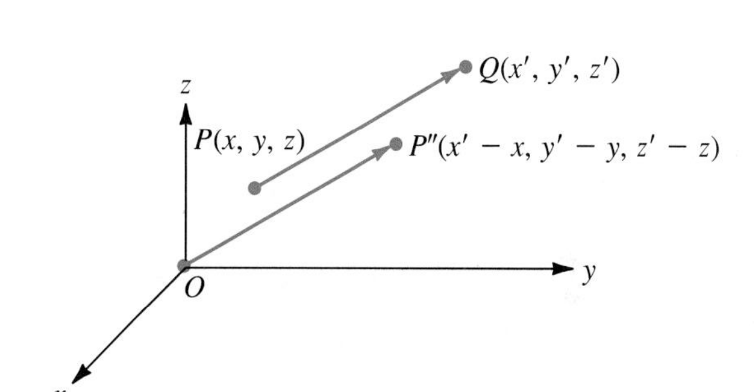

(b) Different directed line segments representing the same vector.

If $\mathbf{u} = \begin{bmatrix} u_1 \\ u_2 \\ u_3 \end{bmatrix}$ and $\mathbf{v} = \begin{bmatrix} v_1 \\ v_2 \\ v_3 \end{bmatrix}$ are vectors in R^3 and c is a scalar, then the **sum** $\mathbf{u} + \mathbf{v}$ and the **scalar multiple** $c\mathbf{u}$ are defined, respectively, as

$$\mathbf{u} + \mathbf{v} = \begin{bmatrix} u_1 + v_1 \\ u_2 + v_2 \\ u_3 + v_3 \end{bmatrix} \quad \text{and} \quad c\mathbf{u} = \begin{bmatrix} cu_1 \\ cu_2 \\ cu_3 \end{bmatrix}$$

The sum is shown in Figure 2.13, which resembles Figure 2.5, and the scalar multiple is shown in Figure 2.14, which resembles Figure 2.8.

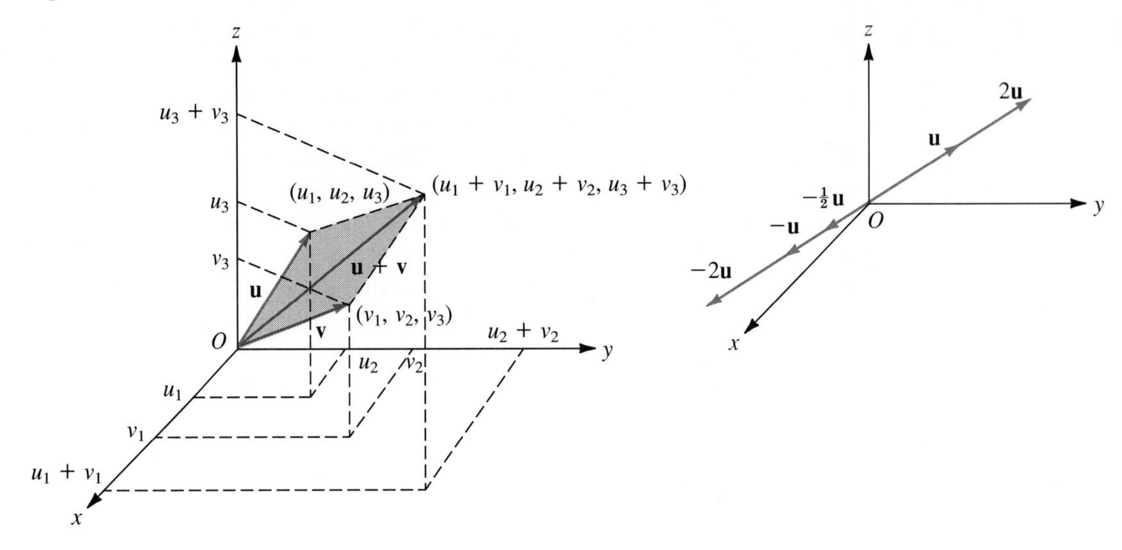

Figure 2.13 Vector addition.

Figure 2.14 Scalar multiplication.

Example 4. Let

$$\mathbf{u} = \begin{bmatrix} 2 \\ 3 \\ -1 \end{bmatrix} \quad \text{and} \quad \mathbf{v} = \begin{bmatrix} 3 \\ -4 \\ 2 \end{bmatrix}.$$

Compute: (a) $\mathbf{u} + \mathbf{v}$; (b) $-2\mathbf{u}$; (c) $3\mathbf{u} - 2\mathbf{v}$.

Solution:

(a) $$\mathbf{u} + \mathbf{v} = \begin{bmatrix} 2 + 3 \\ 3 + (-4) \\ -1 + 2 \end{bmatrix} = \begin{bmatrix} 5 \\ -1 \\ 1 \end{bmatrix}.$$

(b) $$-2\mathbf{u} = \begin{bmatrix} -2(2) \\ -2(3) \\ -2(-1) \end{bmatrix} = \begin{bmatrix} -4 \\ -6 \\ 2 \end{bmatrix}.$$

(c) $$3\mathbf{u} - 2\mathbf{v} = \begin{bmatrix} 3(2) \\ 3(3) \\ 3(-1) \end{bmatrix} - \begin{bmatrix} 2(3) \\ 2(-4) \\ 2(2) \end{bmatrix} = \begin{bmatrix} 0 \\ 17 \\ -7 \end{bmatrix}.$$ ■

The zero vector in R^3 is denoted by $\mathbf{0}$, where

$$\mathbf{0} = \begin{bmatrix} 0 \\ 0 \\ 0 \end{bmatrix}.$$

The vector $\mathbf{0}$ has the property that if $\mathbf{u}$ is any vector in R^3, then

$$\mathbf{u} + \mathbf{0} = \mathbf{u}.$$

The negative of the vector $\mathbf{u} = \begin{bmatrix} u_1 \\ u_2 \\ u_3 \end{bmatrix}$ is the vector $-\mathbf{u} = \begin{bmatrix} -u_1 \\ -u_2 \\ -u_3 \end{bmatrix}$ and

$$\mathbf{u} + (-\mathbf{u}) = \mathbf{0}.$$

Observe that we have defined a vector in the plane as an ordered pair of real numbers or as a 2×1 matrix. Similarly, a vector in space is an ordered triple of real numbers or a 3×1 matrix. However, in physics we often treat a vector as a directed line segment. Thus we have three very different representations of a vector, and one can then ask why all three are legitimately valid. That is, why are we justified in referring to an ordered pair of real numbers, a 2×1 matrix, and a directed line segment by the same name, "vector"?

To answer this question, we first observe that, mathematically speaking, the only thing that concerns us is the behavior of the object we call "vector." It turns out that all three objects behave, from an algebraic point of view, in exactly the same manner. Moreover, many other objects that arise naturally in applied problems behave, algebraically speaking, as do the above-mentioned objects. To a mathematician this is a perfect situation. For we can now abstract those features that all such objects have in common (i.e., those properties that make them all behave alike) and define a new structure. The great advantage of doing this is that we can now talk about properties of all such objects at the same time without having to refer to any one object in particular. This, of course, is much more efficient than studying the properties of each object separately. For example, the following theorem summarizes the properties of addition and scalar multiplication for vectors in the plane and in space. Moreover, this theorem will serve as the model for the generalization of the set of all vectors in the plane or in space to a more abstract setting.

Theorem 2.1. *If* **u**, **v**, *and* **w** *are vectors in* R^2 *or* R^3 *and* c *and* d *are real scalars, then*

(a) $\mathbf{u} + \mathbf{v} = \mathbf{v} + \mathbf{u}$.
(b) $\mathbf{u} + (\mathbf{v} + \mathbf{w}) = (\mathbf{u} + \mathbf{v}) + \mathbf{w}$.
(c) $\mathbf{u} + \mathbf{0} = \mathbf{0} + \mathbf{u} = \mathbf{u}$.
(d) $\mathbf{u} + (-\mathbf{u}) = \mathbf{0}$.
(e) $c(\mathbf{u} + \mathbf{v}) = c\mathbf{u} + c\mathbf{v}$.
(f) $(c + d)\mathbf{u} = c\mathbf{u} + d\mathbf{u}$.
(g) $c(d\mathbf{u}) = (cd)\mathbf{u}$.
(h) $1\mathbf{u} = \mathbf{u}$.

Proof: (a) Suppose that **u** and **v** are vectors in R^2 so that

$$\mathbf{u} = \begin{bmatrix} u_1 \\ u_2 \end{bmatrix} \quad \text{and} \quad \mathbf{v} = \begin{bmatrix} v_1 \\ v_2 \end{bmatrix}.$$

Then

$$\mathbf{u} + \mathbf{v} = \begin{bmatrix} u_1 + v_1 \\ u_2 + v_2 \end{bmatrix} \quad \text{and} \quad \mathbf{v} + \mathbf{u} = \begin{bmatrix} v_1 + u_1 \\ v_2 + u_2 \end{bmatrix}.$$

Since the components of **u** and **v** are real numbers, $u_1 + v_1 = v_1 + u_1$ and $u_2 + v_2 = v_2 + u_2$. Therefore,

$$\mathbf{u} + \mathbf{v} = \mathbf{v} + \mathbf{u}.$$

A similar proof can be given if **u** and **v** are vectors in R^3.

Property (a) can also be established geometrically, as shown in Figure 2.15. The proofs of the remaining properties will be left as exercises. Remember, they can all be proved by either an algebraic or a geometric approach. ■

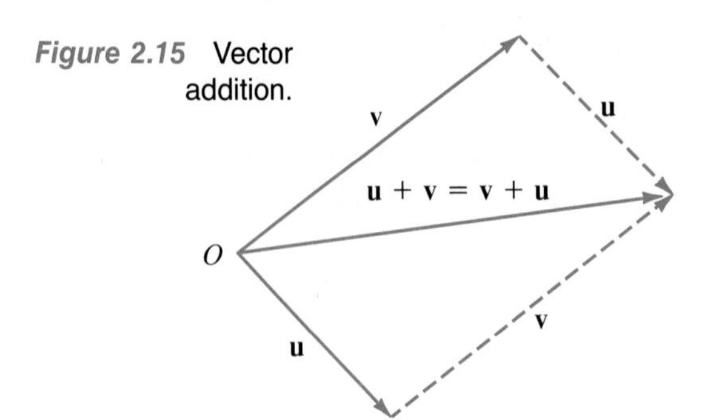

Figure 2.15 Vector addition.

2.1. EXERCISES

1. Sketch a directed line segment in R^2 representing each of the following vectors.

(a) $\mathbf{u} = \begin{bmatrix} -2 \\ 3 \end{bmatrix}$. (b) $\mathbf{v} = \begin{bmatrix} 3 \\ 4 \end{bmatrix}$.

(c) $\mathbf{w} = \begin{bmatrix} -3 \\ -3 \end{bmatrix}$. (d) $\mathbf{z} = \begin{bmatrix} 0 \\ -3 \end{bmatrix}$.

2. Determine the head of the vector $\begin{bmatrix} -2 \\ 5 \end{bmatrix}$ whose tail is at $(-3, 2)$. Make a sketch.

3. Determine the tail of the vector $\begin{bmatrix} 2 \\ 6 \end{bmatrix}$ whose head is at $(1, 2)$. Make a sketch.

4. Determine the tail of the vector $\begin{bmatrix} 2 \\ 4 \\ -1 \end{bmatrix}$ whose head is at $(3, -2, 2)$.

5. For what values of a and b are the vectors $\begin{bmatrix} a-b \\ 2 \end{bmatrix}$ and $\begin{bmatrix} 4 \\ a+b \end{bmatrix}$ equal?

6. For what values of a, b, and c are the vectors $\begin{bmatrix} 2a-b \\ a-2b \\ 6 \end{bmatrix}$ and $\begin{bmatrix} -2 \\ 2 \\ a+b-2c \end{bmatrix}$ equal?

In Exercises 7 and 8, determine the components of the vector $\overrightarrow{PQ}$.

7. (a) $P(1, 2)$, $Q(3, 5)$.
(b) $P(-2, 2, 3)$, $Q(-3, 5, 2)$.

8. (a) $P(-1, 0)$, $Q(-3, -4)$.
(b) $P(1, 1, 2)$, $Q(1, -2, -4)$.

In Exercises 9 and 10, find a vector with tail at the origin representing the vector $\overrightarrow{PQ}$.

9. (a) $P(-1, 2)$, $Q(3, 5)$.
(b) $P(1, 1, -2)$, $Q(3, 4, 5)$.

10. (a) $P(2, -3)$, $Q(-2, 4)$.
(b) $P(-2, -3, 4)$, $Q(0, 0, 1)$.

11. Compute $\mathbf{u} + \mathbf{v}$, $\mathbf{u} - \mathbf{v}$, $2\mathbf{u}$, and $3\mathbf{u} - 2\mathbf{v}$ if:

(a) $\mathbf{u} = \begin{bmatrix} 2 \\ 3 \end{bmatrix}$, $\mathbf{v} = \begin{bmatrix} -2 \\ 5 \end{bmatrix}$.

(b) $\mathbf{u} = \begin{bmatrix} 0 \\ 3 \end{bmatrix}$, $\mathbf{v} = \begin{bmatrix} 3 \\ 2 \end{bmatrix}$.

(c) $\mathbf{u} = \begin{bmatrix} 2 \\ 6 \end{bmatrix}$, $\mathbf{v} = \begin{bmatrix} 3 \\ 2 \end{bmatrix}$.

12. Compute $\mathbf{u} + \mathbf{v}$, $2\mathbf{u} - \mathbf{v}$, $3\mathbf{u} - 2\mathbf{v}$, and $\mathbf{0} - 3\mathbf{v}$ if:

(a) $\mathbf{u} = \begin{bmatrix} 1 \\ 2 \\ 3 \end{bmatrix}$, $\mathbf{v} = \begin{bmatrix} 2 \\ 0 \\ 1 \end{bmatrix}$.

(b) $\mathbf{u} = \begin{bmatrix} 2 \\ -1 \\ 4 \end{bmatrix}$, $\mathbf{v} = \begin{bmatrix} 1 \\ 2 \\ -3 \end{bmatrix}$.

(c) $\mathbf{u} = \begin{bmatrix} 1 \\ 0 \\ -1 \end{bmatrix}$, $\mathbf{v} = \begin{bmatrix} -1 \\ 1 \\ 4 \end{bmatrix}$.

13. Let

$$\mathbf{u} = \begin{bmatrix} 2 \\ 3 \\ -1 \end{bmatrix}, \quad \mathbf{v} = \begin{bmatrix} -1 \\ 2 \\ 4 \end{bmatrix}, \quad \mathbf{w} = \begin{bmatrix} 0 \\ 1 \\ -1 \end{bmatrix},$$

$c = -2$, and $d = 3$.

Compute:
(a) $\mathbf{u} + \mathbf{v}$.
(b) $c\mathbf{u} + d\mathbf{w}$.
(c) $\mathbf{u} + \mathbf{v} + \mathbf{w}$.
(d) $c\mathbf{u} + d\mathbf{v} + \mathbf{w}$.

14. Let

$$\mathbf{x} = \begin{bmatrix} 1 \\ 2 \end{bmatrix}, \quad \mathbf{y} = \begin{bmatrix} -3 \\ 4 \end{bmatrix},$$

$$\mathbf{z} = \begin{bmatrix} r \\ 4 \end{bmatrix}, \quad \text{and} \quad \mathbf{u} = \begin{bmatrix} -2 \\ s \end{bmatrix}.$$

Find r and s so that:
(a) $\mathbf{z} = 2\mathbf{x}$. (b) $\frac{3}{2}\mathbf{u} = \mathbf{y}$. (c) $\mathbf{z} + \mathbf{u} = \mathbf{x}$.

15. Let

$$\mathbf{x} = \begin{bmatrix} 1 \\ -2 \\ 3 \end{bmatrix}, \quad \mathbf{y} = \begin{bmatrix} -3 \\ 1 \\ 3 \end{bmatrix}, \quad \mathbf{z} = \begin{bmatrix} r \\ -1 \\ s \end{bmatrix},$$

and $\mathbf{u} = \begin{bmatrix} 3 \\ t \\ 2 \end{bmatrix}$.

Find r, s, and t so that:
(a) $\mathbf{z} = \frac{1}{2}\mathbf{x}$. (b) $\mathbf{z} + \mathbf{u} = \mathbf{x}$. (c) $\mathbf{z} - \mathbf{x} = \mathbf{y}$.

16. If possible, find scalars c_1 and c_2 so that

$$c_1 \begin{bmatrix} 1 \\ -2 \end{bmatrix} + c_2 \begin{bmatrix} 3 \\ -4 \end{bmatrix} = \begin{bmatrix} -5 \\ 6 \end{bmatrix}.$$

17. If possible, find scalars c_1, c_2, and c_3 so that

$$c_1 \begin{bmatrix} 1 \\ 2 \\ -3 \end{bmatrix} + c_2 \begin{bmatrix} -1 \\ 1 \\ 1 \end{bmatrix} + c_3 \begin{bmatrix} -1 \\ 4 \\ -1 \end{bmatrix} = \begin{bmatrix} 2 \\ -2 \\ 3 \end{bmatrix}.$$

18. If possible, find scalars c_1 and c_2, not both zero, so that

$$c_1 \begin{bmatrix} 1 \\ 2 \end{bmatrix} + c_2 \begin{bmatrix} 3 \\ 4 \end{bmatrix} = \begin{bmatrix} 0 \\ 0 \end{bmatrix}.$$

19. If possible, find scalars c_1, c_2, and c_3, not all zero, so that

$$c_1 \begin{bmatrix} 1 \\ 2 \\ -1 \end{bmatrix} + c_2 \begin{bmatrix} 1 \\ 3 \\ -2 \end{bmatrix} + c_3 \begin{bmatrix} 3 \\ 7 \\ -4 \end{bmatrix} = \begin{bmatrix} 0 \\ 0 \\ 0 \end{bmatrix}.$$

20. Let

$$\mathbf{i} = \begin{bmatrix} 1 \\ 0 \\ 0 \end{bmatrix}, \quad \mathbf{j} = \begin{bmatrix} 0 \\ 1 \\ 0 \end{bmatrix}, \quad \text{and} \quad \mathbf{k} = \begin{bmatrix} 0 \\ 0 \\ 1 \end{bmatrix}.$$

Find scalars c_1, c_2, and c_3 so that any vector

$\mathbf{u} = \begin{bmatrix} r \\ s \\ t \end{bmatrix}$ can be written as

$$\mathbf{u} = c_1\mathbf{i} + c_2\mathbf{j} + c_3\mathbf{k}.$$

21. Show that if $\mathbf{u}$ is a vector in R^2 or R^3, then $\mathbf{u} + \mathbf{0} = \mathbf{u}$.

22. Show that if $\mathbf{u}$ is a vector in R^2 or R^3, then $(-1)\mathbf{u} = -\mathbf{u}$.

23. Prove parts (b) and (d) through (h) of Theorem 2.1.

24. Determine if the software you use supports graphics. If it does, experiment with plotting vectors in R^2. Usually, you must supply coordinates for the head and tail of the vector and then tell the software to connect these points. The points in Exercises 7(a) and 8(a) can be used in this regard.

25. Assuming that the software you use supports graphics (see Exercise 24), plot the vector

$$\mathbf{v} = \begin{bmatrix} 3 \\ 4 \end{bmatrix}$$

on the same coordinate axes for each of the following.
(a) $\mathbf{v}$ is to have its head at (1, 1).
(b) $\mathbf{v}$ is to have its head at (2, 3).

26. Determine if the software you use supports three-dimensional graphics, that is, plotting points in R^3. If it does, experiment with plotting points and connecting them to form vectors in R^3.

2.2. *Vector Spaces and Subspaces*

Definition 2.4. A **real vector space** is a set V of elements on which we have two operations $\oplus$ and $\odot$ defined with the following properties:

(a) If $\mathbf{u}$ and $\mathbf{v}$ are any elements in V, then $\mathbf{u} \oplus \mathbf{v}$ is in V (i.e., V is closed under the operation $\oplus$).

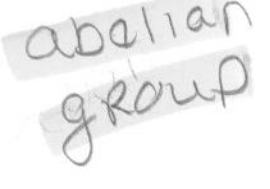

(1) $\mathbf{u} \oplus \mathbf{v} = \mathbf{v} \oplus \mathbf{u}$ for all $\mathbf{u}$, $\mathbf{v}$ in V.

(2) $\mathbf{u} \oplus (\mathbf{v} \oplus \mathbf{w}) = (\mathbf{u} \oplus \mathbf{v}) \oplus \mathbf{w}$ for all $\mathbf{u}$, $\mathbf{v}$, $\mathbf{w}$ in V.

(3) There exists a unique element $\mathbf{0}$ in V such that $\mathbf{u} \oplus \mathbf{0} = \mathbf{0} \oplus \mathbf{u} = \mathbf{u}$ for any $\mathbf{u}$ in V.

(4) For each $\mathbf{u}$ in V there exists a unique $\mathbf{v}$ in V such that $\mathbf{u} \oplus \mathbf{v} = \mathbf{v} \oplus \mathbf{u} = \mathbf{0}$. We denote $\mathbf{v}$ by $-\mathbf{u}$ and call it the negative of $\mathbf{u}$.

(b) If $\mathbf{v}$ is any element in V and c is any real number, then $c \odot \mathbf{u}$ is in V (i.e., V is closed under the operation $\odot$).

(5) $c \odot (\mathbf{u} \oplus \mathbf{v}) = c \odot \mathbf{u} \oplus c \odot \mathbf{v}$ for any $\mathbf{u}$, $\mathbf{v}$ in V and any real number c.

(6) $(c + d) \odot \mathbf{u} = c \odot \mathbf{u} \oplus d \odot \mathbf{u}$ for any $\mathbf{u}$ in V and any real numbers c and d.

(7) $c \odot (d \odot \mathbf{u}) = (cd) \odot \mathbf{u}$ for any $\mathbf{u}$ in V and any real numbers c and d.

(8) $1 \odot \mathbf{u} = \mathbf{u}$ for any $\mathbf{u}$ in V.

The elements of V are called **vectors**; the elements of the set of real numbers R are called **scalars**. The operation $\oplus$ is called **vector addition**; the operation $\odot$ is called **scalar multiplication**. The vector $\mathbf{0}$ is called the **zero vector**. ■

In order to specify a vector space, we must be given a set V and two operations $\oplus$ and $\odot$ satisfying all the properties of the definition. We shall often refer to a real vector space merely as a **vector space.** Thus a "vector" is now an element of a vector space and no longer needs to be interpreted as a directed line segment. In our examples we shall see, however, how this name came about in a natural manner. We now consider some examples of vector spaces, leaving it to the reader to verify that all the properties of Definition 2.4 hold.

Example 1. Consider R^n, the set of all $n \times 1$ matrices $\begin{bmatrix} a_1 \\ a_2 \\ \vdots \\ a_n \end{bmatrix}$ with real entries. Let the operation $\oplus$ be matrix addition and let the operation $\odot$ be multiplication of a matrix by a real number (scalar multiplication).

By the use of the properties of matrices established in Section 1.3, it is not difficult to show that R^n is a vector space by verifying that the properties of Definition 2.4 hold. Thus the matrix $\begin{bmatrix} a_1 \\ a_2 \\ \vdots \\ a_n \end{bmatrix}$, as an element of R^n, is now called a *vector*. We have already discussed R^2 and R^3 in Section 2.1. Although we shall see later that many geometric notions such as length and the angle between vectors can be defined in R^n for $n > 3$, we cannot draw pictures in these cases. ■

Example 2. The set of all $m \times n$ matrices with matrix addition as $\oplus$ and multiplication of a matrix by a real number as $\odot$ is a vector space (verify). We denote this vector space by ${}_mR_n$. ■

Example 3. The set of all real numbers with $\oplus$ as the usual addition of real numbers and $\odot$ as the usual multiplication of real numbers is a vector space (verify). In this case the real numbers play the dual roles of both vectors and scalars. ■

Example 4. Let R_n be the set of all $1 \times n$ matrices $[a_1 \quad a_2 \quad \cdots \quad a_n]$, where we define $\oplus$ by

$$[a_1 \quad a_2 \quad \cdots \quad a_n] \oplus [b_1 \quad b_2 \quad \cdots \quad b_n] = [a_1 + b_1 \quad a_2 + b_2 \quad \cdots \quad a_n + b_n]$$

and $\odot$ by

$$c \odot [a_1 \quad a_2 \quad \cdots \quad a_n] = [ca_1 \quad ca_2 \quad \cdots \quad ca_n].$$

Then R_n is a vector space (verify). This is just a special case of Example 2. ■

Example 5. Another source of examples will be sets of polynomials; therefore, we recall some well-known facts about such functions. A **polynomial** (in t) is a function which is expressible as

$$p(t) = a_0t^n + a_1t^{n-1} + \cdots + a_{n-1}t + a_n,$$

where $a_0, a_1, \ldots, a_n$ are real numbers and n is a nonnegative integer. If $a_0 \neq 0$, then $p(t)$ is said to have **degree n**. Thus the degree of a polynomial is the highest power having a nonzero coefficient; $2t + 1$ has degree 1 and the constant polynomial 3 has degree 0. The **zero polynomial**, defined as $0t^n + 0t^{n-1} + \cdots + 0t + 0$, has no degree. We now let P_n be the set of all polynomials of degree $\leq n$ together with the zero polynomial. If

$$p(t) = a_0t^n + a_1t^{n-1} + \cdots + a_{n-1}t + a_n$$

and

$$q(t) = b_0t^n + b_1t^{n-1} + \cdots + b_{n-1}t + b_n,$$

we define $p(t) \oplus q(t)$ as

$$p(t) \oplus q(t) = (a_0 + b_0)t^n + (a_1 + b_1)t^{n-1} + \cdots + (a_{n-1} + b_{n-1})t + (a_n + b_n).$$

If c is a scalar, we also define $c \odot p(t)$ as

$$c \odot p(t) = (ca_0)t^n + (ca_1)t^{n-1} + \cdots + (ca_{n-1})t + (ca_n).$$

We now show that P_n is a vector space.

Let $p(t)$ and $q(t)$ as above be elements of P_n; that is, they are polynomials of degree $\leq n$ or the zero polynomial. Then the definitions above of the operations $\oplus$ and $\odot$ show that $p(t) \oplus q(t)$ and $c \odot p(t)$, for any scalar c, are polynomials of degree $\leq n$ or the zero polynomial. That is, $p(t) \oplus q(t)$ and $c \odot p(t)$ are in P_n so that (a) and (b) in Definition 2.4 hold. To verify property (1), we observe that

$$q(t) \oplus p(t) = (b_0 + a_0)t^n + (b_1 + a_1)t^{n-1} + \cdots + (b_{n-1} + a_{n-1})t + (a_n + b_n),$$

and since $a_i + b_i = b_i + a_i$ holds for the real numbers, we conclude that $p(t) \oplus q(t) = q(t) \oplus p(t)$. Similarly, we verify property (2). The zero polynomial is the element $\mathbf{0}$ needed in property (3). If $p(t)$ is as given above, then its negative, $-p(t)$, is

$$-a_0t^n - a_1t^{n-1} - \cdots - a_{n-1}t - a_n.$$

We shall now verify property (6) and will leave the verification of the remaining properties to the reader. Thus

$$\begin{aligned}(c + d) \odot p(t) &= (c + d)a_0t^n + (c + d)a_1t^{n-1} + \cdots + (c + d)a_{n-1}t \\ &\quad + (c + d)a_n \\ &= ca_0t^n + da_0t^n + ca_1t^{n-1} + da_1t^{n-1} + \cdots + ca_{n-1}t \\ &\quad + da_{n-1}t + ca_n + da_n \\ &= ca_0t^n + ca_1t^{n-1} + \cdots + ca_{n-1}t + ca_n + da_0t^n \\ &\quad + da_1t^{n-1} + \cdots + da_{n-1}t + da_n \\ &= c(a_0t^n + a_1t^{n-1} + \cdots + a_{n-1}t + a_n) \\ &\quad + d(a_0t^n + a_1t^{n-1} + \cdots + a_{n-1}t + a_n) \\ &= c \odot p(t) \oplus d \odot p(t).\end{aligned}$$ ■

For each natural number n, we have just defined the vector space P_n of all polynomials of degree $\leq n$ together with the zero polynomial. We could also consider the space P of *all* polynomials (of any degree), together with the zero polynomial. Here P is the mathematical union of all the vector spaces P_n. Two polynomials $p(t)$ of degree n and $q(t)$ of degree m are added in P in the same way as they would be added in P_r, where r is the maximum of the two numbers m and n. Then P is a vector space (Exercise 2).

As in the case of ordinary real number arithmetic, in an expression containing both $\odot$ and $\oplus$, the $\odot$ operation is performed first. Moreover, the familiar arithmetic rules when parentheses are encountered, apply in this case also.

Example 6. Let V be the set of all real-valued continuous functions defined on R^1. If f and g are in V, we define $f \oplus g$ by $(f \oplus g)(t) = f(t) + g(t)$. If f is in V and c is a scalar, we define $c \odot f$ by $(c \odot f)(t) = cf(t)$. Then V is a vector space (verify), which is denoted by $C(-\infty, \infty)$. ■

Example 7. Let V be the set of all real numbers with the operations $\mathbf{u} \oplus \mathbf{v} = \mathbf{u} - \mathbf{v}$ ($\oplus$ is ordinary subtraction) and $c \odot \mathbf{u} = c\mathbf{u}$ ($\odot$ is ordinary multiplication). Is V a vector space? If it is not, which properties in Definition 2.4 fail to hold?

Solution: If $\mathbf{u}$ and $\mathbf{v}$ are in V, and c is a scalar, then $\mathbf{u} \oplus \mathbf{v}$ and $c \odot \mathbf{u}$ are in V so that (a) and (b) in Definition 2.4 hold. However, property (1) fails to hold, since

$$\mathbf{u} \oplus \mathbf{v} = \mathbf{u} - \mathbf{v}$$

and

$$\mathbf{v} \oplus \mathbf{u} = \mathbf{v} - \mathbf{u}.$$

Also, we shall let the reader verify that properties (2), (3), and (4) fail to hold. Properties (5), (7), and (8) hold, but property (6) does not hold:

$$(c + d) \odot \mathbf{u} = (c + d)\mathbf{u} = c\mathbf{u} + d\mathbf{u}$$

while

$$c \odot \mathbf{u} \oplus d \odot \mathbf{u} = c\mathbf{u} \oplus d\mathbf{u} = c\mathbf{u} - d\mathbf{u}.$$

Thus V is not a vector space. ■

Example 8. Let V be the set of all ordered triples of real numbers (x, y, z) with the operations

$$(x, y, z) \oplus (x', y', z') = (x', y + y', z + z'); \qquad c \odot (x, y, z) = (cx, cy, cz).$$

It is easy to verify that properties (1), (3), (4), and (6) of Definition 2.4 fail to hold.

For example, if $\mathbf{u} = (x, y, z)$ and $\mathbf{v} = (x', y', z')$, then

$$\mathbf{u} \oplus \mathbf{v} = (x, y, z) \oplus (x', y', z') = (x', y + y', z + z')$$

while

$$\mathbf{v} \oplus \mathbf{u} = (x', y', z') \oplus (x, y, z) = (x, y' + y, z' + z),$$

so property (1) fails to hold. Also,

$$\begin{aligned}(c + d) \odot \mathbf{u} &= (c + d) \odot (x, y, z)\\ &= ((c + d)x, (c + d)y, (c + d)z)\end{aligned}$$

while

$$\begin{aligned}c \odot \mathbf{u} \oplus d \odot \mathbf{u} &= c \odot (x, y, z) \oplus d \odot (x, y, z)\\ &= (cx, cy, cz) \oplus (dx, dy, dz)\\ &= (dx, cy + dy, cz + dz).\end{aligned}$$

so property (6) fails to hold. Thus V is not a vector space. ■

To verify that a given set V with two operations $\oplus$ and $\odot$ is a real vector space, we must show that it satisfies all the properties of Definition 2.4. The first thing to check is whether (a) and (b) hold, for, if either of these fails, we do not have a vector space. If both (a) and (b) hold, it is recommended that (3), the existence of a zero element, be verified next. Naturally, if (3) fails to hold, we do not have a vector space and do not have to check the remaining properties.

Example 9. Let V be the set of all integers; define $\oplus$ as ordinary addition and $\odot$ as ordinary multiplication. Here V is not a vector space because if $\mathbf{u}$ is any vector in V and $c = \sqrt{3}$, then $c \odot \mathbf{u}$ is not in V. Thus (b) fails to hold. ■

The following theorem presents some useful properties common to all vector spaces.

Theorem 2.2. If V is a vector space, then
(a) $0 \odot \mathbf{u} = \mathbf{0}$ *for any vector* $\mathbf{u}$ *in* V.
(b) $c \odot \mathbf{0} = \mathbf{0}$ *for any scalar* c.
(c) *If* $c \odot \mathbf{u} = \mathbf{0}$, *then either* $c = 0$ *or* $\mathbf{u} = \mathbf{0}$.
(d) $(-1) \odot \mathbf{u} = -\mathbf{u}$ *for any* $\mathbf{u}$ *in* V.

Proof: (a) We have

$$0 \odot \mathbf{u} = (0 + 0) \odot \mathbf{u} = 0 \odot \mathbf{u} + 0 \odot \mathbf{u} \tag{1}$$

by (6) of Definition 2.4. Adding $-0 \odot \mathbf{u}$ to both sides of Equation (1), we obtain by (2), (3), and (4) of Definition 2.4,

$$0 \odot \mathbf{u} = \mathbf{0}.$$

(d) $$\begin{aligned}(-1) \odot \mathbf{u} \oplus \mathbf{u} &= (-1) \odot \mathbf{u} \oplus 1 \odot \mathbf{u} \\ &= (-1 + 1) \odot \mathbf{u} = 0 \odot \mathbf{u} = \mathbf{0}.\end{aligned}$$
Thus $(-1) \odot \mathbf{u} = -\mathbf{u}$, by (4) of Definition 2.4.
Parts (b) and (c) are left as exercises. ■

Subspaces

We now begin to analyze the structure of a vector space. First, it is convenient to have a name for a subset of a given vector space that is itself a vector space with respect to the same operations as those in V. Thus we have a definition.

Definition 2.5. Let V be a vector space and W a subset of V. If W is a vector space with respect to the operations in V, then W is called a **subspace** of V. ■

Examples of subspaces of a given vector space occur very frequently. We shall list several of these, leaving the verifications that they are subspaces to the reader. More examples will be found in the exercises.

Example 10. Every vector space has at least two subspaces, itself and the subspace $\{\mathbf{0}\}$ consisting only of the zero vector (recall that $\mathbf{0} \oplus \mathbf{0} = \mathbf{0}$ and $c \odot \mathbf{0} = \mathbf{0}$ in any vector space). These subspaces are called **trivial subspaces**. ■

Example 11. Let P_2 be the set consisting of all polynomials of degree ≤ 2 and the zero polynomial; P_2 is a subset of P, the vector space of all polynomials. It is easy to verify that P_2 is a *subspace of* P. In general, the set P_n consisting of all polynomials of degree $\leq n$ and the zero polynomial is a subspace of P. ■

Example 12. Let V be the set of all polynomials of degree exactly $= 2$; V is a *subset* of P, the vector space of all polynomials, but not a *subspace* of P because the sum of the polynomials $2t^2 + 3t + 1$ and $-2t^2 + t + 2$ is not in V, since it is a polynomial of degree 1. ■

We now pause in our listing of subspaces to develop a labor-saving result. We just noted that to verify that a subset W of a vector space V is a subspace, one must check that (a), (b), and (1) through (8) of Definition 2.4 hold. However, the following theorem says that it is enough to merely check that (a) and (b) hold to verify that a subset W of a vector space V is a subspace.

Theorem 2.3. *Let V be a vector space with operations $\oplus$ and $\odot$ and let W be a nonempty subset of V. Then W is a subspace of V if and only if the following conditions hold:*
(a) *If $\mathbf{u}$ and $\mathbf{v}$ are any vectors in W, then $\mathbf{u} \oplus \mathbf{v}$ is in W.*
(b) *If c is any real number and $\mathbf{u}$ is any vector in W, then $c \odot \mathbf{u}$ is in W.*

(⟹)

Proof: We first show that if W is a subspace of V, then (a) and (b) hold. This follows at once from the observation that if W is a subspace, then it is a vector space and so (a) and (b) of Definition 2.4 hold; these are precisely (a) and (b) of the theorem.

Conversely, suppose that (a) and (b) hold. We wish to show that W is a subspace of V. First, from (b) we have that $(-1) \odot \mathbf{u}$ is in W for any $\mathbf{u}$ in W. From (a) we have that $\mathbf{u} \oplus (-1) \odot \mathbf{u}$ is in W. But $\mathbf{u} \oplus (-1) \odot \mathbf{u} = \mathbf{0}$, so $\mathbf{0}$ is in W. Then $\mathbf{u} \oplus \mathbf{0} = \mathbf{u}$ for any $\mathbf{u}$ in W. Finally, properties (1), (2), (5), (6), (7), and (8) hold in W because they hold in V. Hence W is a subspace of V. ■

Example 13. Let W be the set of all vectors in R^3 of the form $\begin{bmatrix} a \\ b \\ a+b \end{bmatrix}$, where a and b are any real numbers. To verify Theorem 2.3(a) and (b), we let

$$\mathbf{u} = \begin{bmatrix} a_1 \\ b_1 \\ a_1 + b_1 \end{bmatrix} \quad \text{and} \quad \mathbf{v} = \begin{bmatrix} a_2 \\ b_2 \\ a_2 + b_2 \end{bmatrix}$$

be two vectors in W. Then

$$\mathbf{u} \oplus \mathbf{v} = \begin{bmatrix} a_1 + a_2 \\ b_1 + b_2 \\ (a_1 + b_1) + (a_2 + b_2) \end{bmatrix} = \begin{bmatrix} a_1 + a_2 \\ b_1 + b_2 \\ (a_1 + a_2) + (b_1 + b_2) \end{bmatrix}$$

is in W, for W consists of all those vectors whose third entry is the sum of the first two entries. Similarly,

$$c \odot \begin{bmatrix} a_1 \\ b_1 \\ a_1 + b_1 \end{bmatrix} = \begin{bmatrix} ca_1 \\ cb_1 \\ c(a_1 + b_1) \end{bmatrix} = \begin{bmatrix} ca_1 \\ cb_1 \\ ca_1 + cb_1 \end{bmatrix}$$

is in W. Hence W is a subspace of R^3. ■

> Henceforth, we shall usually denote $\mathbf{u} \oplus \mathbf{v}$ and $c \odot \mathbf{u}$ in a vector space V as $\mathbf{u} + \mathbf{v}$ and $c\mathbf{u}$, respectively.

Example 14. A simple way of constructing subspaces in a vector space is as follows. Let $\mathbf{v}_1$ and $\mathbf{v}_2$ be fixed vectors in a vector space V and let W be the set of all vectors in V of the form

$$a_1\mathbf{v}_1 + a_2\mathbf{v}_2,$$

where a_1 and a_2 are any real numbers. To show that W is a subspace of V, we verify properties (a) and (b) of Theorem 2.3. Thus let

$$\mathbf{w}_1 = a_1\mathbf{v}_1 + a_2\mathbf{v}_2 \qquad \text{and} \qquad \mathbf{w}_2 = b_1\mathbf{v}_1 + b_2\mathbf{v}_2$$

be vectors in W. Then

$$\begin{aligned}\mathbf{w}_1 \oplus \mathbf{w}_2 &= (a_1\mathbf{v}_1 + a_2\mathbf{v}_2) + (b_1\mathbf{v}_1 + b_2\mathbf{v}_2)\\ &= (a_1 + b_1)\mathbf{v}_1 + (a_2 + b_2)\mathbf{v}_2,\end{aligned}$$

which is in W. Also, if c is a scalar, then

$$c \odot \mathbf{w}_1 = c(a_1\mathbf{v}_1 + a_2\mathbf{v}_2) = (ca_1)\mathbf{v}_1 + (ca_2)\mathbf{v}_2$$

is in W. Hence W is a subspace of V. ■

The construction carried out in Example 14 for two vectors can easily be performed for more than two vectors. Thus if $S = \{\mathbf{v}_1, \mathbf{v}_2, \ldots, \mathbf{v}_k\}$ is a set of vectors in a vector space V, then the set of all vectors in V of the form

$$a_1\mathbf{v}_1 + a_2\mathbf{v}_2 + \cdots + a_k\mathbf{v}_k$$

is a subspace of V (Exercise 29). We denote this subspace by

$$\text{span } S \qquad \text{or} \qquad \text{span } \{\mathbf{v}_1, \mathbf{v}_2, \ldots, \mathbf{v}_k\}.$$

Example 15. Let W be the set of all 2×3 matrices in ${}_2R_3$ of the form $\begin{bmatrix} a & b & 0 \\ 0 & c & d \end{bmatrix}$, where a, b, c, and d are real numbers. Then

$$W = \text{span}\left\{\begin{bmatrix} 1 & 0 & 0 \\ 0 & 0 & 0 \end{bmatrix}, \begin{bmatrix} 0 & 1 & 0 \\ 0 & 0 & 0 \end{bmatrix}, \begin{bmatrix} 0 & 0 & 0 \\ 0 & 1 & 0 \end{bmatrix}, \begin{bmatrix} 0 & 0 & 0 \\ 0 & 0 & 1 \end{bmatrix}\right\}. \quad ■$$

Another very important example of a subspace is provided by Example 16. Before turning to this example, we discuss a slight change in the notation used to denote a linear system. In Chapter 1, we denoted a linear system consisting of m equations in n unknowns by $AX = B$, where X is an $n \times 1$ matrix and B is an $m \times 1$ matrix. We shall now take the point of view that X is a vector in R^n and B is a vector in R^m and will write these as $\mathbf{x}$ and $\mathbf{b}$, respectively. Thus the linear system $AX = B$ will now be written as $A\mathbf{x} = \mathbf{b}$.

Example 16. If A is an $m \times n$ matrix, the homogeneous system of m equations in n unknowns with coefficient matrix A can be written as

$$A\mathbf{x} = \mathbf{0},$$

where $\mathbf{x}$ is a vector in R^n and $\mathbf{0}$ is the zero vector. Thus the set W of all solutions is a subset of R^n. We now show that W is a subspace of R^n (called the **solution space** of the homogeneous system or the **null space** of the matrix A) by verifying (a) and (b) of Theorem 2.3. Let $\mathbf{x}$ and $\mathbf{y}$ be solutions. Then

$$A\mathbf{x} = \mathbf{0} \quad \text{and} \quad A\mathbf{y} = \mathbf{0}.$$

Now

$$A(\mathbf{x} + \mathbf{y}) = A\mathbf{x} + A\mathbf{y} = \mathbf{0} + \mathbf{0} = \mathbf{0}$$

so $\mathbf{x} + \mathbf{y}$ is a solution. Also, if c is a scalar, then

$$A(c\mathbf{x}) = c(A\mathbf{x}) = c\mathbf{0} = \mathbf{0},$$

so $c\mathbf{x}$ is a solution. Thus W is closed under addition and scalar multiplication of vectors and is therefore a subspace of R^n. ■

It should be noted that the set of all solutions to the linear system $A\mathbf{x} = \mathbf{b}$, $\mathbf{b} \neq \mathbf{0}$, is not a subspace of R^n (see Exercise 27). We leave it as an exercise to show that the subspaces of R^1 are $\{\mathbf{0}\}$ and R^1 itself (Exercise 38). As for R^2, its subspaces are $\{\mathbf{0}\}$, R^2, and any set consisting of all scalar multiples of a nonzero vector (Exercise 39), that is, any line passing through the origin. Exercise 37 in Section 2.4 asks you to show that all the subspaces of R^3 are $\{\mathbf{0}\}$, R^3 itself, and any line or plane passing through the origin.

Lines in R^3 (Optional)

In R^2 a line is determined by specifying its slope and one of its points. In R^3 a line is determined by specifying its direction and one of its points. Let $\mathbf{v} = \begin{bmatrix} a \\ b \\ c \end{bmatrix}$ be a nonzero vector in R^3. Then the line ℓ_0 through the origin and parallel to $\mathbf{v}$ consists of all the points $P(x, y, z)$ whose position vector $\mathbf{u} = \begin{bmatrix} x \\ y \\ z \end{bmatrix}$ is of the form

Figure 2.16 A line in R^3.

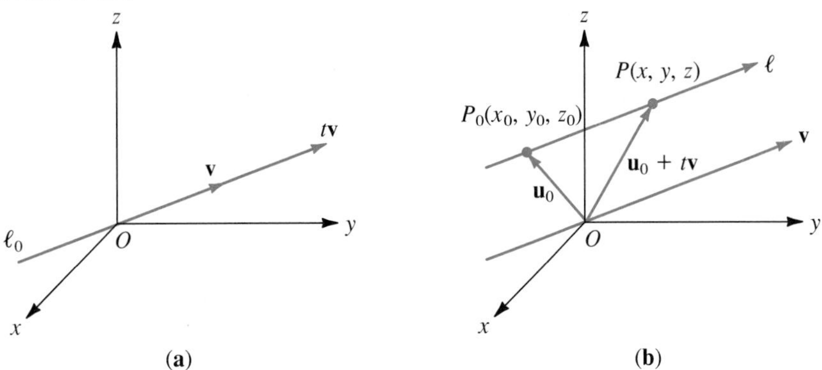

$\mathbf{u} = t\mathbf{v}$, $-\infty < t < \infty$ [Figure 2.16(a)]. It is easy to verify that the line ℓ_0 is a subspace of R^3. Now let $P_0 = (x_0, y_0, z_0)$ be a point in R^3, and let $\mathbf{u}_0 = \begin{bmatrix} x_0 \\ y_0 \\ z_0 \end{bmatrix}$ be the position vector of P_0. Then the line ℓ through P_0 and parallel to $\mathbf{v}$ consists of the points $P(x, y, z)$ whose position vector $\mathbf{u} = \begin{bmatrix} x \\ y \\ z \end{bmatrix}$ satisfies [Figure 2.16(b)]

$$\mathbf{u} = \mathbf{u}_0 + t\mathbf{v}, \qquad -\infty < t < \infty. \tag{2}$$

Equation (2) is called a **parametric equation** of ℓ, since it contains the parameter t, which can be assigned any real number. Equation (2) can also be written in terms of the components as

$$\begin{aligned} x &= x_0 + ta \\ y &= y_0 + tb \qquad -\infty < t < \infty \\ z &= z_0 + tc, \end{aligned} \tag{3}$$

which are called **parametric equations** of ℓ.

Example 17. Parametric equations of the line passing through the point $P_0(-3, 2, 1)$ and parallel to the vector $\mathbf{v} = \begin{bmatrix} 2 \\ -3 \\ 4 \end{bmatrix}$ are

$$\begin{aligned} x &= -3 + 2t \\ y &= 2 - 3t \qquad -\infty < t < \infty. \\ z &= 1 + 4t, \end{aligned}$$

■

Example 18. Find parametric equations of the line ℓ through the points $P_0(2, 3, -4)$ and $P_1(3, -2, 5)$.

Solution: The desired line is parallel to the vector $\mathbf{v} = \overrightarrow{P_0P_1}$. Now

$$\mathbf{v} = \begin{bmatrix} 3 - 2 \\ -2 - 3 \\ 5 - (-4) \end{bmatrix} = \begin{bmatrix} 1 \\ -5 \\ 9 \end{bmatrix}.$$

Since P_0 is on the line, we can write the following parametric equation of ℓ

$$\begin{aligned} x &= 2 + t \\ y &= 3 - 5t \qquad -\infty < t < \infty. \\ z &= -4 + 9t, \end{aligned}$$

Of course, we could have used the point P_1 instead of P_0. In fact, we could use any point on the line in a parametric equation for ℓ. Thus a line can be represented in infinitely many ways in parametric form. ■

2.2. EXERCISES

1. Prove in detail that R^n is a vector space.

2. Show that P, the set of all polynomials, is a vector space.

In Exercises 3 through 7, the given set together with the given operations is not a vector space. List the properties of Definition 2.4 that fail to hold.

3. The set of all positive real numbers with the operations of $\oplus$ as ordinary addition and $\odot$ as ordinary multiplication.

4. The set of all ordered pairs of real numbers with the operations $(x, y) \oplus (x', y') = (x + x', y + y')$ and $r \odot (x, y) = (x, ry)$.

5. The set of all ordered triples of real numbers with the operations $(x, y, z) \oplus (x', y', z') = (x + x', y + y', z + z')$ and $r \odot (x, y, z) = (x, 1, z)$.

6. The set of all 2×1 matrices $\begin{bmatrix} x \\ y \end{bmatrix}$, where $x \leq 0$, with the usual operations in R^2.

7. The set of all ordered pairs of real numbers with the operations $(x, y) \oplus (x', y') = (x + x', y + y')$ and $r \odot (x, y) = (0, 0)$.

8. Let V be the set of all positive real numbers; define $\oplus$ by $\mathbf{u} \oplus \mathbf{v} = \mathbf{uv}$ ($\oplus$ is ordinary multiplication) and define $\odot$ by $c \odot \mathbf{v} = \mathbf{v}^c$. Prove that V is a vector space.

9. Let V be the set of all real-valued continuous functions. If f and g are in V, define $f \oplus g$ by $(f \oplus g)(t) = f(t) + g(t)$. If f is in V, define $c \odot f$ by $(c \odot f)(t) = cf(t)$. Prove that V is a vector space (this is the vector space defined in Example 6).

10. Let V be the set consisting of a single element $\mathbf{0}$. Let $\mathbf{0} \oplus \mathbf{0} = \mathbf{0}$ and $c \odot \mathbf{0} = \mathbf{0}$. Prove that V is a vector space.

11. Consider the differential equation $y'' - y' + 2y = 0$. A solution is a real-valued function f satisfying the equation. Let V be the set of all solutions to the given differential equation; define $\oplus$ and $\odot$ as in Exercise 9. Prove that V is a vector space. (See also Chapter 7.)

12. Let V be the set of all positive real numbers; define $\oplus$ by $\mathbf{u} \oplus \mathbf{v} = \mathbf{uv} - 1$ and $\odot$ by $c \odot \mathbf{v} = \mathbf{v}$. Is V a vector space?

13. Let V be the set of all real numbers; define $\oplus$ by $\mathbf{u} \oplus \mathbf{v} = \mathbf{uv}$ and $\odot$ by $c \odot \mathbf{u} = c + \mathbf{u}$. Is V a vector space?

14. Let V be the set of all real numbers; define $\oplus$ by $\mathbf{u} \oplus \mathbf{v} = 2\mathbf{u} - \mathbf{v}$ and $\odot$ by $c \odot \mathbf{u} = c\mathbf{u}$. Is V a vector space?

15. Prove Theorem 2.2(b) and (c).

16. Prove that P_2 is a subspace of P_3.

In Exercises 17 and 18, which of the given subsets of the vector space, ${}_2R_3$, of all 2×3 matrices are subspaces?

17. The set of all matrices of the form:

(a) $\begin{bmatrix} a & b & c \\ d & 0 & 0 \end{bmatrix}$, where $b = a + c$.

(b) $\begin{bmatrix} a & b & c \\ d & 0 & 0 \end{bmatrix}$, where $c > 0$.

18. The set of all matrices of the form:

(a) $\begin{bmatrix} a & b & c \\ d & e & f \end{bmatrix}$, where $a = -2c$ and $f = 2e + d$.

(b) $\begin{bmatrix} a & b & b \\ b & a & 0 \end{bmatrix}$, where $b = a + 2$.

In Exercises 19 and 20, which of the given subsets of R^3 are subspaces?

19. The set of all vectors of the form:

(a) $\begin{bmatrix} a \\ b \\ 1 \end{bmatrix}$. (b) $\begin{bmatrix} a \\ b \\ a + 2b \end{bmatrix}$. (c) $\begin{bmatrix} a \\ 0 \\ 0 \end{bmatrix}$.

20. The set of all vectors of the form:

(a) $\begin{bmatrix} a \\ b \\ 0 \end{bmatrix}$. (b) $\begin{bmatrix} a \\ b \\ c \end{bmatrix}$, where $a > 0$.

(c) $\begin{bmatrix} a \\ a \\ c \end{bmatrix}$.

In Exercises 21 and 22, which of the given subsets of ${}_2R_2$ are subspaces?

21. The set of all 2×2:
(a) Symmetric matrices.
(b) Singular matrices.

22. The set of all 2×2:
(a) Nonsingular matrices.
(b) Diagonal matrices.

23. Show that P is a subspace of the vector space of all real-valued continuous functions introduced in Exercise 9.

In Exercises 24 and 25, let V be the vector space of all real-valued continuous functions considered in Exercise 9. Which of the given subsets are subspaces of V?

24. The set of all:
(a) Nonnegative functions.
(b) Functions f such that $f(0) = 0$.
(c) Differentiable functions.

25. The set of all:
(a) Constant functions.
(b) Functions f such that $f(0) = 5$.
(c) Functions f such that $|f(t)| \leq 2$ for all t.

26. Prove that the set V of all real-valued functions is a vector space under the operations defined as in Exercise 9. Show also that the space of Exercise 9 is a subspace of V.

27. Show that the set of all solutions to the linear system $A\mathbf{x} = \mathbf{b}$, $\mathbf{b} \neq \mathbf{0}$, is not a subspace of R^n.

28. If A is a nonsingular matrix, what is the null space of A?

29. Let V be a vector space and $S = \{\mathbf{v}_1, \mathbf{v}_2, \ldots, \mathbf{v}_k\}$ be a set of vectors in V. Prove that span S is a subspace of V.

30. Prove that $-(-\mathbf{v}) = \mathbf{v}$.

31. Prove that if $\mathbf{u} \oplus \mathbf{v} = \mathbf{u} \oplus \mathbf{w}$, then $\mathbf{v} = \mathbf{w}$.

32. Prove that if $\mathbf{u} \neq \mathbf{0}$ and $a\mathbf{u} = b\mathbf{u}$, then $a = b$.

33. (a) Show that a line ℓ_0 through the origin of R^3 is a subspace of R^3.
(b) Show that a line ℓ in R^3 not passing through the origin is not a subspace of R^3.

34. State which of the following points are on the line

$$\begin{aligned} x &= 3 + 2t \\ y &= -2 + 3t \qquad -\infty < t < \infty. \\ z &= 4 + 3t, \end{aligned}$$

(a) $(1, 1, 1)$. (b) $(1, -1, 0)$.
(c) $(1, 0, -2)$. (d) $(4, -\frac{1}{2}, \frac{5}{2})$.

35. State which of the following points are on the line

$$\begin{aligned} x &= 4 - 2t \\ y &= -3 + 2t \qquad -\infty < t < \infty. \\ z &= 4 - 5t, \end{aligned}$$

(a) $(0, 1, -6)$. (b) $(1, 2, 3)$.
(c) $(4, -3, 4)$. (d) $(0, 1, -1)$.

36. Find parametric equations of the line through $P_0 = (x_0, y_0, z_0)$ parallel to $\mathbf{v}$.

(a) $P_0 = (3, 4, -2)$, $\mathbf{v} = \begin{bmatrix} 4 \\ -5 \\ 2 \end{bmatrix}$.

(b) $P_0 = (3, 2, 4)$, $\mathbf{v} = \begin{bmatrix} -2 \\ 5 \\ 1 \end{bmatrix}$.

37. Find parametric equations of the line through the given points.
(a) $(2, -3, 1)$, $(4, 2, 5)$.
(b) $(-3, -2, -2)$, $(5, 5, 4)$.

38. Show that the only subspaces of R^1 are $\{\mathbf{0}\}$ and R^1 itself.

39. Show that the only subspaces of R^2 are $\{\mathbf{0}\}$, R^2, and any set consisting of all scalar multiples of a nonzero vector.

40. Determine which of the following subsets of R^2 are subspaces.
(a)

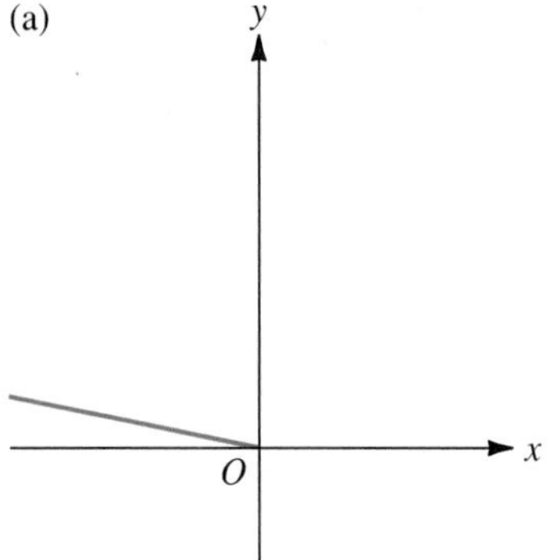

(b)

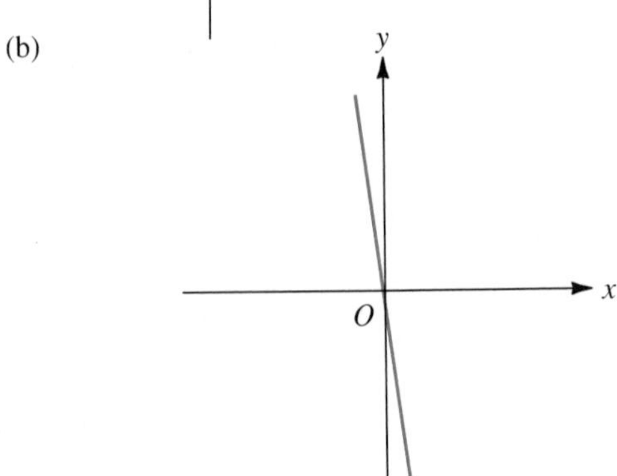

(c)

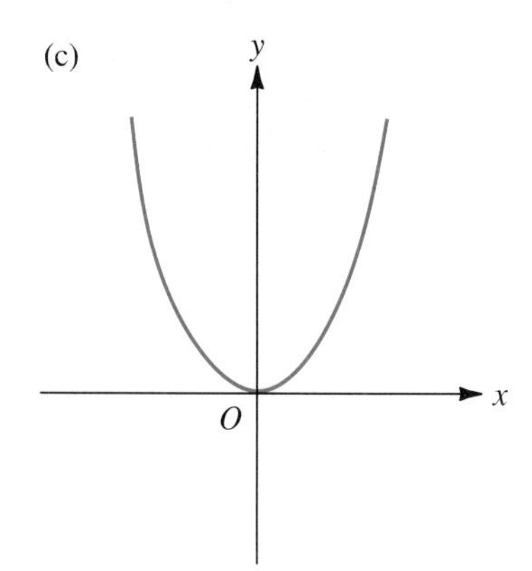

41. Numerical experiments in software *cannot* be used to verify that a set V with two operations $\oplus$ and $\odot$ is a vector space or a subspace. Such a verification must be done ''abstractly'' to take into account all possibilities for elements of V. However, numerical experiments can yield counterexamples which show that V is not a vector space or not a subspace. Use your software to verify that each of the following is *not* a subspace of ${}_2R_2$ with the usual operations of addition of matrices and scalar multiplication.
   (a) The set of symmetric matrices with the (1, 1) entry equal to 3.
   (b) The set of matrices whose first column is $[0 \quad 1]^T$.
   (c) The set of matrices $\begin{bmatrix} a & b \\ c & d \end{bmatrix}$ such that $ad - bc \neq 0$.

42. Example 5 discusses the vector space P_n of polynomials of degree n or less. Operations on polynomials can be performed in linear algebra software by associating a row matrix of size $n + 1$ with polynomial $p(t)$ of P_n. The row matrix consists of the coefficients of $p(t)$ using the association

$$p(t) = a_0t^n + a_1t^{n-1} + \cdots + a_{n-1}t + a_n \rightarrow [a_0 \quad a_1 \quad \cdots \quad a_n].$$

If any term of $p(t)$ is explicitly missing, a zero is used for its coefficient. Then the addition of polynomials corresponds to matrix addition and multiplication of a polynomial by a scalar corresponds to scalar multiplication of matrices. In your software perform the following operations on polynomials, using the matrix association as just described. Let $n = 3$ and

$$p(t) = 2t^3 + 5t^2 + t - 2, \quad q(t) = t^3 + 3t + 5.$$

(a) $p(t) + q(t)$. (b) $5p(t)$. (c) $3p(t) - 4q(t)$.

2.3. *Linear Independence*

So far we have defined a mathematical system called a real vector space and noted some of its properties. We further observe that the only real vector space having a finite number of vectors in it is the vector space whose only vector is $\mathbf{0}$, for if $\mathbf{v} \neq \mathbf{0}$ is in a vector space V, then $c \odot \mathbf{v}$ is in V, where c is any real number, and so V has infinitely many vectors in it. However, in this section and the following one we show that each vector space V studied here has a finite number of vectors that completely describe V. We now turn to a formulation of these ideas. Remember that we will denote $\mathbf{u} \oplus \mathbf{v}$ and $c \odot \mathbf{u}$ in a vector space V as $\mathbf{u} + \mathbf{v}$ and $c\mathbf{u}$, respectively.

Definition 2.6. Let $S = \{\mathbf{v}_1, \mathbf{v}_2, \ldots, \mathbf{v}_k\}$ be a set of vectors in a vector space V. A vector $\mathbf{v}$ in V is called a **linear combination** of the vectors in S if

$$\mathbf{v} = a_1\mathbf{v}_1 + a_2\mathbf{v}_2 + \cdots + a_k\mathbf{v}_k$$

for some real numbers $a_1, a_2, \ldots, a_k$. ■

In Figure 2.17 we show the vector $\mathbf{v}$ in R^2 or R^3 as a linear combination of the vectors $\mathbf{v}_1$ and $\mathbf{v}_2$.

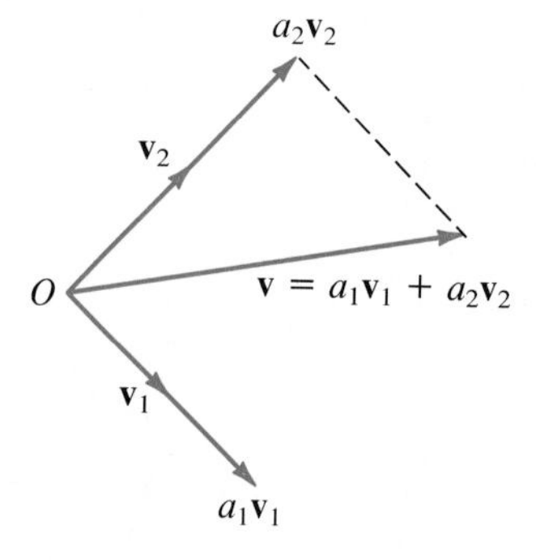

Figure 2.17 A linear combination of two vectors.

Example 1. In R^3 let

$$\mathbf{v}_1 = \begin{bmatrix} 1 \\ 2 \\ 1 \end{bmatrix}, \quad \mathbf{v}_2 = \begin{bmatrix} 1 \\ 0 \\ 2 \end{bmatrix}, \quad \text{and} \quad \mathbf{v}_3 = \begin{bmatrix} 1 \\ 1 \\ 0 \end{bmatrix}.$$

The vector

$$\mathbf{v} = \begin{bmatrix} 2 \\ 1 \\ 5 \end{bmatrix}$$

is a linear combination of $\mathbf{v}_1$, $\mathbf{v}_2$, and $\mathbf{v}_3$ if we can find real numbers a_1, a_2, and a_3 so that

$$a_1\mathbf{v}_1 + a_2\mathbf{v}_2 + a_3\mathbf{v}_3 = \mathbf{v}.$$

Substituting for $\mathbf{v}$, $\mathbf{v}_1$, $\mathbf{v}_2$, and $\mathbf{v}_3$, we have

$$a_1\begin{bmatrix} 1 \\ 2 \\ 1 \end{bmatrix} + a_2\begin{bmatrix} 1 \\ 0 \\ 2 \end{bmatrix} + a_3\begin{bmatrix} 1 \\ 1 \\ 0 \end{bmatrix} = \begin{bmatrix} 2 \\ 1 \\ 5 \end{bmatrix}.$$

Equating corresponding entries leads to the linear system (verify)

$$\begin{aligned} a_1 + a_2 + a_3 &= 2 \\ 2a_1 + a_3 &= 1 \\ a_1 + 2a_2 &= 5. \end{aligned}$$

Solving this linear system by the methods of Chapter 1 gives (verify) $a_1 = 1$, $a_2 = 2$, and $a_3 = -1$, which means that $\mathbf{v}$ is a linear combination of $\mathbf{v}_1$, $\mathbf{v}_2$, and $\mathbf{v}_3$. Thus

$$\mathbf{v} = \mathbf{v}_1 + 2\mathbf{v}_2 - \mathbf{v}_3.$$ ■

Recall that toward the end of Section 2.2 we had observed that the set of all linear combinations of a set of vectors $S = \{\mathbf{v}_1, \mathbf{v}_2, \ldots, \mathbf{v}_k\}$ in a vector space V is a subspace of V, which we call span S. In general, to determine if a specific vector $\mathbf{v}$ belongs to span S, we investigate the consistency of an appropriate linear system.

Example 2. In R^3 let

$$\mathbf{v}_1 = \begin{bmatrix} 2 \\ 1 \\ 2 \end{bmatrix}, \quad \mathbf{v}_2 = \begin{bmatrix} 1 \\ 0 \\ 1 \end{bmatrix}, \quad \mathbf{v}_3 = \begin{bmatrix} 2 \\ 1 \\ 0 \end{bmatrix}, \quad \text{and} \quad \mathbf{v}_4 = \begin{bmatrix} 1 \\ 0 \\ -1 \end{bmatrix}.$$

Determine if the vector

$$\mathbf{v} = \begin{bmatrix} 0 \\ -2 \\ 4 \end{bmatrix}$$

belongs to span $\{\mathbf{v}_1, \mathbf{v}_2, \mathbf{v}_3, \mathbf{v}_4\}$.

Solution: If we can find scalars a_1, a_2, a_3, and a_4 so that

$$a_1\mathbf{v}_1 + a_2\mathbf{v}_2 + a_3\mathbf{v}_3 + a_4\mathbf{v}_4 = \mathbf{v},$$

then $\mathbf{v}$ belongs to span $\{\mathbf{v}_1, \mathbf{v}_2, \mathbf{v}_3, \mathbf{v}_4\}$. Substituting for $\mathbf{v}$, $\mathbf{v}_1$, $\mathbf{v}_2$, $\mathbf{v}_3$, and $\mathbf{v}_4$, we have

$$a_1\begin{bmatrix} 2 \\ 1 \\ 2 \end{bmatrix} + a_2\begin{bmatrix} 1 \\ 0 \\ 1 \end{bmatrix} + a_3\begin{bmatrix} 2 \\ 1 \\ 0 \end{bmatrix} + a_4\begin{bmatrix} 1 \\ 0 \\ -1 \end{bmatrix} = \begin{bmatrix} 0 \\ -2 \\ 4 \end{bmatrix}.$$

Equating corresponding entries leads to the system of linear equations

$$\begin{aligned} 2a_1 + a_2 + 2a_3 + a_4 &= 0 \\ a_1 \qquad + a_3 \qquad &= -2 \\ 2a_1 + a_2 \qquad - a_4 &= 4. \end{aligned}$$

To investigate whether or not this system of linear equations is consistent, we form the augmented matrix and transform it to row echelon form. For this linear system we obtain the matrix in row echelon form (verify),

$$\left[\begin{array}{cccc|c} 1 & \frac{1}{2} & 1 & \frac{1}{2} & 0 \\ 0 & 1 & 0 & 1 & 4 \\ 0 & 0 & 1 & 1 & -2 \end{array}\right],$$

which indicates that the system is consistent, and back substitution gives us $a_1 = a_4$, $a_2 = 4 - a_4$, $a_3 = -2 - a_4$, and a_4 may be chosen arbitrarily. For example, if we set $a_4 = 1$, we obtain $a_1 = 1$, $a_2 = 3$, $a_3 = -3$, and $a_4 = 1$. Hence $\mathbf{v}$ is in span $\{\mathbf{v}_1, \mathbf{v}_2, \mathbf{v}_3, \mathbf{v}_4\}$. ■

Definition 2.7. Let $S = \{\mathbf{v}_1, \mathbf{v}_2, \ldots, \mathbf{v}_k\}$ be a set of vectors in a vector space V. The set S **spans** V, or V is **spanned** by S, if every vector in V is a linear combination of the vectors in S. ■

Again we investigate the consistency of a linear system, but this time for an arbitrary right-hand side.

Example 3. Let V be the vector space R^3. Let

$$\mathbf{v}_1 = \begin{bmatrix} 1 \\ 2 \\ 1 \end{bmatrix}, \quad \mathbf{v}_2 = \begin{bmatrix} 1 \\ 0 \\ 2 \end{bmatrix}, \quad \text{and} \quad \mathbf{v}_3 = \begin{bmatrix} 1 \\ 1 \\ 0 \end{bmatrix}.$$

To find out whether $\{\mathbf{v}_1, \mathbf{v}_2, \mathbf{v}_3\}$ spans V, we pick any vector $\mathbf{v} = \begin{bmatrix} a \\ b \\ c \end{bmatrix}$ in V (a, b, and c are arbitrary real numbers) and must find out whether there are constants a_1, a_2, and a_3 such that

$$a_1\mathbf{v}_1 + a_2\mathbf{v}_2 + a_3\mathbf{v}_3 = \mathbf{v}.$$

This leads to the linear system (verify)

$$\begin{aligned} a_1 + a_2 + a_3 &= a \\ 2a_1 \qquad + a_3 &= b \\ a_1 + 2a_2 \qquad &= c. \end{aligned}$$

A solution is (verify)

$$a_1 = \frac{-2a + 2b + c}{3}, \quad a_2 = \frac{a - b + c}{3}, \quad a_3 = \frac{4a - b - 2c}{3}.$$

Thus $\{\mathbf{v}_1, \mathbf{v}_2, \mathbf{v}_3\}$ spans V. This is equivalent to saying that

$$\text{span } \{\mathbf{v}_1, \mathbf{v}_2, \mathbf{v}_3\} = R^3. \qquad \blacksquare$$

Example 4. Let V be P_2, the vector space consisting of all polynomials of degree ≤ 2 and the zero polynomial. Let $\mathbf{v}_1 = t^2 + 2t + 1$ and $\mathbf{v}_2 = t^2 + 2$. Does $\{\mathbf{v}_1, \mathbf{v}_2\}$ span V?

Solution: Let $\mathbf{v} = at^2 + bt + c$ be any vector in V, where a, b, and c are any real numbers. We must find out whether there are constants a_1 and a_2 such that

$$a_1\mathbf{v}_1 + a_2\mathbf{v}_2 = \mathbf{v}$$

or

$$a_1(t^2 + 2t + 1) + a_2(t^2 + 2) = at^2 + bt + c.$$

Thus

$$(a_1 + a_2)t^2 + (2a_1)t + (a_1 + 2a_2) = at^2 + bt + c.$$

Now two polynomials agree for all values of t only if the coefficients of respective powers of t agree. Thus we get the linear system

$$\begin{aligned} a_1 + a_2 &= a \\ 2a_1 \qquad &= b \\ a_1 + 2a_2 &= c. \end{aligned}$$

Transforming the augmented matrix of this linear system, we obtain (verify)

$$\left[\begin{array}{cc|c} 1 & 0 & 2a - c \\ 0 & 1 & c - a \\ 0 & 0 & b - 4a + 2c \end{array}\right].$$

If $b - 4a + 2c \neq 0$, then there is no solution. Hence $\{\mathbf{v}_1, \mathbf{v}_2\}$ does not span V. ■

Example 5. Consider the homogeneous linear system $A\mathbf{x} = \mathbf{0}$, where

$$A = \begin{bmatrix} 1 & 1 & 0 & 2 \\ -2 & -2 & 1 & -5 \\ 1 & 1 & -1 & 3 \\ 4 & 4 & -1 & 9 \end{bmatrix}.$$

From Example 16 in Section 2.2, the set of all solutions to $A\mathbf{x} = \mathbf{0}$ forms a subspace of R^4. To determine a spanning set for the solution space of this homogeneous system, we find that the reduced row echelon form of the augmented matrix is (verify)

$$\left[\begin{array}{cccc|c} 1 & 1 & 0 & 2 & 0 \\ 0 & 0 & 1 & -1 & 0 \\ 0 & 0 & 0 & 0 & 0 \\ 0 & 0 & 0 & 0 & 0 \end{array}\right].$$

The general solution is then given by

$$x_1 = -r - 2s, \qquad x_2 = r, \qquad x_3 = s, \qquad x_4 = s,$$

where r and s are any real numbers. In matrix form we have that any member of the solution space is given by

$$x = r\begin{bmatrix} -1 \\ 1 \\ 0 \\ 0 \end{bmatrix} + s\begin{bmatrix} -2 \\ 0 \\ 1 \\ 1 \end{bmatrix}.$$

Hence $\begin{bmatrix} -1 \\ 1 \\ 0 \\ 0 \end{bmatrix}$ and $\begin{bmatrix} -2 \\ 0 \\ 1 \\ 1 \end{bmatrix}$ span the solution space. ■

Definition 2.8. Let $S = \{\mathbf{v}_1, \mathbf{v}_2, \ldots, \mathbf{v}_k\}$ be a set of distinct vectors in a vector space V. Then S is said to be **linearly dependent** if there exist constants $a_1, a_2, \ldots, a_k$, not all zero, such that

$$a_1\mathbf{v}_1 + a_2\mathbf{v}_2 + \cdots + a_k\mathbf{v}_k = \mathbf{0}. \tag{1}$$

Otherwise, S is called **linearly independent**. That is, S is linearly independent if (1) holds only when

$$a_1 = a_2 = \cdots = a_k = 0.$$

■

It should be emphasized that for any set of distinct vectors $S = \{\mathbf{v}_1, \mathbf{v}_2, \ldots, \mathbf{v}_k\}$, (1) always holds if we choose all the scalars $a_1, a_2, \ldots, a_k$ equal to zero. The important point in this definition is whether or not it is possible to satisfy (1) with at least one of the scalars different from zero.

Example 6. Determine whether the vectors

$$\begin{bmatrix} -1 \\ 1 \\ 0 \\ 0 \end{bmatrix} \quad \text{and} \quad \begin{bmatrix} -2 \\ 0 \\ 1 \\ 1 \end{bmatrix}$$

determined in Example 5 as spanning the solution space of $A\mathbf{x} = \mathbf{0}$ are linearly dependent or linearly independent.

Solution: Forming Equation (1),

$$a_1\begin{bmatrix} -1 \\ 1 \\ 0 \\ 0 \end{bmatrix} + a_2\begin{bmatrix} -2 \\ 0 \\ 1 \\ 1 \end{bmatrix} = \begin{bmatrix} 0 \\ 0 \\ 0 \\ 0 \end{bmatrix},$$

we obtain the homogeneous system

$$\begin{aligned} -a_1 - 2a_2 &= 0 \\ a_1 + 0a_2 &= 0 \\ 0a_1 + a_2 &= 0 \\ 0a_1 + a_2 &= 0, \end{aligned}$$

whose only solution is $a_1 = a_2 = 0$ (verify). Hence the given vectors are linearly independent. ■

Example 7. Let V be R_4 and $\mathbf{v}_1 = [1 \quad 0 \quad 1 \quad 2]$, $\mathbf{v}_2 = [0 \quad 1 \quad 1 \quad 2]$, and $\mathbf{v}_3 = [1 \quad 1 \quad 1 \quad 3]$. Is $S = \{\mathbf{v}_1, \mathbf{v}_2, \mathbf{v}_3\}$ linearly dependent or linearly independent?

Solution: We form Equation (1),

$$a_1\mathbf{v}_1 + a_2\mathbf{v}_2 + a_3\mathbf{v}_3 = \mathbf{0},$$

and solve for a_1, a_2, and a_3. The resulting homogeneous system is

$$\begin{aligned} a_1 \qquad\quad + a_3 &= 0 \\ a_2 + a_3 &= 0 \\ a_1 + a_2 + a_3 &= 0 \\ 2a_1 + 2a_2 + 3a_3 &= 0, \end{aligned}$$

which has as its only solution $a_1 = a_2 = a_3 = 0$ (verify), showing that S is linearly independent. ■

Example 8. Consider the vectors

$$\mathbf{v}_1 = \begin{bmatrix} 1 \\ 2 \\ -1 \end{bmatrix}, \quad \mathbf{v}_2 = \begin{bmatrix} 1 \\ -2 \\ 1 \end{bmatrix}, \quad \mathbf{v}_3 = \begin{bmatrix} -3 \\ 2 \\ -1 \end{bmatrix}, \quad \text{and} \quad \mathbf{v}_4 = \begin{bmatrix} 2 \\ 0 \\ 0 \end{bmatrix}$$

in R^3. Is $S = \{\mathbf{v}_1, \mathbf{v}_2, \mathbf{v}_3, \mathbf{v}_4\}$ linearly dependent or linearly independent?

Solution: Setting up Equation (1), we are led to the homogeneous system

$$\begin{aligned} a_1 + a_2 - 3a_3 + 2a_4 &= 0 \\ 2a_1 - 2a_2 + 2a_3 \qquad\quad &= 0 \\ -a_1 + a_2 - a_3 \qquad\quad &= 0, \end{aligned}$$

of three equations in four unknowns. By Theorem 1.12, we are assured of the existence of a nontrivial solution. Hence S is linearly dependent. In fact, two of the infinitely many solutions are

$$\begin{aligned} &a_1 = 1, \quad a_2 = 2, \quad a_3 = 1, \quad a_4 = 0; \\ &a_1 = 1, \quad a_2 = 1, \quad a_3 = 0, \quad a_4 = -1. \end{aligned}$$

■

Example 9. Let V be R_3 and $\mathbf{v}_1 = [1 \quad 0 \quad 0]$, $\mathbf{v}_2 = [0 \quad 1 \quad 0]$, and $\mathbf{v}_3 = [0 \quad 0 \quad 1]$. To find out whether $S = \{\mathbf{v}_1, \mathbf{v}_2, \mathbf{v}_3\}$ is linearly dependent or linearly independent, we form Equation (1),

$$a_1\mathbf{v}_1 + a_2\mathbf{v}_2 + a_3\mathbf{v}_3 = \mathbf{0},$$

and solve for a_1, a_2, and a_3. Since $a_1 = a_2 = a_3 = 0$ (verify), we conclude that S is linearly independent. ■

Example 10. Let V be P_2 and $\mathbf{v}_1 = t^2 + t + 2$, $\mathbf{v}_2 = 2t^2 + t$, $\mathbf{v}_3 = 3t^2 + 2t + 2$. Is $S = \{\mathbf{v}_1, \mathbf{v}_2, \mathbf{v}_3\}$ linearly dependent or linearly independent?

Solution: Forming Equation (1), we have

$$\begin{aligned} a_1 + 2a_2 + 3a_3 &= 0 \\ a_1 + a_2 + 2a_3 &= 0 \\ 2a_1 \qquad + 2a_3 &= 0, \end{aligned}$$

which has infinitely many solutions (verify). A particular solution is $a_1 = 1$, $a_2 = 1$, $a_3 = -1$, so

$$\mathbf{v}_1 + \mathbf{v}_2 - \mathbf{v}_3 = \mathbf{0}.$$

Hence S is linearly dependent. ■

Theorem 2.4. *Let S_1 and S_2 be finite subsets of a vector space and let S_1 be a subset of S_2. Then:*
(a) *If S_1 is linearly dependent, so is S_2.*
(b) *If S_2 is linearly independent, so is S_1.*

Proof: Let

$$S_1 = \{\mathbf{v}_1, \mathbf{v}_2, \ldots, \mathbf{v}_k\} \quad \text{and} \quad S_2 = \{\mathbf{v}_1, \mathbf{v}_2, \ldots, \mathbf{v}_k, \mathbf{v}_{k+1}, \ldots, \mathbf{v}_m\}.$$

We first prove (a). Since S_1 is linearly dependent, there exist $a_1, a_2, \ldots, a_k$, not all zero, such that

$$a_1\mathbf{v}_1 + a_2\mathbf{v}_2 + \cdots + a_k\mathbf{v}_k = \mathbf{0}.$$

Then

$$a_1\mathbf{v}_1 + a_2\mathbf{v}_2 + \cdots + a_k\mathbf{v}_k + 0\mathbf{v}_{k+1} + \cdots + 0\mathbf{v}_m = \mathbf{0}. \tag{2}$$

Since not all the coefficients in (2) are zero, we conclude that S_2 is linearly dependent.

Next, we prove (b). Let S_2 be linearly independent. If S_1 is assumed as linearly dependent, then S_2 is linearly dependent, by (a), a contradiction. Hence S_1 must be linearly independent. ■

We now note that the set $S = \{\mathbf{0}\}$ consisting only of $\mathbf{0}$ is linearly dependent, since $5\mathbf{0} = \mathbf{0}$ and $5 \neq 0$. From this it follows that if S is any set of vectors that contains $\mathbf{0}$, then S must be linearly dependent. Also, a set consisting of a single nonzero vector is linearly independent (verify).

We consider next the meaning of linear dependence in R^2 and R^3. Suppose that $\{\mathbf{v}_1, \mathbf{v}_2\}$ is linearly dependent in R^2. Then there exist scalars a_1 and a_2 not both zero such that $a_1\mathbf{v}_1 + a_2\mathbf{v}_2 = \mathbf{0}$. If $a_1 \neq 0$, then $\mathbf{v}_1 = -(a_2/a_1)\mathbf{v}_2$. If $a_2 \neq 0$, then $\mathbf{v}_2 = -(a_1/a_2)\mathbf{v}_1$. Thus one of the vectors is a multiple of the other. Conversely, suppose that $\mathbf{v}_1 = a\mathbf{v}_2$. Then $1\mathbf{v}_1 - a\mathbf{v}_2 = \mathbf{0}$, and since the coefficients of $\mathbf{v}_1$ and $\mathbf{v}_2$ are not both zero, it follows that $\{\mathbf{v}_1, \mathbf{v}_2\}$ is linearly dependent. Thus $\{\mathbf{v}_1, \mathbf{v}_2\}$ is linearly dependent in R^2 if and only if one of the vectors is a multiple of the other (Figure 2.18). Hence two vectors in R^2 are linearly dependent if and only if they both lie on the same line passing through the origin [Figure 2.18(a)].

Figure 2.18

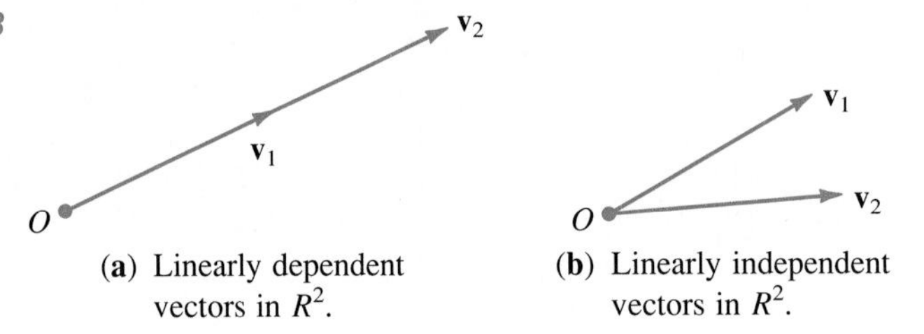

(a) Linearly dependent vectors in R^2.

(b) Linearly independent vectors in R^2.

Suppose now that $\{\mathbf{v}_1, \mathbf{v}_2, \mathbf{v}_3\}$ is a linearly dependent set of vectors in R^3. Then we can write

$$a_1\mathbf{v}_1 + a_2\mathbf{v}_2 + a_3\mathbf{v}_3 = \mathbf{0},$$

where a_1, a_2, and a_3 are not all zero, say $a_2 \neq 0$. Then

$$\mathbf{v}_2 = -\frac{a_1}{a_2}\mathbf{v}_1 - \frac{a_3}{a_2}\mathbf{v}_3,$$

which means that $\mathbf{v}_2$ is in span $\{\mathbf{v}_1, \mathbf{v}_3\}$.

Now span $\{\mathbf{v}_1, \mathbf{v}_3\}$ is either a plane through the origin (when $\{\mathbf{v}_1, \mathbf{v}_3\}$ is linearly independent) or a line through the origin (when $\{\mathbf{v}_1, \mathbf{v}_3\}$ is linearly dependent). Since a line through the origin always lies in a plane through the origin, we conclude that $\mathbf{v}_1$, $\mathbf{v}_2$, and $\mathbf{v}_3$ all lie in the same plane through the origin. Conversely, suppose that $\mathbf{v}_1$, $\mathbf{v}_2$, and $\mathbf{v}_3$ all lie in the same plane through the origin.

Then either all three vectors are the zero vector, or all three vectors lie on the same line through the origin, or all three vectors lie in a plane through the origin spanned by two vectors, say $\mathbf{v}_1$ and $\mathbf{v}_3$. Thus in all these cases $\mathbf{v}_2$ is a linear combination of $\mathbf{v}_1$ and $\mathbf{v}_3$:

$$\mathbf{v}_2 = a_1\mathbf{v}_1 + a_3\mathbf{v}_3.$$

Then

$$a_1\mathbf{v}_1 - 1\mathbf{v}_2 + a_3\mathbf{v}_3 = \mathbf{0},$$

which means that $\{\mathbf{v}_1, \mathbf{v}_2, \mathbf{v}_3\}$ is linearly dependent. Hence three vectors in R^3 are linearly dependent if and only if they all lie in the same plane passing through the origin [Figure 2.19(a) and (b)].

Figure 2.19

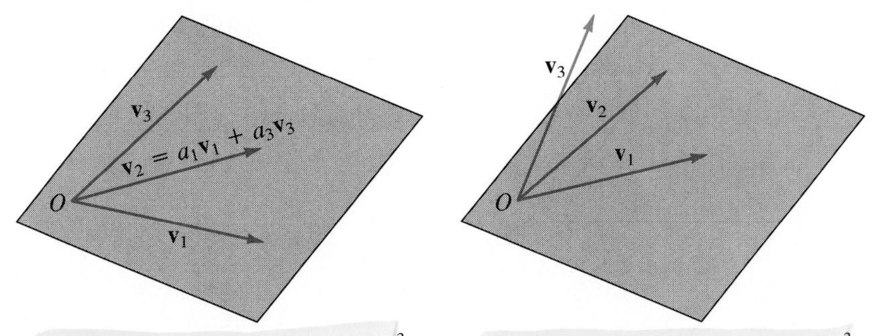

(**a**) Linearly dependent vectors in R^3. (**b**) Linearly independent vectors in R^3.

Theorem 2.5. *Let $S = \{\mathbf{v}_1, \mathbf{v}_2, \ldots, \mathbf{v}_n\}$ be a set of nonzero vectors in a vector space V. Then S is linearly dependent if and only if one of the vectors $\mathbf{v}_j$ is a linear combination of the preceding vectors in S.*

Proof: If $\mathbf{v}_j$ is a linear combination of the preceding vectors,

$$\mathbf{v}_j = a_1\mathbf{v}_1 + a_2\mathbf{v}_2 + \cdots + a_{j-1}\mathbf{v}_{j-1},$$

then

$$a_1\mathbf{v}_1 + a_2\mathbf{v}_2 + \cdots + a_{j-1}\mathbf{v}_{j-1} + (-1)\mathbf{v}_j + 0\mathbf{v}_{j+1} + \cdots + 0\mathbf{v}_n = \mathbf{0}.$$

Since at least one coefficient, -1, is nonzero, we conclude that S is linearly dependent.

Conversely, let S be linearly dependent. Then there exist scalars, $a_1, a_2, \ldots, a_n$, not all zero, such that

$$a_1\mathbf{v}_1 + a_2\mathbf{v}_2 + \cdots + a_n\mathbf{v}_n = \mathbf{0}.$$

Now let j be the largest subscript for which $a_j \neq 0$. If $j > 1$, then

$$\mathbf{v}_j = -\left(\frac{a_1}{a_j}\right)\mathbf{v}_1 - \left(\frac{a_2}{a_j}\right)\mathbf{v}_2 - \cdots - \left(\frac{a_{j-1}}{a_j}\right)\mathbf{v}_{j-1}.$$

If $j = 1$, then $a_1\mathbf{v}_1 = \mathbf{0}$, which implies that $\mathbf{v}_1 = \mathbf{0}$, a contradiction to the hypothesis that none of the vectors in S is the zero vector. Thus one of the vectors in S is a linear combination of the preceding vectors in S. ■

Example 11. Let $V = R_3$ and $\mathbf{v}_1 = [1 \quad 2 \quad -1]$, $\mathbf{v}_2 = [1 \quad -2 \quad 1]$, $\mathbf{v}_3 = [-3 \quad 2 \quad -1]$, and $\mathbf{v}_4 = [2 \quad 0 \quad 0]$. We find (verify) that

$$\mathbf{v}_1 + 2\mathbf{v}_2 + \mathbf{v}_3 + 0\mathbf{v}_4 = \mathbf{0},$$

so $\mathbf{v}_3 = -\mathbf{v}_1 - 2\mathbf{v}_2$. Also,

$$\mathbf{v}_1 + \mathbf{v}_2 + 0\mathbf{v}_3 - \mathbf{v}_4 = \mathbf{0},$$

so $\mathbf{v}_4 = \mathbf{v}_1 + \mathbf{v}_2 + 0\mathbf{v}_3$. ■

We observe that Theorem 2.5 does not say that *every* vector $\mathbf{v}$ is a linear combination of the preceding vectors in S. Thus, in Example 11, $\mathbf{v}_1 + 2\mathbf{v}_2 + \mathbf{v}_3 + 0\mathbf{v}_4 = \mathbf{0}$. We cannot solve, in this equation, for $\mathbf{v}_4$ as a linear combination of $\mathbf{v}_1$, $\mathbf{v}_2$, and $\mathbf{v}_3$, since its coefficient is zero. However, in this same example we have $\mathbf{v}_1 + \mathbf{v}_2 + 0\mathbf{v}_3 - \mathbf{v}_4 = \mathbf{0}$. In this equation we *can* solve for $\mathbf{v}_4$, since its coefficient is nonzero.

We can also prove that if $S = \{\mathbf{v}_1, \mathbf{v}_2, \ldots, \mathbf{v}_k\}$ is a set of vectors in a vector space V, then S is linearly dependent if and only if one of the vectors in S is a linear combination of all the other vectors in S (see Exercise 21). Thus, in Example 11,

$$\mathbf{v}_1 = -\mathbf{v}_3 - 2\mathbf{v}_2, \qquad \mathbf{v}_2 = -\tfrac{1}{2}\mathbf{v}_1 - \tfrac{1}{2}\mathbf{v}_3, \qquad \mathbf{v}_1 = \mathbf{v}_4 - \mathbf{v}_2,$$

and

$$\mathbf{v}_2 = \mathbf{v}_4 - \mathbf{v}_1.$$

The following result will be used in Section 2.4 as well as in several other places. Suppose that $S = \{\mathbf{v}_1, \mathbf{v}_2, \ldots, \mathbf{v}_n\}$ spans a vector space V and $\mathbf{v}_j$ is a linear combination of the preceding vectors in S. Then the set $S_1 = \{\mathbf{v}_1, \mathbf{v}_2, \ldots, \mathbf{v}_{j-1}, \mathbf{v}_{j+1}, \ldots, \mathbf{v}_n\}$, consisting of S with $\mathbf{v}_j$ deleted, also spans V. To show this result, observe that if $\mathbf{v}$ is any vector in V, then, since S spans V, we can find scalars $a_1, a_2, \ldots, a_n$ such that

$$\mathbf{v} = a_1\mathbf{v}_1 + a_2\mathbf{v}_2 + \cdots + a_{j-1}\mathbf{v}_{j-1} + a_j\mathbf{v}_j + a_{j+1}\mathbf{v}_{j+1} + \cdots + a_n\mathbf{v}_n.$$

Now if

$$\mathbf{v}_j = b_1\mathbf{v}_1 + b_2\mathbf{v}_2 + \cdots + b_{j-1}\mathbf{v}_{j-1},$$

then

$$\begin{aligned}\mathbf{v} = a_1\mathbf{v}_1 + a_2\mathbf{v}_2 + \cdots + a_{j-1}\mathbf{v}_{j-1} + a_j(b_1\mathbf{v}_1 + b_2\mathbf{v}_2 + \cdots + b_{j-1}\mathbf{v}_{j-1})\\ + a_{j+1}\mathbf{v}_{j+1} + \cdots + a_n\mathbf{v}_n = c_1\mathbf{v}_1 + c_2\mathbf{v}_2\\ + \cdots + c_{j-1}\mathbf{v}_{j-1} + c_{j+1}\mathbf{v}_{j+1} + \cdots + c_n\mathbf{v}_n,\end{aligned}$$

which means that span $S_1 = V$.

Example 12. Consider the set of vectors $S = \{\mathbf{v}_1, \mathbf{v}_2, \mathbf{v}_3, \mathbf{v}_4\}$ in R^4, where

$$\mathbf{v}_1 = \begin{bmatrix}1\\1\\0\\0\end{bmatrix}, \quad \mathbf{v}_2 = \begin{bmatrix}1\\0\\1\\0\end{bmatrix}, \quad \mathbf{v}_3 = \begin{bmatrix}0\\1\\1\\0\end{bmatrix}, \quad \text{and} \quad \mathbf{v}_4 = \begin{bmatrix}2\\1\\1\\0\end{bmatrix},$$

and let W = span S. Since

$$\mathbf{v}_4 = \mathbf{v}_1 + \mathbf{v}_2,$$

we conclude that W = span S_1, where $S_1 = \{\mathbf{v}_1, \mathbf{v}_2, \mathbf{v}_3\}$. ■

2.3. EXERCISES

1. Which of the following vectors in R^3 are linear combinations of

$$\mathbf{v}_1 = \begin{bmatrix}4\\2\\-3\end{bmatrix}, \quad \mathbf{v}_2 = \begin{bmatrix}2\\1\\-2\end{bmatrix}, \quad \text{and} \quad \mathbf{v}_3 = \begin{bmatrix}-2\\-1\\0\end{bmatrix}?$$

(a) $\begin{bmatrix}1\\1\\1\end{bmatrix}$. (b) $\begin{bmatrix}4\\2\\-6\end{bmatrix}$.

(c) $\begin{bmatrix} -2 \\ -1 \\ 1 \end{bmatrix}$. (d) $\begin{bmatrix} -1 \\ 2 \\ 3 \end{bmatrix}$.

2. Which of the following vectors in R_4 are linear combinations of

$\mathbf{v}_1 = [1 \quad 2 \quad 1 \quad 0]$, $\mathbf{v}_2 = [4 \quad 1 \quad -2 \quad 3]$,
$\mathbf{v}_3 = [1 \quad 2 \quad 6 \quad -5]$, $\mathbf{v}_4 = [-2 \quad 3 \quad -1 \quad 2]$?

(a) $[3 \quad 6 \quad 3 \quad 0]$. (b) $[1 \quad 0 \quad 0 \quad 0]$.
(c) $[3 \quad 6 \quad -2 \quad 5]$. (d) $[0 \quad 0 \quad 0 \quad 1]$.

3. In each part, determine whether the given vector $p(t)$ in P_2 belongs to span $\{p_1(t), p_2(t), p_3(t)\}$, where

$$p_1(t) = t^2 + 2t + 1, \quad p_2(t) = t^2 + 3, \quad \text{and} \quad p_3(t) = t - 1.$$

(a) $p(t) = t^2 + t + 2$.
(b) $p(t) = 2t^2 + 2t + 3$.
(c) $p(t) = -t^2 + t - 4$.
(d) $p(t) = -2t^2 + 3t + 1$.

4. In each part, determine whether the given vector A in ${}_2R_2$ belongs to span $\{A_1, A_2, A_3\}$, where

$$A_1 = \begin{bmatrix} 1 & -1 \\ 0 & 3 \end{bmatrix}, \quad A_2 = \begin{bmatrix} 1 & 1 \\ 0 & 2 \end{bmatrix}, \quad \text{and} \quad A_3 = \begin{bmatrix} 2 & 2 \\ -1 & 1 \end{bmatrix}.$$

(a) $A = \begin{bmatrix} 5 & 1 \\ -1 & 9 \end{bmatrix}$.

(b) $A = \begin{bmatrix} -3 & -1 \\ 3 & 2 \end{bmatrix}$.

(c) $A = \begin{bmatrix} 3 & -2 \\ 3 & 2 \end{bmatrix}$. (d) $A = \begin{bmatrix} 1 & 0 \\ 2 & 1 \end{bmatrix}$.

5. Which of the following sets of vectors span R_2?
(a) $\{[1 \quad 2], [-1 \quad 1]\}$.
(b) $\{[0 \quad 0], [1 \quad 1], [-2 \quad -2]\}$.
(c) $\{[1 \quad 3], [2 \quad -3], [0 \quad 2]\}$.
(d) $\{[2 \quad 4], [-1 \quad 2]\}$.

6. Which of the following sets of vectors span R^4?

(a) $\left\{ \begin{bmatrix} 1 \\ -1 \\ 2 \\ 0 \end{bmatrix}, \begin{bmatrix} 0 \\ 1 \\ 1 \\ 1 \end{bmatrix} \right\}$.

(b) $\left\{ \begin{bmatrix} 3 \\ 2 \\ 1 \\ 0 \end{bmatrix}, \begin{bmatrix} 1 \\ 2 \\ -1 \\ 0 \end{bmatrix}, \begin{bmatrix} 0 \\ 0 \\ 0 \\ 1 \end{bmatrix} \right\}$.

(c) $\left\{ \begin{bmatrix} 3 \\ 2 \\ -1 \\ 2 \end{bmatrix}, \begin{bmatrix} 4 \\ 0 \\ 0 \\ 2 \end{bmatrix}, \begin{bmatrix} 3 \\ 2 \\ -1 \\ 2 \end{bmatrix}, \begin{bmatrix} 5 \\ 6 \\ -3 \\ 2 \end{bmatrix}, \begin{bmatrix} 0 \\ 4 \\ -2 \\ -1 \end{bmatrix} \right\}$.

(d) $\left\{ \begin{bmatrix} 1 \\ 1 \\ 0 \\ 0 \end{bmatrix}, \begin{bmatrix} 1 \\ 2 \\ -1 \\ 1 \end{bmatrix}, \begin{bmatrix} 0 \\ 0 \\ 1 \\ -1 \end{bmatrix}, \begin{bmatrix} 2 \\ 1 \\ 2 \\ -1 \end{bmatrix} \right\}$.

7. Which of the following sets of vectors span R_4?
(a) $\{[1 \quad 0 \quad 0 \quad 1], [0 \quad 1 \quad 0 \quad 0], [1 \quad 1 \quad 1 \quad 1], [1 \quad 1 \quad 1 \quad 0]\}$.
(b) $\{[1 \quad 2 \quad 1 \quad 0], [1 \quad 1 \quad -1 \quad 0], [0 \quad 0 \quad 0 \quad 1]\}$.
(c) $\{[6 \quad 4 \quad -2 \quad 4], [2 \quad 0 \quad 0 \quad 1], [3 \quad 2 \quad -1 \quad 2], [5 \quad 6 \quad -3 \quad 2], [0 \quad 4 \quad -2 \quad -1]\}$.
(d) $\{[1 \quad 1 \quad 0 \quad 0], [1 \quad 2 \quad -1 \quad 1], [0 \quad 0 \quad 1 \quad 1], [2 \quad 1 \quad 2 \quad 1]\}$.

8. Which of the following sets of polynomials span P_2?
(a) $\{t^2 + 1, t^2 + t, t + 1\}$.
(b) $\{t^2 + 1, t - 1, t^2 + t\}$.
(c) $\{t^2 + 2, 2t^2 - t + 1, t + 2, t^2 + t + 4\}$.
(d) $\{t^2 + 2t - 1, t^2 - 1\}$.

9. Do the polynomials $t^3 + 2t + 1$, $t^2 - t + 2$, $t^3 + 2$, $-t^3 + t^2 - 5t + 2$ span P_3?

10. Does the set $S = \left\{\begin{bmatrix}1 & 1\\0 & 0\end{bmatrix}, \begin{bmatrix}0 & 0\\1 & 1\end{bmatrix}, \begin{bmatrix}1 & 0\\0 & 1\end{bmatrix}, \begin{bmatrix}0 & 1\\1 & 1\end{bmatrix}\right\}$ span ${}_2R_2$?

11. Find a set of vectors spanning the solution space of $A\mathbf{x} = \mathbf{0}$, where

$$A = \begin{bmatrix}1 & 0 & 1 & 0\\1 & 2 & 3 & 1\\2 & 1 & 3 & 1\\1 & 1 & 2 & 1\end{bmatrix}.$$

12. Find a set of vectors spanning the null space of

$$A = \begin{bmatrix}1 & 1 & 2 & -1\\2 & 3 & 6 & -2\\-2 & 1 & 2 & 2\\0 & -2 & -4 & 0\end{bmatrix}.$$

13. Let $\mathbf{x}_1 = \begin{bmatrix}2\\-1\\1\end{bmatrix}$, $\mathbf{x}_2 = \begin{bmatrix}4\\-7\\-1\end{bmatrix}$, $\mathbf{x}_3 = \begin{bmatrix}1\\2\\2\end{bmatrix}$ belong to the solution space of $A\mathbf{x} = \mathbf{0}$. Is $\{\mathbf{x}_1, \mathbf{x}_2, \mathbf{x}_3\}$ linearly independent?

14. Let $\mathbf{x}_1 = \begin{bmatrix}1\\2\\0\\1\end{bmatrix}$, $\mathbf{x}_2 = \begin{bmatrix}1\\0\\-1\\1\end{bmatrix}$, $\mathbf{x}_3 = \begin{bmatrix}1\\6\\2\\0\end{bmatrix}$ belong to the null space of A. Is $\{\mathbf{x}_1, \mathbf{x}_2, \mathbf{x}_3\}$ linearly independent?

15. Which of the following sets of vectors in R_3 are linearly dependent? For those which are, express one vector as a linear combination of the rest.

(a) $\{[1\ \ 1\ \ 0], [0\ \ 2\ \ 3], [1\ \ 2\ \ 3], [3\ \ 6\ \ 6]\}$.

(b) $\{[1\ \ 1\ \ 0], [3\ \ 4\ \ 2]\}$.

(c) $\{[1\ \ 1\ \ 0], [0\ \ 2\ \ 3], [1\ \ 2\ \ 3], [0\ \ 0\ \ 0]\}$.

16. Consider the vector space ${}_2R_2$. Follow the directions of Exercise 15.

(a) $\left\{\begin{bmatrix}1 & 1\\2 & 1\end{bmatrix}, \begin{bmatrix}1 & 0\\0 & 2\end{bmatrix}, \begin{bmatrix}0 & 3\\2 & 1\end{bmatrix}, \begin{bmatrix}4 & 6\\8 & 6\end{bmatrix}\right\}$.

(b) $\left\{\begin{bmatrix}1 & 1\\1 & 1\end{bmatrix}, \begin{bmatrix}1 & 0\\0 & 2\end{bmatrix}, \begin{bmatrix}0 & 1\\0 & 2\end{bmatrix}\right\}$.

(c) $\left\{\begin{bmatrix}1 & 1\\1 & 1\end{bmatrix}, \begin{bmatrix}2 & 3\\1 & 2\end{bmatrix}, \begin{bmatrix}3 & 1\\2 & 1\end{bmatrix}, \begin{bmatrix}2 & 2\\1 & 1\end{bmatrix}\right\}$.

17. Consider the vector space P_2. Follow the directions of Exercise 15.

(a) $\{t^2 + 1, t - 2, t + 3\}$.

(b) $\{2t^2 + t, t^2 + 3, t\}$.

(c) $\{2t^2 + t + 1, 3t^2 + t - 5, t + 13\}$.

18. Let V be the vector space of all real-valued continuous functions. Follow the directions of Exercise 15.

(a) $\{\cos t, \sin t, e^t\}$. (b) $\{t, e^t, \sin t\}$.

(c) $\{t^2, t, e^t\}$. (d) $(\cos^2 t, \sin^2 t, \cos 2t\}$.

19. Consider the vector space R^3. Follow the directions of Exercise 15.

(a) $\left\{\begin{bmatrix}1\\0\\0\end{bmatrix}, \begin{bmatrix}0\\1\\1\end{bmatrix}, \begin{bmatrix}1\\2\\-1\end{bmatrix}\right\}$.

(b) $\left\{\begin{bmatrix}1\\1\\-1\end{bmatrix}, \begin{bmatrix}0\\1\\1\end{bmatrix}, \begin{bmatrix}1\\1\\1\end{bmatrix}, \begin{bmatrix}1\\2\\-2\end{bmatrix}\right\}$.

(c) $\left\{\begin{bmatrix}1\\0\\0\end{bmatrix}, \begin{bmatrix}2\\1\\1\end{bmatrix}, \begin{bmatrix}-1\\2\\1\end{bmatrix}\right\}$.

20. Let $\mathbf{u}$ and $\mathbf{v}$ be nonzero vectors in a vector space V. Show that $\{\mathbf{u}, \mathbf{v}\}$ is linearly dependent if and only if there is a scalar k such that $\mathbf{v} = k\mathbf{u}$. Equivalently, $\{\mathbf{u}, \mathbf{v}\}$ is linearly independent if and only if one of the vectors is not a multiple of the other.

21. Let $S = \{\mathbf{v}_1, \mathbf{v}_2, \ldots, \mathbf{v}_k\}$ be a set of vectors in a vector space V. Prove that S is linearly dependent if and only if one of the vectors in S is a linear combination of all the other vectors in S.

22. Suppose that $S = \{\mathbf{v}_1, \mathbf{v}_2, \mathbf{v}_3\}$ is a linearly independent set of vectors in a vector space V. Prove that $T = \{\mathbf{w}_1, \mathbf{w}_2, \mathbf{w}_3\}$ is also linearly independent, where $\mathbf{w}_1 = \mathbf{v}_1 + \mathbf{v}_2 + \mathbf{v}_3$, $\mathbf{w}_2 = \mathbf{v}_2 + \mathbf{v}_3$, and $\mathbf{w}_3 = \mathbf{v}_3$.

23. Let A be an $m \times n$ matrix in reduced row echelon form. Prove that the nonzero rows of A, viewed as vectors in R_n, form a linearly independent set of vectors.

24. For what values of c is the set of vectors $\{t + 3, 2t + c^2 + 2\}$ in P_1 linearly independent?

25. Let $S = \{\mathbf{u}_1, \mathbf{u}_2, \ldots, \mathbf{u}_k\}$ be a set of vectors in a vector space, and let $T = \{\mathbf{v}_1, \mathbf{v}_2, \ldots, \mathbf{v}_m\}$ where each $\mathbf{v}_i$, $i = 1, 2, \ldots, m$, is a linear combination of the vectors in S. Prove that

$$\mathbf{w} = b_1\mathbf{v}_1 + b_2\mathbf{v}_2 + \cdots + b_m\mathbf{v}_m$$

is a linear combination of the vectors in S.

26. Let S_1 and S_2 be finite subsets of a vector space and let S_1 be a subset of S_2. If S_2 is linearly dependent, why or why not is S_1 linearly dependent? Give an example.

27. Let S_1 and S_2 be finite subsets of a vector space and let S_1 be a subset of S_2. If S_1 is linearly independent, why or why not is S_2 linearly independent? Give an example.

28. A linear combination of vectors $\mathbf{v}_1, \mathbf{v}_2, \ldots, \mathbf{v}_k$ in R^n with coefficients $a_1, \ldots, a_k$ is given algebraically, as in Definition 2.6, by

$$\mathbf{v} = a_1\mathbf{v}_1 + a_2\mathbf{v}_2 + \cdots + a_k\mathbf{v}_k.$$

In software we can compute such a linear combination of columns by a matrix multiplication $\mathbf{v} = A\mathbf{c}$, where

$$A = \begin{bmatrix} \mathbf{v}_1 & \mathbf{v}_2 & \cdots & \mathbf{v}_k \end{bmatrix} \quad \text{and} \quad \mathbf{c} = \begin{bmatrix} a_1 \\ a_2 \\ \vdots \\ a_k \end{bmatrix}.$$

That is, matrix A has $\text{col}_j(A) = \mathbf{v}_j$ for $j = 1, 2, \ldots, k$. Experiment in your software with such linear combinations.

(a) Using $\mathbf{v}_1, \mathbf{v}_2, \mathbf{v}_3$ from Example 1, compute

$$5\mathbf{v}_1 - 2\mathbf{v}_2 + 4\mathbf{v}_3.$$

(b) Using $\mathbf{v}_j$, $j = 1, \ldots, 4$ in Example 2, compute

$$3\mathbf{v}_1 - \mathbf{v}_2 + 4\mathbf{v}_3 + 2\mathbf{v}_4.$$

(See also Exercise 26 in Section 1.2.)

29. In Exercise 28, suppose that the vectors were in R_n. Devise a procedure for linear combinations of vectors in R_n that uses matrix multiplication.

30. Determine if your software has a command for finding the null space (see Example 16 in Section 2.2) of a matrix A. If it does, use it on the matrix A in Example 5 and compare the command's output with the results in Example 5. To experiment further, use Exercises 11 and 12.

31. As noted after Example 1, to determine if a specific vector $\mathbf{v}$ belongs to a span S, we investigate the consistency of an appropriate nonhomogeneous linear system $A\mathbf{x} = \mathbf{b}$. In addition, to determine if a set of vectors is linearly independent or not, we investigate the null space of an appropriate homogeneous system $A\mathbf{x} = \mathbf{0}$. These investigations can be performed computationally using a command for reduced row echelon form if available. (See Exercise 27 in Section 1.5.) We summarize the use of a reduced row echelon form command in these cases as follows. Let RREF(C) represent the reduced row echelon form of matrix C.

(i) $\mathbf{v}$ belongs to span S provided that RREF $([A \mid \mathbf{b}])$ contains no row of the form $[0 \ \cdots \ 0 \mid *]$, where $*$ represents a nonzero number.

(ii) The set of vectors is linearly independent if RREF$([A \mid \mathbf{0}])$ contains only rows from an identity matrix and possibly rows of all zeros.

Experiment in your software with this approach using the data given in Examples 1, 2, 6, 7, 8, and 10.

32. **Warning:** The strategy given in Exercise 31 assumes the computations are performed using exact arithmetic. Most software uses a model of exact arithmetic called floating point arithmetic, hence the use of reduced row echelon form may yield incorrect results in these cases. Computationally, the ''line between'' linear independence and linear dependence may be blurred. Experiment in your software with the use of reduced row echelon form for each of the following sets of vectors in R^2. Are they linearly independent or linearly dependent? Compare the theoretical answer with the computational answer from your software.

(a) $\begin{bmatrix} 1 \\ 0 \end{bmatrix}, \begin{bmatrix} 1 \\ 1 \times 10^{-5} \end{bmatrix}.$

(b) $\begin{bmatrix} 1 \\ 0 \end{bmatrix}, \begin{bmatrix} 1 \\ 1 \times 10^{-10} \end{bmatrix}.$

(c) $\begin{bmatrix} 1 \\ 0 \end{bmatrix}, \begin{bmatrix} 1 \\ 1 \times 10^{-16} \end{bmatrix}.$

2.4. *Basis and Dimension*

In this section we continue our study of the structure of a vector space V by determining a set of vectors in V that completely describes V.

Basis

Definition 2.9. A set of vectors $S = \{\mathbf{v}_1, \mathbf{v}_2, \ldots, \mathbf{v}_k\}$ in a vector space V is called a **basis** for V if S spans V and S is linearly independent. ■

Example 1. Let $V = R^3$ and $S = \left\{ \begin{bmatrix} 1 \\ 0 \\ 0 \end{bmatrix}, \begin{bmatrix} 0 \\ 1 \\ 0 \end{bmatrix}, \begin{bmatrix} 0 \\ 0 \\ 1 \end{bmatrix} \right\}$. Then S is a basis for R^3, called the **natural basis** for R^3. One can easily see how to generalize this to obtain the natural basis for R^n. Similarly,

$$S = \{[1 \quad 0 \quad 0], [0 \quad 1 \quad 0], [0 \quad 0 \quad 1]\}$$

is the natural basis for R_3. ■

The natural basis for R^n is denoted by $\mathbf{e}_1, \mathbf{e}_2, \ldots, \mathbf{e}_n$, where

$$\mathbf{e}_i = \begin{bmatrix} 0 \\ 0 \\ \vdots \\ 1 \\ 0 \\ \vdots \\ 0 \end{bmatrix} \leftarrow i\text{th row};$$

that is, $\mathbf{e}_i$ is an $n \times 1$ matrix with a 1 in the ith row and zeros elsewhere.

The natural basis for R^3 is also often denoted by

$$\mathbf{i} = \begin{bmatrix} 1 \\ 0 \\ 0 \end{bmatrix}, \quad \mathbf{j} = \begin{bmatrix} 0 \\ 1 \\ 0 \end{bmatrix}, \quad \text{and} \quad \mathbf{k} = \begin{bmatrix} 0 \\ 0 \\ 1 \end{bmatrix}.$$

These vectors are shown in Figure 2.20. Thus any vector $\mathbf{v} = \begin{bmatrix} a_1 \\ a_2 \\ a_3 \end{bmatrix}$ in R^3 can be written as

$$\mathbf{v} = a_1\mathbf{i} + a_2\mathbf{j} + a_3\mathbf{k}.$$

Figure 2.20 The natural basis for R^3.

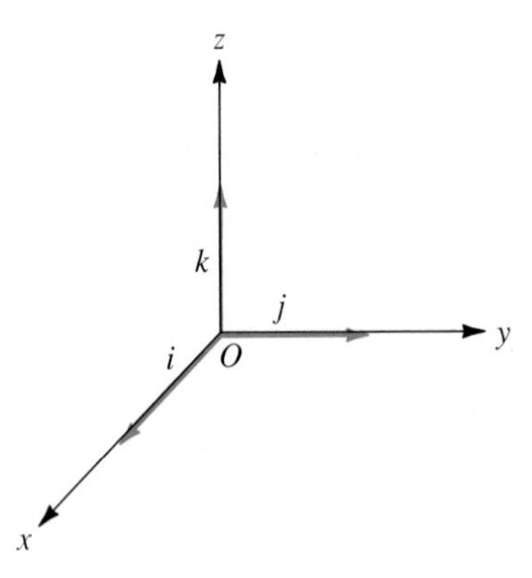

Example 2. Show that the set $S = \{t^2 + 1, t - 1, 2t + 2\}$ is a basis for the vector space P_2.

Solution: To show this, we must show that S spans V and is linearly independent. To show that it spans V, we take any vector in V, that is, a polynomial

$at^2 + bt + c$, and must find constants a_1, a_2, and a_3 such that

$$\begin{aligned} at^2 + bt + c &= a_1(t^2 + 1) + a_2(t - 1) + a_3(2t + 2) \\ &= a_1t^2 + (a_2 + 2a_3)t + (a_1 - a_2 + 2a_3). \end{aligned}$$

Since two polynomials agree for all values of t only if the coefficients of respective powers of t agree, we get the linear system

$$\begin{aligned} a_1 \qquad\qquad &= a \\ a_2 + 2a_3 &= b \\ a_1 - a_2 + 2a_3 &= c. \end{aligned}$$

Solving, we have

$$a_1 = a, \qquad a_2 = \frac{a + b - c}{2}, \qquad a_3 = \frac{c + b - a}{4}.$$

Hence S spans V.

To illustrate this result, suppose that we are given the vector $2t^2 + 6t + 13$. Here $a = 2$, $b = 6$, and $c = 13$. Substituting in the above expressions for a_1, a_2, and a_3, we find that

$$a_1 = 2, \qquad a_2 = -\tfrac{5}{2}, \qquad a_3 = \tfrac{17}{4}.$$

Hence

$$2t^2 + 6t + 13 = 2(t^2 + 1) - \tfrac{5}{2}(t - 1) + \tfrac{17}{4}(2t + 2).$$

To show that S is linearly independent, we form

$$a_1(t^2 + 1) + a_2(t - 1) + a_3(2t + 2) = 0.$$

Then

$$a_1t^2 + (a_2 + 2a_3)t + (a_1 - a_2 + 2a_3) = 0.$$

Again, this can hold for all values of t only if

$$\begin{aligned} a_1 \qquad\qquad &= 0 \\ a_2 + 2a_3 &= 0 \\ a_1 - a_2 + 2a_3 &= 0. \end{aligned}$$

The only solution to this homogeneous system is $a_1 = a_2 = a_3 = 0$, which implies that S is linearly independent. Thus S is a basis for P_2. ■

Example 3. Show that the set $S = \{\mathbf{v}_1, \mathbf{v}_2, \mathbf{v}_3, \mathbf{v}_4\}$, where $\mathbf{v}_1 = [1 \quad 0 \quad 1 \quad 0]$, $\mathbf{v}_2 = [0 \quad 1 \quad -1 \quad 2]$, $\mathbf{v}_3 = [0 \quad 2 \quad 2 \quad 1]$, and $\mathbf{v}_4 = [1 \quad 0 \quad 0 \quad 1]$, is a basis for R_4.

Solution: To show that S is linearly independent, we form the equation

$$a_1\mathbf{v}_1 + a_2\mathbf{v}_2 + a_3\mathbf{v}_3 + a_4\mathbf{v}_4 = \mathbf{0}$$

and solve for a_1, a_2, a_3, and a_4. Substituting for $\mathbf{v}_1$, $\mathbf{v}_2$, $\mathbf{v}_3$, and $\mathbf{v}_4$, we obtain the linear system

$$\begin{aligned} a_1 \qquad\qquad\qquad + a_4 &= 0 \\ a_2 + 2a_3 \qquad &= 0 \\ a_1 - a_2 + 2a_3 \qquad &= 0 \\ 2a_2 + a_3 + a_4 &= 0, \end{aligned}$$

which has as its only solution $a_1 = a_2 = a_3 = a_4 = 0$ (verify), showing that S is linearly independent.

To show that S spans R_4, we let $\mathbf{v} = [a \quad b \quad c \quad d]$ be any vector in R_4. We now seek constants a_1, a_2, a_3, and a_4 such that

$$a_1\mathbf{v}_1 + a_2\mathbf{v}_2 + a_3\mathbf{v}_3 + a_4\mathbf{v}_4 = \mathbf{v}.$$

Substituting for $\mathbf{v}_1$, $\mathbf{v}_2$, $\mathbf{v}_3$, $\mathbf{v}_4$, and $\mathbf{v}$, we find a solution for a_1, a_2, a_3, and a_4 (verify) to the resulting linear system. Hence S spans R_4 and is a basis for R_4. ■

We now establish some results that will tell about the number of vectors in a basis, compare two different bases, and give properties of bases. First, we observe that if $\{\mathbf{v}_1, \mathbf{v}_2, \ldots, \mathbf{v}_k\}$ is a basis for a vector space V, then $\{c\mathbf{v}_1, \mathbf{v}_2, \ldots, \mathbf{v}_k\}$ is also a basis when $c \neq 0$ (Exercise 29). Thus a real vector space always has infinitely many bases.

Theorem 2.6. *If $S = \{\mathbf{v}_1, \mathbf{v}_2, \ldots, \mathbf{v}_n\}$ is a basis for a vector space V, then every vector in V can be written in one and only one way as a linear combination of the vectors in S.*

Proof: First, every vector $\mathbf{v}$ in V can be written as a linear combination of the vectors in S because S spans V. Now let

$$\mathbf{v} = a_1\mathbf{v}_1 + a_2\mathbf{v}_2 + \cdots + a_n\mathbf{v}_n$$

and

$$\mathbf{v} = b_1\mathbf{v}_1 + b_2\mathbf{v}_2 + \cdots + b_n\mathbf{v}_n.$$

We must show that $a_i = b_i$ for $i = 1, 2, \ldots, n$. We have

$$\mathbf{0} = \mathbf{v} - \mathbf{v} = (a_1 - b_1)\mathbf{v}_1 + (a_2 - b_2)\mathbf{v}_2 + \cdots + (a_n - b_n)\mathbf{v}_n.$$

Since S is linearly independent, we conclude that

$$a_i - b_i = 0 \qquad \text{for } i = 1, 2, \ldots, n. \qquad \blacksquare$$

We can also prove (Exercise 38) that if $S = \{\mathbf{v}_1, \mathbf{v}_2, \ldots, \mathbf{v}_n\}$ is a set of nonzero vectors in a vector space V such that every vector in V can be written in one and only one way as a linear combination of the vectors in S, then S is a basis for V.

Even though a nonzero vector space contains an infinite number of elements, a vector space with a finite basis is in a sense completely described by a finite number of vectors, namely, by those vectors in the basis.

Theorem 2.7. *If $S = \{\mathbf{v}_1, \mathbf{v}_2, \ldots, \mathbf{v}_n\}$ is a set of vectors spanning a vector space V, then S contains a basis T for V.*

Proof: If S itself is linearly independent, then S is a basis for V. If S is linearly dependent, then some $\mathbf{v}_j$ is a linear combination of the preceding vectors in S (Theorem 2.5). We now delete $\mathbf{v}_j$ from S, getting a subset S_1 of S. Then, by the remarks made at the end of Section 2.3, we conclude that $S_1 = \{\mathbf{v}_1, \mathbf{v}_2, \ldots, \mathbf{v}_{j-1}, \mathbf{v}_{j+1}, \ldots, \mathbf{v}_n\}$ also spans V.

If S_1 is linearly independent, then S_1 is a basis. If S_1 is linearly dependent, delete a vector of S_1 that is a linear combination of the preceding vectors of S_1 and get a new set S_2 which spans V. Continuing, since S is a finite set, we find a subset T of S that is linearly independent and spans V. The set T is a basis for V.

Alternate Constructive Proof: If S is linearly independent, then since S already spans V, we conclude that S is a basis for V. To see if S is linearly dependent or is linearly independent, we solve

$$a_1\mathbf{v}_1 + a_2\mathbf{v}_2 + \cdots + a_n\mathbf{v}_n = \mathbf{0} \tag{1}$$

for $a_1, a_2, \ldots, a_n$. Equation (1) leads to a homogeneous system in the n unknowns $a_1, a_2, \ldots, a_n$, which is solved by first transforming the augmented matrix of the system to reduced row echelon form. Suppose that S is linearly dependent so that the system has a nontrivial solution. This means that the unknowns corresponding to the leading 1's in the reduced row echelon form matrix can be solved for in terms of other unknowns. The latter unknowns can be assigned arbitrary real values. Without loss of generality, we may assume that $a_1, a_2, \ldots, a_k$ are the unknowns corresponding to the leading 1's. Thus $a_1, a_2, \ldots, a_k$ can be expressed in terms of $a_{k+1}, a_{k+2}, \ldots, a_n$; for example,

$$a_1 = b_1a_{k+1} + b_2a_{k+2} + \cdots + b_{n-k}a_n. \tag{2}$$

Letting $a_{k+1} = 1, a_{k+2} = 0, \ldots, a_n = 0$ in Equation (1), we see that $\mathbf{v}_{k+1}$ is a linear combination of $\mathbf{v}_1, \mathbf{v}_2, \ldots, \mathbf{v}_k$. By the remark at the end of Section 2.3, the set of vectors obtained from S by deleting $\mathbf{v}_{k+1}$ spans V. Similarly, letting $a_{k+1} = 0, a_{k+2} = 1, a_{k+3} = 0, \ldots, a_n = 0$, we find that $\mathbf{v}_{k+2}$ is a linear combination of $\mathbf{v}_1, \mathbf{v}_2, \ldots, \mathbf{v}_k$ and the set of vectors obtained from S by deleting $\mathbf{v}_{k+1}$ and $\mathbf{v}_{k+2}$ spans V. Continuing in this manner, $\mathbf{v}_{k+3}, \mathbf{v}_{k+4}, \ldots, \mathbf{v}_n$ are linear combinations of $\mathbf{v}_1, \mathbf{v}_2, \ldots, \mathbf{v}_k$, so it follows that $\{\mathbf{v}_1, \mathbf{v}_2, \ldots, \mathbf{v}_k\}$ spans V.

We next show that $\{\mathbf{v}_1, \mathbf{v}_2, \ldots, \mathbf{v}_k\}$ is linearly independent. Consider the equation

$$a_1\mathbf{v}_1 + a_2\mathbf{v}_2 + \cdots + a_k\mathbf{v}_k = \mathbf{0}. \tag{3}$$

If we can find values for $a_1, a_2, \ldots, a_k$ satisfying Equation (3), then these values will also satisfy the following equation:

$$a_1\mathbf{v}_1 + a_2\mathbf{v}_2 + \cdots + a_k\mathbf{v}_k + 0\mathbf{v}_{k+1} + 0\mathbf{v}_{k+2} + \cdots + 0\mathbf{v}_n = \mathbf{0}. \tag{4}$$

On the other hand, observe that every solution to Equation (4) is a solution to Equation (1) in which the $n - k$ arbitrary constants $a_{k+1}, a_{k+2}, \ldots, a_n$ have been set equal to zero. Since each of the constants $a_1, a_2, \ldots, a_k$ can be written in terms of $a_{k+1}, a_{k+2}, \ldots, a_n$ [see Equation (2)] and these have been chosen to be zero to satisfy Equation (1), it follows from Equation (2) that $a_1 = 0, a_2 = 0, \ldots, a_k = 0$. Thus the only choice of constants that satisfies Equation (3) is $a_1, a_2, \ldots, a_k$ all zero; hence $\{\mathbf{v}_1, \mathbf{v}_2, \ldots, \mathbf{v}_k\}$ is linearly independent and is therefore a basis for V. ■

The alternate proof of Theorem 2.7 leads to a simple procedure for finding a subset of a spanning set that is a basis. Let $S = \{\mathbf{v}_1, \mathbf{v}_2, \ldots, \mathbf{v}_n\}$. To find a subset of S that is a basis for $V = \text{span } S$, proceed as follows:

STEP 1. Form Equation (1):

$$a_1\mathbf{v}_1 + a_2\mathbf{v}_2 + \cdots + a_n\mathbf{v}_n = \mathbf{0}.$$

STEP 2. Construct the augmented matrix associated with the homogeneous system of Equation (1), and transform it to reduced row echelon form.

STEP 3. The vectors corresponding to the columns containing the leading 1's form a basis for $V = \text{span } S$. Thus, if $S = \{\mathbf{v}_1, \mathbf{v}_2, \ldots, \mathbf{v}_6\}$ and the leading 1's occur in columns 1, 3, and 4, then $\{\mathbf{v}_1, \mathbf{v}_3, \mathbf{v}_4\}$ is a basis for span S.

REMARK. In step 2 of the procedure above, it is sufficient to transform the augmented matrix to row echelon form.

Example 4. Let $V = R_3$ and $S = \{\mathbf{v}_1, \mathbf{v}_2, \mathbf{v}_3, \mathbf{v}_4, \mathbf{v}_5\}$, where $\mathbf{v}_1 = [1 \quad 0 \quad 1]$, $\mathbf{v}_2 = [0 \quad 1 \quad 1]$, $\mathbf{v}_3 = [1 \quad 1 \quad 2]$, $\mathbf{v}_4 = [1 \quad 2 \quad 1]$, and $\mathbf{v}_5 = [-1 \quad 1 \quad -2]$. We find that S spans R_3 (verify) and we now wish to find a subset of S that is a basis for R_3. Using the procedure just developed, we proceed as follows.

STEP 1. $a_1[1 \quad 0 \quad 1] + a_2[0 \quad 1 \quad 1] + a_3[1 \quad 1 \quad 2] + a_4[1 \quad 2 \quad 1] + a_5[-1 \quad 1 \quad -2] = [0 \quad 0 \quad 0]$.

STEP 2. Equating corresponding components, we obtain the homogeneous system

$$\begin{aligned} a_1 \phantom{{}+a_2} + a_3 + a_4 - a_5 &= 0 \\ a_2 + a_3 + 2a_4 + a_5 &= 0 \\ a_1 + a_2 + 2a_3 + a_4 - 2a_5 &= 0. \end{aligned}$$

The reduced row echelon form of the associated augmented matrix is (verify)

$$\left[\begin{array}{ccccc|c} 1 & 0 & 1 & 0 & -2 & 0 \\ 0 & 1 & 1 & 0 & -1 & 0 \\ 0 & 0 & 0 & 1 & 1 & 0 \end{array}\right].$$

STEP 3. The leading 1's appear in columns 1, 2, and 4, so $\{\mathbf{v}_1, \mathbf{v}_2, \mathbf{v}_4\}$ is a basis for R_3. ■

REMARK. In the alternate proof of Theorem 2.7, the order of the vectors in the original spanning set S determines which basis for V is obtained. If, for example, we consider Example 4, where $S = \{\mathbf{w}_1, \mathbf{w}_2, \mathbf{w}_3, \mathbf{w}_4, \mathbf{w}_5\}$ with $\mathbf{w}_1 = \mathbf{v}_5$, $\mathbf{w}_2 = \mathbf{v}_4$, $\mathbf{w}_3 = \mathbf{v}_3$, $\mathbf{w}_4 = \mathbf{v}_2$, and $\mathbf{w}_5 = \mathbf{v}_1$, then the reduced row echelon form of the augmented matrix is (verify)

$$\left[\begin{array}{ccccc|c} 1 & 0 & 0 & 1 & -1 & 0 \\ 0 & 1 & 0 & -1 & 1 & 0 \\ 0 & 0 & 1 & 2 & -1 & 0 \end{array}\right].$$

It now follows that $\{\mathbf{w}_1, \mathbf{w}_2, \mathbf{w}_3\} = \{\mathbf{v}_5, \mathbf{v}_4, \mathbf{v}_3\}$ is a basis for R_3.

We are now about to establish a major result (Corollary 2.1) of this section, which will tell us about the number of vectors in two different bases.

Theorem 2.8. *If $S = \{\mathbf{v}_1, \mathbf{v}_2, \ldots, \mathbf{v}_n\}$ is a basis for a vector space V and $T = \{\mathbf{w}_1, \mathbf{w}_2, \ldots, \mathbf{w}_r\}$ is a linearly independent set of vectors in V, then $r \leq n$.*

Proof: Let $T_1 = \{\mathbf{w}_1, \mathbf{v}_1, \ldots, \mathbf{v}_n\}$. Since S spans V, so does T_1. Since $\mathbf{w}_1$ is a linear combination of the vectors in S, we find that T_1 is linearly dependent. Then, by Theorem 2.5, some $\mathbf{v}_j$ is a linear combination of the preceding vectors in T_1. Delete that particular vector $\mathbf{v}_j$.

Let $S_1 = \{\mathbf{w}_1, \mathbf{v}_1, \ldots, \mathbf{v}_{j-1}, \mathbf{v}_{j+1}, \ldots, \mathbf{v}_n\}$. Note that S_1 spans V. Next, let $T_2 = \{\mathbf{w}_2, \mathbf{w}_1, \mathbf{v}_1, \ldots, \mathbf{v}_{j-1}, \mathbf{v}_{j+1}, \ldots, \mathbf{v}_n\}$. Then T_2 is linearly dependent and some vector in T_2 is a linear combination of the preceding vectors in T_2. Since T is linearly independent, this vector cannot be $\mathbf{w}_1$, so it is $\mathbf{v}_i$, $i \neq j$. Repeat this process over and over. Each time there is a new $\mathbf{w}$ vector available from the set T, it is possible to discard one of the $\mathbf{v}$ vectors from the set S. Thus the number r of $\mathbf{w}$ vectors must be no greater than the number n of $\mathbf{v}$ vectors. That is, $r \leq n$. ■

Corollary 2.1. *If $S = \{\mathbf{v}_1, \mathbf{v}_2, \ldots, \mathbf{v}_n\}$ and $T = \{\mathbf{w}_1, \mathbf{w}_2, \ldots, \mathbf{w}_m\}$ are bases for a vector space V, then $n = m$.*

Proof: Since S is a basis and T is linearly independent, Theorem 2.8 implies that $m \leq n$. Similarly, we obtain $n \leq m$ because T is a basis and S is linearly independent. Hence $n = m$. ■

Dimension

Although a vector space has many bases, all bases have the same number of vectors. We can then make the following definition.

Definition 2.10. The **dimension** of a nonzero vector space V is the number of vectors in a basis for V. We often write **dim** V for the dimension of V. We also define the dimension of the trivial vector space $\{\mathbf{0}\}$ to be zero. ■

Example 5. The set $S = \{t^2, t, 1\}$ is a basis for P_2, so dim $P_2 = 3$. ■

Example 6. Let V be the subspace of R_3 spanned by $S = \{\mathbf{v}_1, \mathbf{v}_2, \mathbf{v}_3\}$, where $\mathbf{v}_1 = [0 \quad 1 \quad 1]$, $\mathbf{v}_2 = [1 \quad 0 \quad 1]$, and $\mathbf{v}_3 = [1 \quad 1 \quad 2]$. Thus every vector in V is of the form

$$a_1\mathbf{v}_1 + a_2\mathbf{v}_2 + a_3\mathbf{v}_3,$$

where a_1, a_2, and a_3 are arbitrary real numbers. We find that S is linearly dependent, and $\mathbf{v}_3 = \mathbf{v}_1 + \mathbf{v}_2$ (verify). Thus $S_1 = \{\mathbf{v}_1, \mathbf{v}_2\}$ also spans V. Since S_1 is linearly independent (verify), we conclude that it is a basis for V. Hence dim $V = 2$. ■

Corollary 2.2. *If vector space V has dimension n, then a largest linearly independent subset of vectors in V contains n vectors and is a basis for V.*

Proof: Let $S = \{\mathbf{v}_1, \mathbf{v}_2, \ldots, \mathbf{v}_k\}$ be a largest linearly independent subset of V. If span $S \neq V$, then there exists a vector $\mathbf{v}$ in V that cannot be written as a linear combination of $\mathbf{v}_1, \mathbf{v}_2, \ldots, \mathbf{v}_k$. It follows by Theorem 2.5 that $\{\mathbf{v}_1, \mathbf{v}_2, \ldots, \mathbf{v}_k, \mathbf{v}\}$ is a linearly independent set of vectors. However, this contradicts the assumption that S is a largest linearly independent subset of V. Hence span $S = V$, which implies that set S is a basis for V and $k = n$ by Corollary 2.1. ■

Corollary 2.3. *If vector space V has dimension n, then a smallest set of vectors that spans V contains n vectors and is a basis for V.*

Proof: Exercise 32. ■

Although Corollaries 2.2 and 2.3 are theoretically of considerable importance, they can be computationally awkward.

From the results above, we can make the following observations. If V has dimension n, then any set of $n + 1$ vectors in V is necessarily linearly dependent; also any set of $n - 1$ vectors in V cannot span V. More generally, we can establish the following results.

Corollary 2.4. *If vector space V has dimension n, then any subset of $m > n$ vectors must be linearly dependent.*

Proof: Exercise 33. ■

Corollary 2.5. *If vector space V has dimension n, then any subset of $m < n$ vectors cannot span V.*

Proof: Exercise 34. ■

It is easy to see that the set $\{\mathbf{0}\}$ is linearly dependent. This is why in Definition 2.10 we defined the dimension of the trivial vector space $\{\mathbf{0}\}$ to be zero.

Thus R^3 has dimension 3, R_2 has dimension 2, and R^n and R_n both have dimension n. Similarly, P_3 has dimension 4 because $\{t^3, t^2, t, 1\}$ is a basis for P_3. In general, P_n has dimension $n + 1$. A vector space that has a basis consisting of a finite number of vectors is called a **finite-dimensional vector space**. All vector spaces considered henceforth in this book are finite dimensional. However, we point out that there are vector spaces with bases consisting of infinitely many vectors (called **infinite-dimensional vector spaces**) which are extremely important in mathematics and physics; their study lies beyond the scope of this book. The vector space P of all polynomials is an infinite-dimensional vector space (Exercise 30).

In Section 2.2, it was an exercise to show that the subspaces of R^2 are $\{\mathbf{0}\}$, R^2 itself, and any line passing through the origin. We can now establish this result using the material developed in this section. First, we have $\{\mathbf{0}\}$ and R^2, the trivial subspaces of dimensions 0 and 2, respectively. The subspace V of R^2 spanned by a vector $\mathbf{v} \neq \mathbf{0}$ is a one-dimensional subspace of R^2; V is a line through the origin. Thus the subspaces of R^2 are $\{\mathbf{0}\}$, R^2, and all the lines through the origin. In a similar way, Exercise 37 asks you to show that the subspaces of R^3 are $\{\mathbf{0}\}$, R^3 itself, and all lines and planes passing through the origin. We now prove a theorem that we shall have occasion to use several times in constructing a basis containing a given set of linearly independent vectors.

Theorem 2.9. *If S is a linearly independent set of vectors in a finite-dimensional vector space V, then there is a basis T for V, which contains S.*

Proof: Let $S = \{\mathbf{v}_1, \mathbf{v}_2, \ldots, \mathbf{v}_m\}$ be a linearly independent set of vectors in the n-dimensional vector space V, where $m < n$. Now let $\{\mathbf{w}_1, \mathbf{w}_2, \ldots, \mathbf{w}_n\}$ be a basis for V and let $S_1 = \{\mathbf{v}_1, \mathbf{v}_2, \ldots, \mathbf{v}_m, \mathbf{w}_1, \mathbf{w}_2, \ldots, \mathbf{w}_n\}$. Since S_1 spans V, it contains, by Theorem 2.7, a basis T for V. Recall that T is obtained by deleting from S_1 every vector that is a linear combination of the preceding vectors. Since S is linearly independent, none of the $\mathbf{v}_i$ can be linear combinations of other $\mathbf{v}_j$ and thus are not deleted. Hence T will contain S. ■

Example 7. To find a basis for R_3 that contains the vector $\mathbf{v} = [1 \quad 0 \quad 1]$, we use Theorem 2.9 as follows. First, let $\{\mathbf{e}_1', \mathbf{e}_2', \mathbf{e}_3'\}$ be the natural basis for R_3, where

$$\mathbf{e}_1' = [1 \quad 0 \quad 0],\ \mathbf{e}_2' = [0 \quad 1 \quad 0],\ \text{and } \mathbf{e}_3' = [0 \quad 0 \quad 1].$$

Let

$$S_1 = \{[1 \quad 0 \quad 1], [1 \quad 0 \quad 0], [0 \quad 1 \quad 0], [0 \quad 0 \quad 1]\}.$$

Since $\mathbf{e}_1'$ is not a linear combination of $\mathbf{v}$, we retain $\mathbf{e}_1'$. Now we check whether $\mathbf{e}_2' = [0 \quad 1 \quad 0]$ is a linear combination of $\mathbf{v}$ and $\mathbf{e}_1'$. Since the answer is no, we retain $\mathbf{e}_2'$. Since $\mathbf{e}_3'$ is a linear combination of $\mathbf{v}$, $\mathbf{e}_1'$, and $\mathbf{e}_2'$ (verify), we delete it. Thus $\{\mathbf{v}, \mathbf{e}_1', \mathbf{e}_2'\}$ is a basis for R_3. ■

It can be shown (Exercise 35) that if W is a subspace of a finite-dimensional vector space V, then W is finite-dimensional and $\dim W \leq \dim V$.

As defined earlier, a given set S of vectors in a vector space V is a basis for V if it spans V and is linearly independent. However, if we are given the *additional* information that the dimension of V is n, we need only verify one of the two conditions. This is the content of the following theorem.

Theorem 2.10. *Let V be an n-dimensional vector space.*

(a) *If $S = \{\mathbf{v}_1, \mathbf{v}_2, \ldots, \mathbf{v}_n\}$ is a linearly independent set of vectors in V, then S is a basis for V.*

(b) *If $S = \{\mathbf{v}_1, \mathbf{v}_2, \ldots, \mathbf{v}_n\}$ spans V, then S is a basis for V.*

Proof: Exercise. ■

As a particular application of Theorem 2.10, we have the following. To determine if a subset S of R^n (R_n) is a basis for R^n (R_n), first count the number of elements in S. If S has n elements, we can use either part (a) or part (b) of Theorem 2.10 to determine whether S is or is not a basis. If S does not have n elements, it is not a basis for R^n (R_n). (Why?) The same line of reasoning applies to any vector space or subspace whose *dimension is known*.

Example 8. In Example 4, since dim $R_3 = 3$ and the set S contains five vectors, we conclude by Theorem 2.10 that S is not a basis for R_3. In Example 3, since dim $R_4 = 4$ and the set S contains four vectors, it is possible for S to be a basis for R_4. If S is linearly independent *or* spans R_4, it is a basis; otherwise, it is not a basis. Thus we need check only one of the conditions in Theorem 2.10, not both. ■

Definition 2.11. Let S be a set of vectors in a vector space V. A subset T of S is called a **maximal independent subset** of S if T is a linearly independent set of vectors, and if there is no linearly independent subset of S having more vectors than T does. ■

Example 9. Let V be R^3 and consider the set $S = \{\mathbf{v}_1, \mathbf{v}_2, \mathbf{v}_3, \mathbf{v}_4\}$, where

$$\mathbf{v}_1 = \begin{bmatrix} 1 \\ 0 \\ 0 \end{bmatrix}, \quad \mathbf{v}_2 = \begin{bmatrix} 0 \\ 1 \\ 0 \end{bmatrix}, \quad \mathbf{v}_3 = \begin{bmatrix} 0 \\ 0 \\ 1 \end{bmatrix}, \quad \text{and} \quad \mathbf{v}_4 = \begin{bmatrix} 1 \\ 1 \\ 1 \end{bmatrix}.$$

Maximal independent subsets of S are

$$\{\mathbf{v}_1, \mathbf{v}_2, \mathbf{v}_3\}, \quad \{\mathbf{v}_1, \mathbf{v}_2, \mathbf{v}_4\}, \quad \{\mathbf{v}_1, \mathbf{v}_3, \mathbf{v}_4\}, \quad \text{and} \quad \{\mathbf{v}_2, \mathbf{v}_3, \mathbf{v}_4\}.$$ ■

Theorem 2.11. *Let S be a finite subset of the vector space V that spans V. A maximal independent subset T of S is a basis for V.*

Proof: Exercise. ■

We conclude this section with one more important example of the properties of the solution space of the homogeneous system $A\mathbf{x} = \mathbf{0}$. This example will be used in an essential manner on a number of occasions throughout the book, especially in Chapter 6.

Example 10. Find a basis for the solution space V of the homogeneous system

$$\begin{bmatrix} 1 & 2 & 0 & 3 & 1 \\ 2 & 3 & 0 & 3 & 1 \\ 1 & 1 & 2 & 2 & 1 \\ 3 & 5 & 0 & 6 & 2 \\ 2 & 3 & 2 & 5 & 2 \end{bmatrix} \begin{bmatrix} x_1 \\ x_2 \\ x_3 \\ x_4 \\ x_5 \end{bmatrix} = \begin{bmatrix} 0 \\ 0 \\ 0 \\ 0 \\ 0 \end{bmatrix}.$$

Solution: Gauss–Jordan reduction leads to the equivalent system (verify)

$$\begin{bmatrix} 1 & 0 & 0 & -3 & -1 \\ 0 & 1 & 0 & 3 & 1 \\ 0 & 0 & 1 & 1 & \frac{1}{2} \\ 0 & 0 & 0 & 0 & 0 \\ 0 & 0 & 0 & 0 & 0 \end{bmatrix} \begin{bmatrix} x_1 \\ x_2 \\ x_3 \\ x_4 \\ x_5 \end{bmatrix} = \begin{bmatrix} 0 \\ 0 \\ 0 \\ 0 \\ 0 \end{bmatrix},$$

whose general solution is

$$\mathbf{x} = \begin{bmatrix} 3s + t \\ -3s - t \\ -s - \frac{1}{2}t \\ s \\ t \end{bmatrix},$$

where s and t are any real numbers. We can write

$$\mathbf{x} = s \begin{bmatrix} 3 \\ -3 \\ -1 \\ 1 \\ 0 \end{bmatrix} + t \begin{bmatrix} 1 \\ -1 \\ -\frac{1}{2} \\ 0 \\ 1 \end{bmatrix}. \tag{5}$$

Since s and t can take on any values, letting them first be 1 and 0, and then 0 and 1, we get as solutions

$$\mathbf{x}_1 = \begin{bmatrix} 3 \\ -3 \\ -1 \\ 1 \\ 0 \end{bmatrix} \quad \text{and} \quad \mathbf{x}_2 = \begin{bmatrix} 1 \\ -1 \\ -\frac{1}{2} \\ 0 \\ 1 \end{bmatrix}.$$

From (5) it is clear that $\{\mathbf{x}_1, \mathbf{x}_2\}$ spans V, and since $\{\mathbf{x}_1, \mathbf{x}_2\}$ is linearly independent (verify), it is a basis for V. Thus the dimension of V is 2. ■

It can be shown that the dimension of the solution space of the homogeneous system $A\mathbf{x} = \mathbf{0}$ is equal to the number of arbitrary constants in the general solution. Of course, this dimension is the same as the dimension of the null space of A.

2.4. EXERCISES

1. Which of the following sets of vectors are bases for R^2?

(a) $\left\{\begin{bmatrix}1\\3\end{bmatrix}, \begin{bmatrix}1\\-1\end{bmatrix}\right\}$. (b) $\left\{\begin{bmatrix}0\\0\end{bmatrix}, \begin{bmatrix}1\\2\end{bmatrix}, \begin{bmatrix}2\\4\end{bmatrix}\right\}$.

(c) $\left\{\begin{bmatrix}1\\2\end{bmatrix}, \begin{bmatrix}2\\-3\end{bmatrix}, \begin{bmatrix}3\\2\end{bmatrix}\right\}$. (d) $\left\{\begin{bmatrix}1\\3\end{bmatrix}, \begin{bmatrix}-2\\6\end{bmatrix}\right\}$.

2. Which of the following sets of vectors are bases for R^3?

(a) $\left\{\begin{bmatrix}1\\2\\0\end{bmatrix}, \begin{bmatrix}0\\1\\-1\end{bmatrix}\right\}$.

(b) $\left\{\begin{bmatrix}1\\1\\-1\end{bmatrix}, \begin{bmatrix}2\\3\\4\end{bmatrix}, \begin{bmatrix}4\\1\\-1\end{bmatrix}, \begin{bmatrix}0\\1\\-1\end{bmatrix}\right\}$.

(c) $\left\{\begin{bmatrix}3\\2\\2\end{bmatrix}, \begin{bmatrix}-1\\2\\1\end{bmatrix}, \begin{bmatrix}0\\1\\0\end{bmatrix}\right\}$.

(d) $\left\{\begin{bmatrix}1\\0\\0\end{bmatrix}, \begin{bmatrix}0\\2\\-1\end{bmatrix}, \begin{bmatrix}3\\4\\1\end{bmatrix}, \begin{bmatrix}0\\1\\0\end{bmatrix}\right\}$.

3. Which of the following sets of vectors are bases for R_4?

(a) $\{[1\ \ 0\ \ 0\ \ 1], [0\ \ 1\ \ 0\ \ 0], [1\ \ 1\ \ 1\ \ 1], [0\ \ 1\ \ 1\ \ 1]\}$.

(b) $\{[1\ \ -1\ \ 0\ \ 2], [3\ \ -1\ \ 2\ \ 1], [1\ \ 0\ \ 0\ \ 1]\}$.

(c) $\{[-2\ \ 4\ \ 6\ \ 4], [0\ \ 1\ \ 2\ \ 0], [-1\ \ 2\ \ 3\ \ 2], [-3\ \ 2\ \ 5\ \ 6], [-2\ \ -1\ \ 0\ \ 4]\}$.

(d) $\{[0\ \ 0\ \ 1\ \ 1], [-1\ \ 1\ \ 1\ \ 2], [1\ \ 1\ \ 0\ \ 0], [2\ \ 1\ \ 2\ \ 1]\}$.

4. Which of the following sets of vectors are bases for P_2?

(a) $\{-t^2 + t + 2, 2t^2 + 2t + 3, 4t^2 - 1\}$.

(b) $\{t^2 + 2t - 1, 2t^2 + 3t - 2\}$.

(c) $\{t^2 + 1, 3t^2 + 2t, 3t^2 + 2t + 1, 6t^2 + 6t + 3\}$.

(d) $\{3t^2 + 2t + 1, t^2 + t + 1, t^2 + 1\}$.

5. Which of the following sets of vectors are bases for P_3?

(a) $\{t^3 + 2t^2 + 3t, 2t^3 + 1, 6t^3 + 8t^2 + 6t + 4, t^3 + 2t^2 + t + 1\}$.

(b) $\{t^3 + t^2 + 1, t^3 - 1, t^3 + t^2 + t\}$.

(c) $\{t^3 + t^2 + t + 1, t^3 + 2t^2 + t + 3, 2t^3 + t^2 + 3t + 2, t^3 + t^2 + 2t + 2\}$.

(d) $\{t^3 - t, t^3 + t^2 + 1, t - 1\}$.

6. Show that the set of matrices

$$\left\{\begin{bmatrix}1 & 1\\0 & 0\end{bmatrix}, \begin{bmatrix}0 & 0\\1 & 1\end{bmatrix}, \begin{bmatrix}1 & 0\\0 & 1\end{bmatrix}, \begin{bmatrix}0 & 1\\1 & 1\end{bmatrix}\right\}$$

forms a basis for the vector space ${}_2R_2$.

In Exercises 7 and 8, determine which of the given subsets forms a basis for R^3. Express the vector $\begin{bmatrix}2\\1\\3\end{bmatrix}$ as a linear combination of the vectors in each subset that is a basis.

7. (a) $\left\{\begin{bmatrix}1\\1\\1\end{bmatrix}, \begin{bmatrix}1\\2\\3\end{bmatrix}, \begin{bmatrix}0\\1\\0\end{bmatrix}\right\}$.

(b) $\left\{\begin{bmatrix}1\\2\\3\end{bmatrix}, \begin{bmatrix}2\\1\\3\end{bmatrix}, \begin{bmatrix}0\\0\\0\end{bmatrix}\right\}$.

8. (a) $\left\{\begin{bmatrix}2\\1\\3\end{bmatrix}, \begin{bmatrix}1\\2\\1\end{bmatrix}, \begin{bmatrix}1\\1\\4\end{bmatrix}, \begin{bmatrix}1\\5\\1\end{bmatrix}\right\}$.

(b) $\left\{\begin{bmatrix}1\\1\\2\end{bmatrix}, \begin{bmatrix}2\\2\\0\end{bmatrix}, \begin{bmatrix}3\\4\\-1\end{bmatrix}\right\}$.

In Exercises 9 and 10, determine which of the given subsets form a basis for P_2. Express $5t^2 - 3t + 8$ as a linear combination of the vectors in each subset that is a basis.

9. (a) $\{t^2 + t, t - 1, t + 1\}$.
(b) $\{t^2 + 1, t - 1\}$.

10. (a) $\{t^2 + t, t^2, t^2 + 1\}$.
(b) $\{t^2 + 1, t^2 - t + 1\}$.

11. Find a basis for the subspace W of R^3 spanned by

$$\left\{\begin{bmatrix}1\\2\\2\end{bmatrix}, \begin{bmatrix}3\\2\\1\end{bmatrix}, \begin{bmatrix}11\\10\\7\end{bmatrix}, \begin{bmatrix}7\\6\\4\end{bmatrix}\right\}.$$

What is the dimension of W?

12. Find a basis for the subspace W of R_4 spanned by the set of vectors

$$\{[1 \;\; 1 \;\; 0 \;\; -1], [0 \;\; 1 \;\; 2 \;\; 1], [1 \;\; 0 \;\; 1 \;\; -1], [1 \;\; 1 \;\; -6 \;\; -3], [-1 \;\; -5 \;\; 1 \;\; 0]\}.$$

What is dim W?

13. Let W be the subspace of P_3 spanned by

$$\{t^3 + t^2 - 2t + 1, t^2 + 1, t^3 - 2t, 2t^3 + 3t^2 - 4t + 3\}.$$

Find a basis for W. What is the dimension of W?

14. Let W be the subspace of the space of all continuous real-valued functions spanned by $\{\cos^2 t, \sin^2 t, \cos 2t\}$. Find a basis for W. What is the dimension of W?

In Exercises 15 and 16, find a basis for the given subspaces of R^3 and R^4.

15. (a) All vectors of the form $\begin{bmatrix}a\\b\\c\end{bmatrix}$, where $b = a + c$.

(b) All vectors of the form $\begin{bmatrix}a\\b\\c\end{bmatrix}$, where $b = a$.

16. (a) All vectors of the form $\begin{bmatrix}a\\b\\c\end{bmatrix}$, where $a = 0$.

(b) All vectors of the form $\begin{bmatrix}a+c\\a-b\\b+c\\-a+b\end{bmatrix}$.

In Exercises 17 and 18, find the dimensions of the given subspaces of R_4.

17. (a) All vectors of the form $[a \;\; b \;\; c \;\; d]$, where $d = a + b$.
(b) All vectors of the form $[a \;\; b \;\; c \;\; d]$, where $c = a - b$ and $d = a + b$.

18. (a) All vectors of the form $[a \;\; b \;\; c \;\; d]$, where $a = b$.
(b) All vectors of the form $[a + c \;\; a - b \;\; b + c \;\; -a + b]$.

19. Find the dimensions of the subspaces of R^2 spanned by the vectors in Exercise 1.

20. Find the dimensions of the subspaces of R^3 spanned by the vectors in Exercise 2.

21. Find the dimensions of the subspaces of R_4 spanned by the vectors in Exercise 3.

22. Find a basis for R^3 that includes

(a) The vector $\begin{bmatrix}1\\0\\2\end{bmatrix}$.

(b) The vectors $\begin{bmatrix}1\\0\\2\end{bmatrix}$ and $\begin{bmatrix}0\\1\\3\end{bmatrix}$.

23. Find a basis for P_3 that includes the vectors $t^3 + t$ and $t^2 - t$.

24. Find a basis for ${}_2R_3$. What is the dimension of ${}_2R_3$? Generalize to ${}_mR_n$.

25. Find a basis for the solution space V of the homogeneous system

$$\begin{bmatrix} 1 & 2 & 1 & 2 & 1 \\ 1 & 2 & 2 & 1 & 2 \\ 2 & 4 & 3 & 3 & 3 \\ 0 & 0 & 1 & -1 & -1 \end{bmatrix} \begin{bmatrix} x_1 \\ x_2 \\ x_3 \\ x_4 \\ x_5 \end{bmatrix} = \begin{bmatrix} 0 \\ 0 \\ 0 \\ 0 \end{bmatrix}.$$

What is the dimension of V?

26. Find the dimension of the solution space of the homogeneous system

$$\begin{bmatrix} 1 & 0 & 2 \\ 2 & 1 & 3 \\ 3 & 1 & 2 \end{bmatrix} \begin{bmatrix} x_1 \\ x_2 \\ x_3 \end{bmatrix} = \begin{bmatrix} 0 \\ 0 \\ 0 \end{bmatrix}.$$

27. Find a basis for the solution space V of the homogeneous system

$$\begin{bmatrix} 1 & 2 & 2 & -1 & 1 \\ 0 & 2 & 2 & -2 & -1 \\ 2 & 6 & 2 & -4 & 1 \\ 1 & 4 & 0 & -3 & 0 \end{bmatrix} \begin{bmatrix} x_1 \\ x_2 \\ x_3 \\ x_4 \\ x_5 \end{bmatrix} = \begin{bmatrix} 0 \\ 0 \\ 0 \\ 0 \end{bmatrix}.$$

What is the dimension of V?

28. Determine the dimension of the null space of

$$A = \begin{bmatrix} 1 & -1 & 2 & 1 & 0 \\ 2 & 0 & 1 & -1 & 3 \\ 5 & -1 & 3 & 0 & 3 \\ 4 & -2 & 5 & 1 & 3 \\ 1 & 3 & -4 & -5 & 6 \end{bmatrix}.$$

29. Prove that if $\{\mathbf{v}_1, \mathbf{v}_2, \ldots, \mathbf{v}_k\}$ is a basis for a vector space V, then $\{c\mathbf{v}_1, \mathbf{v}_2, \ldots, \mathbf{v}_k\}$ for $c \neq 0$, is also a basis for V.

30. Prove that the vector space P of all polynomials is not finite-dimensional. [*Hint:* Suppose that $\{p_1(t), p_2(t), \ldots, p_k(t)\}$ is a finite basis for P. Let $d_j =$ degree $p_j(t)$. Establish a contradiction.]

31. Let V be an n-dimensional vector space. Show that any $n + 1$ vectors in V form a linearly dependent set.

32. Prove Corollary 2.3.

33. Prove Corollary 2.4.

34. Prove Corollary 2.5.

35. Show that if W is a subspace of a finite-dimensional vector space V, then W is finite-dimensional and $\dim W \leq \dim V$.

36. Show that if W is a subspace of a finite-dimensional vector space V and $\dim W = \dim V$, then $W = V$.

37. Prove that the subspaces of R^3 are $\{\mathbf{0}\}$, R^3 itself, and any line or plane passing through the origin.

38. Let $S = \{\mathbf{v}_1, \mathbf{v}_2, \ldots, \mathbf{v}_n\}$ be a set of nonzero vectors in a vector space V such that every vector in V can be written in one and only one way as a linear combination of the vectors in S. Prove that S is a basis for V.

39. Prove Theorem 2.10.

40. Prove Theorem 2.11.

2.5. *Coordinates and Isomorphisms*

Coordinates

If V is an n-dimensional vector space, we know that V has a basis S with n vectors in it; so far we have not paid any attention to the order of the vectors in S. However, in the discussion of this section we speak of an **ordered basis** $S = \{\mathbf{v}_1, \mathbf{v}_2, \ldots, \mathbf{v}_n\}$ for V; thus $S_1 = \{\mathbf{v}_2, \mathbf{v}_1, \ldots, \mathbf{v}_n\}$ is a different ordered basis for V.

If $S = \{\mathbf{v}_1, \mathbf{v}_2, \ldots, \mathbf{v}_n\}$ is an ordered basis for the n-dimensional vector space V, then every vector $\mathbf{v}$ in V can be uniquely expressed in the form

$$\mathbf{v} = a_1\mathbf{v}_1 + a_2\mathbf{v}_2 + \cdots + a_n\mathbf{v}_n,$$

where $a_1, a_2, \ldots, a_n$ are real numbers. We shall refer to

$$[\mathbf{v}]_S = \begin{bmatrix} a_1 \\ a_2 \\ \vdots \\ a_n \end{bmatrix}$$

as the **coordinate vector of v with respect to the ordered basis** S. The entries of $[\mathbf{v}]_S$ are called the **coordinates of v with respect to** S. Theorem 2.6 implies that $[\mathbf{v}]_S$ is unique.

Example 1. Consider the vector space P_1 of all polynomials of degree ≤ 1 and the zero polynomial and let $S = \{\mathbf{v}_1, \mathbf{v}_2\}$ be an ordered basis for P_1, where $\mathbf{v}_1 = t$ and $\mathbf{v}_2 = 1$. If $\mathbf{v} = p(t) = 5t - 2$, then $[\mathbf{v}]_S = \begin{bmatrix} 5 \\ -2 \end{bmatrix}$ is the coordinate vector of $\mathbf{v}$ with respect to the ordered basis S. On the other hand, if $T = \{t + 1, t - 1\}$ is the ordered basis, we have $5t - 2 = \frac{3}{2}(t + 1) + \frac{7}{2}(t - 1)$, which implies that

$$[\mathbf{v}]_T = \begin{bmatrix} \frac{3}{2} \\ \frac{7}{2} \end{bmatrix}. \qquad \blacksquare$$

Notice that the coordinate vector $[\mathbf{v}]_S$ depends upon the order in which the vectors in S are listed; a change in the order of this listing may change the coordinates of $\mathbf{v}$ with respect to S.

Example 2. Consider the vector space R^3 and let $S = \{\mathbf{v}_1, \mathbf{v}_2, \mathbf{v}_3\}$ be an ordered basis for R^3, where

$$\mathbf{v}_1 = \begin{bmatrix} 1 \\ 1 \\ 0 \end{bmatrix}, \quad \mathbf{v}_2 = \begin{bmatrix} 2 \\ 0 \\ 1 \end{bmatrix}, \quad \text{and} \quad \mathbf{v}_3 = \begin{bmatrix} 0 \\ 1 \\ 2 \end{bmatrix}.$$

If

$$\mathbf{v} = \begin{bmatrix} 1 \\ 1 \\ -5 \end{bmatrix},$$

compute $[\mathbf{v}]_S$.

Solution: To find $[\mathbf{v}]_S$ we need to find the constants a_1, a_2, and a_3 such that

$$a_1\mathbf{v}_1 + a_2\mathbf{v}_2 + a_3\mathbf{v}_3 = \mathbf{v},$$

which leads to the linear system whose augmented matrix is (verify)

$$\left[\begin{array}{ccc|c} 1 & 2 & 0 & 1 \\ 1 & 0 & 1 & 1 \\ 0 & 1 & 2 & -5 \end{array}\right] \tag{1}$$

or equivalently

$$\left[\mathbf{v}_1 \quad \mathbf{v}_2 \quad \mathbf{v}_3 \;\vdots\; \mathbf{v}\right].$$

Transforming the matrix in (1) to reduced row echelon form, we obtain the solution (verify)

$$a_1 = 3, \qquad a_2 = -1, \qquad a_3 = -2,$$

so

$$[\mathbf{v}]_S = \begin{bmatrix} 3 \\ -1 \\ -2 \end{bmatrix}.$$

■

Let $S = \{\mathbf{v}_1, \mathbf{v}_2, \ldots, \mathbf{v}_n\}$ be an ordered basis for a vector space V and let $\mathbf{v}$ and $\mathbf{w}$ be vectors in V such that

$$[\mathbf{v}]_S = [\mathbf{w}]_S.$$

We shall now show that $\mathbf{v} = \mathbf{w}$. Let

$$[\mathbf{v}]_S = \begin{bmatrix} a_1 \\ a_2 \\ \vdots \\ a_n \end{bmatrix}, \qquad [\mathbf{w}]_S = \begin{bmatrix} b_1 \\ b_2 \\ \vdots \\ b_n \end{bmatrix}.$$

Since $[\mathbf{v}]_S = [\mathbf{w}]_S$, we have

$$a_i = b_i, \qquad i = 1, 2, \ldots, n.$$

Then

$$\mathbf{v} = a_1\mathbf{v}_1 + a_2\mathbf{v}_2 + \cdots + a_n\mathbf{v}_n$$

and

$$\begin{aligned} \mathbf{w} &= b_1\mathbf{v}_1 + b_2\mathbf{v}_2 + \cdots + b_n\mathbf{v}_n \\ &= a_1\mathbf{v}_1 + a_2\mathbf{v}_2 + \cdots + a_n\mathbf{v}_n \\ &= \mathbf{v}. \end{aligned}$$

The choice of an ordered basis and the consequent assignment of a coordinate vector for every $\mathbf{v}$ in V enables us to "picture" the vector space. We illustrate this notion by using Example 1. Choose a fixed point O in the plane R^2 and draw any two arrows $\mathbf{w}_1$ and $\mathbf{w}_2$ from O which depict the basis vectors t and 1 in the ordered basis $S = \{t, 1\}$ for P_2 (see Figure 2.21). The directions of $\mathbf{w}_1$ and $\mathbf{w}_2$ determine two lines, which we call the $\boldsymbol{x_1}$**-** and $\boldsymbol{x_2}$**-axes**, respectively. The positive direction on the x_1-axis is in the direction of $\mathbf{w}_1$; the negative direction on the x_1-axis is along $-\mathbf{w}_1$. Similarly, the positive direction on the x_2-axis is in the direction of $\mathbf{w}_2$; the negative direction on the x_2-axis is along $-\mathbf{w}_2$. The lengths of $\mathbf{w}_1$ and $\mathbf{w}_2$ determine the scales on the x_1- and x_2-axes, respectively. If $\mathbf{v}$ is a vector in P_1, we can write $\mathbf{v}$, uniquely, as $\mathbf{v} = a_1\mathbf{w}_1 + a_2\mathbf{w}_2$. We now mark off a segment of length $|a_1|$ on the x_1-axis (in the positive direction if a_1 is positive and in the negative direction if a_1 is negative) and draw a line through the end point of this segment parallel to $\mathbf{w}_2$. Similarly, mark off a segment of length $|a_2|$ on the x_2-axis (in the positive direction if a_2 is positive and in the negative direction if a_2 is negative) and draw a line through the end point of this segment parallel to $\mathbf{w}_1$. We draw a directed line segment from O to the point of intersection of these two lines. This directed line segment represents $\mathbf{v}$.

Figure 2.21

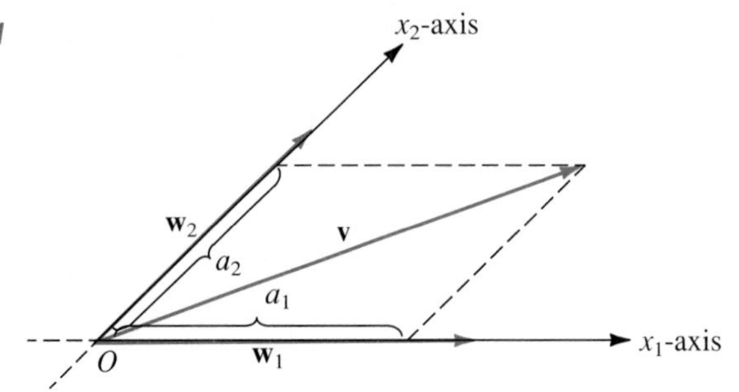

Isomorphism

If $\mathbf{v}$ and $\mathbf{w}$ are vectors in an n-dimensional vector space V with an ordered basis $S = \{\mathbf{v}_1, \mathbf{v}_2, \ldots, \mathbf{v}_n\}$, then we can write $\mathbf{v}$ and $\mathbf{w}$ uniquely, as

$$\mathbf{v} = a_1\mathbf{v}_1 + a_2\mathbf{v}_2 + \cdots + a_n\mathbf{v}_n, \qquad \mathbf{w} = b_1\mathbf{v}_1 + b_2\mathbf{v}_2 + \cdots + b_n\mathbf{v}_n.$$

Thus with $\mathbf{v}$ and $\mathbf{w}$, we associate $[\mathbf{v}]_S$ and $[\mathbf{w}]_S$, respectively, elements in R^n:

$$\begin{aligned}\mathbf{v} &\longrightarrow [\mathbf{v}]_S \\ \mathbf{w} &\longrightarrow [\mathbf{w}]_S.\end{aligned}$$

The sum $\mathbf{v} + \mathbf{w} = (a_1 + b_1)\mathbf{v}_1 + (a_2 + b_2)\mathbf{v}_2 + \cdots + (a_n + b_n)\mathbf{v}_n$, which means that with $\mathbf{v} + \mathbf{w}$ we associate the vector

$$[\mathbf{v} + \mathbf{w}]_S = \begin{bmatrix} (a_1 + b_1) \\ (a_2 + b_2) \\ \vdots \\ (a_n + b_n) \end{bmatrix} = [\mathbf{v}]_S + [\mathbf{w}]_S.$$

Therefore,

$$\mathbf{v} + \mathbf{w} \longrightarrow [\mathbf{v} + \mathbf{w}]_S = [\mathbf{v}]_S + [\mathbf{w}]_S.$$

That is, when we add $\mathbf{v}$ and $\mathbf{w}$ in V, we add their associated vectors $[\mathbf{v}]_S$ and $[\mathbf{w}]_S$ to obtain the vector in R^n associated with $\mathbf{v} + \mathbf{w}$.

Similarly, if c is a real number, then

$$c\mathbf{v} = (ca_1)\mathbf{v}_1 + (ca_2)\mathbf{v}_2 + \cdots + (ca_n)\mathbf{v}_n,$$

which implies that

$$[c\mathbf{v}]_S = \begin{bmatrix} ca_1 \\ ca_2 \\ \vdots \\ ca_n \end{bmatrix} = c[\mathbf{v}]_S.$$

Therefore,

$$c\mathbf{v} \longrightarrow [c\mathbf{v}]_S = c[\mathbf{v}]_S,$$

and thus when $\mathbf{v}$ is multiplied by a scalar c, we multiply $[\mathbf{v}]_S$ by c to obtain the vector in R^n associated with $c\mathbf{v}$.

This discussion suggests that, from an algebraic point of view, V and R^n behave ''rather similarly.'' We now clarify this notion.

Definition 2.12. Let V be a real vector space with operations $\oplus$ and $\odot$, and let W be a real vector space with operations $\boxplus$ and $\boxdot$. A one-to-one function L mapping V onto W is called an **isomorphism** of V onto W if:
(a) $L(\mathbf{v} \oplus \mathbf{w}) = L(\mathbf{v}) \boxplus L(\mathbf{w})$ for $\mathbf{v}$, $\mathbf{w}$ in V.
(b) $L(c \odot \mathbf{v}) = c \boxdot L(\mathbf{v})$ for $\mathbf{v}$ in V, c a real number.
In this case we say that ***V* is isomorphic to *W***. ■

REMARK. A function L mapping a vector space V onto a vector space W satisfying properties (a) and (b) of Definition 2.12 is called a linear transformation. These functions will be studied in depth in Chapter 4. Thus an isomorphism of a vector space V onto a vector space W is a linear transformation that is one-to-one and onto.

Recall that L is **one-to-one** if $L(\mathbf{v}_1) = L(\mathbf{v}_2)$, for $\mathbf{v}_1$, $\mathbf{v}_2$ in V, implies that $\mathbf{v}_1 = \mathbf{v}_2$. Also, L is **onto** if for each $\mathbf{w}$ in W there is at least one $\mathbf{v}$ in V for which $L(\mathbf{v}) = \mathbf{w}$.† Thus the mapping $L: R^3 \rightarrow R^2$ defined by

$$L\left(\begin{bmatrix} a_1 \\ a_2 \\ a_3 \end{bmatrix}\right) = \begin{bmatrix} a_1 + a_2 \\ a_1 \end{bmatrix}$$

† See Appendix A for further discussion of one-to-one and onto functions.

is onto. To see this, suppose that $\mathbf{w} = \begin{bmatrix} b_1 \\ b_2 \end{bmatrix}$; we seek $\mathbf{v} = \begin{bmatrix} a_1 \\ a_2 \\ a_3 \end{bmatrix}$ such that

$$L(\mathbf{v}) = \begin{bmatrix} a_1 + a_2 \\ a_1 \end{bmatrix} = \mathbf{w} = \begin{bmatrix} b_1 \\ b_2 \end{bmatrix}.$$

Thus we obtain the solution: $a_1 = b_2$, $a_2 = b_1 - b_2$, and a_3 is arbitrary. However, L is not one-to-one, for if $\mathbf{v}_1 = \begin{bmatrix} 1 \\ 2 \\ 3 \end{bmatrix}$ and $\mathbf{v}_2 = \begin{bmatrix} 1 \\ 2 \\ 4 \end{bmatrix}$, then

$$L(\mathbf{v}_1) = L(\mathbf{v}_2) = \begin{bmatrix} 3 \\ 1 \end{bmatrix} \text{ but } \mathbf{v}_1 \neq \mathbf{v}_2. \qquad \blacksquare$$

As a result of Theorem 2.13 below we can replace the expressions "V is isomorphic to W" and "W is isomorphic to V" by "V and W are isomorphic."

Isomorphic vector spaces differ only in the nature of their elements; their algebraic properties are identical. That is, if the vector spaces V and W are isomorphic, under the isomorphism L, then for each $\mathbf{v}$ in V there is a unique $\mathbf{w}$ in W so that $L(\mathbf{v}) = \mathbf{w}$ and, conversely, for each $\mathbf{w}$ in W there is a unique $\mathbf{v}$ in V so that $L(\mathbf{v}) = \mathbf{w}$. If we now replace each element of V by its image under L and replace the operations $\oplus$ and $\odot$ of V by $\boxplus$ and $\boxdot$, respectively, we get precisely W. The most important example of isomorphic vector spaces is given in the following theorem.

Theorem 2.12. IF V is an n-dimensional real vector space, then V is isomorphic to R^n.

Proof: Let $S = \{\mathbf{v}_1, \mathbf{v}_2, \ldots, \mathbf{v}_n\}$ be an ordered basis for V, and let $L\colon V \to R^n$ be defined by

$$L(\mathbf{v}) = [\mathbf{v}]_S = \begin{bmatrix} a_1 \\ a_2 \\ \vdots \\ a_n \end{bmatrix},$$

where $\mathbf{v} = a_1\mathbf{v}_1 + a_2\mathbf{v}_2 + \cdots + a_n\mathbf{v}_n$.

We show that L is an isomorphism. First, L is one-to-one. Let

$$[\mathbf{v}]_S = \begin{bmatrix} a_1 \\ a_2 \\ \vdots \\ a_n \end{bmatrix} \quad \text{and} \quad [\mathbf{w}]_S = \begin{bmatrix} b_1 \\ b_2 \\ \vdots \\ b_n \end{bmatrix}$$

and suppose that $L(\mathbf{v}) = L(\mathbf{w})$. Then $[\mathbf{v}]_S = [\mathbf{w}]_S$, and from our earlier remarks it follows that $\mathbf{v} = \mathbf{w}$.

Next, L is onto, for if $\mathbf{w} = \begin{bmatrix} b_1 \\ b_2 \\ \vdots \\ b_n \end{bmatrix}$ is a given vector in R^n and

$$\mathbf{v} = b_1\mathbf{v}_1 + b_2\mathbf{v}_2 + \cdots + b_n\mathbf{v}_n,$$

then $L(\mathbf{v}) = \mathbf{w}$.

Finally, L satisfies Definition 2.12(a) and (b). Let $\mathbf{v}$ and $\mathbf{w}$ be vectors in V such that $[\mathbf{v}]_S = \begin{bmatrix} a_1 \\ a_2 \\ \vdots \\ a_n \end{bmatrix}$ and $[\mathbf{w}]_S = \begin{bmatrix} b_1 \\ b_2 \\ \vdots \\ b_n \end{bmatrix}$. Then

$$L(\mathbf{v} + \mathbf{w}) = [\mathbf{v} + \mathbf{w}]_S = [\mathbf{v}]_S + [\mathbf{w}]_S = L(\mathbf{v}) + L(\mathbf{w})$$

and

$$L(c\mathbf{v}) = [c\mathbf{v}]_S = c[\mathbf{v}]_S = cL(\mathbf{v}),$$

as we saw before. Hence V and R^n are isomorphic. ■

Another example of isomorphism is given by the vector spaces discussed in the review section at the beginning of this chapter: R^2, the vector space of directed line segments emanating from a point in the plane and the vector space of all ordered pairs of real numbers. There is a corresponding isomorphism for R^3.

Some important properties of isomorphisms are given in Theorem 2.13.

Theorem 2.13. (a) *Every vector space V is isomorphic to itself.*
(b) *If V is isomorphic to W, then W is isomorphic to V.*
(c) *If U is isomorphic to V and V is isomorphic to W, then U is isomorphic to W.*

Proof: Exercise. [Parts (a) and (c) are not difficult to show; (b) is slightly harder and will essentially be proved in Theorem 4.6.] ■

The following theorem shows that all vector spaces of the same dimension are, algebraically speaking, alike, and conversely, that isomorphic vector spaces have the same dimensions.

Theorem 2.14. *Two finite-dimensional vector spaces are isomorphic if and only if their dimensions are equal.*

Proof: Let V and W be n-dimensional vector spaces. Then V and R^n are isomorphic and W and R^n are isomorphic. From Theorem 2.13 it follows that V and W are isomorphic.

Conversely, let V and W be isomorphic finite-dimensional vector spaces; let $L: V \to W$ be an isomorphism. Assume that $\dim V = n$, and let $S = \{\mathbf{v}_1, \mathbf{v}_2, \ldots, \mathbf{v}_n\}$ be a basis for V.

We now prove that the set $T = \{L(\mathbf{v}_1), L(\mathbf{v}_2), \ldots, L(\mathbf{v}_n)\}$ is a basis for W. First, T spans W. If $\mathbf{w}$ is any vector in W, then $\mathbf{w} = L(\mathbf{v})$ for some $\mathbf{v}$ in V. Now $\mathbf{v} = a_1\mathbf{v}_1 + a_2\mathbf{v}_2 + \cdots + a_n\mathbf{v}_n$, where the a_i are uniquely determined real numbers, so

$$\begin{aligned} L(\mathbf{v}) &= L(a_1\mathbf{v}_1 + a_2\mathbf{v}_2 + \cdots + a_n\mathbf{v}_n) = L(a_1\mathbf{v}_1) + L(a_2\mathbf{v}_2) + \cdots \\ &\quad + L(a_n\mathbf{v}_n) = a_1L(\mathbf{v}_1) + a_2L(\mathbf{v}_2) + \cdots + a_nL(\mathbf{v}_n). \end{aligned}$$

Thus T spans W.

Now suppose that

$$a_1L(\mathbf{v}_1) + a_2L(\mathbf{v}_2) + \cdots + a_nL(\mathbf{v}_n) = \mathbf{0}_W.$$

Then $L(a_1\mathbf{v}_1 + a_2\mathbf{v}_2 + \cdots + a_n\mathbf{v}_n) = \mathbf{0}_W$. From Exercise 7, $L(\mathbf{0}_V) = \mathbf{0}_W$ so $L(a_1\mathbf{v}_1 + a_2\mathbf{v}_2 + \cdots + a_n\mathbf{v}_n) = \mathbf{0}_W$. Since L is one-to-one, we get $a_1\mathbf{v}_1 + a_2\mathbf{v}_2 + \cdots + a_n\mathbf{v}_n = \mathbf{0}_V$. Since S is linearly independent, we conclude that $a_1 = a_2 = \cdots = a_n = 0$, which means that T is linearly independent. Hence T is a basis for W and $\dim W = n$. ■

It is easy to show, as a consequence of Theorem 2.14, that the spaces R^n and R^m are isomorphic if and only if $n = m$ (see Exercise 8).

We can now establish the converse of Theorem 2.12, as follows.

Corollary 2.6. *If V is a finite-dimensional vector space that is isomorphic to R^n, then $\dim V = n$.*

Proof: This result follows from Theorem 2.14. ■

If $L: V \to W$ is an isomorphism, then since L is a one-to-one onto mapping, it has an inverse L^{-1}. (This will be shown in Theorem 4.6.) It is not difficult to show that $L^{-1}: W \to V$ is also an isomorphism. (This will also be essentially shown in Theorem 4.6.) Moreover, if $S = \{\mathbf{v}_1, \mathbf{v}_2, \ldots, \mathbf{v}_n\}$ is a basis for V, then one can prove that $T = L(S) = \{L(\mathbf{v}_1), L(\mathbf{v}_2), \ldots, L(\mathbf{v}_n)\}$ is a basis for W, as we have seen in the proof of Theorem 2.14.

As an example of isomorphism, we note that the vector spaces P_3, R_4, and R^4 are all isomorphic, since each has dimension four.

We have shown in this section that the idea of a finite-dimensional vector space, which at first seemed fairly abstract, is not so mysterious. In fact, such a vector space does not differ much from R^n.

Transition Matrices

We now look at the relationship between two coordinate vectors for the same vector $\mathbf{v}$ with respect to different bases. Thus let $S = \{\mathbf{v}_1, \mathbf{v}_2, \ldots, \mathbf{v}_n\}$ and $T = \{\mathbf{w}_1, \mathbf{w}_2, \ldots, \mathbf{w}_n\}$ be two ordered bases for the n-dimensional vector space V. If $\mathbf{v}$ is any vector in V, then

$$\mathbf{v} = c_1\mathbf{w}_1 + c_2\mathbf{w}_2 + \cdots + c_n\mathbf{w}_n \quad \text{and} \quad [\mathbf{v}]_T = \begin{bmatrix} c_1 \\ c_2 \\ \vdots \\ c_n \end{bmatrix}.$$

Then

$$\begin{aligned} [\mathbf{v}]_S &= [c_1\mathbf{w}_1 + c_2\mathbf{w}_2 + \cdots + c_n\mathbf{w}_n]_S \\ &= [c_1\mathbf{w}_1]_S + [c_2\mathbf{w}_2]_S + \cdots + [c_n\mathbf{w}_n]_S \\ &= c_1[\mathbf{w}_1]_S + c_2[\mathbf{w}_2]_S + \cdots + c_n[\mathbf{w}_n]_S \end{aligned}$$

Let the coordinate vector of $\mathbf{w}_j$ with respect to S be denoted by

$$[\mathbf{w}_j]_S = \begin{bmatrix} a_{1j} \\ a_{2j} \\ \vdots \\ a_{nj} \end{bmatrix}.$$

Then

$$[\mathbf{v}]_S = c_1 \begin{bmatrix} a_{11} \\ a_{21} \\ \vdots \\ a_{n1} \end{bmatrix} + c_2 \begin{bmatrix} a_{12} \\ a_{22} \\ \vdots \\ a_{n2} \end{bmatrix} + \cdots + c_n \begin{bmatrix} a_{1n} \\ a_{2n} \\ \vdots \\ a_{nn} \end{bmatrix}$$

$$= \begin{bmatrix} a_{11} & a_{12} & \cdots & a_{1n} \\ a_{21} & a_{22} & \cdots & a_{2n} \\ \vdots & \vdots & & \vdots \\ a_{n1} & a_{n2} & \cdots & a_{nn} \end{bmatrix} \begin{bmatrix} c_1 \\ c_2 \\ \vdots \\ c_n \end{bmatrix}$$

or

$$[\mathbf{v}]_S = P[\mathbf{v}]_T, \tag{2}$$

where

$$P = \begin{bmatrix} a_{11} & a_{12} & \cdots & a_{1n} \\ a_{21} & a_{22} & \cdots & a_{2n} \\ \vdots & \vdots & & \vdots \\ a_{n1} & a_{n2} & \cdots & a_{nn} \end{bmatrix} = \begin{bmatrix} [\mathbf{w}_1]_S & [\mathbf{w}_2]_S \cdots [\mathbf{w}_n]_S \end{bmatrix}$$

is called the **transition matrix from the T-basis to the S-basis**.

Thus, to find the transition matrix from the T-basis to the S-basis, we first compute the coordinate vector of each member of the T-basis with respect to the S-basis. Forming a matrix with these vectors as columns arranged in their natural order, we obtain the transition matrix. Equation (2) says that the coordinate vector of $\mathbf{v}$ with respect to the basis S is the transition matrix P times the coordinate vector of $\mathbf{v}$ with respect to the basis T.

Example 3. Let V be R^3 and let $S = \{\mathbf{v}_1, \mathbf{v}_2, \mathbf{v}_3\}$ and $T = \{\mathbf{w}_1, \mathbf{w}_2, \mathbf{w}_3\}$ be ordered bases for R^3, where

$$\mathbf{v}_1 = \begin{bmatrix} 2 \\ 0 \\ 1 \end{bmatrix}, \quad \mathbf{v}_2 = \begin{bmatrix} 1 \\ 2 \\ 0 \end{bmatrix}, \quad \mathbf{v}_3 = \begin{bmatrix} 1 \\ 1 \\ 1 \end{bmatrix}$$

and

$$\mathbf{w}_1 = \begin{bmatrix} 6 \\ 3 \\ 3 \end{bmatrix}, \quad \mathbf{w}_2 = \begin{bmatrix} 4 \\ -1 \\ 3 \end{bmatrix}, \quad \mathbf{w}_3 = \begin{bmatrix} 5 \\ 5 \\ 2 \end{bmatrix}.$$

(a) Compute the transition matrix P from the T-basis to the S-basis.

(b) Verify Equation (2) for $\mathbf{v} = \begin{bmatrix} 4 \\ -9 \\ 5 \end{bmatrix}$.

Solution: (a) To compute P, we need to find a_1, a_2, a_3 such that

$$a_1\mathbf{v}_1 + a_2\mathbf{v}_2 + a_3\mathbf{v}_3 = \mathbf{w}_1,$$

which leads to a linear system of three equations in three unknowns, whose augmented matrix is

$$\left[\mathbf{v}_1 \quad \mathbf{v}_2 \quad \mathbf{v}_3 \;\vdots\; \mathbf{w}_1\right].$$

That is, the augmented matrix is

$$\left[\begin{array}{ccc:c} 2 & 1 & 1 & 6 \\ 0 & 2 & 1 & 3 \\ 1 & 0 & 1 & 3 \end{array}\right].$$

Similarly, we need to find b_1, b_2, b_3 and c_1, c_2, c_3 such that

$$\begin{aligned} b_1\mathbf{v}_1 + b_2\mathbf{v}_2 + b_3\mathbf{v}_3 &= \mathbf{w}_2 \\ c_1\mathbf{v}_1 + c_2\mathbf{v}_2 + c_3\mathbf{v}_3 &= \mathbf{w}_3. \end{aligned}$$

These vector equations lead to two linear systems, each of three equations in three unknowns, whose augmented matrices are

$$\left[\mathbf{v}_1 \quad \mathbf{v}_2 \quad \mathbf{v}_3 \;\vdots\; \mathbf{w}_2\right] \quad \text{and} \quad \left[\mathbf{v}_1 \quad \mathbf{v}_2 \quad \mathbf{v}_3 \;\vdots\; \mathbf{w}_3\right],$$

or specifically,

$$\left[\begin{array}{ccc:c} 2 & 1 & 1 & 4 \\ 0 & 2 & 1 & -1 \\ 1 & 0 & 1 & 3 \end{array}\right] \quad \text{and} \quad \left[\begin{array}{ccc:c} 2 & 1 & 1 & 5 \\ 0 & 2 & 1 & 5 \\ 1 & 0 & 1 & 2 \end{array}\right].$$

Since the coefficient matrix of all three linear systems is $[\mathbf{v}_1 \quad \mathbf{v}_2 \quad \mathbf{v}_3]$, we can transform the three augmented matrices to reduced row echelon form simultaneously by transforming the partitioned matrix

$$[\mathbf{v}_1 \quad \mathbf{v}_2 \quad \mathbf{v}_3 \mid \mathbf{w}_1 \mid \mathbf{w}_2 \mid \mathbf{w}_3]$$

to reduced row echelon form. Thus we transform

$$\left[\begin{array}{ccc|c|c|c} 2 & 1 & 1 & 6 & 4 & 5 \\ 0 & 2 & 1 & 3 & -1 & 5 \\ 1 & 0 & 1 & 3 & 3 & 2 \end{array}\right]$$

to reduced row echelon form, obtaining (verify)

$$\left[\begin{array}{ccc|c|c|c} 1 & 0 & 0 & 2 & 2 & 1 \\ 0 & 1 & 0 & 1 & -1 & 2 \\ 0 & 0 & 1 & 1 & 1 & 1 \end{array}\right],$$

which implies that the transition matrix from the T-basis to the S-basis is

$$P = \begin{bmatrix} 2 & 2 & 1 \\ 1 & -1 & 2 \\ 1 & 1 & 1 \end{bmatrix}.$$

(b) If $\mathbf{v} = \begin{bmatrix} 4 \\ -9 \\ 5 \end{bmatrix}$, then expressing $\mathbf{v}$ in terms of the T-basis, we have (verify)

$$\mathbf{v} = \begin{bmatrix} 4 \\ -9 \\ 5 \end{bmatrix} = 1\begin{bmatrix} 6 \\ 3 \\ 3 \end{bmatrix} + 2\begin{bmatrix} 4 \\ -1 \\ 3 \end{bmatrix} - 2\begin{bmatrix} 5 \\ 5 \\ 2 \end{bmatrix}.$$

So $[\mathbf{v}]_T = \begin{bmatrix} 1 \\ 2 \\ -2 \end{bmatrix}$. Then

$$[\mathbf{v}]_S = P[\mathbf{v}]_T = \begin{bmatrix} 2 & 2 & 1 \\ 1 & -1 & 2 \\ 1 & 1 & 1 \end{bmatrix}\begin{bmatrix} 1 \\ 2 \\ -2 \end{bmatrix} = \begin{bmatrix} 4 \\ -5 \\ 1 \end{bmatrix}.$$

If we compute $[\mathbf{v}]_S$ directly, we find that

$$\mathbf{v} = \begin{bmatrix} 4 \\ -9 \\ 5 \end{bmatrix} = 4\begin{bmatrix} 2 \\ 0 \\ 1 \end{bmatrix} - 5\begin{bmatrix} 1 \\ 2 \\ 0 \end{bmatrix} + 1\begin{bmatrix} 1 \\ 1 \\ 1 \end{bmatrix}, \text{ so } [\mathbf{v}]_S = \begin{bmatrix} 4 \\ -5 \\ 1 \end{bmatrix}.$$

Hence

$$[\mathbf{v}]_S = P[\mathbf{v}]_T. \qquad \blacksquare$$

We next want to show that the transition matrix P from the T-basis to the S-basis is nonsingular. Suppose that $P[\mathbf{v}]_T = \mathbf{0}_{R^n}$ for some $\mathbf{v}$ in V. From Equation (2) we have

$$P[\mathbf{v}]_T = [\mathbf{v}]_S = \mathbf{0}_{R^n}.$$

If $\mathbf{v} = b_1\mathbf{v}_1 + b_2\mathbf{v}_2 + \cdots + b_n\mathbf{v}_n$, then

$$\begin{bmatrix} b_1 \\ b_2 \\ \vdots \\ b_n \end{bmatrix} = [\mathbf{v}]_S = \mathbf{0}_{R^n} = \begin{bmatrix} 0 \\ 0 \\ \vdots \\ 0 \end{bmatrix},$$

so

$$\mathbf{v} = 0\mathbf{v}_1 + 0\mathbf{v}_2 + \cdots + 0\mathbf{v}_n = \mathbf{0}_V.$$

Hence $[\mathbf{0}]_T = \mathbf{0}_{R^n}$. Thus the homogeneous system $P\mathbf{x} = \mathbf{0}$ has only the trivial solution; it then follows from Theorem 1.17 that P is nonsingular. Of course, we then also have

$$[\mathbf{v}]_T = P^{-1}[\mathbf{v}]_S.$$

That is, P^{-1} is then the transition matrix from the S-basis to the T-basis; the jth column of P^{-1} is $[\mathbf{v}_j]_T$.

Example 4. Let S and T be the ordered bases for R^3 defined in Example 3. Compute the transition matrix Q from the S-basis to the T-basis directly.

Solution: Q is the matrix whose columns are the solution vectors to the linear systems obtained from the vector equations

$$\begin{aligned} a_1\mathbf{w}_1 + a_2\mathbf{w}_2 + a_3\mathbf{w}_3 &= \mathbf{v}_1 \\ b_1\mathbf{w}_1 + b_2\mathbf{w}_2 + b_3\mathbf{w}_3 &= \mathbf{v}_2 \\ c_1\mathbf{w}_1 + c_2\mathbf{w}_2 + c_3\mathbf{w}_3 &= \mathbf{v}_3. \end{aligned}$$

As in Example 3, we can solve these linear systems simultaneously by transforming the partitioned matrix

$$\left[\mathbf{w}_1 \quad \mathbf{w}_2 \quad \mathbf{w}_3 \;\vdots\; \mathbf{v}_1 \;\vdots\; \mathbf{v}_2 \;\vdots\; \mathbf{v}_3\right]$$

to reduced row echelon form. That is, we transform

$$\left[\begin{array}{rrr|r|r|r} 6 & 4 & 5 & 2 & 1 & 1 \\ 3 & -1 & 5 & 0 & 2 & 1 \\ 3 & 3 & 2 & 1 & 0 & 1 \end{array}\right]$$

to reduced row echelon form, obtaining (verify)

$$\left[\begin{array}{rrr|r|r|r} 1 & 0 & 0 & \frac{3}{2} & \frac{1}{2} & -\frac{5}{2} \\ 0 & 1 & 0 & -\frac{1}{2} & -\frac{1}{2} & \frac{3}{2} \\ 0 & 0 & 1 & -1 & 0 & 2 \end{array}\right],$$

so

$$Q = \begin{bmatrix} \frac{3}{2} & \frac{1}{2} & -\frac{5}{2} \\ -\frac{1}{2} & -\frac{1}{2} & \frac{3}{2} \\ -1 & 0 & 2 \end{bmatrix}.$$

Computing P^{-1}, we find (verify) that $Q = P^{-1}$. ■

Example 5. Let V be P_1 and let $S = \{\mathbf{v}_1, \mathbf{v}_2\}$ and $T = \{\mathbf{w}_1, \mathbf{w}_2\}$ be ordered bases for P_2, where

$$\mathbf{v}_1 = t, \qquad \mathbf{v}_2 = t - 3, \qquad \mathbf{w}_1 = t - 1, \qquad \mathbf{w}_2 = t + 1.$$

(a) Compute the transition matrix P from the T-basis to the S-basis.
(b) Verify Equation (2) for $\mathbf{v} = 5t + 1$.
(c) Compute the transition matrix Q from the S-basis to the T-basis and show that $Q = P^{-1}$.

Solution: (a) To compute P, we need to solve the vector equations

$$a_1\mathbf{v}_1 + a_2\mathbf{v}_2 = \mathbf{w}_1$$
$$b_1\mathbf{v}_1 + b_2\mathbf{v}_2 = \mathbf{w}_2$$

simultaneously by transforming the resulting partitioned matrix (verify)

$$\left[\begin{array}{cc|c|c} 1 & 1 & 1 & 1 \\ 0 & -3 & -1 & 1 \end{array}\right]$$

to reduced row echelon form. The result is

$$\left[\begin{array}{cc|c|c} 1 & 0 & \frac{2}{3} & \frac{4}{3} \\ 0 & 1 & \frac{1}{3} & -\frac{1}{3} \end{array}\right],$$

so

$$P = \begin{bmatrix} \frac{2}{3} & \frac{4}{3} \\ \frac{1}{3} & -\frac{1}{3} \end{bmatrix}.$$

(b) If $\mathbf{v} = 5t + 1$, then expressing $\mathbf{v}$ in terms of T, we have (verify)

$$\mathbf{v} = 5t + 1 = 2(t - 1) + 3(t + 1),$$

so $[\mathbf{v}]_T = \begin{bmatrix} 2 \\ 3 \end{bmatrix}$. Then

$$[\mathbf{v}]_S = P[\mathbf{v}]_T = \begin{bmatrix} \frac{2}{3} & \frac{4}{3} \\ \frac{1}{3} & -\frac{1}{3} \end{bmatrix}\begin{bmatrix} 2 \\ 3 \end{bmatrix} = \begin{bmatrix} \frac{16}{3} \\ -\frac{1}{3} \end{bmatrix}.$$

Computing $[\mathbf{v}]_S$ directly, we find that

$$\mathbf{v} = 5t + 1 = \tfrac{16}{3}t - \tfrac{1}{3}(t - 3), \qquad \text{so} \quad [\mathbf{v}]_S = \begin{bmatrix} \frac{16}{3} \\ -\frac{1}{3} \end{bmatrix}.$$

Hence

$$[\mathbf{v}]_S = P[\mathbf{v}]_T.$$

(c) The transition matrix Q from the S-basis to the T-basis is obtained (verify) by transforming the partitioned matrix

$$\left[\begin{array}{rr|r|r} 1 & 1 & 1 & 1 \\ -1 & 1 & 0 & -3 \end{array}\right]$$

to reduced row echelon form, obtaining (verify)

$$\left[\begin{array}{rr|r|r} 1 & 0 & \frac{1}{2} & 2 \\ 0 & 1 & \frac{1}{2} & -1 \end{array}\right].$$

Hence

$$Q = \begin{bmatrix} \frac{1}{2} & 2 \\ \frac{1}{2} & -1 \end{bmatrix}.$$

Computing P^{-1}, we find (verify) that $Q = P^{-1}$. ■

2.5. EXERCISES

1. Let $S = \left\{ \begin{bmatrix} 1 \\ 2 \end{bmatrix}, \begin{bmatrix} 0 \\ 1 \end{bmatrix} \right\}$ and $T = \left\{ \begin{bmatrix} 1 \\ 1 \end{bmatrix}, \begin{bmatrix} 2 \\ 3 \end{bmatrix} \right\}$ be ordered bases for R^2. Let $\mathbf{v} = \begin{bmatrix} 1 \\ 5 \end{bmatrix}$ and $\mathbf{w} = \begin{bmatrix} 5 \\ 4 \end{bmatrix}$.

(a) Find the coordinate vectors of $\mathbf{v}$ and $\mathbf{w}$ with respect to the basis T.

(b) What is the transition matrix P from the T- to the S-basis?

(c) Find the coordinate vectors of $\mathbf{v}$ and $\mathbf{w}$ with respect to S using P.

(d) Find the coordinate vectors of $\mathbf{v}$ and $\mathbf{w}$ with respect to S directly.

(e) Find the transition matrix Q from the S- to the T-basis.

(f) Find the coordinate vectors of $\mathbf{v}$ and $\mathbf{w}$ with respect to T using Q. Compare the answers with those of (a).

2. Let $S = \left\{ \begin{bmatrix} 1 \\ 0 \\ 1 \end{bmatrix}, \begin{bmatrix} -1 \\ 0 \\ 0 \end{bmatrix}, \begin{bmatrix} 0 \\ 1 \\ 2 \end{bmatrix} \right\}$ and $T = \left\{ \begin{bmatrix} -1 \\ 1 \\ 0 \end{bmatrix}, \begin{bmatrix} 1 \\ 2 \\ -1 \end{bmatrix}, \begin{bmatrix} 0 \\ 1 \\ 0 \end{bmatrix} \right\}$ be ordered bases for R^3. Let $\mathbf{v} = \begin{bmatrix} 1 \\ 3 \\ 8 \end{bmatrix}$ and $\mathbf{w} = \begin{bmatrix} -1 \\ 8 \\ -2 \end{bmatrix}$. Follow the directions of Exercise 1.

3. Let $S = \{t^2 + 1, t - 2, t + 3\}$ and $T = \{2t^2 + t, t^2 + 3, t\}$ be ordered bases for P_2. Let $\mathbf{v} = 8t^2 - 4t + 6$ and $\mathbf{w} = 7t^2 - t + 9$. Follow the directions of Exercise 1.

4. Let $S = \{[1\ \ 1\ \ 1], [1\ \ 2\ \ 3], [1\ \ 0\ \ 1]\}$ and $T = \{[0\ \ 1\ \ 1], [1\ \ 0\ \ 0], [1\ \ 0\ \ 1]\}$ be ordered bases for R_3. Let $\mathbf{v} = [-1\ \ 4\ \ 5]$ and $\mathbf{w} = [2\ \ 0\ \ -6]$. Follow the directions of Exercise 1.

5. Let

$$S = \left\{\begin{bmatrix} 1 & 0 \\ 0 & 0 \end{bmatrix}, \begin{bmatrix} 0 & 1 \\ 1 & 0 \end{bmatrix}, \begin{bmatrix} 0 & 2 \\ 0 & 1 \end{bmatrix}, \begin{bmatrix} 0 & 0 \\ 1 & 1 \end{bmatrix}\right\}$$

and

$$T = \left\{\begin{bmatrix} 1 & 1 \\ 0 & 0 \end{bmatrix}, \begin{bmatrix} 0 & 0 \\ 1 & 0 \end{bmatrix}, \begin{bmatrix} 0 & 0 \\ 0 & 1 \end{bmatrix}, \begin{bmatrix} 1 & 0 \\ 0 & 0 \end{bmatrix}\right\}$$

be ordered bases for ${}_2R_2$. Let $\mathbf{v} = \begin{bmatrix} 1 & 1 \\ 1 & 1 \end{bmatrix}$ and $\mathbf{w} = \begin{bmatrix} 1 & 2 \\ -2 & 1 \end{bmatrix}$. Follow the directions of Exercise 1.

6. Prove parts (a) and (c) of Theorem 2.13.

7. Let $L: V \to W$ be an isomorphism of vector space V onto vector space W.
(a) Prove that $L(\mathbf{0}_V) = \mathbf{0}_W$.
(b) Show that $L(\mathbf{v} - \mathbf{w}) = L(\mathbf{v}) - L(\mathbf{w})$.

8. Prove that R^n and R^m are isomorphic if and only if $n = m$.

9. Find an isomorphism $L: R_n \to R^n$.

10. Find an isomorphism $L: P_2 \to R^3$.

11. (a) Show that ${}_2R_2$ is isomorphic to R^4.
(b) What is dim ${}_2R_2$?

12. Let V be the subspace of the vector space of all real-valued continuous functions that has basis $S = \{e^t, e^{-t}\}$. Show that V and R^2 are isomorphic.

13. Let V be the subspace of the vector space of all real-valued functions that is *spanned* by the set $S = \{\cos^2 t, \sin^2 t, \cos 2t\}$. Show that V and R_2 are isomorphic.

14. The set $S = \left\{\begin{bmatrix} 1 \\ -1 \\ 1 \end{bmatrix}, \begin{bmatrix} 0 \\ 0 \\ 1 \end{bmatrix}, \begin{bmatrix} 1 \\ 1 \\ 0 \end{bmatrix}\right\}$ is an ordered basis for R^3. Find $\mathbf{v}$ if $[\mathbf{v}]_s$ is:

(a) $\begin{bmatrix} 2 \\ -1 \\ 3 \end{bmatrix}$. (b) $\begin{bmatrix} 0 \\ 0 \\ 1 \end{bmatrix}$. (c) $\begin{bmatrix} 0 \\ 2 \\ 1 \end{bmatrix}$.

15. The set $S = \{t^2 + 1, t + 1, t^2 + t\}$ is an ordered basis for P_2. Find $\mathbf{v}$ if $[\mathbf{v}]_S$ is:

(a) $\begin{bmatrix} 3 \\ -1 \\ -2 \end{bmatrix}$. (b) $\begin{bmatrix} 1 \\ 0 \\ 0 \end{bmatrix}$. (c) $\begin{bmatrix} 2 \\ 0 \\ -1 \end{bmatrix}$.

16. If the vector $\mathbf{v}$ in P_2 has the coordinate vector $\begin{bmatrix} 1 \\ 2 \\ 3 \end{bmatrix}$ with respect to the ordered basis $T = \{t^2, t - 1, 1\}$, what is $[\mathbf{v}]_S$ if $S = \{t^2 + t + 1, t + 1, 1\}$?

17. Let V and W be isomorphic vector spaces. Prove that if V_1 is a subspace of V, then V_1 is isomorphic to a subspace W_1 of W.

18. Suppose that $S = \{\mathbf{v}_1, \mathbf{v}_2, \mathbf{v}_3\}$ is a basis for R^3, where

$$\mathbf{v}_1 = \begin{bmatrix} 1 \\ 0 \\ 1 \end{bmatrix}, \quad \mathbf{v}_2 = \begin{bmatrix} 1 \\ 1 \\ 0 \end{bmatrix}, \quad \mathbf{v}_3 = \begin{bmatrix} 0 \\ 0 \\ 1 \end{bmatrix}.$$

Let $T = \{\mathbf{w}_1, \mathbf{w}_2, \mathbf{w}_3\}$. Suppose that the transition matrix from T to S is

$$P = \begin{bmatrix} 1 & 1 & 2 \\ 2 & 1 & 1 \\ -1 & -1 & 1 \end{bmatrix}.$$

Determine T.

2.6. The Rank of a Matrix

In this section we obtain another effective method for finding a basis for a vector space V spanned by a given set of vectors $S = \{\mathbf{v}_1, \mathbf{v}_2, \ldots, \mathbf{v}_k\}$. In the Alternate Constructive Proof to Theorem 2.7, we develop a technique for choosing a basis for V that is a subset of S. The method to be developed in this section produces a basis for V that is not guaranteed to be a subset of S. We shall also attach a unique number to a matrix A that we later show gives us information about the dimension of the solution space of a homogeneous system with coefficient matrix A.

Definition 2.13. Let

$$A = \begin{bmatrix} a_{11} & a_{12} & \cdots & a_{1n} \\ a_{21} & a_{22} & \cdots & a_{2n} \\ \vdots & \vdots & & \vdots \\ a_{m1} & a_{m2} & \cdots & a_{mn} \end{bmatrix}$$

be an $m \times n$ matrix. The rows of A, considered as vectors in R_n, span a subspace of R_n, called the **row space** of A. Similarly, the columns of A, considered as vectors in R^m, span a subspace of R^m called the **column space** of A. ■

Theorem 2.15. *If A and B are two $m \times n$ row (column) equivalent matrices, then the row (column) spaces of A and B are equal.*

Proof: If A and B are row equivalent, then the rows of B are obtained from the rows of A by a finite number of the three elementary row operations. Thus each row of B is a linear combination of the rows of A. Hence the row space of B is contained in the row space of A. If we apply the inverse elementary row operations to B, we get A, so the row space of A is contained in the row space of B. Hence the row spaces of A and B are identical. The proof for the column spaces is similar. ■

We can use this theorem to find a basis for a subspace spanned by a given set of vectors. We illustrate this method with the following example.

Example 1. Find a basis for the subspace V of R_3 spanned by $S = \{\mathbf{v}_1, \mathbf{v}_2, \mathbf{v}_3, \mathbf{v}_4, \mathbf{v}_5\}$, where

$$\mathbf{v}_1 = [1 \quad -2 \quad 1], \qquad \mathbf{v}_2 = [-1 \quad 1 \quad 1], \qquad \mathbf{v}_3 = [1 \quad -3 \quad 3],$$
$$\mathbf{v}_4 = [3 \quad -5 \quad 1], \qquad \text{and} \qquad \mathbf{v}_5 = [1 \quad -4 \quad 5].$$

Solution: Note that V is the row space of the matrix A whose rows are the given vectors,

$$A = \begin{bmatrix} 1 & -2 & 1 \\ -1 & 1 & 1 \\ 1 & -3 & 3 \\ 3 & -5 & 1 \\ 1 & -4 & 5 \end{bmatrix}.$$

Using elementary row operations, we find that A is row equivalent to the matrix (verify)

$$B = \begin{bmatrix} 1 & 0 & -3 \\ 0 & 1 & -2 \\ 0 & 0 & 0 \\ 0 & 0 & 0 \\ 0 & 0 & 0 \end{bmatrix},$$

which is in reduced row echelon form. The row spaces of A and B are identical and a basis for the row space of B consists of

$$\mathbf{w}_1 = \begin{bmatrix} 1 & 0 & -3 \end{bmatrix} \quad \text{and} \quad \mathbf{w}_2 = \begin{bmatrix} 0 & 1 & -2 \end{bmatrix}.$$

(See Exercise 23 in Section 2.3.) Hence $\{\mathbf{w}_1, \mathbf{w}_2\}$ is also a basis for V. ■

It is not necessary to find a matrix B in reduced row echelon form that is row equivalent to A. All that is required is that we have a matrix B that is row equivalent to A and such that we can easily obtain a basis for the row space of B. Often one does not have to reduce A all the way to reduced row echelon form to get such a matrix B. It is easy to show that if A is row equivalent to a matrix B that is in row echelon form, then the nonzero rows of B form a basis for the row space of A.

Of course, the basis that has been obtained by the procedure used in Example 1 produced a basis that is not a subset of the given spanning set. The method used in Example 4 of Section 2.4 does produce a basis that is a subset of the spanning set. However, the basis for a subspace V of R^n that is obtained by the procedure used in Example 1 is analogous, in its simplicity, to the natural basis for R^n. Thus if

$$\mathbf{v} = \begin{bmatrix} a_1 \\ a_2 \\ \vdots \\ a_n \end{bmatrix}$$

is a vector in V and $\{\mathbf{v}_1, \mathbf{v}_2, \ldots, \mathbf{v}_k\}$ is a basis for V obtained by the method of Example 1 where the leading 1's occur in columns $j_1, j_2, \ldots, j_k$, then it can be shown (Exercise 26) that

$$\mathbf{v} = a_{j_1}\mathbf{v}_1 + a_{j_2}\mathbf{v}_2 + \cdots + a_{j_k}\mathbf{v}_k.$$

Example 2. Let V be the subspace of Example 1. Given that the vector

$$\mathbf{v} = [5 \quad -13 \quad 11]$$

is in V, write $\mathbf{v}$ as a linear combination of the basis determined in Example 1.

Solution: We have $\mathbf{v} = 5\mathbf{w}_1 - 13\mathbf{w}_2$ (verify). ■

Example 3. Let V be the subspace of P_4 spanned by $S = \{\mathbf{v}_1, \mathbf{v}_2, \mathbf{v}_3, \mathbf{v}_4\}$, where $\mathbf{v}_1 = t^4 + t^2 + 2t + 1$, $\mathbf{v}_2 = t^4 + t^2 + 2t + 2$, $\mathbf{v}_3 = 2t^4 + t^3 + t + 2$, and $\mathbf{v}_4 = t^4 + t^3 - t^2 - t$. Find a basis for V.

Solution: Since P_4 is isomorphic to R_5 under the isomorphism L defined by

$$L(at^4 + bt^3 + ct^2 + dt + e) = [a \quad b \quad c \quad d \quad e],$$

then $L(V)$ is isomorphic to a subspace W of R_5 (see Exercise 17 of Section 2.5). The subspace W is spanned by $\{L(\mathbf{v}_1), L(\mathbf{v}_2), L(\mathbf{v}_3), L(\mathbf{v}_4)\}$, as we have seen in the proof of Theorem 2.14. We now find a basis for W by proceeding as in Example 1. Thus W is the row space of the matrix

$$A = \begin{bmatrix} 1 & 0 & 1 & 2 & 1 \\ 1 & 0 & 1 & 2 & 2 \\ 2 & 1 & 0 & 1 & 2 \\ 1 & 1 & -1 & -1 & 0 \end{bmatrix},$$

and A is row equivalent to (verify)

$$B = \begin{bmatrix} 1 & 0 & 1 & 2 & 0 \\ 0 & 1 & -2 & -3 & 0 \\ 0 & 0 & 0 & 0 & 1 \\ 0 & 0 & 0 & 0 & 0 \end{bmatrix}.$$

A basis for W is therefore $T = \{\mathbf{w}_1, \mathbf{w}_2, \mathbf{w}_3\}$, where $\mathbf{w}_1 = [1 \quad 0 \quad 1 \quad 2 \quad 0]$, $\mathbf{w}_2 = [0 \quad 1 \quad -2 \quad -3 \quad 0]$, and $\mathbf{w}_3 = [0 \quad 0 \quad 0 \quad 0 \quad 1]$. A basis for V is then

$$\{L^{-1}(\mathbf{w}_1), L^{-1}(\mathbf{w}_2), L^{-1}(\mathbf{w}_3)\} = \{t^4 + t^2 + 2t,\ t^3 - 2t^2 - 3t,\ 1\}. \qquad \blacksquare$$

Definition 2.14. The dimension of the row (column) space of A is called the **row (column) rank** of A. ■

If A and B are row equivalent, then row rank A = row rank B and if A and B are column equivalent, then column rank A = column rank B. Therefore, if we start out with an $m \times n$ matrix A and find a matrix B in reduced row echelon form that is row equivalent to A, then A and B have equal row ranks. But the row rank of B is clearly the number of nonzero rows. Thus we have a good method for finding the row rank of a given matrix A.

Example 4. To determine the row rank of the matrix

$$A = \begin{bmatrix} 1 & 2 & 3 & 1 & 2 \\ 2 & 1 & 2 & 3 & 1 \\ 3 & 3 & 5 & 4 & 3 \\ 1 & -1 & -1 & 2 & -1 \end{bmatrix},$$

we find that A is row equivalent to (verify)

$$B = \begin{bmatrix} 1 & 0 & \frac{1}{3} & \frac{5}{3} & 0 \\ 0 & 1 & \frac{4}{3} & -\frac{1}{3} & 1 \\ 0 & 0 & 0 & 0 & 0 \\ 0 & 0 & 0 & 0 & 0 \end{bmatrix},$$

which means that row rank A = row rank $B = 2$. ■

We may also conclude that if A is an $m \times n$ matrix and P is a nonsingular $m \times m$ matrix, then row rank (PA) = row rank A, for A and PA are row equivalent (Exercise 21 in Section 1.6). Similarly, if Q is a nonsingular $n \times n$ matrix, then column rank (AQ) = column rank A. Moreover, since dimension $R_n = n$, we see that row rank $A \leq n$. Also, since the row space of A is spanned by m vectors, row rank $A \leq m$. Thus row rank $A \leq$ minimum $\{m, n\}$.

We now prove the following important theorem.

Theorem 2.16. The row rank and column rank of the $m \times n$ matrix $A = [a_{ij}]$ are equal.

Proof: Let $\mathbf{v}_1, \mathbf{v}_2, \ldots, \mathbf{v}_m$ be the row vectors of A, where

$$\mathbf{v}_i = [a_{i1} \quad a_{i2} \quad \cdots \quad a_{in}], \qquad i = 1, 2, \ldots, m.$$

Let row rank $A = r$ and let the set of vectors $\{\mathbf{w}_1, \mathbf{w}_2, \ldots, \mathbf{w}_r\}$ form a basis for the row space of A, where $\mathbf{w}_i = [b_{i1} \quad b_{i2} \quad \cdots \quad b_{in}]$ for $i = 1, 2, \ldots, r$. Now each of the row vectors is a linear combination of $\mathbf{w}_1, \mathbf{w}_2, \ldots, \mathbf{w}_r$:

$$\begin{aligned}
\mathbf{v}_1 &= c_{11}\mathbf{w}_1 + c_{12}\mathbf{w}_2 + \cdots + c_{1r}\mathbf{w}_r \\
\mathbf{v}_2 &= c_{21}\mathbf{w}_1 + c_{22}\mathbf{w}_2 + \cdots + c_{2r}\mathbf{w}_r \\
&\vdots \\
\mathbf{v}_m &= c_{m1}\mathbf{w}_1 + c_{m2}\mathbf{w}_2 + \cdots + c_{mr}\mathbf{w}_r,
\end{aligned}$$

where the c_{ij} are uniquely determined real numbers. Recalling that two matrices are equal if and only if the corresponding entries are equal, we equate the entries of these vector equations to get

$$\begin{aligned}
a_{1j} &= c_{11}b_{1j} + c_{12}b_{2j} + \cdots + c_{1r}b_{rj} \\
a_{2j} &= c_{21}b_{1j} + c_{22}b_{2j} + \cdots + c_{2r}b_{rj} \\
&\vdots \\
a_{mj} &= c_{m1}b_{1j} + c_{m2}b_{2j} + \cdots + c_{mr}b_{rj}
\end{aligned}$$

or

$$\begin{bmatrix} a_{1j} \\ a_{2j} \\ \vdots \\ a_{mj} \end{bmatrix} = b_{1j}\begin{bmatrix} c_{11} \\ c_{21} \\ \vdots \\ c_{m1} \end{bmatrix} + b_{2j}\begin{bmatrix} c_{12} \\ c_{22} \\ \vdots \\ c_{m2} \end{bmatrix} + \cdots + b_{rj}\begin{bmatrix} c_{1r} \\ c_{2r} \\ \vdots \\ c_{mr} \end{bmatrix}$$

for $j = 1, 2, \ldots, n$.

Since every column of A is a linear combination of r vectors, the dimension of the column space of A is at most r, or column rank $A \leq r =$ row rank A. Similarly, we get row rank $A \leq$ column rank A. Hence the row and column ranks of A are equal.

Alternate Proof: Let $\mathbf{x}_1, \mathbf{x}_2, \ldots, \mathbf{x}_n$ denote the columns of A. To determine the dimension of the column space of A, we use the procedure in the alternate proof of Theorem 2.7. Thus we consider the equation

$$c_1\mathbf{x}_1 + c_2\mathbf{x}_2 + \cdots + c_n\mathbf{x}_n = \mathbf{0}.$$

We now transform the augmented matrix, $[A \mid \mathbf{0}]$, of this homogeneous system to reduced row echelon form. The vectors corresponding to the columns containing the leading 1's form a basis for the column space of A. Thus the column rank of A is the number of leading 1's. But this number is also the number of nonzero rows in the reduced row echelon form matrix that is row equivalent to A, so it is the row rank of A. Thus row rank A = column rank A. ■

Example 5. Let A be the matrix of Example 4. The column space of A is the subspace V of R^4 spanned by the columns of A:

$$\mathbf{v}_1 = \begin{bmatrix} 1 \\ 2 \\ 3 \\ 1 \end{bmatrix}, \quad \mathbf{v}_2 = \begin{bmatrix} 2 \\ 1 \\ 3 \\ -1 \end{bmatrix}, \quad \mathbf{v}_3 = \begin{bmatrix} 3 \\ 2 \\ 5 \\ -1 \end{bmatrix},$$

$$\mathbf{v}_4 = \begin{bmatrix} 1 \\ 3 \\ 4 \\ 2 \end{bmatrix}, \quad \text{and} \quad \mathbf{v}_5 = \begin{bmatrix} 2 \\ 1 \\ 3 \\ -1 \end{bmatrix}.$$

Writing the columns of A as row vectors, we obtain the matrix

$$A^T = \begin{bmatrix} 1 & 2 & 3 & 1 \\ 2 & 1 & 3 & -1 \\ 3 & 2 & 5 & -1 \\ 1 & 3 & 4 & 2 \\ 2 & 1 & 3 & -1 \end{bmatrix},$$

where A^T is the transpose of A (see Section 1.2). Transforming A^T to reduced row echelon form, we obtain (verify)

$$C = \begin{bmatrix} 1 & 0 & 1 & -1 \\ 0 & 1 & 1 & 1 \\ 0 & 0 & 0 & 0 \\ 0 & 0 & 0 & 0 \\ 0 & 0 & 0 & 0 \end{bmatrix}.$$

Thus $\{[1 \quad 0 \quad 1 \quad -1], [0 \quad 1 \quad 1 \quad 1]\}$ is a basis for the row space of A^T, and

$$\left\{\begin{bmatrix} 1 \\ 0 \\ 1 \\ -1 \end{bmatrix}, \begin{bmatrix} 0 \\ 1 \\ 1 \\ 1 \end{bmatrix}\right\}$$

is a basis for the column space of A. Hence column rank $A = 2$. ■

Since the row and column ranks of a matrix are equal, we now merely refer to the **rank** of a matrix. Note that rank $I_n = n$. Theorem 1.21 states that A is equivalent to B if and only if there exist nonsingular matrices P and Q such that $B = PAQ$. If A is equivalent to B, then rank A = rank B, for rank B = rank (PAQ) = rank (PA) = rank A.

We also recall from Section 1.7 that if A is an $m \times n$ matrix, then A is equivalent to a matrix $C = \begin{bmatrix} I_r & O \\ O & O \end{bmatrix}$. Now rank A = rank $C = r$. We use these facts to establish the result that if A and B are $m \times n$ matrices of equal rank, then A and B are equivalent. Thus let rank $A = r$ = rank B. Then there exist nonsingular matrices P_1, Q_1, P_2, and Q_2 such that

$$P_1AQ_1 = \begin{bmatrix} I_r & O \\ O & O \end{bmatrix} = P_2BQ_2.$$

Then $P_2^{-1}P_1AQ_1Q_2^{-1} = B$. Letting $P = P_2^{-1}P_1$ and $Q = Q_1Q_2^{-1}$, we find that P and Q are nonsingular and $B = PAQ$. Hence A and B are equivalent.

At this point we can show that an $n \times n$ matrix is nonsingular if and only if its rank is n. We first prove Theorem 2.17.

Theorem 2.17. *If A is an $n \times n$ matrix, then rank $A = n$ if and only if A is row equivalent to I_n.*

Proof: If rank $A = n$, then A is row equivalent to a matrix B in reduced row echelon form, and rank $B = n$. Since rank $B = n$, we conclude that B has no zero rows, and this implies (by Exercise 3 of Section 1.5) that $B = I_n$. Hence A is row equivalent to I_n.

Conversely, if A is row equivalent to I_n, then rank A = rank $I_n = n$. ■

Corollary 2.7. *A is nonsingular if and only if rank $A = n$.*

Proof: This follows from Theorem 2.17 and Corollary 1.3. ■

From a practical point of view, this result is not too useful, since most of the time we want to know not only whether A is nonsingular but also its inverse. The method developed in Chapter 1 enables us to find A^{-1}, if it exists, and tells us if it does not exist. Thus we do not have to learn first if A^{-1} exists and then go through another procedure to obtain it.

Corollary 2.8. *The homogeneous system* $A\mathbf{x} = \mathbf{0}$ *where* A *is* $n \times n$, *has a nontrivial solution if and only if rank* $A < n$.

Proof: This follows from Corollary 2.7 and from the fact that $A\mathbf{x} = \mathbf{0}$ has a nontrivial solution if and only if A is singular (Theorem 1.17). ■

If A is an $m \times n$ matrix, it can be shown (see Supplementary Exercise 20) that the dimension of null space of $A = n -$ rank A. That is,

$$\begin{aligned}\dim(\text{solution space of } A\mathbf{x} = \mathbf{0}) &= n - \text{rank } A \\ &= \text{number of columns of } A - \text{rank } A.\end{aligned}$$

Example 6. Let

$$A = \begin{bmatrix} 1 & 2 & 0 \\ 0 & 1 & 3 \\ 2 & 1 & 3 \end{bmatrix}.$$

Then if we transform A to reduced row echelon form B, we find that $B = I_3$ (verify). Thus rank $A = 3$, so the dimension of the null space of A is $3 - 3 = 0$. That is, the dimension of the solution space of $A\mathbf{x} = \mathbf{0}$ is zero, which means that $A\mathbf{x} = \mathbf{0}$ has only the trivial solution. ■

Example 7. Let

$$A = \begin{bmatrix} 1 & 2 & 0 & 1 \\ 1 & 1 & -3 & 0 \\ 1 & 3 & 3 & 2 \\ -1 & -1 & 3 & 1 \end{bmatrix}.$$

Then A is row equivalent to (verify)

$$B = \begin{bmatrix} 1 & 0 & 6 & 3 \\ 0 & 1 & 3 & 1 \\ 0 & 0 & 0 & 0 \\ 0 & 0 & 0 & 0 \end{bmatrix},$$

so rank $A = 2$. It then follows that the dimension of the solution space of $A\mathbf{x} = \mathbf{0}$ is

$$4 - 2 = 2.$$

Hence the homogeneous system $A\mathbf{x} = \mathbf{0}$ has a nontrivial solution. ■

Our final application of rank is to linear systems. If we consider the linear system $A\mathbf{x} = \mathbf{b}$ of m equations in n unknowns, where $A = [a_{ij}]$, then we observe that the system can also be written as the equation (verify)

$$x_1\begin{bmatrix} a_{11} \\ a_{21} \\ \vdots \\ a_{m1} \end{bmatrix} + x_2\begin{bmatrix} a_{12} \\ a_{22} \\ \vdots \\ a_{m2} \end{bmatrix} + \cdots + x_n\begin{bmatrix} a_{1n} \\ a_{2n} \\ \vdots \\ a_{mn} \end{bmatrix} = \begin{bmatrix} b_1 \\ b_2 \\ \vdots \\ b_m \end{bmatrix}.$$

Thus $A\mathbf{x} = \mathbf{b}$ is consistent (has a solution) if and only if $\mathbf{b}$ is a linear combination of the columns of A, that is, if and only if $\mathbf{b}$ belongs to the column space of A. This means that if $A\mathbf{x} = \mathbf{b}$ has a solution, then rank A = rank $[A \mid \mathbf{b}]$, where $[A \mid \mathbf{b}]$ is the augmented matrix of the linear system. Conversely, if rank $[A \mid \mathbf{b}]$ = rank A, then $\mathbf{b}$ is in the column space of A, which means that the linear system has a solution. Thus $A\mathbf{x} = \mathbf{b}$ is consistent if and only if rank A = rank $[A \mid \mathbf{b}]$. This result, although of interest, is not of great computational value, since we usually are interested in finding a solution rather than in merely knowing whether or not a solution exists.

Example 8. Consider the linear system

$$\begin{bmatrix} 2 & 1 & 3 \\ 1 & -2 & 2 \\ 0 & 1 & 3 \end{bmatrix}\begin{bmatrix} x_1 \\ x_2 \\ x_3 \end{bmatrix} = \begin{bmatrix} 1 \\ 2 \\ 3 \end{bmatrix}.$$

Since rank A = rank $[A \mid \mathbf{b}] = 3$ (verify), the linear system has a solution. ■

Example 9. The linear system

$$\begin{bmatrix} 1 & 2 & 3 \\ 1 & -3 & 4 \\ 2 & -1 & 7 \end{bmatrix}\begin{bmatrix} x_1 \\ x_2 \\ x_3 \end{bmatrix} = \begin{bmatrix} 4 \\ 5 \\ 6 \end{bmatrix}$$

has no solution because rank $A = 2$ and rank $[A \mid \mathbf{b}] = 3$ (verify). ■

Let $S = \{\mathbf{v}_1, \mathbf{v}_2, \ldots, \mathbf{v}_n\}$ be a set of n vectors in R_n, and let A be the matrix whose jth row is $\mathbf{v}_j$. It is not difficult to show (Exercise 21) that S is linearly independent if and only if rank $A = n$. Similarly, let $S = \{\mathbf{v}_1, \mathbf{v}_2, \ldots, \mathbf{v}_n\}$ be a set of n vectors in R^n, and let A be the matrix whose jth column is $\mathbf{v}_j$. It can then be shown (Exercise 22) that S is linearly independent if and only if rank $A = n$.

The following statements are equivalent for an $n \times n$ matrix A:

1. A is nonsingular.
2. $A\mathbf{x} = \mathbf{0}$ has only the trivial solution.
3. A is row (column) equivalent to I_n.
4. The system $A\mathbf{x} = \mathbf{b}$ has a unique solution for every vector $\mathbf{b}$ in R^n.
5. A is a product of elementary matrices.
6. A has rank n.
7. The rows (columns) of A form a linearly independent set of vectors in R_n (R^n).

2.6. EXERCISES

1. Find a basis for the subspace V of R^3 spanned by

$$S = \left\{ \begin{bmatrix} 1 \\ 2 \\ 3 \end{bmatrix}, \begin{bmatrix} 2 \\ 1 \\ 4 \end{bmatrix}, \begin{bmatrix} -1 \\ -1 \\ 2 \end{bmatrix}, \begin{bmatrix} 0 \\ 1 \\ 2 \end{bmatrix}, \begin{bmatrix} 1 \\ 1 \\ 1 \end{bmatrix} \right\}$$

and write each of the following vectors in terms of the basis vectors:

(a) $\begin{bmatrix} 3 \\ 4 \\ 12 \end{bmatrix}$. (b) $\begin{bmatrix} 3 \\ 2 \\ 2 \end{bmatrix}$. (c) $\begin{bmatrix} 1 \\ 2 \\ 6 \end{bmatrix}$.

2. Find a basis for the subspace of P_3 spanned by

$$S = \{t^3 + t^2 + 2t + 1,\ t^3 - 3t + 1,\ t^2 + t + 2,\ t + 1,\ t^3 + 1\}.$$

3. Find a basis for the subspace of ${}_2R_2$ spanned by

$$S = \left\{ \begin{bmatrix} 1 & 2 \\ 1 & 1 \end{bmatrix}, \begin{bmatrix} 2 & 1 \\ 3 & 1 \end{bmatrix}, \begin{bmatrix} 0 & 2 \\ 1 & 2 \end{bmatrix}, \begin{bmatrix} 3 & 2 \\ 1 & 4 \end{bmatrix}, \begin{bmatrix} 5 & 0 \\ 0 & -1 \end{bmatrix} \right\}.$$

4. Find a basis for the subspace of R_2 spanned by

$$S = \{[1 \quad 2], [2 \quad 3], [3 \quad 1], [-4 \quad 3]\}.$$

In Exercises 5 and 6, find the row and column ranks of the given matrices.

5. (a) $\begin{bmatrix} 1 & 2 & 3 & 2 & 1 \\ 3 & 1 & -5 & -2 & 1 \\ 7 & 8 & -1 & 2 & 5 \end{bmatrix}$.

(b) $\begin{bmatrix} 1 & 3 & 2 & 0 & 0 & 1 \\ 2 & 1 & -5 & 1 & 2 & 0 \\ 3 & 2 & 5 & 1 & -2 & 1 \\ 5 & 8 & 9 & 1 & -2 & 2 \\ 9 & 9 & 4 & 2 & 0 & 2 \end{bmatrix}$.

6. (a) $\begin{bmatrix} 1 & 2 & 3 & 2 & 1 \\ 0 & 5 & 4 & 0 & -1 \\ 2 & -1 & 2 & 4 & 3 \end{bmatrix}$.

(b) $\begin{bmatrix} 1 & 1 & -1 & 2 & 0 \\ 2 & -4 & 0 & 1 & 1 \\ 5 & -1 & -3 & 7 & 1 \\ 3 & -9 & 1 & 0 & 2 \end{bmatrix}$.

7. Let A be an $m \times n$ matrix in row echelon form. Prove that rank A = the number of nonzero rows of A.

8. For each of the following matrices, verify Theorem 2.16 by computing the row and column ranks.

(a) $\begin{bmatrix} 1 & 2 & 3 \\ -1 & 2 & 1 \\ 3 & 1 & 2 \end{bmatrix}$. (b) $\begin{bmatrix} 1 & -2 & -1 \\ 2 & -1 & 3 \\ 7 & -8 & 3 \end{bmatrix}$.

(c) $\begin{bmatrix} 1 & -2 & -1 \\ 2 & -1 & 3 \\ 7 & -8 & 3 \\ 5 & -7 & 0 \end{bmatrix}$.

In Exercises 9 and 10, compute the rank of the given matrix.

9. (a) $\begin{bmatrix} 1 & -1 & 2 & 3 \\ 2 & 6 & -8 & 1 \\ 5 & 3 & -2 & 10 \end{bmatrix}$.

(b) $\begin{bmatrix} 1 & 2 & 0 & 3 \\ 3 & 2 & -1 & 0 \\ 2 & -1 & 0 & 1 \end{bmatrix}$.

10. (a) $\begin{bmatrix} 1 & 3 & -2 & 4 \\ -1 & 4 & -5 & 10 \\ 3 & 2 & 1 & -2 \\ 3 & -5 & 8 & -16 \end{bmatrix}$.

(b) $\begin{bmatrix} 1 & 1 & 1 & 1 \\ 2 & -1 & 0 & 0 \\ 0 & 1 & -1 & 2 \\ 1 & 1 & -1 & 2 \end{bmatrix}$.

11. Which of the following matrices are equivalent?

$$A = \begin{bmatrix} 1 & 2 & 1 & 3 \\ 2 & 1 & -4 & -5 \\ 7 & 8 & -5 & -1 \\ 10 & 14 & -2 & 3 \end{bmatrix},$$

$$B = \begin{bmatrix} 1 & 2 & 1 & 3 \\ 2 & 1 & -4 & -5 \\ 1 & 1 & 0 & 0 \\ 0 & 0 & 1 & 1 \end{bmatrix},$$

$$C = \begin{bmatrix} 1 & 5 & 1 & 3 \\ 2 & 1 & 2 & 1 \\ -3 & 0 & 1 & 0 \\ 4 & 7 & -4 & 3 \end{bmatrix},$$

$$D = \begin{bmatrix} 1 & 2 & -4 & 3 \\ 4 & 7 & -4 & 1 \\ 7 & 12 & -4 & -1 \\ 2 & 3 & 4 & -5 \end{bmatrix},$$

$$E = \begin{bmatrix} 4 & 3 & -1 & -5 \\ -2 & -6 & -7 & 10 \\ -2 & -3 & -2 & 5 \\ 0 & -6 & -10 & 10 \end{bmatrix}.$$

In Exercises 12 and 13, determine which of the given linear systems are consistent by comparing the ranks of the coefficients and augmented matrices.

12. (a) $\begin{bmatrix} 1 & 2 & 5 & -2 \\ 2 & 3 & -2 & 4 \\ 5 & 1 & 0 & 2 \end{bmatrix} \begin{bmatrix} x_1 \\ x_2 \\ x_3 \\ x_4 \end{bmatrix} = \begin{bmatrix} 0 \\ 0 \\ 0 \end{bmatrix}$.

(b) $\begin{bmatrix} 1 & 2 & 5 & -2 \\ 2 & 3 & -2 & 4 \\ 5 & 1 & 0 & 2 \end{bmatrix} \begin{bmatrix} x_1 \\ x_2 \\ x_3 \\ x_4 \end{bmatrix} = \begin{bmatrix} -1 \\ -13 \\ 3 \end{bmatrix}$.

13. (a) $\begin{bmatrix} 1 & -2 & -3 & 4 \\ 4 & -1 & -5 & 6 \\ 2 & 3 & 1 & -2 \end{bmatrix} \begin{bmatrix} x_1 \\ x_2 \\ x_3 \\ x_4 \end{bmatrix} = \begin{bmatrix} 1 \\ 2 \\ 2 \end{bmatrix}$.

(b) $\begin{bmatrix} 1 & 1 & 1 \\ 1 & -1 & 1 \\ 5 & 1 & 5 \end{bmatrix} \begin{bmatrix} x_1 \\ x_2 \\ x_3 \end{bmatrix} = \begin{bmatrix} 6 \\ 2 \\ 5 \end{bmatrix}.$

In Exercises 14 and 15, use Corollary 2.7 to find which of the given matrices are nonsingular.

14. (a) $\begin{bmatrix} 1 & 2 & -3 \\ -1 & 2 & 3 \\ 0 & 8 & 0 \end{bmatrix}.$

(b) $\begin{bmatrix} 1 & 2 & -3 \\ -1 & 2 & 3 \\ 0 & 1 & 1 \end{bmatrix}.$

15. (a) $\begin{bmatrix} 1 & 1 & 2 \\ -1 & 3 & 4 \\ -5 & 7 & 8 \end{bmatrix}.$

(b) $\begin{bmatrix} 1 & 1 & 4 & -1 \\ 1 & 2 & 3 & 2 \\ -1 & 3 & 2 & 1 \\ -2 & 6 & 12 & -4 \end{bmatrix}.$

In Exercises 16 and 17, use Corollary 2.8 to find which of the given homogeneous systems have a nontrivial solution.

16. (a) $\begin{bmatrix} 1 & 2 & 3 \\ 0 & 1 & 0 \\ 1 & 0 & 3 \end{bmatrix} \begin{bmatrix} x_1 \\ x_2 \\ x_3 \end{bmatrix} = \begin{bmatrix} 0 \\ 0 \\ 0 \end{bmatrix}.$

(b) $\begin{bmatrix} 1 & 1 & 2 & -1 \\ 1 & 3 & -1 & 2 \\ 1 & 1 & 1 & 3 \\ 1 & 2 & 1 & 1 \end{bmatrix} \begin{bmatrix} x_1 \\ x_2 \\ x_3 \\ x_4 \end{bmatrix} = \begin{bmatrix} 0 \\ 0 \\ 0 \\ 0 \end{bmatrix}.$

17. (a) $\begin{bmatrix} 1 & 2 & -1 \\ 2 & -1 & 3 \\ 5 & -4 & 3 \end{bmatrix} \begin{bmatrix} x_1 \\ x_2 \\ x_3 \end{bmatrix} = \begin{bmatrix} 0 \\ 0 \\ 0 \end{bmatrix}.$

(b) $\begin{bmatrix} 1 & 2 & 3 \\ -1 & 2 & -1 \\ 1 & 0 & 3 \end{bmatrix} \begin{bmatrix} x_1 \\ x_2 \\ x_3 \end{bmatrix} = \begin{bmatrix} 0 \\ 0 \\ 0 \end{bmatrix}.$

In Exercises 18 and 19, find rank A by obtaining a matrix of the form $\begin{bmatrix} I_r & O \\ O & O \end{bmatrix}$ that is equivalent to A.

18. (a) $A = \begin{bmatrix} 1 & 1 & -2 \\ 1 & 2 & 3 \\ 0 & 1 & 3 \end{bmatrix}.$

(b) $A = \begin{bmatrix} 1 & 1 & -2 & 0 & 0 \\ 1 & 2 & 3 & 6 & 7 \\ 2 & 1 & 3 & 6 & 5 \end{bmatrix}.$

19. (a) $A = \begin{bmatrix} 1 & 1 & -2 \\ 1 & 2 & 3 \\ 3 & 4 & -1 \end{bmatrix}.$

(b) $A = \begin{bmatrix} 1 & -1 & 2 & 3 \\ 2 & 2 & 0 & 1 \\ 1 & -5 & 6 & 8 \\ 4 & 0 & 4 & 6 \end{bmatrix}.$

20. Let A be a 3×5 matrix.
(a) Give *all* possible values for the rank of A.
(b) If the rank of A is 3, what is the dimension of its column space?
(c) If the rank of A is 3, what is the dimension of the solution space of the homogeneous system $A\mathbf{x} = \mathbf{0}$?

21. Let $S = \{\mathbf{v}_1, \mathbf{v}_2, \ldots, \mathbf{v}_n\}$ be a set of n vectors in R_n, and let A be the matrix whose jth row is $\mathbf{v}_j$. Show that S is linearly independent if and only if rank $A = n$.

22. Let $S = \{\mathbf{v}_1, \mathbf{v}_2, \ldots, \mathbf{v}_n\}$ be a set of n vectors in R^n, and let A be the matrix whose jth column is $\mathbf{v}_j$. Show that S is linearly independent if and only if rank $A = n$.

23. Let A be an $n \times n$ matrix. Show that the homogeneous system $A\mathbf{x} = \mathbf{0}$ has a nontrivial solution if and only if the columns of A are linearly dependent.

24. Let A be an $n \times n$ matrix. Show that if rank $A = n$, then the columns of A are linearly independent.

25. Let A be an $n \times n$ matrix. Prove that if the rows of A are linearly independent, then the columns of A span R^n.

26. Let $S = \{\mathbf{v}_1, \mathbf{v}_2, \ldots, \mathbf{v}_k\}$ be a basis for a subspace V of R_n that is obtained by the method of Example 1. If

$$\mathbf{v} = \begin{bmatrix} a_1 & a_2 & \cdots & a_n \end{bmatrix}$$

belongs to V and the leading 1's in the reduced row echelon form from the method in Example 1 occur in columns $j_1, j_2, \ldots, j_k$, then show that

$$\mathbf{v} = a_{j_1}\mathbf{v}_1 + a_{j_2}\mathbf{v}_2 + \cdots + a_{j_k}\mathbf{v}_k.$$

27. Determine if your software has a command for computing the rank of a matrix. If it does, experiment with the command on matrices A in Examples 4 and 5 and Exercises 9 and 10.

28. Assuming that exact arithmetic is used, rank A is the number of nonzero rows in the reduced row echelon form of A. Compare the results using your rank command and the reduced row echelon form approach on the following matrices:

$$A = \begin{bmatrix} 1 & 1 \\ 0 & 1 \times 10^{-j} \end{bmatrix}, \qquad j = 5,\ 10,\ 16.$$

(See Exercise 32 in Section 2.3.)

Supplementary Exercises

1. Let $C[a, b]$ denote the set of all real-valued continuous functions defined on $[a, b]$. If f and g are in $C[a, b]$, we define $f \oplus g$ by $(f \oplus g)(t) = f(t) + g(t)$, for t in $[a, b]$. If f is in $C[a, b]$ and c is a scalar, we define $c \odot f$ by $(c \odot f)(t) = c\,f(t)$, for t in $[a, b]$.
(a) Show that $C[a, b]$ is a real vector space.
(b) Let $W(k)$ be the set of all functions in $C[a, b]$ with $f(a) = k$. For what values of k will $W(k)$ be a subspace of $C[a, b]$?
(c) Let $t_1, t_2, \ldots, t_n$ be a fixed set of points in $[a, b]$. Show that the subset of all functions f in $C[a, b]$ that have roots at $t_1, t_2, \ldots, t_n$, that is, $f(t_i) = 0$ for $i = 1, 2, \ldots, n$, forms a subspace.

2. In R^4, let W be the subset of all vectors

$$\mathbf{v} = \begin{bmatrix} a_1 \\ a_2 \\ a_3 \\ a_4 \end{bmatrix}$$

that satisfy $a_4 - a_3 = a_2 - a_1$.
(a) Show that W is a subspace of R^4.
(b) Show that $S = \left\{ \begin{bmatrix} 1 \\ 0 \\ 0 \\ -1 \end{bmatrix}, \begin{bmatrix} 0 \\ 1 \\ 0 \\ 1 \end{bmatrix}, \begin{bmatrix} 1 \\ 1 \\ 1 \\ 1 \end{bmatrix}, \begin{bmatrix} 0 \\ 0 \\ 1 \\ 1 \end{bmatrix} \right\}$ spans W.
(c) Find a subset of S that is a basis for W.
(d) Express $\mathbf{v} = \begin{bmatrix} 0 \\ 4 \\ 2 \\ 6 \end{bmatrix}$ as a linear combination of the basis obtained in part (c).

3. Consider $\mathbf{v}_1 = \begin{bmatrix} 1 \\ 1 \\ -2 \\ 1 \end{bmatrix}$, $\mathbf{v}_2 = \begin{bmatrix} 1 \\ 5 \\ 2 \\ -1 \end{bmatrix}$, and $\mathbf{v}_3 = \begin{bmatrix} 3 \\ 0 \\ 2 \\ 1 \end{bmatrix}$. Determine whether the vector $\mathbf{v}$ belongs to span $\{\mathbf{v}_1, \mathbf{v}_2, \mathbf{v}_3\}$.

(a) $\mathbf{v} = \begin{bmatrix} 0 \\ 7 \\ 4 \\ 0 \end{bmatrix}$. (b) $\mathbf{v} = \begin{bmatrix} 5 \\ 6 \\ 2 \\ 1 \end{bmatrix}$. (c) $\mathbf{v} = \begin{bmatrix} 3 \\ -8 \\ -6 \\ 5 \end{bmatrix}$.

4. Let A be a fixed $n \times n$ matrix and let the set of all $n \times n$ matrices B such that $AB = BA$ be denoted by $C(A)$. Is $C(A)$ a subspace of ${}_nR_n$?

5. Let W and U be subspaces of vector space V.
(a) Show that $W \cup U$, the set of all vectors $\mathbf{v}$ that are either in W or in U is not always a subspace of V.
(b) Show that $W \cap U$, the set of all vectors $\mathbf{v}$ that are in both W and U, is a subspace of V.

6. Prove that a subspace W of R^3 coincides with R^3 if and only if it contains the vectors $\begin{bmatrix} 1 \\ 0 \\ 0 \end{bmatrix}$, $\begin{bmatrix} 0 \\ 1 \\ 0 \end{bmatrix}$, and $\begin{bmatrix} 0 \\ 0 \\ 1 \end{bmatrix}$.

7. Let A be a fixed $m \times n$ matrix and define W to be the subset of all $m \times 1$ matrices $\mathbf{b}$ in R^m for which the linear system $A\mathbf{x} = \mathbf{b}$ has a solution.
(a) Is W a subspace of R^m?
(b) What is the relationship between W and the column space of A?

8. Consider vector space R_2.
(a) For what values of m and b will all vectors of the form $[x \quad mx + b]$ be a subspace of R_2?
(b) For what value of r will the set of all vectors of the form $[x \quad rx^2]$ be a subspace of R_2?

9. For what values of a is the vector $\begin{bmatrix} a^2 \\ -3a \\ -2 \end{bmatrix}$ in

$$\text{span}\left\{\begin{bmatrix} 1 \\ 2 \\ 3 \end{bmatrix}, \begin{bmatrix} 0 \\ 1 \\ 1 \end{bmatrix}, \begin{bmatrix} 1 \\ 3 \\ 4 \end{bmatrix}\right\}?$$

10. For what values of a is the vector $\begin{bmatrix} a^2 \\ a \\ 1 \end{bmatrix}$ in

$$\text{span}\left\{\begin{bmatrix} 1 \\ 2 \\ 3 \end{bmatrix}, \begin{bmatrix} 1 \\ 1 \\ 1 \end{bmatrix}, \begin{bmatrix} 0 \\ 1 \\ 2 \end{bmatrix}\right\}?$$

11. For what values of k will the set S form a basis for R^6?

$$S = \left\{\begin{bmatrix} 1 \\ 2 \\ -2 \\ 1 \\ 1 \\ 1 \end{bmatrix}, \begin{bmatrix} 0 \\ 2 \\ 0 \\ 0 \\ 0 \\ 0 \end{bmatrix}, \begin{bmatrix} 0 \\ 5 \\ k \\ 1 \\ 0 \\ 0 \end{bmatrix}, \begin{bmatrix} 0 \\ 2 \\ 1 \\ k \\ 0 \\ 0 \end{bmatrix}, \begin{bmatrix} 0 \\ 1 \\ -2 \\ 4 \\ -3 \\ 1 \end{bmatrix}, \begin{bmatrix} 0 \\ 3 \\ 1 \\ 1 \\ 1 \\ 0 \end{bmatrix}\right\}.$$

12. Consider the subspace of R^4 given by $W =$

$$\text{span}\left\{\begin{bmatrix} 1 \\ 1 \\ 1 \\ 1 \end{bmatrix}, \begin{bmatrix} 3 \\ 2 \\ 1 \\ 1 \end{bmatrix}, \begin{bmatrix} 1 \\ 2 \\ 3 \\ 1 \end{bmatrix}, \begin{bmatrix} 0 \\ 2 \\ 4 \\ 1 \end{bmatrix}\right\}.$$

(a) Determine a subset S of the spanning set that is a basis for W.
(b) Find a basis T for W that is not a subset of the spanning set.
(c) Find the coordinate vector of $\mathbf{v} = \begin{bmatrix} 3 \\ -1 \\ -5 \\ 0 \end{bmatrix}$ with respect to each of the bases from parts (a) and (b).

13. Prove that if $S = \{\mathbf{v}_1, \mathbf{v}_2, \ldots, \mathbf{v}_k\}$ is a basis for a subspace W of vector space V, then there is a basis for V that includes the set S. (*Hint:* Use Theorem 2.9.)

14. Let $V = \text{span}\,\{\mathbf{v}_1, \mathbf{v}_2\}$, where

$$\mathbf{v}_1 = \begin{bmatrix} 1 \\ 0 \\ 2 \end{bmatrix} \quad \text{and} \quad \mathbf{v}_2 = \begin{bmatrix} 1 \\ 1 \\ 1 \end{bmatrix}.$$

Find a basis S for R^3 that includes $\mathbf{v}_1$ and $\mathbf{v}_2$. (*Hint:* Use the technique developed in the Alternate Constructive Proof of Theorem 2.7.)

15. Describe the set of all vectors $\mathbf{b}$ in R^3 for which the linear system $A\mathbf{x} = \mathbf{b}$ is consistent.

(a) $A = \begin{bmatrix} 1 & -2 & 1 & 0 \\ 2 & 1 & 1 & 2 \\ 1 & -7 & 2 & -2 \end{bmatrix}$.

(b) $A = \begin{bmatrix} 1 & 2 & 1 \\ 1 & 3 & 1 \\ 2 & 4 & 3 \end{bmatrix}$.

16. Find a basis for the solution space of the homogeneous system $(\lambda I_3 - A)\mathbf{x} = \mathbf{0}$ for the given scalar λ and given matrix A.

(a) $\lambda = 1$, $A = \begin{bmatrix} 0 & 0 & 1 \\ 1 & 0 & -3 \\ 0 & 1 & 3 \end{bmatrix}$.

(b) $\lambda = 3$, $A = \begin{bmatrix} 1 & 1 & -2 \\ -1 & 2 & 1 \\ 0 & 1 & -1 \end{bmatrix}$.

17. Show that rank A = rank A^T, for any $m \times n$ matrix A.

18. Let A and B be $m \times n$ matrices that are row equivalent.
(a) Prove that rank A = rank B.
(b) Prove that for $\mathbf{x}$ in R^n, $A\mathbf{x} = \mathbf{0}$ if and only if $B\mathbf{x} = \mathbf{0}$.

19. Let A be $m \times n$ and B be $n \times k$.
(a) Prove that rank $(AB) \leq \min\{\text{rank } A, \text{rank } B\}$.
(b) Find A and B such that rank $(AB) < \min\{\text{rank } A, \text{rank } B\}$.
(c) If $k = n$ and B is nonsingular, prove that rank (AB) = rank A.
(d) If $m = n$ and A is nonsingular, prove that rank (AB) = rank B.
(e) For nonsingular matrices P and Q, what is rank (PAQ)?

20. For an $m \times n$ matrix A, let the set of all vectors $\mathbf{x}$ in R^n such that $A\mathbf{x} = \mathbf{0}$ be denoted by NS (A), which in Example 16 of Section 2.2 has been shown to be a subspace of R^n, called the null space of A.
(a) Prove that rank A + dim NS $(A) = n$.
(b) For $m = n$, prove that A is nonsingular if and only if dim NS $(A) = 0$.

21. Let A be an $m \times n$ matrix and B a nonsingular $m \times m$ matrix. Prove that NS (BA) = NS (A) (see Exercise 20).

22. Find dim NS (A) (see Exercise 20) for each of the following matrices.

(a) $A = \begin{bmatrix} 1 & 2 & 1 \\ 2 & 0 & 3 \\ 0 & 4 & -1 \end{bmatrix}$.

(b) $A = \begin{bmatrix} 1 & 2 & 4 & 1 \\ 2 & 1 & 1 & 1 \\ -1 & 4 & 10 & 1 \end{bmatrix}$.

23. Any nonsingular 3×3 matrix P represents a transition matrix from some ordered basis $T = \{\mathbf{w}_1, \mathbf{w}_2, \mathbf{w}_3\}$ to some other ordered basis $S = \{\mathbf{v}_1, \mathbf{v}_2, \mathbf{v}_3\}$. Let

$$P = \begin{bmatrix} 1 & 1 & 1 \\ 0 & 2 & 1 \\ 1 & 0 & 3 \end{bmatrix}.$$

(a) If $\mathbf{v}_1 = \begin{bmatrix} 1 \\ 1 \\ 0 \end{bmatrix}$, $\mathbf{v}_2 = \begin{bmatrix} 1 \\ 0 \\ 1 \end{bmatrix}$, and $\mathbf{v}_3 = \begin{bmatrix} 1 \\ 1 \\ 1 \end{bmatrix}$, find T.

(b) If $\mathbf{w}_1 = \begin{bmatrix} 1 \\ 0 \\ 2 \end{bmatrix}$, $\mathbf{w}_2 = \begin{bmatrix} 2 \\ 0 \\ 1 \end{bmatrix}$, and $\mathbf{w}_3 = \begin{bmatrix} 1 \\ 1 \\ 0 \end{bmatrix}$, find S.

24. In Supplementary Exercise 18 for Chapter 1, we defined the outer product of two $n \times 1$ column matrices X and Y as XY^T. Determine the rank of an outer product.

25. Suppose that A is an $n \times n$ matrix and that there is no nonzero vector $\mathbf{x}$ in R^n such that $A\mathbf{x} = \mathbf{x}$. Show that $A - I_n$ is nonsingular.

26. Let A be an $m \times n$ matrix. Prove that if A^TA is nonsingular, then rank $A = n$.

27. Prove or disprove by finding a counterexample.
(a) rank $(A + B) \leq \max\{\text{rank } A, \text{rank } B\}$.
(b) rank $(A + B) \geq \min\{\text{rank } A, \text{rank } B\}$.
(c) rank $(A + B)$ = rank A + rank B.

3 Inner Product Spaces[†]

As we noted in Chapter 2, when physicists talk about vectors in R^2 and R^3, they usually refer to objects that have magnitude and direction. However, thus far in our study of vector spaces we have refrained from discussing these notions. In this chapter we deal with magnitude and direction in a vector space.

3.1. *The Standard Inner Product on* R^2 *and* R^3

Length

In this section we discuss the notions of magnitude and direction in R^2 and R^3 and, in the next section, generalize these to R^n. We consider R^2 and R^3 with the usual Cartesian coordinate system. The **length** or **magnitude** of the vector $\mathbf{v} = \begin{bmatrix} v_1 \\ v_2 \end{bmatrix}$ in R^2, denoted by $\|\mathbf{v}\|$, is by the Pythagorean theorem (see Figure 3.1)

$$\|\mathbf{v}\| = \sqrt{v_1^2 + v_2^2}. \tag{1}$$

Example 1. If $\mathbf{v} = \begin{bmatrix} 2 \\ -5 \end{bmatrix}$, then, by Equation (1),

$$\|\mathbf{v}\| = \sqrt{(2)^2 + (-5)^2} = \sqrt{4 + 25} = \sqrt{29}. \qquad \blacksquare$$

[†] This chapter may also be covered after Section 6.1 and before Section 6.2, which is where it is used.

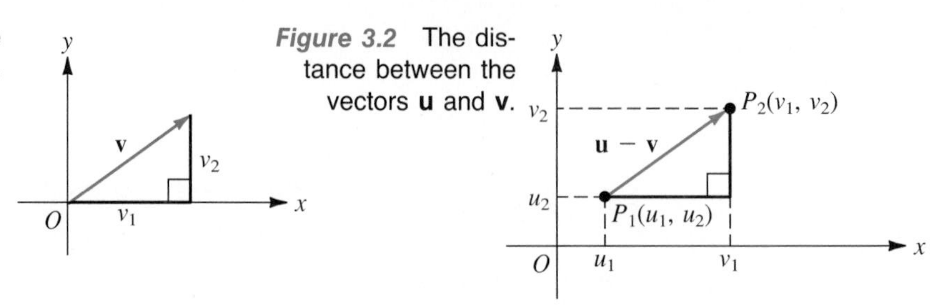

Figure 3.1 The length of **v**.

Figure 3.2 The distance between the vectors **u** and **v**.

If $\mathbf{u} = \begin{bmatrix} u_1 \\ u_2 \end{bmatrix}$ and $\mathbf{v} = \begin{bmatrix} v_1 \\ v_2 \end{bmatrix}$ are vectors in R^2, then the **distance** between **u** and **v** is defined as $\|\mathbf{u} - \mathbf{v}\| = \left\|\begin{matrix} u_1 - v_1 \\ u_2 - v_2 \end{matrix}\right\|$. Thus

$$\|\mathbf{u} - \mathbf{v}\| = \sqrt{(u_1 - v_1)^2 + (u_2 - v_2)^2}. \tag{2}$$

Equation (2) also gives the distance between the points $P_1(u_1, u_2)$ and $P_2(v_1, v_2)$, as shown in Figure 3.2.

Example 2. Compute the distance between the vectors

$$\mathbf{u} = \begin{bmatrix} -1 \\ 5 \end{bmatrix} \quad \text{and} \quad \mathbf{v} = \begin{bmatrix} 3 \\ 2 \end{bmatrix}.$$

Solution: By Equation (2), the distance between **u** and **v** is

$$\|\mathbf{u} - \mathbf{v}\| = \sqrt{(-1 - 3)^2 + (5 - 2)^2} = \sqrt{4^2 + 3^2} = \sqrt{25} = 5. \quad \blacksquare$$

Now let $\mathbf{v} = \begin{bmatrix} v_1 \\ v_2 \\ v_3 \end{bmatrix}$ be a vector in R^3. Using the Pythagorean theorem twice (see Figure 3.3), we can obtain the **length** of **v**, also denoted $\|\mathbf{v}\|$, as

$$\|\mathbf{v}\| = \sqrt{c^2 + v_3^2} = \sqrt{(\sqrt{v_1^2 + v_2^2})^2 + v_3^2} = \sqrt{v_1^2 + v_2^2 + v_3^2}. \tag{3}$$

It follows from Equation (3) that the zero vector has length zero. It is easy to show that the zero vector is the only vector whose length is zero.

If $\mathbf{u} = \begin{bmatrix} u_1 \\ u_2 \\ u_3 \end{bmatrix}$ and $\mathbf{v} = \begin{bmatrix} v_1 \\ v_2 \\ v_3 \end{bmatrix}$, then $\mathbf{u} - \mathbf{v} = \begin{bmatrix} u_1 - v_1 \\ u_2 - v_2 \\ u_3 - v_3 \end{bmatrix}$. Thus

Figure 3.3 The length of **v**.

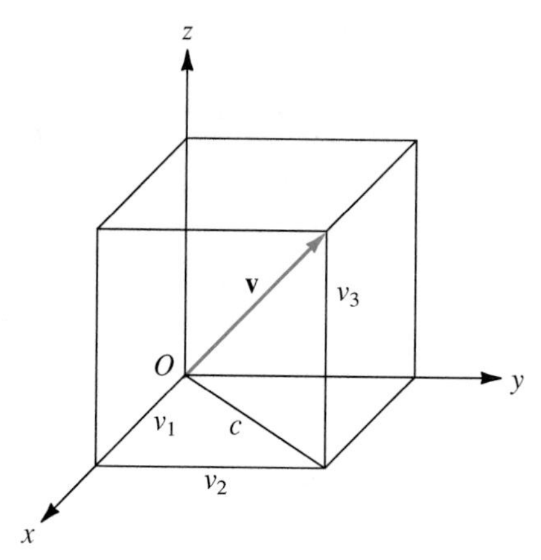

$$\|\mathbf{u} - \mathbf{v}\| = \sqrt{(u_1 - v_1)^2 + (u_2 - v_2)^2 + (u_3 - v_3)^2}. \tag{4}$$

This is the **distance** between the vectors **u** and **v**.

Of course, Equation (4) also gives the distance between two points P_1 and P_2 in R^3 with respective coordinates (u_1, u_2, u_3) and (v_1, v_2, v_3).

Example 3. Compute the length of the vector

$$\mathbf{v} = \begin{bmatrix} 1 \\ 2 \\ 3 \end{bmatrix}.$$

Solution: By Equation (3), the length of **v** is

$$\|\mathbf{v}\| = \sqrt{1^2 + 2^2 + 3^2} = \sqrt{14}.$$ ■

Example 4. Compute the distance between the vectors

$$\mathbf{u} = \begin{bmatrix} 1 \\ 2 \\ 3 \end{bmatrix} \quad \text{and} \quad \mathbf{v} = \begin{bmatrix} -4 \\ 3 \\ 5 \end{bmatrix}.$$

Solution: By Equation (4), the distance between **u** and **v** is

$$\|\mathbf{u} - \mathbf{v}\| = \sqrt{(1 + 4)^2 + (2 - 3)^2 + (3 - 5)^2} = \sqrt{30}.$$ ■

Angle Between Two Vectors

The direction of a vector in R^2 is given by specifying its angle of inclination or slope. The direction of a vector $\mathbf{v}$ in R^3 is specified by giving the cosines of the angles that the vector $\mathbf{v}$ makes with the positive x-, y- and z-axes (see Figure 3.4); these are called **direction cosines**.

Figure 3.4

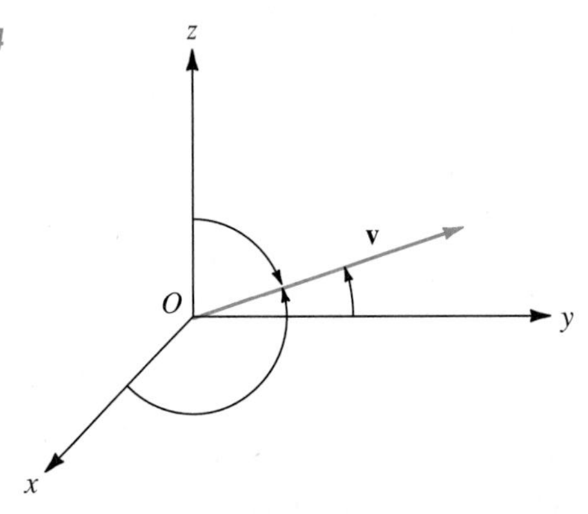

Instead of dealing with the special problem of finding the cosines of these angles for a vector in R^3, or the angle of inclination for a vector in R^2, we consider the more general problem of determining the **angle** θ, $0 \leq \theta \leq \pi$, between two nonzero vectors in R^2 or R^3. As shown in Figure 3.5, let

$$\mathbf{u} = \begin{bmatrix} u_1 \\ u_2 \\ u_3 \end{bmatrix} \quad \text{and} \quad \mathbf{v} = \begin{bmatrix} v_1 \\ v_2 \\ v_3 \end{bmatrix}$$

be two vectors in R^3. Using the law of cosines, we have

$$\|\mathbf{v} - \mathbf{u}\|^2 = \|\mathbf{u}\|^2 + \|\mathbf{v}\|^2 - 2\|\mathbf{u}\|\,\|\mathbf{v}\| \cos \theta.$$

Hence

$$\begin{aligned} \cos \theta &= \frac{\|\mathbf{u}\|^2 + \|\mathbf{v}\|^2 - \|\mathbf{v} - \mathbf{u}\|^2}{2\|\mathbf{u}\|\,\|\mathbf{v}\|} = \frac{(u_1^2 + u_2^2 + u_3^2) + (v_1^2 + v_2^2 + v_3^2)}{2\|\mathbf{u}\|\,\|\mathbf{v}\|} \\ &\quad - \frac{(v_1 - u_1)^2 + (v_2 - u_2)^2 + (v_3 - u_3)^2}{2\|\mathbf{u}\|\,\|\mathbf{v}\|} \\ &= \frac{u_1 v_1 + u_2 v_2 + u_3 v_3}{\|\mathbf{u}\|\,\|\mathbf{v}\|} \end{aligned}$$

Figure 3.5

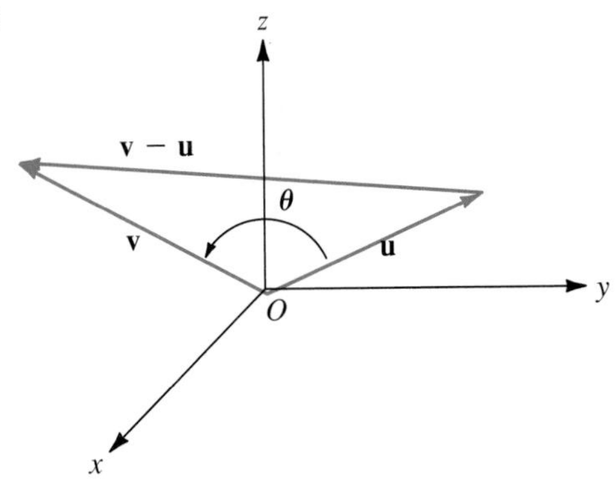

Thus

$$\cos\,\theta = \frac{u_1v_1 + u_2v_2 + u_3v_3}{\|\mathbf{u}\|\;\|\mathbf{v}\|}. \tag{5}$$

In a similar way, if $\mathbf{u} = \begin{bmatrix} u_1 \\ u_2 \end{bmatrix}$ and $\mathbf{v} = \begin{bmatrix} v_1 \\ v_2 \end{bmatrix}$ are nonzero vectors in R^2 and θ is the angle between $\mathbf{u}$ and $\mathbf{v}$, then

$$\cos\,\theta = \frac{u_1v_1 + u_2v_2}{\|\mathbf{u}\|\;\|\mathbf{v}\|}. \tag{6}$$

The zero vector in R^2 or R^3 has no specific direction. The law of cosines expression above is true if $\mathbf{v} \neq \mathbf{0}$ and $\mathbf{u} = \mathbf{0}$ for any angle θ. Thus the zero vector can be assigned any direction.

Example 5. Let $\mathbf{u} = \begin{bmatrix} 1 \\ 1 \\ 0 \end{bmatrix}$ and $\mathbf{v} = \begin{bmatrix} 0 \\ 1 \\ 1 \end{bmatrix}$. The angle θ between $\mathbf{u}$ and $\mathbf{v}$ is determined by

$$\cos\,\theta = \frac{(1)(0) + (1)(1) + (0)(1)}{\sqrt{1^2 + 1^2 + 0^2}\,\sqrt{0^2 + 1^2 + 1^2}} = \frac{1}{2}.$$

Since $0 \le \theta \le \pi$, $\theta = 60°$. ■

The cosine of an angle between two nonzero vectors in R^2 or R^3 can be expressed in terms of a very useful function, which we now define.

Definition 3.1. Let $\mathbf{u} = \begin{bmatrix} u_1 \\ u_2 \end{bmatrix}$ and $\mathbf{v} = \begin{bmatrix} v_1 \\ v_2 \end{bmatrix}$ be vectors in R^2. The **standard inner product** or **dot product** on R^2 is defined as the number

$$u_1v_1 + u_2v_2$$

and is denoted by

$$\mathbf{u} \cdot \mathbf{v}.$$

Let $\mathbf{u} = \begin{bmatrix} u_1 \\ u_2 \\ u_3 \end{bmatrix}$ and $\mathbf{v} = \begin{bmatrix} v_1 \\ v_2 \\ v_3 \end{bmatrix}$ be vectors in R^3. The **standard inner product** or **dot product** on R^3 is defined as the number

$$u_1v_1 + u_2v_2 + u_3v_3$$

and is also denoted by $(\mathbf{u}, \mathbf{v})$ or by $\mathbf{u} \cdot \mathbf{v}$. ■

Example 6. If $\mathbf{u} = \begin{bmatrix} 2 \\ 3 \\ 2 \end{bmatrix}$ and $\mathbf{v} = \begin{bmatrix} 4 \\ 2 \\ -1 \end{bmatrix}$, then

$$\begin{aligned} \mathbf{u} \cdot \mathbf{v} &= (2)(4) + (3)(2) + (2)(-1) \\ &= 12. \end{aligned}$$

■

If we examine Equations (1) and (3), we see that if $\mathbf{v}$ is a vector in R^2 or R^3, then

$$\|\mathbf{v}\| = \sqrt{\mathbf{v} \cdot \mathbf{v}}. \tag{7}$$

Using Definition 3.1, we can also write Equations (5) and (6) for the cosine of the angle θ between two nonzero vectors $\mathbf{u}$ and $\mathbf{v}$ in R^2 or R^3 as

$$\cos\theta = \frac{\mathbf{u} \cdot \mathbf{v}}{\|\mathbf{u}\|\,\|\mathbf{v}\|}, \qquad 0 \leq \theta \leq \pi. \tag{8}$$

We also observe that two vectors $\mathbf{u}$ and $\mathbf{v}$ in R^2 or R^3 are **orthogonal**, or **perpendicular**, if and only if $\mathbf{u} \cdot \mathbf{v} = 0$.

Example 7. The vectors $\mathbf{u} = \begin{bmatrix} 2 \\ -4 \end{bmatrix}$ and $\mathbf{v} = \begin{bmatrix} 4 \\ 2 \end{bmatrix}$ are orthogonal, since

$$\mathbf{u} \cdot \mathbf{v} = (2)(4) + (-4)(2) = 0.$$

See Figure 3.6. ■

Figure 3.6

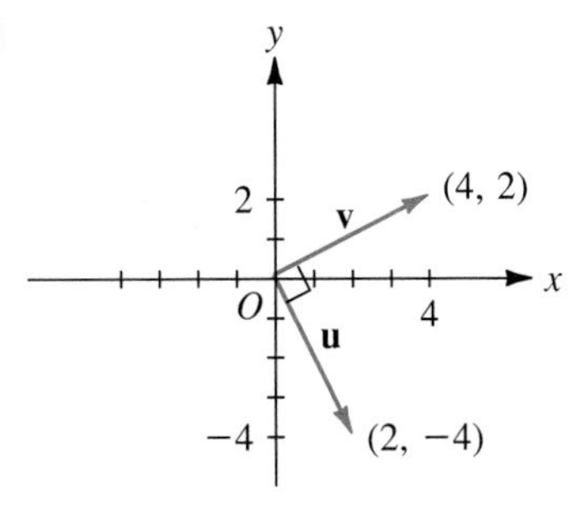

We note the following properties of the standard inner product on R^2 and R^3 that will motivate our next section.

Theorem 3.1. *Let* **u**, **v**, *and* **w** *be vectors in* R^2 *or* R^3, *and let c be a scalar. The standard inner product on* R^2 *and* R^3 *has the following properties:*

(a) $\mathbf{u} \cdot \mathbf{u} > 0$ *if* $\mathbf{u} \neq \mathbf{0}$; $\mathbf{u} \cdot \mathbf{u} = 0$ *if and only if* $\mathbf{u} = \mathbf{0}$.
(b) $\mathbf{u} \cdot \mathbf{v} = \mathbf{v} \cdot \mathbf{u}$.
(c) $(\mathbf{u} + \mathbf{v}) \cdot \mathbf{w} = \mathbf{u} \cdot \mathbf{w} + \mathbf{v} \cdot \mathbf{w}$.
(d) $c\mathbf{u} \cdot \mathbf{v} = c(\mathbf{u} \cdot \mathbf{v})$, *for any real scalar c.*

Proof: Exercise. ■

Unit Vectors

A **unit vector** in R^2 or R^3 is a vector whose length is 1. If **x** is any nonzero vector, then the vector

$$\mathbf{u} = \frac{1}{\|\mathbf{x}\|}\mathbf{x}$$

is a unit vector in the direction of **x** (Exercise 32).

Example 8. Let $\mathbf{x} = \begin{bmatrix} -3 \\ 4 \end{bmatrix}$. Then

$$\|\mathbf{x}\| = \sqrt{(-3)^2 + 4^2} = 5.$$

Hence the vector $\mathbf{u} = \frac{1}{5}\begin{bmatrix} -3 \\ 4 \end{bmatrix} = \begin{bmatrix} -\frac{3}{5} \\ \frac{4}{5} \end{bmatrix}$ is a unit vector, since

$$\|\mathbf{u}\| = \sqrt{\left(-\frac{3}{5}\right)^2 + \left(\frac{4}{5}\right)^2} = \sqrt{\frac{9 + 16}{25}} = 1.$$

Also, $\mathbf{u}$ lies in the direction of $\mathbf{x}$ (Figure 3.7). ■

Figure 3.7

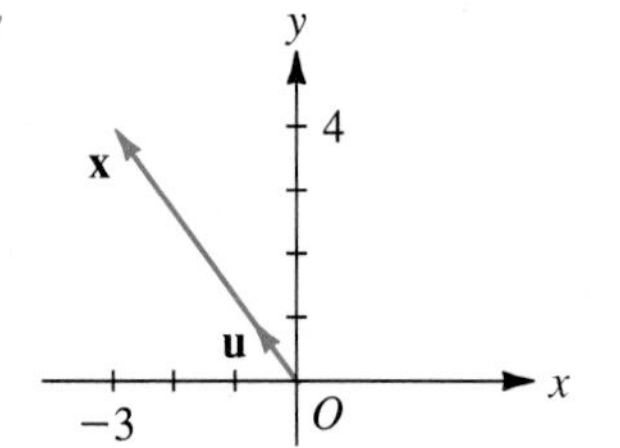

Figure 3.8 Unit vectors in R^2

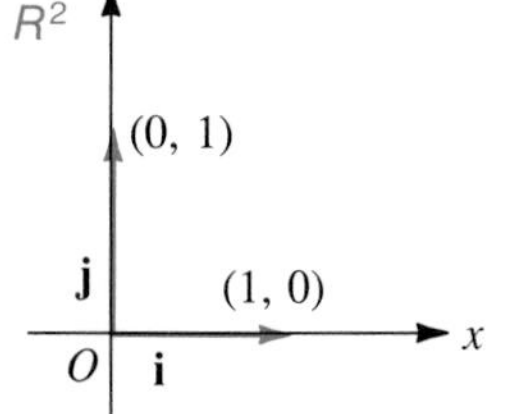

There are two unit vectors in R^2 that are of special importance. These are $\mathbf{i} = \begin{bmatrix} 1 \\ 0 \end{bmatrix}$ and $\mathbf{j} = \begin{bmatrix} 0 \\ 1 \end{bmatrix}$, the unit vectors along the positive x- and y-axes, respectively, shown in Figure 3.8. Observe that $\mathbf{i}$ and $\mathbf{j}$ are orthogonal. Since $\mathbf{i}$ and $\mathbf{j}$ form the natural basis for R^2, every vector in R^2 can be written as a linear combination of the orthogonal vectors $\mathbf{i}$ and $\mathbf{j}$.

Similarly, the vectors in the natural basis for R^3,

$$\mathbf{i} = \begin{bmatrix} 1 \\ 0 \\ 0 \end{bmatrix}, \quad \mathbf{j} = \begin{bmatrix} 0 \\ 1 \\ 0 \end{bmatrix}, \quad \text{and} \quad \mathbf{k} = \begin{bmatrix} 0 \\ 0 \\ 1 \end{bmatrix},$$

are unit vectors that are mutually orthogonal.

When several forces act on a body, we can find a single force, called the **resultant force**, having the equivalent effect. The resultant force can be determined using vectors. The following example illustrates the method.

Example 9. Suppose that a force of 12 pounds is applied to an object along the negative x-axis and a force of 5 pounds is applied to the object along the positive y-axis. Find the magnitude and direction of the resultant force.

Solution: In Figure 3.9 we have represented the force along the negative x-axis by the vector $\overrightarrow{OA}$ and the force along the positive y-axis by the vector $\overrightarrow{OB}$. The resultant force is the vector $\overrightarrow{OC} = \overrightarrow{OA} + \overrightarrow{OB}$. Thus the magnitude of the resultant force is 13 pounds and its direction is as indicated in the figure. ■

Figure 3.9

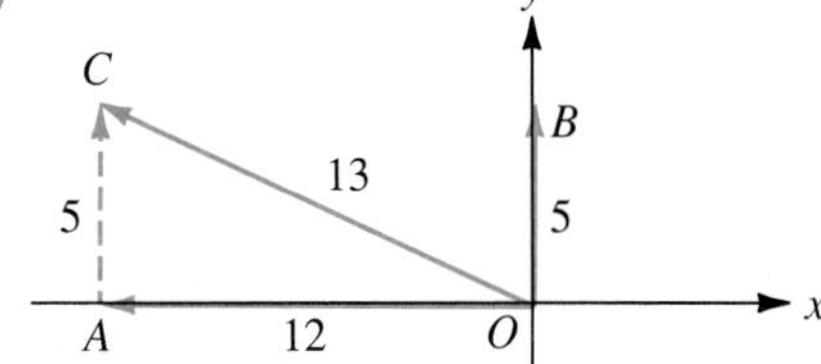

Figure 3.10

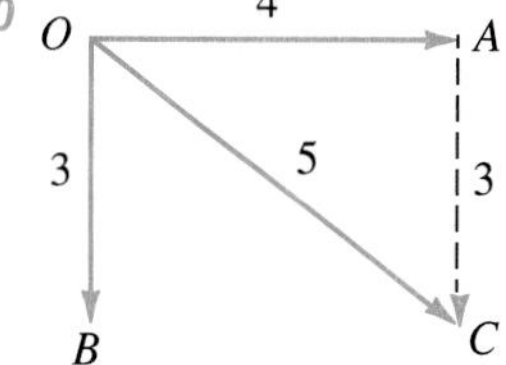

Vectors are also used in physics to deal with velocity problems, as the following example illustrates.

Example 10. Suppose that a boat is traveling east across a river at the rate of 4 miles per hour while the river's current is flowing south at a rate of 3 miles per hour. Find the resultant velocity of the boat.

Solution: In Figure 3.10 we have represented the velocity of the boat by the vector $\overrightarrow{OA}$ and the velocity of the river's current by the vector $\overrightarrow{OB}$. The resultant velocity is the vector $\overrightarrow{OC} = \overrightarrow{OA} + \overrightarrow{OB}$. Thus the magnitude of the resultant velocity is 5 miles per hour and its direction is as indicated in the figure. ■

3.1. EXERCISES

In Exercises 1 and 2, find the length of each vector.

1. (a) $\begin{bmatrix}1\\0\end{bmatrix}$. (b) $\begin{bmatrix}0\\0\end{bmatrix}$. (c) $\begin{bmatrix}1\\2\end{bmatrix}$.

2. (a) $\begin{bmatrix}0\\-2\\0\end{bmatrix}$. (b) $\begin{bmatrix}-1\\-3\\-4\end{bmatrix}$. (c) $\begin{bmatrix}1\\-2\\4\end{bmatrix}$.

In Exercises 3 and 4, compute $\|\mathbf{u} - \mathbf{v}\|$.

3. (a) $\mathbf{u} = \begin{bmatrix}1\\0\end{bmatrix}$, $\mathbf{v} = \begin{bmatrix}1\\1\end{bmatrix}$.

(b) $\mathbf{u} = \begin{bmatrix}0\\0\end{bmatrix}$, $\mathbf{v} = \begin{bmatrix}1\\-1\end{bmatrix}$.

4. (a) $\mathbf{u} = \begin{bmatrix}1\\2\\3\end{bmatrix}$, $\mathbf{v} = \begin{bmatrix}4\\5\\6\end{bmatrix}$.

(b) $\mathbf{u} = \begin{bmatrix}-1\\-2\\-3\end{bmatrix}$, $\mathbf{v} = \begin{bmatrix}-4\\-5\\-6\end{bmatrix}$.

In Exercises 5 and 6, find the distance between **u** and **v**.

5. (a) $\mathbf{u} = \begin{bmatrix}1\\2\end{bmatrix}$, $\mathbf{v} = \begin{bmatrix}-4\\-5\end{bmatrix}$.

(b) $\mathbf{u} = \begin{bmatrix}1\\2\end{bmatrix}$, $\mathbf{v} = \begin{bmatrix}4\\-5\end{bmatrix}$.

6. (a) $\mathbf{u} = \begin{bmatrix}-1\\-2\\-3\end{bmatrix}$, $\mathbf{v} = \begin{bmatrix}4\\5\\6\end{bmatrix}$.

(b) $\mathbf{u} = \begin{bmatrix}0\\1\\-1\end{bmatrix}$, $\mathbf{v} = \begin{bmatrix}1\\2\\0\end{bmatrix}$.

In Exercises 7 and 8, find $\mathbf{u} \cdot \mathbf{v}$.

7. (a) $\mathbf{u} = \begin{bmatrix}0\\0\end{bmatrix}$, $\mathbf{v} = \begin{bmatrix}0\\0\end{bmatrix}$.

(b) $\mathbf{u} = \begin{bmatrix}3\\0\end{bmatrix}$, $\mathbf{v} = \begin{bmatrix}1\\2\end{bmatrix}$.

(c) $\mathbf{u} = \begin{bmatrix}2\\-3\end{bmatrix}$, $\mathbf{v} = \begin{bmatrix}1\\-2\end{bmatrix}$.

8. (a) $\mathbf{u} = \begin{bmatrix}0\\0\\-3\end{bmatrix}$, $\mathbf{v} = \begin{bmatrix}-4\\5\\6\end{bmatrix}$.

(b) $\mathbf{u} = \begin{bmatrix}1\\3\\2\end{bmatrix}$, $\mathbf{v} = \begin{bmatrix}2\\3\\4\end{bmatrix}$.

(c) $\mathbf{u} = \begin{bmatrix}1\\-1\\1\end{bmatrix}$, $\mathbf{v} = \begin{bmatrix}1\\2\\-1\end{bmatrix}$.

9. For each pair of vectors **u** and **v** in Exercise 5, find the cosine of the angle θ between **u** and **v**.

10. For each pair of vectors in Exercise 6, find the cosine of the angle θ between **u** and **v**.

11. For each of the following vectors **v**, find the direction cosines (the cosines of the angle between **v** and the positive x-, y-, and z-axes).

(a) $\mathbf{v} = \begin{bmatrix}1\\0\\0\end{bmatrix}$. (b) $\mathbf{v} = \begin{bmatrix}1\\3\\2\end{bmatrix}$.

(c) $\mathbf{u} = \begin{bmatrix}-1\\-2\\-3\end{bmatrix}$. (d) $\mathbf{u} = \begin{bmatrix}4\\-3\\2\end{bmatrix}$.

12. Let P and Q be the points in R^3 with respective coordinates $(3, -1, 2)$ and $(4, 2, -3)$. Find the length of the segment PQ.

13. Prove Theorem 3.1.

14. Verify Theorem 3.1 for

$$\mathbf{u} = \begin{bmatrix} 1 \\ 2 \\ 3 \end{bmatrix}, \quad \mathbf{v} = \begin{bmatrix} -2 \\ 4 \\ 3 \end{bmatrix},$$

$$\mathbf{w} = \begin{bmatrix} 0 \\ 3 \\ -2 \end{bmatrix}, \quad \text{and} \quad c = -3.$$

15. Show that in R^2:
(a) $\mathbf{i} \cdot \mathbf{i} = \mathbf{j} \cdot \mathbf{j} = 1$. (b) $\mathbf{i} \cdot \mathbf{j} = 0$.

16. Show that in R^3:
(a) $\mathbf{i} \cdot \mathbf{i} = \mathbf{j} \cdot \mathbf{j} = \mathbf{k} \cdot \mathbf{k} = 1$.
(b) $\mathbf{i} \cdot \mathbf{j} = \mathbf{i} \cdot \mathbf{k} = \mathbf{j} \cdot \mathbf{k} = 0$.

17. Which of the vectors $\mathbf{v}_1 = \begin{bmatrix} 1 \\ 2 \end{bmatrix}$, $\mathbf{v}_2 = \begin{bmatrix} 0 \\ 1 \end{bmatrix}$,
$\mathbf{v}_3 = \begin{bmatrix} -2 \\ -4 \end{bmatrix}$, $\mathbf{v}_4 = \begin{bmatrix} -2 \\ 1 \end{bmatrix}$, $\mathbf{v}_5 = \begin{bmatrix} 2 \\ 4 \end{bmatrix}$, and
$\mathbf{v}_6 = \begin{bmatrix} -6 \\ 3 \end{bmatrix}$ are
(a) Orthogonal? (b) In the same direction?
(c) In opposite directions?

18. Which of the vectors $\mathbf{v}_1 = \begin{bmatrix} 1 \\ -1 \\ -2 \end{bmatrix}$, $\mathbf{v}_2 = \begin{bmatrix} 3 \\ -1 \\ 2 \end{bmatrix}$,

$$\mathbf{v}_3 = \begin{bmatrix} 2 \\ 4 \\ -1 \end{bmatrix}, \mathbf{v}_4 = \begin{bmatrix} \frac{1}{2} \\ 0 \\ \frac{1}{4} \end{bmatrix}, \mathbf{v}_5 = \begin{bmatrix} \frac{1}{2} \\ -\frac{1}{2} \\ -1 \end{bmatrix}, \mathbf{v}_6 = \begin{bmatrix} -\frac{2}{3} \\ -\frac{4}{3} \\ \frac{1}{3} \end{bmatrix}$$

are
(a) Orthogonal? (b) In the same direction?
(c) In opposite directions?

19. (Optional) Which of the following pairs of lines are perpendicular?

(a) $x = 2 + 2t$, $y = -3 - 3t$, $z = 4 + 4t$ and $x = 2 + t$, $y = 4 - t$, $z = 5 - t$.

(b) $x = 3 - t$, $y = 4 + t$, $z = 2 + 2t$ and $x = 2t$, $y = 3 - 2t$, $z = 4 + 2t$.

20. (Optional) Find parametric equations of the line passing through $(3, -1, -3)$ and perpendicular to the line passing through $(3, -2, 4)$ and $(0, 3, 5)$.

21. A ship is being pushed by a tugboat with a force of 300 pounds along the negative y-axis while another tugboat is pushing along the negative x-axis with a force of 400 pounds. Find the magnitude and sketch the direction of the resultant force.

22. Suppose that an airplane is flying with an airspeed of 260 kilometers per hour while a wind is blowing to the west at 100 kilometers per hour. Indicate on a figure the approximate direction that the plane must follow to result in a flight directly south. What will be the resultant speed?

23. Let points A, B, C, and D in R^3 have respective coordinates $(1, 2, 3)$, $(-2, 3, 5)$, $(0, 3, 6)$, and $(3, 2, 4)$. Prove that $ABCD$ is a parallelogram.

24. Find c so that the vector $\mathbf{v} = \begin{bmatrix} 1 \\ c \end{bmatrix}$ is orthogonal to
$\mathbf{w} = \begin{bmatrix} 2 \\ -1 \end{bmatrix}$.

25. Find c so that the vector $\mathbf{v} = \begin{bmatrix} 2 \\ c \\ 3 \end{bmatrix}$ is orthogonal to
$\mathbf{w} = \begin{bmatrix} 1 \\ -2 \\ 1 \end{bmatrix}$.

26. If possible, find a, b, and c so that $\mathbf{v} = \begin{bmatrix} a \\ b \\ c \end{bmatrix}$ is orthogonal to both $\mathbf{w} = \begin{bmatrix} 1 \\ 2 \\ 1 \end{bmatrix}$ and $\mathbf{x} = \begin{bmatrix} 1 \\ -1 \\ 1 \end{bmatrix}$.

27. If possible, find a and b so that $\mathbf{v} = \begin{bmatrix} a \\ b \\ 2 \end{bmatrix}$ is orthogonal to both $\mathbf{w} = \begin{bmatrix} 2 \\ 1 \\ 1 \end{bmatrix}$ and $\mathbf{x} = \begin{bmatrix} 1 \\ 0 \\ 1 \end{bmatrix}$.

28. Show that the only vector $\mathbf{x}$ in R^2 or R^3 that is orthogonal to every other vector is the zero vector.

29. Prove that if $\mathbf{v}$, $\mathbf{w}$, and $\mathbf{x}$ are in R^2 or R^3 and $\mathbf{v}$ is orthogonal to both $\mathbf{w}$ and $\mathbf{x}$, then $\mathbf{v}$ is orthogonal to every vector in span $\{\mathbf{w}, \mathbf{x}\}$.

30. Let $\mathbf{u}$ be a fixed vector in R^2 (R^3). Prove that the set V of all vectors $\mathbf{v}$ in R^2 (R^3) such that $\mathbf{u}$ and $\mathbf{v}$ are orthogonal is a subspace of R^2 (R^3).

31. Prove that if c is a scalar and $\mathbf{v}$ is a vector in R^2 or R^3, then $\|c\mathbf{v}\| = |c|\ \|\mathbf{v}\|$.

32. Show that if $\mathbf{x}$ is a nonzero vector in R^2 or R^3, then $\mathbf{u} = \dfrac{1}{\|\mathbf{x}\|}\mathbf{x}$ is a unit vector in the direction of $\mathbf{x}$.

33. Let $S = \{\mathbf{v}_1, \mathbf{v}_2, \mathbf{v}_3\}$ be a set of nonzero vectors in R^3 such that any two vectors in S are orthogonal. Prove that S is linearly independent.

34. Prove that for any vectors $\mathbf{u}$, $\mathbf{v}$, and $\mathbf{w}$ in R^2 or R^3 we have

$$\mathbf{u} \cdot (\mathbf{v} + \mathbf{w}) = \mathbf{u} \cdot \mathbf{v} + \mathbf{u} \cdot \mathbf{w}.$$

35. Prove that for any vectors $\mathbf{u}$, $\mathbf{v}$, and $\mathbf{w}$ in R^2 or R^3 and any scalar c we have:
(a) $(\mathbf{u} + c\mathbf{v}) \cdot \mathbf{w} = \mathbf{u} \cdot \mathbf{w} + c(\mathbf{v} \cdot \mathbf{w})$.
(b) $\mathbf{u} \cdot (c\mathbf{v}) = c(\mathbf{u} \cdot \mathbf{v})$.
(c) $(\mathbf{u} + \mathbf{v}) \cdot (c\mathbf{w}) = c(\mathbf{u} \cdot \mathbf{w}) + c(\mathbf{v} \cdot \mathbf{w})$.

36. Prove that the diagonals of a rectangle are of equal length. [*Hint:* Take the vertices of the rectangle as $(0, 0)$, $(0, b)$, $(a, 0)$, and (a, b).]

37. Prove that the angles at the base of an isosceles triangle are equal.

38. Prove that a parallelogram is a rhombus, a parallelogram with four equal sides, if and only if its diagonals are orthogonal.

39. To compute the dot product of a pair of vectors $\mathbf{u}$ and $\mathbf{v}$ in R^2 or R^3, use the matrix product operation in your software as follows. Let U and V be column matrices for vectors $\mathbf{u}$ and $\mathbf{v}$, respectively. Then $\mathbf{u} \cdot \mathbf{v}$ is the product of U^T and V (or V^T and U). Experiment with the vectors in Exercises 7 and 8.

40. Determine if there is a command in your software to compute the length of a vector. If there is, use it on the vector in Example 3 and then compute the distance between vectors in Example 4.

41. Assuming that your software has a command to compute the length of a vector (see Exercise 40), determine a unit vector in the direction of $\mathbf{v}$ for each of the following.

(a) $\mathbf{v} = \begin{bmatrix} 2 \\ 4 \end{bmatrix}$. (b) $\mathbf{v} = \begin{bmatrix} 7 \\ 1 \\ 0 \end{bmatrix}$. (c) $\mathbf{v} = \begin{bmatrix} 1 \\ 2 \\ -1 \end{bmatrix}$

42. Referring to Exercise 39, how could your software check for orthogonal vectors?

3.2. Cross Product in R^3 (Optional)

In this section we discuss an operation that is meaningful only in R^3. Despite this limitation, it has a number of important applications, some of which we discuss in this section. Suppose that $\mathbf{u} = u_1\mathbf{i} + u_2\mathbf{j} + u_3\mathbf{k}$ and $\mathbf{v} = v_1\mathbf{i} + v_2\mathbf{j} + v_3\mathbf{k}$ and that we want to find a vector $\mathbf{w} = \begin{bmatrix} x \\ y \\ z \end{bmatrix}$ orthogonal (perpendicular) to both $\mathbf{u}$ and $\mathbf{v}$. Thus we want $\mathbf{u} \cdot \mathbf{w} = 0$ and $\mathbf{v} \cdot \mathbf{w} = 0$, which leads to the linear system

$$\begin{aligned} u_1x + u_2y + u_3z &= 0 \\ v_1x + v_2y + v_3z &= 0. \end{aligned} \tag{1}$$

It can be shown that

$$\mathbf{w} = \begin{bmatrix} u_2v_3 - u_3v_2 \\ u_3v_1 - u_1v_3 \\ u_1v_2 - u_2v_1 \end{bmatrix}$$

is a solution to (1) (verify). Of course, we can also write $\mathbf{w}$ as

$$w = (u_2v_3 - u_3v_2)\mathbf{i} + (u_3v_1 - u_1v_3)\mathbf{j} + (u_1v_2 - u_2v_1)\mathbf{k}. \tag{2}$$

This vector is called the **cross product** of $\mathbf{u}$ and $\mathbf{v}$ and is denoted by $\mathbf{u} \times \mathbf{v}$. Note that the cross product, $\mathbf{u} \times \mathbf{v}$, is a vector, while the dot product, $\mathbf{u} \cdot \mathbf{v}$, is a scalar or number. Although the cross product is not defined on R^n if $n \neq 3$, it has many applications; we shall use it when we study planes in R^3.

Example 1. Let $\mathbf{u} = 2\mathbf{i} + \mathbf{j} + 2\mathbf{k}$ and $\mathbf{v} = 3\mathbf{i} - \mathbf{j} - 3\mathbf{k}$. From (2),

$$\mathbf{u} \times \mathbf{v} = -\mathbf{i} + 12\mathbf{j} - 5\mathbf{k}.$$

■

Let $\mathbf{u}$, $\mathbf{v}$, and $\mathbf{w}$ be vectors in R^3 and c a scalar. The cross product operation satisfies the following properties, whose verification we leave to the reader:

(a) $\mathbf{u} \times \mathbf{v} = -(\mathbf{v} \times \mathbf{u})$.
(b) $\mathbf{u} \times (\mathbf{v} + \mathbf{w}) = \mathbf{u} \times \mathbf{v} + \mathbf{u} \times \mathbf{w}$.
(c) $(\mathbf{u} + \mathbf{v}) \times \mathbf{w} = \mathbf{u} \times \mathbf{w} + \mathbf{v} \times \mathbf{w}$.

(d) $c(\mathbf{u} \times \mathbf{v}) = (c\mathbf{u}) \times \mathbf{v} = \mathbf{u} \times (c\mathbf{v})$.
(e) $\mathbf{u} \times \mathbf{u} = \mathbf{0}$.
(f) $\mathbf{0} \times \mathbf{u} = \mathbf{u} \times \mathbf{0} = \mathbf{0}$.
(g) $\mathbf{u} \times (\mathbf{v} \times \mathbf{w}) = (\mathbf{u} \cdot \mathbf{w})\mathbf{v} - (\mathbf{u} \cdot \mathbf{v})\mathbf{w}$.
(h) $(\mathbf{u} \times \mathbf{v}) \times \mathbf{w} = (\mathbf{w} \cdot \mathbf{u})\mathbf{v} - (\mathbf{w} \cdot \mathbf{v})\mathbf{u}$.

Example 2. It follows from (2) that

$$\mathbf{i} \times \mathbf{i} = \mathbf{j} \times \mathbf{j} = \mathbf{k} \times \mathbf{k} = \mathbf{0},$$
$$\mathbf{i} \times \mathbf{j} = \mathbf{k}, \qquad \mathbf{j} \times \mathbf{k} = \mathbf{i}, \qquad \mathbf{k} \times \mathbf{i} = \mathbf{j}.$$

Also,

$$\mathbf{j} \times \mathbf{i} = -\mathbf{k}, \qquad \mathbf{k} \times \mathbf{j} = -\mathbf{i}, \qquad \mathbf{i} \times \mathbf{k} = -\mathbf{j}.$$

These can be remembered by the method indicated in Figure 3.11. Moving around the circle in a clockwise direction, we see that the cross product of two vectors taken in the indicated order is the third vector; moving in a counterclockwise direction, we see that the cross product taken in the indicated order is the negative of the third vector. The cross product of a vector with itself is the zero vector. ■

Figure 3.11

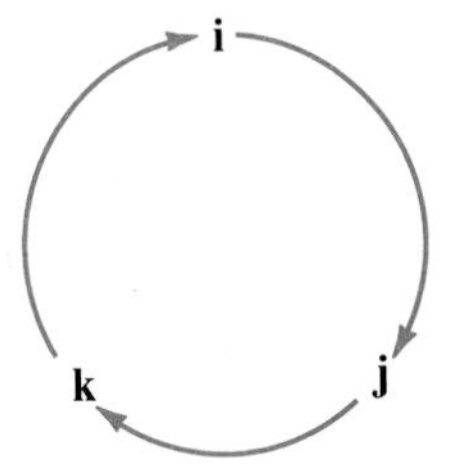

Although many of the familiar properties of the real numbers hold for the cross product, it should be noted that two important properties do not hold. The commutative law does not hold, since $\mathbf{u} \times \mathbf{v} = -(\mathbf{v} \times \mathbf{u})$. Also, the associative law does not hold, since $\mathbf{i} \times (\mathbf{i} \times \mathbf{j}) = \mathbf{i} \times \mathbf{k} = -\mathbf{j}$ while $(\mathbf{i} \times \mathbf{i}) \times \mathbf{j} = \mathbf{0} \times \mathbf{j} = \mathbf{0}$.

We shall now take a closer look at the geometric properties of the cross product. First, we observe the following two additional properties of the cross product, whose proofs we leave to the reader:

$$(\mathbf{u} \times \mathbf{v}) \cdot \mathbf{w} = \mathbf{u} \cdot (\mathbf{v} \times \mathbf{w}) \qquad \text{Exercise 7(a)} \qquad (3)$$

$$\mathbf{u} \times (\mathbf{v} \times \mathbf{w}) = (\mathbf{u} \cdot \mathbf{w})\mathbf{v} - (\mathbf{u} \cdot \mathbf{v})\mathbf{w} \qquad \text{Exercise 7(b)} \qquad (4)$$

Example 3. Let $\mathbf{u}$ and $\mathbf{v}$ be as in Example 1, and let $\mathbf{w} = \mathbf{i} + 2\mathbf{j} + 3\mathbf{k}$. Then

$$\mathbf{u} \times \mathbf{v} = -\mathbf{i} + 12\mathbf{j} - 5\mathbf{k} \quad \text{and} \quad (\mathbf{u} \times \mathbf{v}) \cdot \mathbf{w} = 8$$
$$\mathbf{v} \times \mathbf{w} = 3\mathbf{i} - 12\mathbf{j} + 7\mathbf{k} \quad \text{and} \quad \mathbf{u} \cdot (\mathbf{v} \times \mathbf{w}) = 8,$$

which illustrates Equation (3). Also,

$$\mathbf{u} \times (\mathbf{v} \times \mathbf{w}) = 31\mathbf{i} - 8\mathbf{j} - 27\mathbf{k},$$
$$\mathbf{u} \cdot \mathbf{w} = 10, \qquad \mathbf{u} \cdot \mathbf{v} = -1,$$
$$(\mathbf{u} \cdot \mathbf{w})\mathbf{v} = 30\mathbf{i} - 10\mathbf{j} - 30\mathbf{k}, \qquad (\mathbf{u} \cdot \mathbf{v})\mathbf{w} = -\mathbf{i} - 2\mathbf{j} - 3\mathbf{k}.$$

Hence

$$(\mathbf{u} \cdot \mathbf{w})\mathbf{v} - (\mathbf{u} \cdot \mathbf{v})\mathbf{w} = 31\mathbf{i} - 8\mathbf{j} - 27\mathbf{k},$$

which illustrates Equation (4). ■

From Equation (3) and properties (a) and (e) of the cross product operation, we have

$$(\mathbf{u} \times \mathbf{v}) \cdot \mathbf{v} = \mathbf{u} \cdot (\mathbf{v} \times \mathbf{v}) = \mathbf{u} \cdot \mathbf{0} = 0 \tag{5}$$

$$\begin{aligned}(\mathbf{u} \times \mathbf{v}) \cdot \mathbf{u} &= -(\mathbf{v} \times \mathbf{u}) \cdot \mathbf{u} = -\mathbf{v} \cdot (\mathbf{u} \times \mathbf{u}) \\ &= -\mathbf{v} \cdot \mathbf{0} = 0.\end{aligned} \tag{6}$$

From (5) and (6) and the definition for orthogonality of vectors given in Section 3.1, it follows that if $\mathbf{u} \times \mathbf{v} \neq \mathbf{0}$, then $\mathbf{u} \times \mathbf{v}$ is orthogonal to both $\mathbf{u}$ and $\mathbf{v}$ and hence to the plane determined by them. It can be shown that the direction of $\mathbf{u} \times \mathbf{v}$ is such that $\mathbf{u}$, $\mathbf{v}$ and $\mathbf{u} \times \mathbf{v}$ form a right-handed coordinate system. That is, if θ is the angle between $\mathbf{u}$ and $\mathbf{v}$, then the direction of $\mathbf{u} \times \mathbf{v}$ is that in which a right-hand screw perpendicular to the plane of $\mathbf{u}$ and $\mathbf{v}$ will move if we rotate it through the angle θ from $\mathbf{u}$ to $\mathbf{v}$ (Figure 3.12).

Figure 3.12

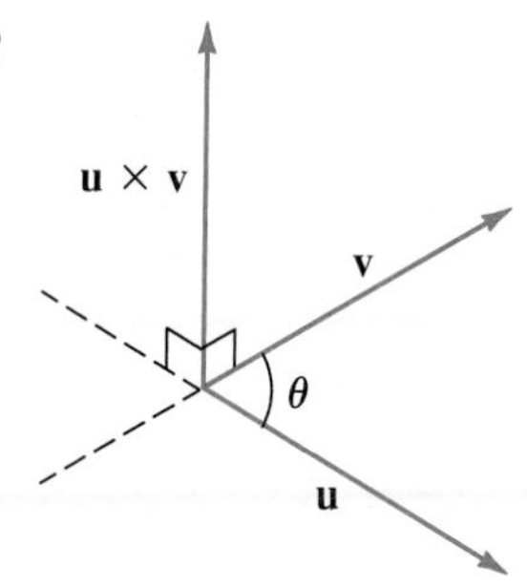

The magnitude of $\mathbf{u} \times \mathbf{v}$ can be determined as follows. From Equation (7) of Section 3.1, we have

$$
\begin{aligned}
\|\mathbf{u} \times \mathbf{v}\|^2 &= (\mathbf{u} \times \mathbf{v}) \cdot (\mathbf{u} \times \mathbf{v}) \\
&= \mathbf{u} \cdot [\mathbf{v} \times (\mathbf{u} \times \mathbf{v})], \qquad \text{by (3)} \\
&= \mathbf{u} \cdot [(\mathbf{v} \cdot \mathbf{v})\mathbf{u} - (\mathbf{v} \cdot \mathbf{u})\mathbf{v}] \qquad \text{by (4)} \\
&= (\mathbf{u} \cdot \mathbf{u})(\mathbf{v} \cdot \mathbf{v}) - (\mathbf{v} \cdot \mathbf{u})(\mathbf{v} \cdot \mathbf{u}) \qquad \text{by (d) and (b) of Theorem 3.1} \\
&= \|\mathbf{u}\|^2 \|\mathbf{v}\|^2 - (\mathbf{u} \cdot \mathbf{v})^2 \qquad \text{by Equation (7) of Section 3.1 and (b) of Theorem 3.1.}
\end{aligned}
$$

From Equation (8) of Section 3.1 it follows that

$$\mathbf{u} \cdot \mathbf{v} = \|\mathbf{u}\| \, \|\mathbf{v}\| \cos \theta,$$

where θ is the angle between $\mathbf{u}$ and $\mathbf{v}$. Hence

$$
\begin{aligned}
\|\mathbf{u} \times \mathbf{v}\|^2 &= \|\mathbf{u}\|^2 \|\mathbf{v}\|^2 - \|\mathbf{u}\|^2 \|\mathbf{v}\|^2 \cos^2 \theta \\
&= \|\mathbf{u}\|^2 \|\mathbf{v}\|^2 (1 - \cos^2 \theta) \\
&= \|\mathbf{u}\|^2 \|\mathbf{v}\|^2 \sin^2 \theta.
\end{aligned}
$$

Taking square roots, we obtain

$$\|\mathbf{u} \times \mathbf{v}\| = \|\mathbf{u}\| \, \|\mathbf{v}\| \sin \theta. \tag{7}$$

Note that we do not have $|\sin \theta|$, since $\sin \theta$ is nonnegative for $0 \le \theta \le \pi$. It follows that the vectors $\mathbf{u}$ and $\mathbf{v}$ are parallel if and only if $\mathbf{u} \times \mathbf{v} = \mathbf{0}$ (Exercise 9).

We now consider several applications of cross product.

Area of a Triangle

Consider the triangle with vertices $\mathbf{u}_1$, $\mathbf{u}_2$, and $\mathbf{u}_3$ (Figure 3.13), where we now denote a point in R^3 as a vector. The area of this triangle is $\frac{1}{2}bh$, where b is the base and h is the height. If we take the segment between $\mathbf{u}_1$ and $\mathbf{u}_2$ to be the base, then

$$b = \|\mathbf{u}_2 - \mathbf{u}_1\|.$$

The height h is given by

$$h = \|\mathbf{u}_3 - \mathbf{u}_1\| \sin \theta.$$

Figure 3.13

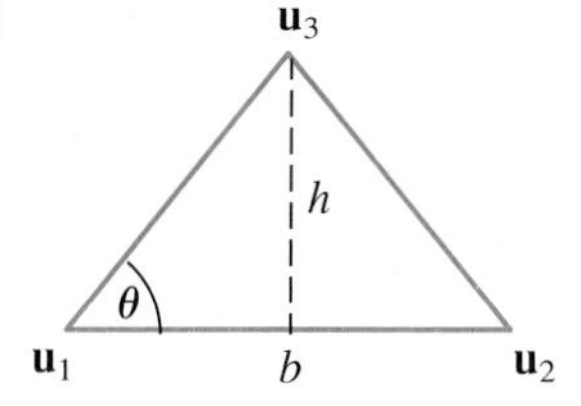

Hence, by (7), the area A_T of the triangle is

$$\begin{aligned} A_T &= \tfrac{1}{2}\|\mathbf{u}_2 - \mathbf{u}_1\|\,\|\mathbf{u}_3 - \mathbf{u}_1\| \sin\theta \\ &= \tfrac{1}{2}\|(\mathbf{u}_2 - \mathbf{u}_1) \times (\mathbf{u}_3 - \mathbf{u}_1)\|. \end{aligned}$$

Example 4. Find the area of the triangle with vertices $\mathbf{u}_1 = (2, 2, 4)$, $\mathbf{u}_2 = (-1, 0, 5)$, and $\mathbf{u}_3 = (3, 4, 3)$.

Solution: We have

$$\begin{aligned} \mathbf{u}_2 - \mathbf{u}_1 &= -3\mathbf{i} - 2\mathbf{j} + \mathbf{k}, \\ \mathbf{u}_3 - \mathbf{u}_1 &= \mathbf{i} + 2\mathbf{j} - \mathbf{k}. \end{aligned}$$

Then

$$\begin{aligned} A_T &= \tfrac{1}{2}\|(-3\mathbf{i} - 2\mathbf{j} + \mathbf{k}) \times (\mathbf{i} + 2\mathbf{j} - \mathbf{k})\| = \tfrac{1}{2}\|-2\mathbf{j} - 4\mathbf{k}\| \\ &= \|-\mathbf{j} - 2\mathbf{k}\| = \sqrt{5}. \end{aligned}$$

■

Area of a Parallelogram

The area A_P of the parallelogram with adjacent sides $\mathbf{u}_2 - \mathbf{u}_1$ and $\mathbf{u}_3 - \mathbf{u}_1$ (Figure 3.14) is $2A_T$, so

$$A_P = \|(\mathbf{u}_2 - \mathbf{u}_1) \times (\mathbf{u}_3 - \mathbf{u}_1)\|.$$

Figure 3.14

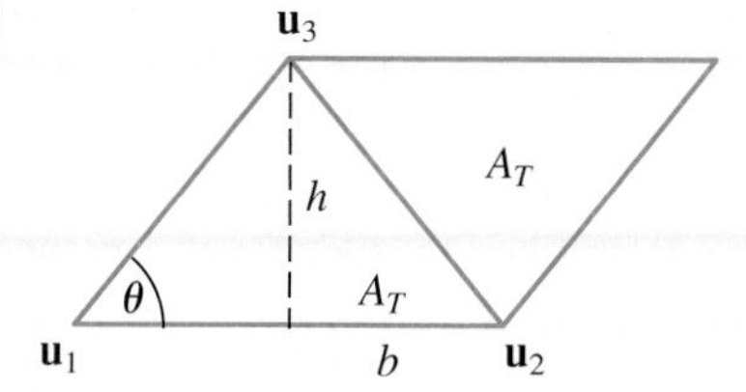

Example 5. If $\mathbf{u}_1$, $\mathbf{u}_2$, and $\mathbf{u}_3$ are as in Example 4, then the area of the parallelogram with adjacent sides $\mathbf{u}_2 - \mathbf{u}_1$ and $\mathbf{u}_3 - \mathbf{u}_1$ is $2\sqrt{5}$ (verify). ■

Volume of a Parallelepiped

Consider the parallelepiped with a vertex at the origin and edges **u**, **v**, and **w** (Figure 3.15). The volume V of the parallelepiped is the product of the area of the face containing **v** and **w** and the distance h from this face to the face parallel to it. Now

$$h = \|\mathbf{u}\| \, |\cos \theta|,$$

where θ is the angle between **u** and $\mathbf{v} \times \mathbf{w}$, and the area of the face determined by **v** and **w** is $\|\mathbf{v} \times \mathbf{w}\|$. Hence

$$V = \|\mathbf{v} \times \mathbf{w}\| \, \|\mathbf{u}\| \, |\cos \theta| = |\mathbf{u} \cdot (\mathbf{v} \times \mathbf{w})|.$$

Figure 3.15

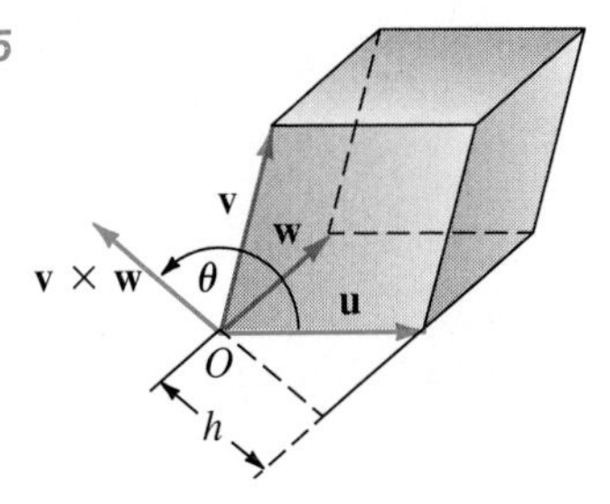

Example 6. Consider the parallelepiped with a vertex at the origin and edges $\mathbf{u} = \mathbf{i} - 2\mathbf{j} + 3\mathbf{k}$, $\mathbf{v} = \mathbf{i} + 3\mathbf{j} + \mathbf{k}$, and $\mathbf{w} = 2\mathbf{i} + \mathbf{j} + 2\mathbf{k}$. Then

$$\mathbf{v} \times \mathbf{w} = 5\mathbf{i} - 5\mathbf{k}.$$

Hence $\mathbf{u} \cdot (\mathbf{v} \times \mathbf{w}) = -10$. Thus the volume V is given by

$$V = |\mathbf{u} \cdot (\mathbf{v} \times \mathbf{w})| = |-10| = 10.$$

■

Planes

A plane in R^3 can be determined by specifying a point in it and a **normal** to it, that is, a vector perpendicular to it.

To obtain an equation of the plane passing through the point $P_0(x_0, y_0, z_0)$ and having the nonzero vector $\mathbf{v} = a\mathbf{i} + b\mathbf{j} + c\mathbf{k}$ as a normal, we proceed as follows. A point $P(x, y, z)$ lies in the plane if and only if the vector $\overrightarrow{P_0P}$ is perpendicular to $\mathbf{v}$ (Figure 3.16). Thus $P(x, y, z)$ lies in the plane if and only if

$$\mathbf{v} \cdot \overrightarrow{P_0P} = 0. \tag{8}$$

Since

$$\overrightarrow{P_0P} = (x - x_0)\mathbf{i} + (y - y_0)\mathbf{j} + (z - z_0)\mathbf{k},$$

we can write (8) as

$$a(x - x_0) + b(y - y_0) + c(z - z_0) = 0. \tag{9}$$

Figure 3.16

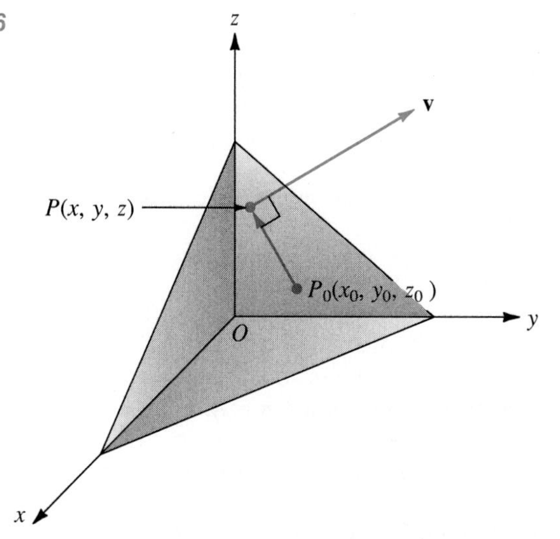

Example 7. Find an equation of the plane passing through the point $(3, 4, -3)$ and perpendicular to the vector $\mathbf{v} = 5\mathbf{i} - 2\mathbf{j} + 4\mathbf{k}$.

Solution: Substituting in (9), we obtain the equation of the plane as

$$5(x - 3) - 2(y - 4) + 4(z + 3) = 0. \quad \blacksquare$$

A plane is also determined by three noncollinear points in it, as we show in the following example.

Example 8. Find an equation of the plane passing through the points $P_1(2, -2, 1)$, $P_2(-1, 0, 3)$, and $P_3(5, -3, 4)$.

Solution: The nonparallel vectors $\overrightarrow{P_1P_2} = -3\mathbf{i} + 2\mathbf{j} + 2\mathbf{k}$ and $\overrightarrow{P_1P_3} = 3\mathbf{i} - \mathbf{j} + 3\mathbf{k}$ lie in the plane, since the points P_1, P_2, and P_3 lie in the plane. The vector

$$\mathbf{v} = \overrightarrow{P_1P_2} \times \overrightarrow{P_1P_3} = 8\mathbf{i} + 15\mathbf{j} - 3\mathbf{k}$$

is then perpendicular to both $\overrightarrow{P_1P_2}$ and $\overrightarrow{P_1P_3}$ and is thus a normal to the plane. Using the vector $\mathbf{v}$ and the point $P_1(2, -2, 1)$ in (9), we obtain

$$8(x - 2) + 15(y + 2) - 3(z - 1) = 0$$

as an equation of the plane. ■

If we multiply out and simplify, (9) can be rewritten as

$$ax + by + cz + d = 0. \tag{10}$$

Example 9. Equation (10) of the plane in Example 8 can be rewritten as

$$8x + 15y - 3z + 17 = 0. \tag{11}$$

It is not difficult to show (Exercise 24) that the graph of an equation of the form given in (10), where a, b, c, and d are constants (with a, b, and c not all zero), is a plane with normal $\mathbf{v} = a\mathbf{i} + b\mathbf{j} + c\mathbf{k}$; moreover, if $d = 0$, it is a two-dimensional subspace of R^3.

Example 10. An alternate solution to Example 8 is as follows. Let the equation of the desired plane be

$$ax + by + cz + d = 0, \tag{12}$$

where a, b, c, and d are to be determined. Since P_1, P_2, and P_3 lie in the plane, their coordinates satisfy (12). Thus we obtain the linear system

$$\begin{aligned} 2a - 2b + c + d &= 0 \\ -a \qquad + 3c + d &= 0 \\ 5a - 3b + 4c + d &= 0. \end{aligned}$$

Solving this system, we have (verify)

$$a = \tfrac{8}{17}r, \qquad b = \tfrac{15}{17}r, \qquad c = -\tfrac{3}{17}r, \qquad \text{and} \qquad d = r,$$

where r is any real number. Letting $r = 17$, we obtain

$$a = 8, \qquad b = 15, \qquad c = -3, \qquad \text{and} \qquad d = 17,$$

which yields (11) as in the first solution. ■

Example 11. Find parametric equations of the line of intersection of the planes

$$\pi_1\text{: } 2x + 3y - 2z + 4 = 0 \qquad \text{and} \qquad \pi_2\text{: } x - y + 2z + 3 = 0.$$

Solution: Solving the linear system consisting of the equations of π_1 and π_2, we obtain (verify)

$$\begin{aligned} x &= -\tfrac{13}{5} - \tfrac{4}{5}t \\ y &= \tfrac{2}{5} + \tfrac{6}{5}t \qquad -\infty < t < \infty \\ z &= 0 + t. \end{aligned}$$

as the parametric equations of the line ℓ of intersection of the planes (see Figure 3.17). ■

Figure 3.17

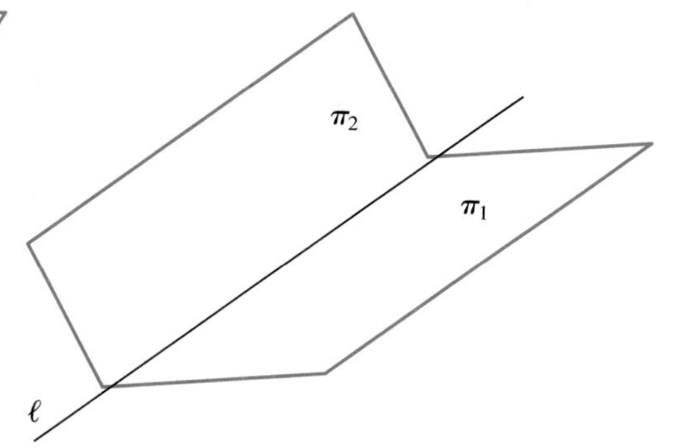

As we have indicated, the cross product cannot be generalized to R^n. However, we can generalize the notions of length, direction, and standard inner product to R^n in the natural manner, but there are some things to be checked. For example, if we define the cosine of the angle θ, between two nonzero vectors **u**

and $\mathbf{v}$ in R^n as

$$\cos\theta = \frac{\mathbf{u}\cdot\mathbf{v}}{\|\mathbf{u}\|\,\|\mathbf{v}\|},$$

we must check that $-1 \le \cos\theta \le 1$; otherwise, it would be misleading to call this fraction $\cos\theta$. Rather than verify this property for R^n at this point, we obtain this result in our next section, where we formulate the notion of inner product in any real vector space.

3.2. EXERCISES

1. Compute $\mathbf{u} \times \mathbf{v}$.
(a) $\mathbf{u} = 2\mathbf{i} + 3\mathbf{j} + 4\mathbf{k}$, $\mathbf{v} = -\mathbf{i} + 3\mathbf{j} - \mathbf{k}$.
(b) $\mathbf{u} = \mathbf{i} + \mathbf{k}$, $\mathbf{v} = 2\mathbf{i} + 3\mathbf{j} - \mathbf{k}$.
(c) $\mathbf{u} = \mathbf{i} - \mathbf{j} + 2\mathbf{k}$, $\mathbf{v} = 3\mathbf{i} - 4\mathbf{j} + \mathbf{k}$.
(d) $\mathbf{u} = \begin{bmatrix} 2 \\ -1 \\ 1 \end{bmatrix}$, $\mathbf{v} = -2\mathbf{u}$.

2. Compute $\mathbf{u} \times \mathbf{v}$.
(a) $\mathbf{u} = \mathbf{i} - \mathbf{j} + 2\mathbf{k}$, $\mathbf{v} = 3\mathbf{i} + \mathbf{j} + 2\mathbf{k}$.
(b) $\mathbf{u} = 2\mathbf{i} + \mathbf{j} - 2\mathbf{k}$, $\mathbf{v} = \mathbf{i} + 3\mathbf{k}$.
(c) $\mathbf{u} = 2\mathbf{j} + \mathbf{k}$, $\mathbf{v} = 3\mathbf{u}$.
(d) $\mathbf{u} = \begin{bmatrix} 4 \\ 0 \\ -2 \end{bmatrix}$, $\mathbf{v} = \begin{bmatrix} 0 \\ 2 \\ -1 \end{bmatrix}$.

3. Let $\mathbf{u} = \mathbf{i} + 2\mathbf{j} - 3\mathbf{k}$, $\mathbf{v} = 2\mathbf{i} + 3\mathbf{j} + \mathbf{k}$, $\mathbf{w} = 2\mathbf{i} - \mathbf{j} + 2\mathbf{k}$, and $c = -3$. Verify properties (a) through (h) for the cross product operation.

4. Prove properties (a) through (h) for the cross product operation.

5. Let $\mathbf{u} = 2\mathbf{i} - \mathbf{j} + 3\mathbf{k}$, $\mathbf{v} = 3\mathbf{i} + \mathbf{j} - \mathbf{k}$, and $\mathbf{w} = 3\mathbf{i} + \mathbf{j} + 2\mathbf{k}$.

(a) Verify Equation (3).
(b) Verify Equation (4).

6. Verify that each of the cross products $\mathbf{u} \times \mathbf{v}$ in Exercise 1 is orthogonal to both $\mathbf{u}$ and $\mathbf{v}$.

7. (a) Show that $(\mathbf{u} \times \mathbf{v}) \cdot \mathbf{w} = \mathbf{u} \cdot (\mathbf{v} \times \mathbf{w})$.
(b) Show that

$$\mathbf{u} \times (\mathbf{v} \times \mathbf{w}) = (\mathbf{u}\cdot\mathbf{w})\mathbf{v} - (\mathbf{u}\cdot\mathbf{v})\mathbf{w}.$$

8. Verify Equation (7) for the pairs of vectors in Exercise 1.

9. Show that $\mathbf{u}$ and $\mathbf{v}$ are parallel if and only if $\mathbf{u} \times \mathbf{v} = \mathbf{0}$.

10. Show that $\|\mathbf{u} \times \mathbf{v}\|^2 + (\mathbf{u}\cdot\mathbf{v})^2 = \|\mathbf{u}\|^2\,\|\mathbf{v}\|^2$.

11. Prove the **Jacobi identity**

$$(\mathbf{u} \times \mathbf{v}) \times \mathbf{w} + (\mathbf{v} \times \mathbf{w}) \times \mathbf{u} + (\mathbf{w} \times \mathbf{u}) \times \mathbf{v} = \mathbf{0}.$$

12. Find the area of the triangle with vertices $\mathbf{u}_1 = (1, -2, 3)$, $\mathbf{u}_2 = (-3, 1, 4)$, and $\mathbf{u}_3 = (0, 4, 3)$.

13. Find the area of the triangle with vertices $\mathbf{u}_1$, $\mathbf{u}_2$, and $\mathbf{u}_3$, where $\mathbf{u}_2 - \mathbf{u}_1 = 2\mathbf{i} + 3\mathbf{j} - \mathbf{k}$ and $\mathbf{u}_3 - \mathbf{u}_1 = \mathbf{i} + 2\mathbf{j} + 2\mathbf{k}$.

14. Find the area of the parallelogram with adjacent sides $\mathbf{u}_2 - \mathbf{u}_1 = \mathbf{i} + 3\mathbf{j} - 2\mathbf{k}$ and $\mathbf{u}_3 - \mathbf{u}_1 = 3\mathbf{i} - \mathbf{j} - \mathbf{k}$.

15. Find the volume of the parallelepiped with a vertex at the origin and edges $\mathbf{u} = 2\mathbf{i} - \mathbf{j}$, $\mathbf{v} = \mathbf{i} - 2\mathbf{j} - 2\mathbf{k}$, and $\mathbf{w} = 3\mathbf{i} - \mathbf{j} + \mathbf{k}$.

16. Repeat Exercise 15 for $\mathbf{u} = \mathbf{i} - 2\mathbf{j} + 4\mathbf{k}$, $\mathbf{v} = 3\mathbf{i} + 4\mathbf{j} + \mathbf{k}$, and $\mathbf{w} = -\mathbf{i} + \mathbf{j} + \mathbf{k}$.

17. State which of the following points are on the plane

$$3(x - 2) + 2(y + 3) - 4(z - 4) = 0.$$

(a) $(0, -2, 3)$. (b) $(1, -2, 3)$.

18. Find an equation of the plane passing through the given point and perpendicular to the given vector.
(a) $(0, 2, -3)$, $3\mathbf{i} - 2\mathbf{j} + 4\mathbf{k}$.
(b) $(-1, 3, 2)$, $\mathbf{j} - 3\mathbf{k}$.

19. Find an equation of the plane passing through the given points.
(a) $(0, 1, 2)$, $(3, -2, 5)$, $(2, 3, 4)$.
(b) $(2, 3, 4)$, $(-1, -2, 3)$, $(-5, -4, 2)$.

20. Find parametric equations of the line of intersection of the given planes.
(a) $2x + 3y - 4z + 5 = 0$ and $-3x + 2y + 5z + 6 = 0$.
(b) $3x - 2y - 5z + 4 = 0$ and $2x + 3y + 4z + 8 = 0$.

21. Find an equation of the plane through $(-2, 3, 4)$ and perpendicular to the line through $(4, -2, 5)$ and $(0, 2, 4)$.

22. Find the point of intersection of the line

$$x = 2 - 3t$$
$$y = 4 + 2t$$
$$z = 3 - 5t$$

and the plane $2x + 3y + 4z + 8 = 0$.

23. Find a line passing through $(-2, 5, -3)$ and perpendicular to the plane $2x - 3y + 4z + 7 = 0$.

24. (a) Show that the graph of an equation of the form given in (10), with a, b, and c not all zero, is a plane with normal $\mathbf{v} = a\mathbf{i} + b\mathbf{j} + c\mathbf{k}$.
(b) Show that the set of all points on the plane $ax + by + cz = 0$ is a subspace of R^3.
(c) Find a basis for the plane $2x - 3y + 4z = 0$.

3.3. *Inner Product Spaces*

In this section we use the properties of the standard inner product or dot product on R^3 listed in Theorem 3.1 as our foundation for generalizing the notion of the inner product to any real vector space. Here V is an arbitrary real vector space, not necessarily finite dimensional.

Definition 3.2. Let V be any real vector space. An **inner product** on V is a function that assigns to each ordered pair of vectors $\mathbf{u}$, $\mathbf{v}$ of V a real number $(\mathbf{u}, \mathbf{v})$ satisfying:
(a) $(\mathbf{u}, \mathbf{u}) > 0$ for $\mathbf{u} \neq \mathbf{0}_V$; $(\mathbf{u}, \mathbf{u}) = 0$ if and only if $\mathbf{u} = \mathbf{0}_V$.
(b) $(\mathbf{v}, \mathbf{u}) = (\mathbf{u}, \mathbf{v})$ for any $\mathbf{u}$, $\mathbf{v}$ in V.
(c) $(\mathbf{u} + \mathbf{v}, \mathbf{w}) = (\mathbf{u}, \mathbf{w}) + (\mathbf{v}, \mathbf{w})$ for any $\mathbf{u}$, $\mathbf{v}$, $\mathbf{w}$ in V.
(d) $(c\mathbf{u}, \mathbf{v}) = c(\mathbf{u}, \mathbf{v})$ for $\mathbf{u}$, $\mathbf{v}$ in V and c a real scalar. ■

From these properties it follows that $(\mathbf{u}, c\mathbf{v}) = c(\mathbf{u}, \mathbf{v})$ because $(\mathbf{u}, c\mathbf{v}) = (c\mathbf{v}, \mathbf{u}) = c(\mathbf{v}, \mathbf{u}) = c(\mathbf{u}, \mathbf{v})$.

Example 1. In Section 3.1 we have defined the standard inner product or dot product on R^3 by

$$(\mathbf{u}, \mathbf{v}) = \mathbf{u} \cdot \mathbf{v} = u_1v_1 + u_2v_2 + u_3v_3$$

where $\mathbf{u} = \begin{bmatrix} u_1 \\ u_2 \\ u_3 \end{bmatrix}$ and $\mathbf{v} = \begin{bmatrix} v_1 \\ v_2 \\ v_3 \end{bmatrix}$. ■

Example 2. We can define the **standard inner product** on R^n by defining $(\mathbf{u}, \mathbf{v})$ for $\mathbf{u} = \begin{bmatrix} u_1 \\ u_2 \\ \vdots \\ u_n \end{bmatrix}$ and $\mathbf{v} = \begin{bmatrix} v_1 \\ v_2 \\ \vdots \\ v_n \end{bmatrix}$ in R^n, as

$$(\mathbf{u}, \mathbf{v}) = u_1v_1 + u_2v_2 + \cdots + u_nv_n.$$

Thus, if

$$\mathbf{u} = \begin{bmatrix} 1 \\ -2 \\ 3 \\ 4 \end{bmatrix} \quad \text{and} \quad \mathbf{v} = \begin{bmatrix} 3 \\ 2 \\ -2 \\ 1 \end{bmatrix}$$

are vectors in R^4, then

$$\begin{aligned} (\mathbf{u}, \mathbf{v}) &= (1)(3) + (-2)(2) + (3)(-2) + (4)(1) \\ &= -3. \end{aligned}$$

Of course, one has to verify that this function satisfies the properties of Definition 3.2. ■

Example 3. Let V be any finite-dimensional vector space and let $S = \{\mathbf{u}_1, \mathbf{u}_2, \ldots, \mathbf{u}_n\}$ be an ordered basis for V. If

$$\mathbf{v} = a_1\mathbf{u}_1 + a_2\mathbf{u}_2 + \cdots + a_n\mathbf{u}_n$$

and

$$\mathbf{w} = b_1\mathbf{u}_1 + b_2\mathbf{u}_2 + \cdots + b_n\mathbf{u}_n,$$

we define

$$(\mathbf{v}, \mathbf{w}) = ([\mathbf{v}]_S, [\mathbf{w}]_S) = a_1b_1 + a_2b_2 + \cdots + a_nb_n.$$

It is not difficult to verify that this defines an inner product on V. This definition of $(\mathbf{v}, \mathbf{w})$ as an inner product on V uses the standard inner product on R^n. ■

Example 3 shows that we can define an inner product on any finite-dimensional vector space. Of course, if we change the basis for V in Example 3, we obtain a different inner product.

Example 4. Let $\mathbf{u} = \begin{bmatrix} u_1 \\ u_2 \end{bmatrix}$ and $\mathbf{v} = \begin{bmatrix} v_1 \\ v_2 \end{bmatrix}$ be vectors in R^2. We define

$$(\mathbf{u}, \mathbf{v}) = u_1v_1 - u_2v_1 - u_1v_2 + 3u_2v_2.$$

Show that this gives an inner product on R^2.

Solution: We have

$$\begin{aligned}(\mathbf{u}, \mathbf{u}) &= u_1^2 - 2u_1u_2 + 3u_2^2 = u_1^2 - 2u_1u_2 + u_2^2 + 2u_2^2 \\ &= (u_1 - u_2)^2 + 2u_2^2 > 0 \text{ if } \mathbf{u} \neq \mathbf{0}.\end{aligned}$$

Moreover, if $(\mathbf{u}, \mathbf{u}) = 0$, then $u_1 = u_2$ and $u_2 = 0$, so $\mathbf{u} = \mathbf{0}$. Conversely, if $\mathbf{u} = \mathbf{0}$, then $(\mathbf{u}, \mathbf{u}) = 0$. We can also verify (see Exercise 2) the remaining three properties of Definition 3.2. This inner product is, of course, not the standard inner product on R^2. ■

Example 4 shows that on one vector space we may have many different inner products, since we also have the standard inner product on R^2.

Example 5. Let V be the vector space of all continuous real-valued functions on the unit interval $[0, 1]$. For f and g in V, we let $(f, g) = \int_0^1 f(t)g(t)\, dt$. We now verify that this is an inner product on V, that is, that the properties of Definition 3.2 are satisfied.

Using results from calculus, we have for $f \neq 0$, the zero function,

$$(f, f) = \int_0^1 (f(t))^2\, dt > 0.$$

Moreover, if $(f, f) = 0$, then $f = 0$. Conversely, if $f = 0$, then $(f, f) = 0$. Also,

$$(f, g) = \int_0^1 f(t)g(t)\, dt = \int_0^1 g(t)f(t)\, dt = (g, f).$$

Next,

$$\begin{aligned}(f + g, h) &= \int_0^1 (f(t) + g(t))h(t)\, dt = \int_0^1 f(t)h(t)\, dt + \int_0^1 g(t)h(t)\, dt \\ &= (f, h) + (g, h).\end{aligned}$$

Finally,

$$(cf, g) = \int_0^1 (cf(t))g(t)\, dt = c \int_0^1 f(t)g(t)\, dt = c(f, g).$$

Thus, if f and g are the functions defined by $f(t) = t + 1$, $g(t) = 2t + 3$, then

$$(f, g) = \int_0^1 (t + 1)(2t + 3)\, dt = \int_0^1 (2t^2 + 5t + 3)\, dt = \tfrac{37}{6}.$$ ■

Example 6. Let $V = R_2$; if $\mathbf{u} = [u_1 \quad u_2]$ and $\mathbf{v} = [v_1 \quad v_2]$ are vectors in V, we define $(\mathbf{u}, \mathbf{v}) = u_1v_1 - u_2v_1 - u_1v_2 + 5u_2v_2$. The verification that this function is an inner product is entirely analogous to the verification required in Example 4. ■

Example 7. Let $V = P$; if $p(t)$ and $q(t)$ are polynomials in P, we define

$$(p(t), q(t)) = \int_0^1 p(t)q(t)\, dt.$$

The verification that this function is an inner product is identical to the verification given for Example 5. ■

We now show that every inner product on a finite-dimensional vector space V is completely determined, in terms of a given basis, by a certain matrix.

Theorem 3.2. *Let $S = \{\mathbf{u}_1, \mathbf{u}_2, \ldots, \mathbf{u}_n\}$ be an ordered basis for a finite-dimensional vector space V and assume that we are given an inner product on V. Let $c_{ij} = (\mathbf{u}_i, \mathbf{u}_j)$ and $C = [c_{ij}]$. Then*

(a) *C is a symmetric matrix.*
(b) *C determines* $(\mathbf{v}, \mathbf{w})$ *for every* $\mathbf{v}$ *and* $\mathbf{w}$ *in* V.

Proof: (a) Exercise.
(b) If $\mathbf{v}$ and $\mathbf{w}$ are in V, then

$$\mathbf{v} = a_1\mathbf{u}_1 + a_2\mathbf{u}_2 + \cdots + a_n\mathbf{u}_n$$
$$\mathbf{w} = b_1\mathbf{u}_1 + b_2\mathbf{u}_2 + \cdots + b_n\mathbf{u}_n,$$

which implies that

$$[\mathbf{v}]_S = \begin{bmatrix} a_1 \\ a_2 \\ \vdots \\ a_n \end{bmatrix} \quad \text{and} \quad [\mathbf{w}]_S = \begin{bmatrix} b_1 \\ b_2 \\ \vdots \\ b_n \end{bmatrix}.$$

The inner product $(\mathbf{v}, \mathbf{w})$ can then be expressed as

$$\begin{aligned}
(\mathbf{v}, \mathbf{w}) &= \left(\sum_{i=1}^{n} a_i\mathbf{u}_i, \mathbf{w} \right) = \sum_{i=1}^{n} a_i(\mathbf{u}_i, \mathbf{w}) \\
&= \sum_{i=1}^{n} a_i \left(\mathbf{u}_i, \sum_{j=1}^{n} b_j\mathbf{u}_j \right) \\
&= \sum_{i=1}^{n} a_i \sum_{j=1}^{n} b_j(\mathbf{u}_i, \mathbf{u}_j) = \sum_{i=1}^{n} \sum_{j=1}^{n} a_ib_j(\mathbf{u}_i, \mathbf{u}_j) \\
&= \sum_{i=1}^{n} \sum_{j=1}^{n} a_ic_{ij}b_j \\
&= [\mathbf{v}]_S^T C [\mathbf{w}]_S,
\end{aligned}$$

so

$$(\mathbf{v}, \mathbf{w}) = [\mathbf{v}]_S^T C [\mathbf{w}]_S, \tag{1}$$

which means that C determines $(\mathbf{v}, \mathbf{w})$ for every $\mathbf{v}$ and $\mathbf{w}$ in V. ■

Thus the inner product in Equation (1) is the product of three matrices. We shall next show that the inner product in (1) can also be expressed as a standard inner product on R^n. We first establish the following result for the standard inner product on R^n.

If $A = [a_{ij}]$ is an $n \times n$ matrix and $\mathbf{x}$ and $\mathbf{y}$ are vectors in R^n, then

$$(A\mathbf{x}, \mathbf{y}) = (\mathbf{x}, A^T\mathbf{y}). \tag{2}$$

This will be proved by first observing that if $\mathbf{u}$ and $\mathbf{v}$ are vectors in R^n, then $\mathbf{u}$ and $\mathbf{v}$ can be viewed as $n \times 1$ matrices and then the standard inner product on R^n can be written as (Exercise 37)

$$(\mathbf{u}, \mathbf{v}) = \mathbf{u}^T\mathbf{v}.$$

Equation (3) together with associativity of matrix multiplication and Theorem 1.4(c) can now be used to prove (2):

$$(A\mathbf{x}, \mathbf{y}) = (A\mathbf{x})^T\mathbf{y} = (\mathbf{x}^TA^T)\mathbf{y} = \mathbf{x}^T(A^T\mathbf{y}) = (\mathbf{x}, A^T\mathbf{y}).$$

Using Equation (3), we can now write Equation (1) as

$$(\mathbf{v}, \mathbf{w}) = ([\mathbf{v}]_S, C[\mathbf{w}]_S), \tag{4}$$

where the inner product on the left is in V and the inner product on the right is the standard inner product on R^n. Using Equation (2) and the fact that C is symmetric, we have

$$(\mathbf{v}, \mathbf{w}) = (C[\mathbf{v}]_S, [\mathbf{w}]_S) \tag{5}$$

(verify). Thus C determines $(\mathbf{v}, \mathbf{w})$ for every $\mathbf{v}$ and $\mathbf{w}$ in V. In summary, we have shown that an inner product on a real finite-dimensional vector space V can be computed using the standard inner product in R^n, as in Equation (4) or (5).

The matrix C in Theorem 3.2 is called the **matrix of the inner product with respect to the ordered basis S**. If the inner product is as defined in Example 3, then $C = I_n$ (verify).

There is another important property satisfied by the matrix of an inner product. If $\mathbf{u}$ is a nonzero vector in R^n, then $(\mathbf{u}, \mathbf{u}) > 0$, so letting $\mathbf{x} = [\mathbf{u}]_S$, Equation (1) says that

$$\mathbf{x}^T C\mathbf{x} > 0 \qquad \text{for every nonzero } \mathbf{x} \text{ in } R^n.$$

This property of the matrix of an inner product is so important that we specifically identify such matrices. An $n \times n$ symmetric matrix C with the property that

$\mathbf{x}^T C\mathbf{x} > 0$ for every nonzero vector $\mathbf{x}$ in R^n is called **positive definite**. A positive definite matrix C is nonsingular, for if C is singular, then the homogeneous system $C\mathbf{x} = \mathbf{0}$ has a nontrivial solution $\mathbf{x}_0$. Then $\mathbf{x}_0^T C\mathbf{x}_0 = 0$, contradicting the requirement that $\mathbf{x}^T C\mathbf{x} > 0$ for any nonzero vector $\mathbf{x}$.

If $C = [c_{ij}]$ is an $n \times n$ positive definite matrix, then we can use C to define an inner product on V. Using the same notation as above, we define

$$(\mathbf{v}, \mathbf{w}) = ([\mathbf{v}]_S, C[\mathbf{w}]_S) = \sum_{i=1}^{n} \sum_{j=1}^{n} a_i c_{ij} b_j.$$

It is not difficult to show that this defines an inner product on V (verify). The only gap in the discussion above is that we still do not know when a real symmetric matrix is positive definite, other than trying to verify the definition, which is usually not a fruitful approach. In Section 6.3 (see Theorem 6.12) we provide a characterization of positive definite matrices.

Example 8. Let $C = \begin{bmatrix} 2 & 1 \\ 1 & 2 \end{bmatrix}$. In this case we may verify that C is positive definite as follows.

$$\begin{aligned} \mathbf{x}^T C\mathbf{x} &= \begin{bmatrix} x_1 & x_2 \end{bmatrix} \begin{bmatrix} 2 & 1 \\ 1 & 2 \end{bmatrix} \begin{bmatrix} x_1 \\ x_2 \end{bmatrix} \\ &= 2x_1^2 + 2x_1x_2 + 2x_2^2 \\ &= x_1^2 + x_2^2 + (x_1 + x_2)^2 > 0 \qquad \text{if } \mathbf{x} \neq \mathbf{0}. \end{aligned}$$

We now define an inner product on P_1 whose matrix with respect to the ordered basis $S = \{t, 1\}$ is C. Thus let $p(t) = a_1t + a_2$ and $q(t) = b_1t + b_2$ be any two vectors in P_1. Let $(p(t), q(t)) = 2a_1b_1 + a_2b_1 + a_1b_2 + 2a_2b_2$. We must verify that if $p(t)$ is not the zero polynomial, then $(p(t), p(t)) > 0$; that is, if both a_1 and a_2 are not zero, then $2a_1^2 + 2a_1a_2 + 2a_2^2 > 0$. We now have

$$2a_1^2 + 2a_1a_2 + 2a_2^2 = a_1^2 + a_2^2 + (a_1 + a_2)^2 > 0.$$

The remaining properties are easy to verify. ■

Definition 3.3. A real vector space that has an inner product defined on it is called an **inner product space**. If the space is finite-dimensional, it is called a **Euclidean space**. ■

If V is an inner product space, then by **dimension** of V we mean the dimension of V as a real vector space, and a set S is a **basis** for V if S is a basis for the real vector space V. Examples 5 and 7 are inner product spaces. In an inner product space we define the **length** of a vector $\mathbf{u}$ by

$$\|\mathbf{u}\| = \sqrt{(\mathbf{u}, \mathbf{u})}.$$

This definition of length seems reasonable because at least we have $\|\mathbf{u}\| > 0$ if $\mathbf{u} \neq \mathbf{0}$. We can show (see Exercise 7) that $\|\mathbf{0}\| = 0$.

We shall now prove a result that will enable us to give a worthwhile definition for the cosine of an angle between two nonzero vectors $\mathbf{u}$ and $\mathbf{v}$ in an inner product space V. This result, called the **Cauchy–Schwarz inequality**, has many important applications in mathematics. The proof, although not difficult, is one that is not too natural and does call for a clever start.

Theorem 3.3. (Cauchy†–Schwarz‡ Inequality) *If* $\mathbf{u}$ *and* $\mathbf{v}$ *are any two vectors in an inner product space V, then*

$$(\mathbf{u}, \mathbf{v})^2 \leq \|\mathbf{u}\|^2 \, \|\mathbf{v}\|^2$$

Proof: If $\mathbf{u} = \mathbf{0}$, then $\|\mathbf{u}\| = 0$ and $(\mathbf{u}, \mathbf{v}) = 0$, so the inequality holds. Now suppose that $\mathbf{u}$ is nonzero. Let r be a scalar and consider the vector $r\mathbf{u} + \mathbf{v}$. Since the inner product of a vector with itself is always nonnegative, we have

$$0 \leq (r\mathbf{u} + \mathbf{v}, r\mathbf{u} + \mathbf{v}) = (\mathbf{u}, \mathbf{u})r^2 + 2r(\mathbf{u}, \mathbf{v}) + (\mathbf{v}, \mathbf{v}) = ar^2 + 2br + c,$$

where $a = (\mathbf{u}, \mathbf{u})$, $b = (\mathbf{u}, \mathbf{v})$, and $c = (\mathbf{v}, \mathbf{v})$. If we fix $\mathbf{u}$ and $\mathbf{v}$, then $ar^2 + 2br + c = p(r)$ is a quadratic polynomial in r that is nonnegative for all values of

† Augustin-Louis Cauchy (1789–1857) grew up in a suburb of Paris as a neighbor of several leading mathematicians of the day, attended the École Polytechnique and the École des Ponts et Chaussées, and was for a time a practicing engineer. He was a devout Roman Catholic, with an abiding interest in Catholic charities. He was also strongly devoted to royalty, especially to the Bourbon kings who ruled France after Napoleon's defeat. When Charles X was deposed in 1830, Cauchy voluntarily followed him into exile in Prague.

Cauchy wrote 7 books and more than 700 papers of varying quality, touching all branches of mathematics. He made important contributions to the early theory of determinants, the theory of eigenvalues, the study of ordinary and partial differential equations, the theory of permutation groups, and the foundations of calculus, and he founded the theory of functions of a complex variable.

‡ Hermann Amandus Schwarz (1843–1921) was born in Poland but was educated and taught in Germany. He was a protégé of Karl Weierstrass and of Ernst Eduard Kummer, whose daughter he married. His main contributions to mathematics were in the geometric aspects of analysis, such as conformal mappings and minimal surfaces. In connection with the latter he sought certain numbers associated with differential equations, numbers that have since come to be called eigenvalues. The inequality given above was used in the search for these numbers.

Figure 3.18

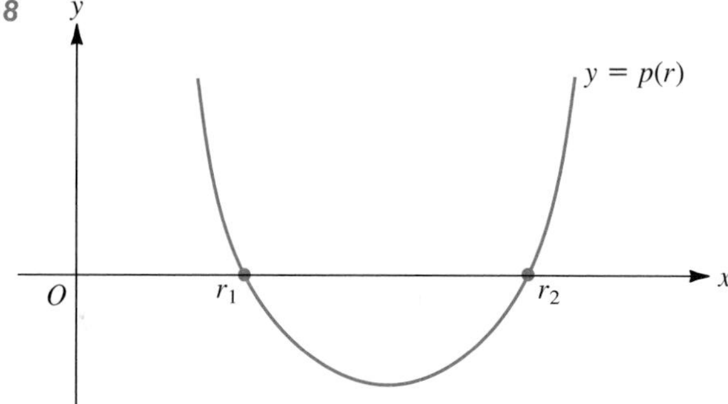

r. This means that $p(r)$ has at most one real root, for if it had two distinct real roots, r_1 and r_2, it would be negative between r_1 and r_2 (Figure 3.18). From the quadratic formula, the roots of $p(r)$ are given by

$$\frac{-b + \sqrt{b^2 - ac}}{a} \quad \text{and} \quad \frac{-b - \sqrt{b^2 - ac}}{a}$$

($a \neq 0$, since $\mathbf{u} \neq \mathbf{0}$). Thus we must have $b^2 - ac \leq 0$, which means that $b^2 \leq ac$. Substituting for a, b, and c, we obtain the desired inequality. ■

Example 9. Let $\mathbf{u} = \begin{bmatrix} 1 \\ 2 \\ -3 \end{bmatrix}$ and $\mathbf{v} = \begin{bmatrix} -3 \\ 2 \\ 2 \end{bmatrix}$ be in the Euclidean space R^3 with the standard inner product. Then $(\mathbf{u}, \mathbf{v}) = -5$, $\|\mathbf{u}\| = \sqrt{14}$, and $\|\mathbf{v}\| = \sqrt{17}$. Therefore, $(\mathbf{u}, \mathbf{v})^2 \leq \|\mathbf{u}\|^2 \|\mathbf{v}\|^2$. ■

If $\mathbf{u}$ and $\mathbf{v}$ are any two nonzero vectors in an inner product space V, the Cauchy–Schwarz inequality can be written as

$$\frac{(\mathbf{u}, \mathbf{v})^2}{\|\mathbf{u}\|^2 \|\mathbf{v}\|^2} \leq 1 \quad \text{or} \quad -1 \leq \frac{(\mathbf{u}, \mathbf{v})}{\|\mathbf{u}\| \|\mathbf{v}\|} \leq 1.$$

It then follows that there is one and only one angle θ such that

$$\cos \theta = \frac{(\mathbf{u}, \mathbf{v})}{\|\mathbf{u}\| \|\mathbf{v}\|} \qquad 0 \leq \theta \leq \pi.$$

We define this angle to be the **angle** between $\mathbf{u}$ and $\mathbf{v}$.

The triangle inequality is an easy consequence of the Cauchy–Schwarz inequality.

Corollary 3.1. (Triangle Inequality) *If* **u** *and* **v** *are any vectors in an inner product space V, then* $\|\mathbf{u} + \mathbf{v}\| \leq \|\mathbf{u}\| + \|\mathbf{v}\|$.

Proof: We have

$$\|\mathbf{u} + \mathbf{v}\|^2 = (\mathbf{u} + \mathbf{v}, \mathbf{u} + \mathbf{v}) = (\mathbf{u}, \mathbf{u}) + 2(\mathbf{u}, \mathbf{v}) + (\mathbf{v}, \mathbf{v})$$
$$= \|\mathbf{u}\|^2 + 2(\mathbf{u}, \mathbf{v}) + \|\mathbf{v}\|^2.$$

The Cauchy–Schwarz inequality states that $(\mathbf{u}, \mathbf{v}) \leq |(\mathbf{u}, \mathbf{v})| \leq \|\mathbf{u}\|\,\|\mathbf{v}\|$, so we get

$$\|\mathbf{u} + \mathbf{v}\|^2 \leq \|\mathbf{u}\|^2 + 2\|\mathbf{u}\|\,\|\mathbf{v}\| + \|\mathbf{v}\|^2 = (\|\mathbf{u}\| + \|\mathbf{v}\|)^2.$$

Taking square roots, we obtain

$$\|\mathbf{u} + \mathbf{v}\| \leq \|\mathbf{u}\| + \|\mathbf{v}\|.$$ ■

We now state the Cauchy–Schwarz inequality for the inner product spaces introduced in several of our examples. In Example 2, if

$$\mathbf{u} = \begin{bmatrix} u_1 \\ u_2 \\ \vdots \\ u_n \end{bmatrix} \quad \text{and} \quad \mathbf{v} = \begin{bmatrix} v_1 \\ v_2 \\ \vdots \\ v_n \end{bmatrix},$$

then

$$(\mathbf{u}, \mathbf{v})^2 = \left(\sum_{i=1}^{n} u_i v_i\right)^2 \leq \left(\sum_{i=1}^{n} u_i^2\right)\left(\sum_{i=1}^{n} v_i^2\right) = \|\mathbf{u}\|^2\,\|\mathbf{v}\|^2.$$

In Example 5, if f and g are continuous functions on $[0, 1]$, then

$$(f, g)^2 = \left(\int_0^1 f(t)g(t)\,dt\right)^2 \leq \left(\int_0^1 f^2(t)\,dt\right)\left(\int_0^1 g^2(t)\,dt\right).$$

Example 10. Let V be the Euclidean space P_2 with inner product defined as in Example 7. If $p(t) = t + 2$, then the length of $p(t)$ is

$$\|p(t)\| = \sqrt{(p(t), p(t))} = \sqrt{\int_0^1 (t+2)^2\, dt} = \sqrt{\frac{19}{3}}.$$

If $q(t) = 2t - 3$, then to find the cosine of the angle θ between $p(t)$ and $q(t)$, we proceed as follows. First,

$$\|q(t)\| = \sqrt{\int_0^1 (2t-3)^2\, dt} = \sqrt{\frac{13}{3}}.$$

Next,

$$(p(t), q(t)) = \int_0^1 (t+2)(2t-3)\, dt = \int_0^1 (2t^2 + t - 6)\, dt = -\frac{29}{6}.$$

Then

$$\cos\theta = \frac{(p(t), q(t))}{|p(t)|\,|q(t)|} = \frac{-29/6}{\sqrt{19/3}\sqrt{13/3}} = \frac{-29}{2\sqrt{(19)(13)}}.$$ ■

Definition 3.4. If V is an inner product space, we define the **distance** between two vectors $\mathbf{u}$ and $\mathbf{v}$ in V as $d(\mathbf{u}, \mathbf{v}) = \|\mathbf{u} - \mathbf{v}\|$. ■

Definition 3.5. Let V be an inner product space. Vectors $\mathbf{u}$ and $\mathbf{v}$ in V are **orthogonal** if $(\mathbf{u}, \mathbf{v}) = 0$. ■

Example 11. Let V be the Euclidean space R^4 with the standard inner product. If

$$\mathbf{u} = \begin{bmatrix} 1 \\ 0 \\ 0 \\ 1 \end{bmatrix} \quad \text{and} \quad \mathbf{v} = \begin{bmatrix} 0 \\ 2 \\ 3 \\ 0 \end{bmatrix},$$

then $(\mathbf{u}, \mathbf{v}) = 0$, so $\mathbf{u}$ and $\mathbf{v}$ are orthogonal. ■

Example 12. Let V be the Euclidean space P_2 considered in Example 10. The vectors t and $t - \frac{2}{3}$ are orthogonal, since

$$\left(t, t - \frac{2}{3}\right) = \int_0^1 t\left(t - \frac{2}{3}\right) dt = \int_0^1 \left(t^2 - \frac{2t}{3}\right) dt = 0.$$ ■

Of course, the vector $\mathbf{0}_V$ in an inner product space V is orthogonal to every vector in V, and two nonzero vectors in V are orthogonal if the angle θ between them is $\pi/2$. Also, the subset of vectors in V orthogonal to a fixed vector in V is a subspace of V (see Exercise 21).

We know from calculus that we can work with any set of coordinate axes for R^3, but that the work becomes less burdensome when we deal with Cartesian coordinates. The comparable notion in an inner product space is that of a basis whose vectors are mutually orthogonal. We now proceed to formulate this idea.

Definition 3.6. Let V be an inner product space. A set S of vectors in V is called **orthogonal** if any two distinct vectors in S are orthogonal. If, in addition, each vector in S is of unit length, then S is called **orthonormal**. ■

We note here that if $\mathbf{x}$ is a nonzero vector in an inner product space, then we can always find a vector of unit length (called a **unit vector**) in the same direction as $\mathbf{x}$; we let $\mathbf{u} = \frac{1}{\|\mathbf{x}\|}\mathbf{x}$. Then

$$\|\mathbf{u}\| = \sqrt{(\mathbf{u}, \mathbf{u})} = \sqrt{\left(\frac{\mathbf{x}}{\|\mathbf{x}\|}, \frac{\mathbf{x}}{\|\mathbf{x}\|}\right)} = \sqrt{\frac{(\mathbf{x}, \mathbf{x})}{\|\mathbf{x}\|\,\|\mathbf{x}\|}} = \sqrt{\frac{\|\mathbf{x}\|^2}{\|\mathbf{x}\|\,\|\mathbf{x}\|}} = 1,$$

and the cosine of the angle between $\mathbf{x}$ and $\mathbf{u}$ is 1, so $\mathbf{x}$ and $\mathbf{u}$ have the same direction.

Example 13. If $\mathbf{x}_1 = \begin{bmatrix} 1 \\ 0 \\ 2 \end{bmatrix}$, $\mathbf{x}_2 = \begin{bmatrix} -2 \\ 0 \\ 1 \end{bmatrix}$, and $\mathbf{x}_3 = \begin{bmatrix} 0 \\ 1 \\ 0 \end{bmatrix}$, then $\{\mathbf{x}_1, \mathbf{x}_2, \mathbf{x}_3\}$ is an orthogonal set (verify). The vectors

$$\mathbf{u}_1 = \begin{bmatrix} \frac{1}{\sqrt{5}} \\ 0 \\ \frac{2}{\sqrt{5}} \end{bmatrix} \quad \text{and} \quad \mathbf{u}_2 = \begin{bmatrix} -\frac{2}{\sqrt{5}} \\ 0 \\ \frac{1}{\sqrt{5}} \end{bmatrix}$$

are unit vectors in the directions of $\mathbf{x}_1$ and $\mathbf{x}_2$, respectively. Since $\mathbf{x}_3$ is also a unit vector, we conclude that $\{\mathbf{x}_1, \mathbf{x}_2, \mathbf{x}_3\}$ is an orthonormal set. ■

Example 14. The natural bases for R^n and R_n are orthonormal sets with respect to the standard inner products on these vector spaces. ■

An important result about orthogonal sets of vectors in an inner product space is the following.

Theorem 3.4. *Let $S = \{\mathbf{u}_1, \mathbf{u}_2, \ldots, \mathbf{u}_n\}$ be a finite orthogonal set of nonzero vectors in an inner product space V. Then S is linearly independent.*

Proof: Suppose that

$$a_1\mathbf{u}_1 + a_2\mathbf{u}_2 + \cdots + a_n\mathbf{u}_n = \mathbf{0}.$$

Then taking the inner product of both sides with $\mathbf{u}_i$, we have

$$(a_1\mathbf{u}_1 + a_2\mathbf{u}_2 + \cdots + a_i\mathbf{u}_i + \cdots + a_n\mathbf{u}_n, \mathbf{u}_i) = (\mathbf{0}, \mathbf{u}_i) = 0.$$

The left-hand side is

$$a_1(\mathbf{u}_1, \mathbf{u}_i) + a_2(\mathbf{u}_2, \mathbf{u}_i) + \cdots + a_i(\mathbf{u}_i, \mathbf{u}_i) + \cdots + a_n(\mathbf{u}_n, \mathbf{u}_i),$$

and since S is orthogonal, this is $a_i(\mathbf{u}_i, \mathbf{u}_i)$. Thus $a_i(\mathbf{u}_i, \mathbf{u}_i) = 0$. Since $\mathbf{u}_i \neq \mathbf{0}$, $(\mathbf{u}_i, \mathbf{u}_i) \neq 0$, so $a_i = 0$. Repeating this for $i = 1, 2, \ldots, n$, we find that $a_1 = a_2 = \cdots = a_n = 0$, so S is linearly independent. ■

Example 15. Let V be the vector space of all continuous functions on $[-\pi, \pi]$. For f and g in V, we let $(f, g) = \int_{-\pi}^{\pi} f(t)g(t)\,dt$, which is easily shown to be an inner product on V (see Example 5). Consider the functions

$$1, \cos t, \sin t, \cos 2t, \sin 2t, \ldots, \cos nt, \sin nt, \ldots, \tag{6}$$

which are clearly in V. The relationships

$$\int_{-\pi}^{\pi} \cos nt\,dt = \int_{-\pi}^{\pi} \sin nt\,dt = \int_{-\pi}^{\pi} \sin nt \cos nt\,dt = 0$$

$$\int_{-\pi}^{\pi} \cos mt \cos nt\,dt = \int_{-\pi}^{\pi} \sin mt \sin nt\,dt = 0 \qquad \text{if } m \neq n$$

demonstrate that $(f, g) = 0$ whenever f and g are distinct functions from (6). Hence every finite subset of functions from (6) is an orthogonal set. Theorem 3.4 then implies that any finite subset of functions from (6) is linearly independent. ■

3.3. EXERCISES

1. Verify that the standard inner product on R^n satisfies the properties of Definition 3.2.

2. Verify that the function in Example 4 satisfies the remaining three properties of Definition 3.2.

3. Let V be the real vector space of all $n \times n$ matrices. If A and B are in V, we define $(A, B) = \text{Tr}\,(B^T A)$, where Tr is the trace function defined in Exercise 23 of Section 1.2. Prove that this function is an inner product on V.

4. Verify that the function defined on V in Example 3 is an inner product.

5. Verify that the function defined on R_2 in Example 6 is an inner product.

6. Verify that the function defined on P in Example 7 is an inner product.

7. Let V be an inner product space. Prove the following.
(a) $\|\mathbf{0}\| = 0$.
(b) $(\mathbf{u}, \mathbf{0}) = (\mathbf{0}, \mathbf{u}) = 0$ for any $\mathbf{u}$ in V.
(c) If $(\mathbf{u}, \mathbf{v}) = 0$ for all $\mathbf{v}$ in V, then $\mathbf{u} = \mathbf{0}$.
(d) If $(\mathbf{u}, \mathbf{w}) = (\mathbf{v}, \mathbf{w})$ for all $\mathbf{w}$ in V, then $\mathbf{u} = \mathbf{v}$.
(e) If $(\mathbf{w}, \mathbf{u}) = (\mathbf{w}, \mathbf{v})$ for all $\mathbf{w}$ in V, then $\mathbf{u} = \mathbf{v}$.

In Exercises 8 and 9, let V be Euclidean space R_4 with the standard inner product. Compute $(\mathbf{u}, \mathbf{v})$.

8. (a) $\mathbf{u} = [1 \;\; 3 \;\; -1 \;\; 2]$, $\mathbf{v} = [-1 \;\; 2 \;\; 0 \;\; 1]$.
(b) $\mathbf{u} = [0 \;\; 0 \;\; 1 \;\; 1]$, $\mathbf{v} = [1 \;\; 1 \;\; 0 \;\; 0]$.
(c) $\mathbf{u} = [-2 \;\; 1 \;\; 3 \;\; 4]$, $\mathbf{v} = [3 \;\; 2 \;\; 1 \;\; -2]$.

9. (a) $\mathbf{u} = [1 \;\; 2 \;\; 3 \;\; 4]$, $\mathbf{v} = [-1 \;\; 0 \;\; -1 \;\; -1]$.
(b) $\mathbf{u} = [0 \;\; -1 \;\; 1 \;\; 4]$, $\mathbf{v} = [2 \;\; 0 \;\; -8 \;\; 2]$.
(c) $\mathbf{u} = [0 \;\; 0 \;\; -1 \;\; 2]$, $\mathbf{v} = [2 \;\; 3 \;\; -1 \;\; 0]$.

In Exercises 10 and 11, use the inner product space of continuous functions on $[0, 1]$ defined in Example 5. Find (f, g) for the following.

10. (a) $f(t) = 1 + t$, $g(t) = 2 - t$.
(b) $f(t) = 1$, $g(t) = 3$.
(c) $f(t) = 1$, $g(t) = 3 + 2t$.

11. (a) $f(t) = 3t$, $g(t) = 2t^2$.
(b) $f(t) = t$, $g(t) = e^t$.
(c) $f(t) = \sin t$, $g(t) = \cos t$.

In Exercises 12 and 13, let V be the Euclidean space of Example 4. Compute the length of the given vector.

12. (a) $\begin{bmatrix} 1 \\ 3 \end{bmatrix}$. (b) $\begin{bmatrix} 3 \\ -1 \end{bmatrix}$. (c) $\begin{bmatrix} 1 \\ 0 \end{bmatrix}$.

13. (a) $\begin{bmatrix} 0 \\ -2 \end{bmatrix}$. (b) $\begin{bmatrix} -2 \\ -4 \end{bmatrix}$. (c) $\begin{bmatrix} 2 \\ 2 \end{bmatrix}$.

In Exercises 14 and 15, let V be the inner product space of Example 7. Find the cosine of the angle between each pair of vectors in V.

14. (a) $p(t) = t$, $q(t) = t - 1$.
(b) $p(t) = t$, $q(t) = t$.
(c) $p(t) = 1$, $q(t) = 2t + 3$.

15. (a) $p(t) = 1$, $q(t) = 1$.
(b) $p(t) = t^2$, $q(t) = 2t^3 - \frac{4}{3}t$.
(c) $p(t) = \sin t$, $q(t) = \cos t$.

16. Prove the **parallelogram law** for any two vectors in an inner product space:

$$\|\mathbf{u} + \mathbf{v}\|^2 + \|\mathbf{u} - \mathbf{v}\|^2 = 2\|\mathbf{u}\|^2 + 2\|\mathbf{v}\|^2.$$

17. Let V be an inner product space. Show that $\|c\mathbf{u}\| = |c|\ \|\mathbf{u}\|$ for any vector $\mathbf{u}$ and any scalar c.

18. State the Cauchy–Schwarz inequality for the inner product spaces defined in Example 4, Example 6, and Exercise 3.

19. Let V be an inner product space. Prove that if $\mathbf{u}$ and $\mathbf{v}$ are any vectors in V, then $\|\mathbf{u} + \mathbf{v}\|^2 = \|\mathbf{u}\|^2 + \|\mathbf{v}\|^2$ if and only if $(\mathbf{u}, \mathbf{v}) = 0$, that is, if and only if $\mathbf{u}$ and $\mathbf{v}$ are orthogonal. This result is known as the **Pythagorean theorem**.

20. Let V be the Euclidean space R_4 considered in Exercise 8. Find which of the pairs of vectors listed there are orthogonal.

21. Let V be an inner product space and $\mathbf{u}$ a fixed vector in V. Prove that the set of all vectors in V orthogonal to $\mathbf{u}$ is a subspace of V.

22. For each of the inner products defined in Examples 1, 4, and 6, choose an ordered basis S for the vector space and find the matrix of the inner product with respect to S.

23. Let $C = \begin{bmatrix} 3 & -2 \\ -2 & 3 \end{bmatrix}$. Define an inner product on R_2 whose matrix with respect to the natural ordered basis is C.

24. If V is an inner product space, prove that the distance function of Definition 3.4 satisfies the following properties for all vectors $\mathbf{u}$, $\mathbf{v}$, and $\mathbf{w}$ in V.
(a) $d(\mathbf{u}, \mathbf{v}) \geq 0$.
(b) $d(\mathbf{u}, \mathbf{v}) = 0$ if and only if $\mathbf{u} = \mathbf{v}$.
(c) $d(\mathbf{u}, \mathbf{v}) = d(\mathbf{v}, \mathbf{u})$.
(d) $d(\mathbf{u}, \mathbf{v}) \leq d(\mathbf{u}, \mathbf{w}) + d(\mathbf{w}, \mathbf{v})$.

In Exercises 25 and 26, let V be the inner product space of Example 5. Compute the distance between the given vectors.

25. (a) $\sin t$, $\cos t$.
(b) t, t^2.

26. (a) $2t + 3$, $3t^2 - 1$.
(b) $3t + 1$, 1.

In Exercises 27 and 28, which of the given sets of vectors in R^3, with the standard inner product, are orthogonal, orthonormal, neither?

27. (a) $\left\{ \begin{bmatrix} \frac{1}{\sqrt{2}} \\ 0 \\ \frac{1}{\sqrt{2}} \end{bmatrix}, \begin{bmatrix} \frac{-1}{\sqrt{2}} \\ 0 \\ \frac{1}{\sqrt{2}} \end{bmatrix}, \begin{bmatrix} 0 \\ 1 \\ 0 \end{bmatrix} \right\}$.

(b) $\left\{ \begin{bmatrix} 1 \\ 1 \\ 0 \end{bmatrix}, \begin{bmatrix} 0 \\ 0 \\ 1 \end{bmatrix}, \begin{bmatrix} 0 \\ 1 \\ 0 \end{bmatrix} \right\}$.

(c) $\left\{ \begin{bmatrix} 1 \\ -1 \\ 0 \end{bmatrix}, \begin{bmatrix} 0 \\ 1 \\ 1 \end{bmatrix}, \begin{bmatrix} 0 \\ 0 \\ 1 \end{bmatrix} \right\}$.

28. (a) $\left\{ \begin{bmatrix} 0 \\ 1 \\ -1 \end{bmatrix}, \begin{bmatrix} 0 \\ 1 \\ 1 \end{bmatrix}, \begin{bmatrix} 2 \\ 0 \\ 0 \end{bmatrix} \right\}$.

(b) $\left\{ \begin{bmatrix} -1 \\ 1 \\ 0 \end{bmatrix}, \begin{bmatrix} \frac{1}{\sqrt{3}} \\ \frac{1}{\sqrt{3}} \\ \frac{1}{\sqrt{3}} \end{bmatrix}, \begin{bmatrix} 0 \\ \frac{1}{\sqrt{2}} \\ \frac{-1}{\sqrt{2}} \end{bmatrix} \right\}$.

(c) $\left\{ \begin{bmatrix} \frac{2}{3} \\ \frac{2}{3} \\ \frac{1}{3} \end{bmatrix}, \begin{bmatrix} -\frac{2}{3} \\ \frac{1}{3} \\ \frac{2}{3} \end{bmatrix}, \begin{bmatrix} \frac{1}{3} \\ -\frac{2}{3} \\ \frac{2}{3} \end{bmatrix} \right\}$.

In Exercises 29 and 30, let V be the inner product space of Example 7.

29. Let $p(t) = 3t + 1$ and $q(t) = at$. For what values of a are $p(t)$ and $q(t)$ orthogonal?

30. Let $p(t) = 3t + 1$ and $q(t) = at + b$. For what values of a and b are $p(t)$ and $q(t)$ orthogonal?

In Exercises 31 and 32, let V be the Euclidean space R^3 with the standard inner product.

31. Let $\mathbf{u} = \begin{bmatrix} 1 \\ 1 \\ -2 \end{bmatrix}$ and $\mathbf{v} = \begin{bmatrix} a \\ -1 \\ 2 \end{bmatrix}$. For what values of a are $\mathbf{u}$ and $\mathbf{v}$ orthogonal?

32. Let $\mathbf{u} = \begin{bmatrix} \frac{1}{\sqrt{2}} \\ 0 \\ \frac{1}{\sqrt{2}} \end{bmatrix}$ and $\mathbf{v} = \begin{bmatrix} a \\ -1 \\ -b \end{bmatrix}$. For what values of a and b is $\{\mathbf{u}, \mathbf{v}\}$ an orthonormal set?

33. Let $A = \begin{bmatrix} 1 & 2 \\ 3 & 4 \end{bmatrix}$. Find a 2×2 matrix $B \neq O_2$ such that A and B are orthogonal in the inner product space defined in Exercise 3. Can there be more than one matrix B that is orthogonal to A?

34. Let V be the inner product space in Example 5.
(a) If $p(t) = \sqrt{t}$, find $q(t) = a + bt \neq 0$ such that $p(t)$ and $q(t)$ are orthogonal.
(b) If $p(t) = \sin t$, find $q(t) = a + be^t \neq 0$ such that $p(t)$ and $q(t)$ are orthogonal.

35. Let $C = [c_{ij}]$ be an $n \times n$ positive definite symmetric matrix and let V be an n-dimensional vector space with ordered basis $S = \{\mathbf{u}_1, \mathbf{u}_2, \ldots, \mathbf{u}_n\}$. For $\mathbf{v} = a_1\mathbf{u}_1 + a_2\mathbf{u}_2 + \cdots + a_n\mathbf{u}_n$ and for $\mathbf{w} = b_1\mathbf{u}_1 + b_2\mathbf{u}_2 + \cdots + b_n\mathbf{u}_n$ in V define

$(\mathbf{v}, \mathbf{w}) = \sum_{i=1}^{n} \sum_{j=1}^{n} a_i c_{ij} b_j$. Prove that this defines an inner product on V.

36. If A and B are $n \times n$ matrices, show that $(A\mathbf{u}, B\mathbf{v}) = (\mathbf{u}, A^TB\mathbf{v})$ for any vectors $\mathbf{u}$ and $\mathbf{v}$ in Euclidean space R^n with the standard inner product.

37. In the Euclidean space R^n with the standard inner product, prove that $(\mathbf{u}, \mathbf{v}) = \mathbf{u}^T\mathbf{v}$.

38. Consider Euclidean space R^4 with the standard inner product and let

$$\mathbf{u}_1 = \begin{bmatrix} 1 \\ 0 \\ 0 \\ 1 \end{bmatrix} \quad \text{and} \quad \mathbf{u}_2 = \begin{bmatrix} 0 \\ 1 \\ 0 \\ 1 \end{bmatrix}.$$

(a) Prove that the set W consisting of all vectors in R^4 that are orthogonal to both $\mathbf{u}_1$ and $\mathbf{u}_2$ is a subspace of R^4.
(b) Find a basis for W.

39. Let V be an inner product space. Show that if $\mathbf{v}$ is orthogonal to $\mathbf{w}_1, \mathbf{w}_2, \ldots, \mathbf{w}_k$, then $\mathbf{v}$ is orthogonal to every vector in

$$\text{span } \{\mathbf{w}_1, \mathbf{w}_2, \ldots, \mathbf{w}_k\}.$$

40. Suppose that $\{\mathbf{v}_1, \mathbf{v}_2, \ldots, \mathbf{v}_n\}$ is an orthonormal set in R^n with the standard inner product. Let the matrix A be given by $A = [\mathbf{v}_1 \quad \mathbf{v}_2 \quad \cdots \quad \mathbf{v}_n]$. Show that A is nonsingular and compute its inverse. Give three different examples of such a matrix in R^2 or R^3.

41. Suppose that $\{\mathbf{v}_1, \mathbf{v}_2, \ldots, \mathbf{v}_n\}$ is an orthogonal set in R^n with the standard inner product. Let the matrix A be given by $A = [\mathbf{v}_1 \quad \mathbf{v}_2 \quad \cdots \quad \mathbf{v}_n]$. That is, A is the matrix whose jth column is $\mathbf{v}_j$, $j = 1, 2, \ldots, n$. Prove or disprove: A is nonsingular.

42. If A is nonsingular, prove that A^TA is positive definite.

43. If C is positive definite, and $\mathbf{x} \neq \mathbf{0}$ is such that $C\mathbf{x} = k\mathbf{x}$ for some scalar k, show that $k > 0$.

44. If C is positive definite, show that its diagonal entries are positive.

45. Let C be positive definite and r any scalar. Prove or disprove: rC is positive definite.

46. If B and C are $n \times n$ positive definite matrices, show that $B + C$ is positive definite.

47. Let S be the set of $n \times n$ positive definite matrices. Is S a subspace of ${}_nR_n$?

48. To compute the standard inner product of a pair of vectors $\mathbf{u}$ and $\mathbf{v}$ in R^n, use the matrix product operation in your software as follows. Let U and V be column matrices for vectors $\mathbf{u}$ and $\mathbf{v}$, respectively. Then $(\mathbf{u}, \mathbf{v})$ = the product of U^T and V (or V^T and U). Experiment with the vectors in Example 2 and Exercises 8 and 9 (see Exercise 37).

49. Exercise 40 in Section 3.1 can be generalized to R^n or even R_n in some software. Determine if this is the case for the software that you use.

50. Exercise 41 in Section 3.1 can be generalized to R^n or even R_n in some software. Determine if this is the case for the software that you use.

3.4. *The Gram–Schmidt Process*

In this section we prove that for every Euclidean space V we can obtain a basis S for V such that S is an orthonormal set; such a basis is called an **orthonormal basis**, and the method we use to obtain it is the **Gram–Schmidt process**. From our work with the natural bases for R^2, R^3, and, in general, for R^n, we know that when these bases are present, the computations are kept to a minimum. The reduction in the computational effort is due to the fact that we are dealing with an orthonormal basis. For example, if $S = \{\mathbf{u}_1, \mathbf{u}_2, \ldots, \mathbf{u}_n\}$ is a basis for an n-dimensional Euclidean space V, then if $\mathbf{v}$ is any vector in V, we can write $\mathbf{v}$ as

$$\mathbf{v} = c_1\mathbf{u}_1 + c_2\mathbf{u}_2 + \cdots + c_n\mathbf{u}_n.$$

The coefficients $c_1, c_2, \ldots, c_n$ are obtained by solving a linear system of n equations in n unknowns.

However, if S is orthonormal, we can obtain the same result with much less work. This is the content of the following theorem.

Theorem 3.5. *Let $S = \{\mathbf{u}_1, \mathbf{u}_2, \ldots, \mathbf{u}_n\}$ be an orthonormal basis for a Euclidean space V and let $\mathbf{v}$ be any vector in V. Then*

$$\mathbf{v} = c_1\mathbf{u}_1 + c_2\mathbf{u}_2 + \cdots + c_n\mathbf{u}_n,$$

where

$$c_i = (\mathbf{v}, \mathbf{u}_i), \qquad i = 1, 2, \ldots, n.$$

Proof: Exercise 13. ■

Example 1. Let $S = \{\mathbf{u}_1, \mathbf{u}_2, \mathbf{u}_3\}$ be an orthonormal basis for R^3, where

$$\mathbf{u}_1 = \begin{bmatrix} \frac{2}{3} \\ -\frac{2}{3} \\ \frac{1}{3} \end{bmatrix}, \quad \mathbf{u}_2 = \begin{bmatrix} \frac{2}{3} \\ \frac{1}{3} \\ -\frac{2}{3} \end{bmatrix}, \quad \text{and} \quad \mathbf{u}_3 = \begin{bmatrix} \frac{1}{3} \\ \frac{2}{3} \\ \frac{2}{3} \end{bmatrix}.$$

Write the vector $\mathbf{v} = \begin{bmatrix} 3 \\ 4 \\ 5 \end{bmatrix}$ as a linear combination of the vectors in S.

Solution: We have

$$\mathbf{v} = c_1\mathbf{u}_1 + c_2\mathbf{u}_2 + c_3\mathbf{u}_3.$$

Theorem 3.5 shows that c_1, c_2, and c_3 can be obtained without having to solve a linear system of three equations in three unknowns. Thus

$$c_1 = (\mathbf{v}, \mathbf{u}_1) = 1, \qquad c_2 = (\mathbf{v}, \mathbf{u}_2) = 0, \qquad c_3 = (\mathbf{v}, \mathbf{u}_3) = 6$$

and $\mathbf{v} = \mathbf{u}_1 + 6\mathbf{u}_3$. ■

Theorem 3.6. (Gram†–Schmidt‡ Process) *Let V be an inner product space and $W \neq \{\mathbf{0}\}$ a subspace of V with basis $S = \{\mathbf{u}_1, \mathbf{u}_2, \ldots, \mathbf{u}_n\}$. Then there exists an orthonormal basis $T = \{\mathbf{w}_1, \mathbf{w}_2, \ldots, \mathbf{w}_n\}$ for W.*

Proof: The proof is constructive; that is, we exhibit the desired basis T. However, we first find an orthogonal basis $T^* = \{\mathbf{v}_1, \mathbf{v}_2, \ldots, \mathbf{v}_n\}$ for W.

First, we pick any one of the vectors in S, say $\mathbf{u}_1$, and call it $\mathbf{v}_1$. Thus $\mathbf{v}_1 = \mathbf{u}_1$. We now look for a vector $\mathbf{v}_2$ in the subspace W_1 of W spanned by $\{\mathbf{u}_1, \mathbf{u}_2\}$, which is orthogonal to $\mathbf{v}_1$. Since $\mathbf{v}_1 = \mathbf{u}_1$, W_1 is also the subspace spanned by $\{\mathbf{v}_1, \mathbf{u}_2\}$. Thus $\mathbf{v}_2 = a_1\mathbf{v}_1 + a_2\mathbf{u}_2$. We determine a_1 and a_2, so that $(\mathbf{v}_1, \mathbf{v}_2) = 0$. Now $0 = (\mathbf{v}_2, \mathbf{v}_1) = (a_1\mathbf{v}_1 + a_2\mathbf{u}_2, \mathbf{v}_1) = a_1(\mathbf{v}_1, \mathbf{v}_1) + a_2(\mathbf{u}_2, \mathbf{v}_1)$. Note that $\mathbf{v}_1 \neq \mathbf{0}$ (why?), so $(\mathbf{v}_1, \mathbf{v}_1) \neq 0$. Thus

$$a_1 = -a_2\frac{(\mathbf{u}_2, \mathbf{v}_1)}{(\mathbf{v}_1, \mathbf{v}_1)}.$$

† Jörgen Pederson Gram (1850–1916) was a Danish actuary.

‡ Erhard Schmidt (1876–1959) taught at several leading German universities and was a student of both Hermann Amandus Schwarz and David Hilbert. He contributed importantly to the study of integral equations and partial differential equations and, as part of this study, he introduced the method for finding an orthonormal basis in 1907. In 1908, he wrote a paper on infinitely many linear equations in infinitely many unknowns, in which he founded the theory of Hilbert spaces and in which he again used his method.

We may assign an arbitrary nonzero value to a_2. Thus, letting $a_2 = 1$, we obtain

$$a_1 = -\frac{(\mathbf{u}_2, \mathbf{v}_1)}{(\mathbf{v}_1, \mathbf{v}_1)}.$$

Hence

$$\mathbf{v}_2 = a_1\mathbf{v}_1 + \mathbf{u}_2 = \mathbf{u}_2 - \frac{(\mathbf{u}_2, \mathbf{v}_1)}{(\mathbf{v}_1, \mathbf{v}_1)}\mathbf{v}_1.$$

At this point we have an orthogonal subset $\{\mathbf{v}_1, \mathbf{v}_2\}$ of W (Figure 3.19).

Next, we look for a vector $\mathbf{v}_3$ in the subspace W_2 of W spanned by $\{\mathbf{u}_1, \mathbf{u}_2, \mathbf{u}_3\}$ which is orthogonal to both $\mathbf{v}_1$ and $\mathbf{v}_2$. Of course, W_2 is also the subspace spanned by $\{\mathbf{v}_1, \mathbf{v}_2, \mathbf{u}_3\}$ (why?). Thus $\mathbf{v}_3 = b_1\mathbf{v}_1 + b_2\mathbf{v}_2 + b_3\mathbf{u}_3$. We try to find b_1, b_2, and b_3 so that $(\mathbf{v}_3, \mathbf{v}_1) = 0$ and $(\mathbf{v}_3, \mathbf{v}_2) = 0$. Now

$$0 = (\mathbf{v}_3, \mathbf{v}_1) = (b_1\mathbf{v}_1 + b_2\mathbf{v}_2 + b_3\mathbf{u}_3, \mathbf{v}_1) = b_1(\mathbf{v}_1, \mathbf{v}_1) + b_3(\mathbf{u}_3, \mathbf{v}_1)$$
$$0 = (\mathbf{v}_3, \mathbf{v}_2) = (b_1\mathbf{v}_1 + b_2\mathbf{v}_2 + b_3\mathbf{u}_3, \mathbf{v}_2) = b_2(\mathbf{v}_2, \mathbf{v}_2) + b_3(\mathbf{u}_3, \mathbf{v}_2).$$

Observe that $\mathbf{v}_2 \neq \mathbf{0}$ (why?). Solving for b_1 and b_2, we have

$$b_1 = -b_3\frac{(\mathbf{u}_3, \mathbf{v}_1)}{(\mathbf{v}_1, \mathbf{v}_1)} \quad \text{and} \quad b_2 = -b_3\frac{(\mathbf{u}_3, \mathbf{v}_2)}{(\mathbf{v}_2, \mathbf{v}_2)}.$$

Letting $b_3 = 1$, we have

$$\mathbf{v}_3 = \mathbf{u}_3 - \frac{(\mathbf{u}_3, \mathbf{v}_1)}{(\mathbf{v}_1, \mathbf{v}_1)}\mathbf{v}_1 - \frac{(\mathbf{u}_3, \mathbf{v}_2)}{(\mathbf{v}_2, \mathbf{v}_2)}\mathbf{v}_2.$$

At this point we have an orthogonal subset $\{\mathbf{v}_1, \mathbf{v}_2, \mathbf{v}_3\}$ of W (Figure 3.20).

Figure 3.19

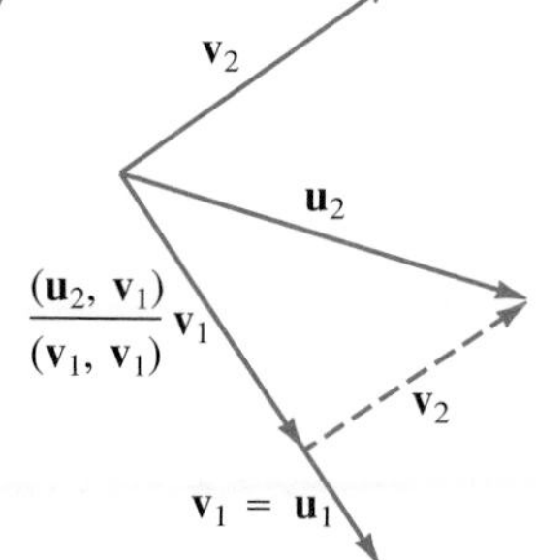

Figure 3.20

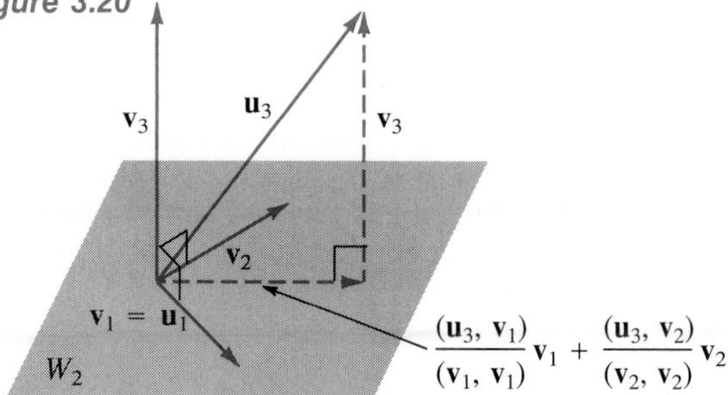

We next seek a vector $\mathbf{v}_4$ in the subspace W_3 spanned by $\{\mathbf{u}_1, \mathbf{u}_2, \mathbf{u}_3, \mathbf{u}_4\}$, and also by $\{\mathbf{v}_1, \mathbf{v}_2, \mathbf{v}_3, \mathbf{u}_4\}$, which is orthogonal to $\mathbf{v}_1, \mathbf{v}_2, \mathbf{v}_3$. We can then write

$$\mathbf{v}_4 = \mathbf{u}_4 - \frac{(\mathbf{u}_4, \mathbf{v}_1)}{(\mathbf{v}_1, \mathbf{v}_1)}\mathbf{v}_1 - \frac{(\mathbf{u}_4, \mathbf{v}_2)}{(\mathbf{v}_2, \mathbf{v}_2)}\mathbf{v}_2 - \frac{(\mathbf{u}_4, \mathbf{v}_3)}{(\mathbf{v}_3, \mathbf{v}_3)}\mathbf{v}_3.$$

Continue in this manner until we have an orthogonal set

$$T^* = \{\mathbf{v}_1, \mathbf{v}_2, \ldots, \mathbf{v}_n\}$$

of n vectors. By Theorem 3.4 we conclude that T^* is a basis for W. If we now let $\mathbf{w}_i = \mathbf{v}_i/\|\mathbf{v}_i\|$ for $i = 1, 2, \ldots, n$, then $T = \{\mathbf{w}_1, \mathbf{w}_2, \ldots, \mathbf{w}_n\}$ is an orthonormal basis for W. ■

Example 2. Consider the basis $S = \{\mathbf{u}_1, \mathbf{u}_2, \mathbf{u}_3\}$ for the Euclidean space R^3 with the standard inner product where

$$\mathbf{u}_1 = \begin{bmatrix} 1 \\ 1 \\ 1 \end{bmatrix}, \quad \mathbf{u}_2 = \begin{bmatrix} -1 \\ 0 \\ -1 \end{bmatrix}, \quad \text{and} \quad \mathbf{u}_3 = \begin{bmatrix} -1 \\ 2 \\ 3 \end{bmatrix}.$$

Transform S to an orthonormal basis $T = \{\mathbf{w}_1, \mathbf{w}_2, \mathbf{w}_3\}$.

Solution: First, let $\mathbf{v}_1 = \mathbf{u}_1$. Then we find that

$$\mathbf{v}_2 = \mathbf{u}_2 - \frac{(\mathbf{u}_2, \mathbf{v}_1)}{(\mathbf{v}_1, \mathbf{v}_1)}\mathbf{v}_1 = \begin{bmatrix} -1 \\ 0 \\ -1 \end{bmatrix} + \left(\frac{2}{3}\right)\begin{bmatrix} 1 \\ 1 \\ 1 \end{bmatrix} = \begin{bmatrix} -\frac{1}{3} \\ \frac{2}{3} \\ -\frac{1}{3} \end{bmatrix}.$$

and

$$\mathbf{v}_3 = \mathbf{u}_3 - \frac{(\mathbf{u}_3, \mathbf{v}_1)}{(\mathbf{v}_1, \mathbf{v}_1)}\mathbf{v}_1 - \frac{(\mathbf{u}_3, \mathbf{v}_2)}{(\mathbf{v}_2, \mathbf{v}_2)}\mathbf{v}_2 = \begin{bmatrix} -1 \\ 2 \\ 3 \end{bmatrix} - \left(\frac{4}{3}\right)\begin{bmatrix} 1 \\ 1 \\ 1 \end{bmatrix} - \left(\frac{\frac{2}{3}}{\frac{6}{9}}\right)\begin{bmatrix} -\frac{1}{3} \\ \frac{2}{3} \\ -\frac{1}{3} \end{bmatrix}$$

$$= \begin{bmatrix} -2 \\ 0 \\ 2 \end{bmatrix}.$$

Then $\{\mathbf{v}_1, \mathbf{v}_2, \mathbf{v}_3\}$ is an orthogonal basis. If we clear the fractions in each $\mathbf{v}_i$ by multiplying by a scalar, the resulting set is also an orthogonal basis. Hence

$\left\{\begin{bmatrix}1\\1\\1\end{bmatrix}, \begin{bmatrix}-1\\2\\-1\end{bmatrix}, \begin{bmatrix}-2\\0\\2\end{bmatrix}\right\}$ is an orthogonal basis for R^3. Multiplying each vector by the reciprocal of its length yields

$\left\{\begin{bmatrix}\frac{1}{\sqrt{3}}\\\frac{1}{\sqrt{3}}\\\frac{1}{\sqrt{3}}\end{bmatrix}, \begin{bmatrix}\frac{-1}{\sqrt{6}}\\\frac{2}{\sqrt{6}}\\\frac{-1}{\sqrt{6}}\end{bmatrix}, \begin{bmatrix}\frac{-1}{\sqrt{2}}\\0\\\frac{1}{\sqrt{2}}\end{bmatrix}\right\}$ which is an orthonormal basis for R^3. ■

Example 3. Let V be the Euclidean space P_3 with the inner product defined in Example 7 of Section 3.3. Let W be the subspace of P_3 having $S = \{t^2, t\}$ as a basis. Find an orthonormal basis for W.

Solution: First, let $\mathbf{u}_1 = t^2$ and $\mathbf{u}_2 = t$. Now let $\mathbf{v}_1 = \mathbf{u}_1 = t^2$. Then

$$\mathbf{v}_2 = \mathbf{u}_2 - \frac{(\mathbf{u}_2, \mathbf{v}_1)}{(\mathbf{v}_1, \mathbf{v}_1)}\mathbf{v}_1 = t - \frac{\frac{1}{4}}{\frac{1}{5}}t^2 = t - \frac{5}{4}t^2,$$

where $(\mathbf{v}_1, \mathbf{v}_1) = \int_0^1 t^2t^2\,dt = \int_0^1 t^4\,dt = \frac{1}{5}$, and $(\mathbf{u}_2, \mathbf{v}_1) = \int_0^1 tt^2\,dt = \int_0^1 t^3\,dt = \frac{1}{4}$. Since $(\mathbf{v}_2, \mathbf{v}_2) = \int_0^1 (t - \frac{5}{4}t^2)^2\,dt = \frac{1}{48}$, $\{\sqrt{5}t^2, \sqrt{48}(t - \frac{5}{4}t^2)\}$ is an orthonormal basis for W. If we choose $\mathbf{u}_1 = t$ and $\mathbf{u}_2 = t^2$, then we obtain (verify) the orthonormal basis $\{\sqrt{3}t, \sqrt{30}(t^2 - \frac{1}{2}t)\}$ for W. ■

In the proof of Theorem 3.6 we have also established the following result. At each stage of the Gram–Schmidt process, the ordered set $\{\mathbf{w}_1, \mathbf{w}_2, \ldots, \mathbf{w}_k\}$ is an orthonormal basis for the subspace spanned by

$$\{\mathbf{u}_1, \mathbf{u}_2, \ldots, \mathbf{u}_k\} \qquad 1 \le k \le n.$$

Also, the final orthonormal basis T depends upon the order of the vectors in the given basis S. Thus, if we change the order of the vectors in S, we might obtain a different orthonormal basis T for W.

We make one final observation with regard to the Gram–Schmidt process. In our proof of Theorem 3.6, we first obtained an orthogonal basis T^* and then normalized all the vectors in T^* to obtain the orthonormal basis T. Of course, an alternative course of action is to normalize each vector as soon as we produce it.

One of the useful consequences of having an orthonormal basis in a Euclidean space V is that an arbitrary inner product on V, when it is expressed in terms of coordinates with respect to the orthonormal basis, behaves like the standard inner product on R^n.

Theorem 3.7. *Let V be an n-dimensional Euclidean space, and let $S = \{\mathbf{u}_1, \mathbf{u}_2, \ldots, \mathbf{u}_n\}$ be an orthonormal basis for V. If $\mathbf{v} = a_1\mathbf{u}_1 + a_2\mathbf{u}_2 + \cdots + a_n\mathbf{u}_n$ and $\mathbf{w} = b_1\mathbf{u}_1 + b_2\mathbf{u}_2 + \cdots + b_n\mathbf{u}_n$, then*

$$(\mathbf{v}, \mathbf{w}) = a_1b_1 + a_2b_2 + \cdots + a_nb_n.$$

Proof: We first compute the matrix $C = [c_{ij}]$ of the given inner product with respect to the ordered basis S. We have

$$\begin{aligned} c_{ij} = (\mathbf{u}_i, \mathbf{u}_j) &= 1 \quad \text{if } i = j \\ &= 0 \quad \text{if } i \neq j. \end{aligned}$$

Hence $C = I_n$, the identity matrix. Now we also know from Equation (1) or Section 3.3 that

$$\begin{aligned} (\mathbf{v}, \mathbf{w}) &= [\mathbf{v}]_S^T C [\mathbf{w}]_S = [\mathbf{v}]_S^T I_n [\mathbf{w}]_S = [\mathbf{v}]_S^T [\mathbf{w}]_S \\ &= \begin{bmatrix} a_1 & a_2 & \cdots & a_n \end{bmatrix} \begin{bmatrix} b_1 \\ b_2 \\ \vdots \\ b_n \end{bmatrix} = a_1b_1 + a_2b_2 + \cdots + a_nb_n, \end{aligned}$$

which establishes the result. ■

The theorem that we just proved has some additional implications. Consider the Euclidean space R_3 with the standard inner product and let W be the subspace with ordered basis $S = \{[2 \quad 1 \quad 1], [1 \quad 1 \quad 2]\}$. Let $\mathbf{u} = [5 \quad 3 \quad 4]$ be a vector in W. Then

$$[5 \quad 3 \quad 4] = 2[2 \quad 1 \quad 1] + 1[1 \quad 1 \quad 2],$$

so $[5 \quad 3 \quad 4]_S = \begin{bmatrix} 2 \\ 1 \end{bmatrix}$. Now the length of $\mathbf{u}$ is $\|\mathbf{u}\| = \sqrt{5^2 + 3^2 + 4^2} = \sqrt{25 + 9 + 16} = \sqrt{50}$. We might expect to compute the length of $\mathbf{u}$ by using the coordinate vector with respect to S, that is, $\|\mathbf{u}\| = \sqrt{2^2 + 1^2} = \sqrt{5}$. Obviously, we have the wrong answer. However, let us transform the given basis S

for W into an orthonormal basis T for W. Using the Gram–Schmidt process, we find that (verify)

$$\{[2 \quad 1 \quad 1], [-\tfrac{4}{6} \quad \tfrac{1}{6} \quad \tfrac{7}{6}]\}$$

is an orthogonal basis for W. It then follows from Exercise 29 that $\{[2 \quad 1 \quad 1], [-4 \quad 1 \quad 7]\}$ is also an orthogonal basis, and so

$$T = \left\{\begin{bmatrix} \frac{2}{\sqrt{6}} & \frac{1}{\sqrt{6}} & \frac{1}{\sqrt{6}} \end{bmatrix}, \begin{bmatrix} \frac{-4}{\sqrt{66}} & \frac{1}{\sqrt{66}} & \frac{7}{\sqrt{66}} \end{bmatrix}\right\}$$

is an orthonormal basis for W. Then the coordinate vector of $\mathbf{u}$ with respect to T is

$$[\mathbf{u}]_T = \begin{bmatrix} \frac{17}{6}\sqrt{6} \\ \frac{\sqrt{66}}{6} \end{bmatrix}.$$

Computing the length of $\mathbf{u}$ using these coordinates, we find that

$$\|\mathbf{u}\|_T = \sqrt{\left(\frac{17}{6}\sqrt{6}\right)^2 + \left(\frac{\sqrt{66}}{6}\right)^2} = \sqrt{\frac{1800}{36}} = \sqrt{50}.$$

It is not difficult to show (Exercise 15) that if T is an orthonormal basis for an inner product space and $[\mathbf{u}]_T = \begin{bmatrix} a_1 \\ a_2 \\ \vdots \\ a_n \end{bmatrix}$, then $\|\mathbf{u}\| = \sqrt{a_1^2 + a_2^2 + \cdots + a_n^2}$.

Projections and Applications (Optional)

Theorem 3.8. *Let W be an m-dimensional subspace of a vector space V with orthonormal basis $S = \{\mathbf{u}_1, \mathbf{u}_2, \ldots, \mathbf{u}_m\}$. Then every vector $\mathbf{u}$ in V can be uniquely written as*

$$\mathbf{u} = \mathbf{v} + \mathbf{w},$$

where $\mathbf{v}$ is in W and $\mathbf{w}$ is orthogonal to every vector in W.

Proof: Let

$$\mathbf{v} = (\mathbf{u}, \mathbf{u}_1)\mathbf{u}_1 + (\mathbf{u}, \mathbf{u}_2)\mathbf{u}_2 + \cdots + (\mathbf{u}, \mathbf{u}_m)\mathbf{u}_m \tag{1}$$

and

$$\mathbf{w} = \mathbf{u} - \mathbf{v}.$$

Since $\mathbf{v}$ is a linear combination of members of S, $\mathbf{v}$ belongs to W. We next show that $\mathbf{w}$ is orthogonal to every vector in W. Thus let

$$\mathbf{z} = c_1\mathbf{u}_1 + c_2\mathbf{u}_2 + \cdots + c_m\mathbf{u}_m$$

be an arbitrary vector in W. Then

$$\begin{aligned}(\mathbf{w}, \mathbf{z}) &= (\mathbf{u} - \mathbf{v}, \mathbf{z}) = (\mathbf{u}, \mathbf{z}) - (\mathbf{v}, \mathbf{z}) \\ &= (\mathbf{u}, c_1\mathbf{u}_1 + c_2\mathbf{u}_2 + \cdots + c_m\mathbf{u}_m) - [(\mathbf{u}, \mathbf{u}_1)\mathbf{u}_1 \\ &\quad + (\mathbf{u}, \mathbf{u}_2)\mathbf{u}_2 + \cdots + (\mathbf{u}, \mathbf{u}_n)\mathbf{u}_m, c_1\mathbf{u}_1 + c_2\mathbf{u}_2 + \cdots \\ &\quad + c_m\mathbf{u}_m] \\ &= c_1(\mathbf{u}, \mathbf{u}_1) + c_2(\mathbf{u}, \mathbf{u}_2) + \cdots + c_m(\mathbf{u}, \mathbf{u}_m) \\ &\quad - c_1(\mathbf{u}, \mathbf{u}_1)(\mathbf{u}_1, \mathbf{u}_1) - c_2(\mathbf{u}, \mathbf{u}_2)(\mathbf{u}_2, \mathbf{u}_2) \\ &\quad - \cdots - c_m(\mathbf{u}, \mathbf{u}_m)(\mathbf{u}_m, \mathbf{u}_m) \\ &= 0,\end{aligned}$$

since $(\mathbf{u}_i, \mathbf{u}_j) = 0$ for $i \neq j$ and $(\mathbf{u}_i, \mathbf{u}_i) = 1$, $1 \leq i \leq m$. We leave it as an exercise to prove that the vectors $\mathbf{v}$ and $\mathbf{w}$ are unique. ■

The vector $\mathbf{v}$ obtained in Equation (1) is called the **orthogonal projection** of $\mathbf{u}$ on W and is denoted by $\text{proj}_W \mathbf{u}$. In Figure 3.21, we illustrate Theorem 3.8 when W is a two-dimensional subspace of R^3 (a plane through the origin).

Figure 3.21

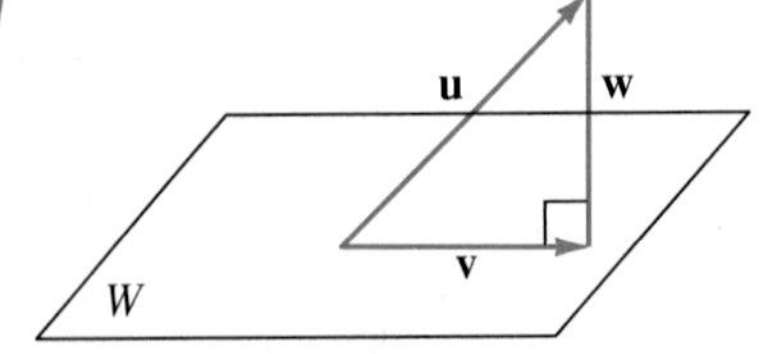

Example 4. Let W be the two-dimensional subspace of R^3 with orthonormal basis $\{\mathbf{u}_1, \mathbf{u}_2\}$, where

$$\mathbf{u}_1 = \begin{bmatrix} \frac{2}{3} \\ \frac{1}{3} \\ -\frac{2}{3} \end{bmatrix} \quad \text{and} \quad \mathbf{u}_2 = \begin{bmatrix} \frac{1}{\sqrt{2}} \\ 0 \\ \frac{1}{\sqrt{2}} \end{bmatrix}.$$

Using the standard inner product on R^3, find the orthogonal projection of

$$\mathbf{u} = \begin{bmatrix} 2 \\ 1 \\ 3 \end{bmatrix}$$

on W and the vector $\mathbf{w}$ that is orthogonal to every vector in W.

Solution: From Equation (1) we have

$$\begin{aligned} \mathbf{v} = \text{proj}_W\, \mathbf{u} &= (\mathbf{u}, \mathbf{u}_1)\mathbf{u}_1 + (\mathbf{u}, \mathbf{u}_2)\mathbf{u}_2 \\ &= -\frac{1}{3}\mathbf{u}_1 + \frac{5}{\sqrt{2}}\mathbf{u}_2 \\ &= \begin{bmatrix} \frac{41}{18} \\ -\frac{1}{9} \\ \frac{49}{18} \end{bmatrix} \end{aligned}$$

and

$$\mathbf{w} = \mathbf{u} - \mathbf{v} = \begin{bmatrix} -\frac{5}{18} \\ \frac{10}{9} \\ \frac{5}{18} \end{bmatrix}. \quad \blacksquare$$

It is clear from Figure 3.20 that the distance from $\mathbf{u}$ to the plane W is given by the length of the vector $\mathbf{w} = \mathbf{u} - \mathbf{v}$, that is, by

$$\|\mathbf{u} - \text{proj}_W\, \mathbf{u}\|.$$

Example 5. Let W be the subspace of R^3 defined in Example 4 and let $\mathbf{u} = \begin{bmatrix} 1 \\ 1 \\ 0 \end{bmatrix}$. Find the distance from $\mathbf{u}$ to W.

Solution: We first compute

$$\begin{aligned} \text{proj}_W \mathbf{u} &= (\mathbf{u}, \mathbf{u}_1)\mathbf{u}_1 + (\mathbf{u}, \mathbf{u}_2)\mathbf{u}_2 \\ &= \mathbf{u}_1 + \frac{1}{\sqrt{2}}\mathbf{u}_2 \\ &= \begin{bmatrix} \frac{7}{6} \\ \frac{1}{3} \\ -\frac{1}{6} \end{bmatrix}. \end{aligned}$$

Then

$$\mathbf{u} - \text{proj}_W \mathbf{u} = \begin{bmatrix} 1 \\ 1 \\ 0 \end{bmatrix} - \begin{bmatrix} \frac{7}{6} \\ \frac{1}{3} \\ -\frac{1}{6} \end{bmatrix} = \begin{bmatrix} -\frac{1}{6} \\ \frac{2}{3} \\ \frac{1}{6} \end{bmatrix}$$

and

$$\|\mathbf{u} - \text{proj}_W \mathbf{u}\| = \sqrt{\frac{1}{36} + \frac{4}{9} + \frac{1}{36}} = \frac{\sqrt{2}}{2},$$

so the distance from $\mathbf{u}$ to W is $\sqrt{2}/2$. ■

In the study of power series in calculus, the functions

$$1, t, t^2, \ldots, t^n, \ldots \tag{2}$$

play a central role. Most of the functions encountered in calculus can be expanded in a power series. The coefficients in Taylor and Maclaurin series expansions involve successive derivatives of the given function evaluated at a point, the center of expansion.

The function $f(t) = |t|$ does not have a Taylor series expansion with center $t_0 = 0$ (a Maclaurin series), because $|t|$ does not have a derivative at $t = 0$. Thus there is no way to compute the coefficients in such an expansion. However, it may be possible to obtain series expansions for such functions as $f(t) = |t|$ by using a different type of expansion. One such important expansion involves the set of functions

$$1,\ \cos t,\ \sin t,\ \cos 2t,\ \sin 2t,\ \ldots,\ \cos nt,\ \sin nt,\ \ldots, \tag{3}$$

which we discussed briefly in Example 15 of Section 3.3. Obtaining a series expansion of $f(t) = |t|$ in terms of the functions in (3) would take us too far afield at this time, but we will show how to obtain an approximation from a finite subset of these functions in the next example.

Example 6. Let V be the vector space of real continuous functions on $[-\pi, \pi]$. Then $W = \text{span}\,\{1, \cos t, \sin t\}$ is a finite-dimensional subspace of V. If f and g belong to V, then $(f, g) = \int_{-\pi}^{\pi} f(t)g(t)\,dt$ defines an inner product on V as in Example 15 of Section 3.3. It also follows that S is an orthogonal set, but Theorem 3.8 requires an orthonormal basis. It is easy to show that

$$\int_{-\pi}^{\pi} 1\,dt = 2\pi, \qquad \int_{-\pi}^{\pi} \cos^2 t\,dt = \pi, \qquad \int_{-\pi}^{\pi} \sin^2 t\,dt = \pi.$$

If we let

$$\mathbf{u}_1 = \frac{1}{\sqrt{2\pi}}, \qquad \mathbf{u}_2 = \frac{1}{\sqrt{\pi}}\cos t, \qquad \text{and} \qquad \mathbf{u}_3 = \frac{1}{\sqrt{\pi}}\sin t,$$

then $T = \{\mathbf{u}_1, \mathbf{u}_2, \mathbf{u}_3\}$ is an orthonormal set in V and $W = \text{span}\,T$ (verify). Using Theorem 3.8, we can compute $\text{proj}_W\,\mathbf{u}$, for $\mathbf{u} = |t|$, as

$$\begin{aligned}\text{proj}_W\,|t| &= \left(|t|, \frac{1}{\sqrt{2\pi}}\right)\frac{1}{\sqrt{2\pi}} + \left(|t|, \frac{1}{\sqrt{\pi}}\cos t\right)\frac{1}{\sqrt{\pi}}\cos t \\ &\quad + \left(|t|, \frac{1}{\sqrt{\pi}}\sin t\right)\frac{1}{\sqrt{\pi}}\sin t.\end{aligned}$$

We have

$$\begin{aligned}\left(|t|, \frac{1}{\sqrt{2\pi}}\right) &= \int_{-\pi}^{\pi} |t|\frac{1}{\sqrt{2\pi}}\,dt = \frac{1}{\sqrt{2\pi}}\int_{-\pi}^{0} -t\,dt + \frac{1}{\sqrt{2\pi}}\int_{0}^{\pi} t\,dt \\ &= \frac{\pi^2}{\sqrt{2\pi}},\end{aligned}$$

$$\left(|t|, \frac{1}{\sqrt{\pi}} \cos t\right) = \int_{-\pi}^{\pi} |t| \frac{1}{\sqrt{\pi}} \cos t \, dt = \frac{1}{\sqrt{\pi}} \int_{-\pi}^{0} - t \cos t \, dt$$
$$+ \frac{1}{\sqrt{\pi}} \int_{0}^{\pi} t \cos t \, dt = -\frac{2}{\sqrt{\pi}} - \frac{2}{\sqrt{\pi}} = -\frac{4}{\sqrt{\pi}},$$

and

$$\left(|t|, \frac{1}{\sqrt{\pi}} \sin t\right) = \int_{-\pi}^{\pi} |t| \frac{1}{\sqrt{\pi}} \sin t \, dt = \frac{1}{\sqrt{\pi}} \int_{-\pi}^{0} - t \sin t \, dt$$
$$+ \frac{1}{\sqrt{\pi}} \int_{0}^{\pi} t \sin t \, dt = -\sqrt{\pi} + \sqrt{\pi} = 0.$$

Then

$$\begin{aligned} \text{proj}_W |t| &= \frac{\pi^2}{\sqrt{2\pi}} \frac{1}{\sqrt{2\pi}} - \frac{4}{\sqrt{\pi}} \frac{1}{\sqrt{\pi}} \cos t \\ &= \frac{\pi}{2} - \frac{4}{\pi} \cos t. \end{aligned}$$

Figure 3.22 contains a graph of both $|t|$ and $\text{proj}_W |t|$. ■

In Example 5 $|\mathbf{u} - \text{proj}_W \mathbf{u}|$ represented the distance in 3-space from $\mathbf{u}$ to the plane W. We can generalize this notion of distance from a vector in V to a

Figure 3.22

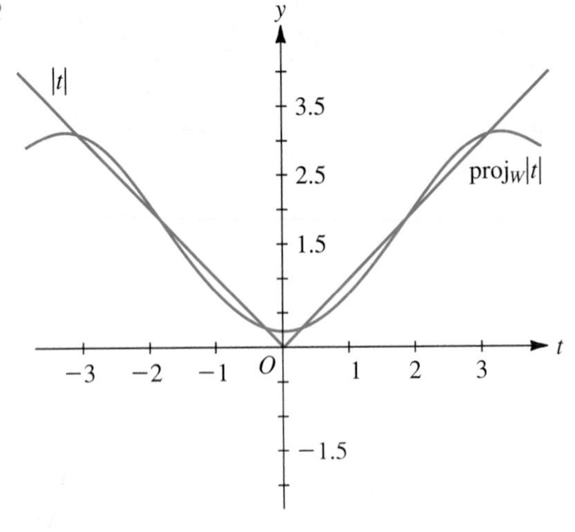

subspace W of V. We can show that the vector in W that is closest to $\mathbf{u}$ is in fact $\text{proj}_W\,\mathbf{u}$, so $\|\mathbf{u} - \text{proj}_W\,\mathbf{u}\|$ represents the distance from $\mathbf{u}$ to W.

Theorem 3.9. *Let W be a finite-dimensional subspace of the inner product space V. Then for vector $\mathbf{u}$ belonging to V, the vector in W closest to $\mathbf{u}$ is $\text{proj}_W\,\mathbf{u}$. That is, $\|\mathbf{u} - \mathbf{v}\|$, for $\mathbf{v}$ belonging to W, is minimized when $\mathbf{v} = \text{proj}_W\,\mathbf{u}$.*

Proof: Let $\mathbf{v}$ be any vector in W. Then

$$\mathbf{u} - \mathbf{v} = (\mathbf{u} - \text{proj}_W\,\mathbf{u}) + (\text{proj}_W\,\mathbf{u} - \mathbf{v}).$$

Since $\mathbf{v}$ and $\text{proj}_W\,\mathbf{u}$ are both in W, $\text{proj}_W\,\mathbf{u} - \mathbf{v}$ is in W. By Theorem 3.8, $\mathbf{u} - \text{proj}_W\,\mathbf{u}$ is orthogonal to every vector in W, so

$$\begin{aligned} \|\mathbf{u} - \mathbf{v}\|^2 &= (\mathbf{u} - \mathbf{v}, \mathbf{u} - \mathbf{v}) \\ &= ((\mathbf{u} - \text{proj}_W\,\mathbf{u}) + (\text{proj}_W\,\mathbf{u} - \mathbf{v}), (\mathbf{u} - \text{proj}_W\,\mathbf{u}) + (\text{proj}_W\,\mathbf{u} - \mathbf{v})) \\ &= \|\mathbf{u} - \text{proj}_W\,\mathbf{u}\|^2 + \|\text{proj}_W\,\mathbf{u} - \mathbf{v}\|^2. \end{aligned}$$

If $\mathbf{v} \neq \text{proj}_W\,\mathbf{u}$, then $\|\text{proj}_W\,\mathbf{u} - \mathbf{v}\|^2$ is positive and

$$\|\mathbf{u} - \mathbf{v}\|^2 > \|\mathbf{u} - \text{proj}_W\,\mathbf{u}\|^2.$$

Thus it follows that $\text{proj}_W\,\mathbf{u}$ is the vector in W that minimizes $\|\mathbf{u} - \mathbf{v}\|^2$ and hence minimizes $\|\mathbf{u} - \mathbf{v}\|$. ■

In Example 4, $\mathbf{v} = \text{proj}_W\,\mathbf{u} = \begin{bmatrix} \frac{41}{18} \\ -\frac{1}{9} \\ \frac{49}{18} \end{bmatrix}$ is the vector in $W = \text{span}\,\{\mathbf{u}_1, \mathbf{u}_2\}$ that is closest to $\mathbf{u} = \begin{bmatrix} 2 \\ 1 \\ 3 \end{bmatrix}$. In Example 6, the function $\dfrac{\pi}{2} - \dfrac{4}{\pi}\cos t$ is the vector in span $\{1, \cos t, \sin t\}$ that is closest to the function $|t|$. Theorem 3.9 is important in the theory of approximation of functions and can be generalized to more abstract settings. Example 6 is a preview of ideas from the study of Fourier series that play an important role in the study of heat distribution and the analysis of sound waves. The study of projections is important in a number of areas in applied mathematics. We illustrate this in the next section by considering the topic of least squares, which provides a technique for dealing with inconsistent linear systems.

3.4. EXERCISES

In this set of exercises, the Euclidean spaces R_n and R^n have the standard inner products on them. Euclidean space P_n has on it the inner product defined in Example 7 of Section 3.3.

1. Use the Gram–Schmidt process to transform the basis $\left\{\begin{bmatrix}1\\2\end{bmatrix}, \begin{bmatrix}-3\\4\end{bmatrix}\right\}$ for the Euclidean space R^2 into:
(a) An orthogonal basis.
(b) An orthonormal basis.

2. Use the Gram–Schmidt process to transform the basis $\left\{\begin{bmatrix}1\\0\\1\end{bmatrix}, \begin{bmatrix}-2\\1\\3\end{bmatrix}\right\}$ for the subspace W of the Euclidean space R^3 into:
(a) An orthogonal basis.
(b) An orthonormal basis.

3. Consider the Euclidean space R_4 and let W be the subspace that has

$$S = \{[1 \quad 1 \quad -1 \quad 0], [0 \quad 2 \quad 0 \quad 1]\}$$

as a basis. Use the Gram–Schmidt process to obtain an orthonormal basis for W.

4. Consider Euclidean space R^3 and let W be the subspace that has basis $S = \left\{\begin{bmatrix}1\\1\\1\end{bmatrix}, \begin{bmatrix}1\\0\\2\end{bmatrix}\right\}$.

Use the Gram–Schmidt process to obtain an orthogonal basis for W.

5. Let $S = \{t, 1\}$ be a basis for a subspace W of the Euclidean space P_2. Find an orthonormal basis for W.

6. Repeat Exercise 5 with $S = \{t + 1, t - 1\}$.

7. Let $S = \{t, \sin 2\pi t\}$ be a basis for a subspace W of the inner product space of Example 5 in Section 3.3. Find an orthonormal basis for W.

8. Let $S = \{t, e^t\}$ be a basis for a subspace W of the inner product space of Example 5 in Section 3.3. Find an orthonormal basis for W.

9. Find an orthonormal basis for the Euclidean space R^3 that contains the vectors

$$\begin{bmatrix}\frac{2}{3}\\-\frac{2}{3}\\\frac{1}{3}\end{bmatrix} \quad \text{and} \quad \begin{bmatrix}\frac{2}{3}\\\frac{1}{3}\\-\frac{2}{3}\end{bmatrix}.$$

10. Use the Gram–Schmidt process to transform the basis

$$\left\{\begin{bmatrix}1\\1\\1\end{bmatrix}, \begin{bmatrix}0\\1\\1\end{bmatrix}, \begin{bmatrix}1\\2\\3\end{bmatrix}\right\}$$

for the Euclidean space R^3 into an orthonormal basis for R^3.

11. Use the Gram–Schmidt process to construct an orthonormal basis for the subspace W of the Euclidean space R^3 *spanned* by

$$\left\{\begin{bmatrix}1\\1\\1\end{bmatrix}, \begin{bmatrix}2\\2\\2\end{bmatrix}, \begin{bmatrix}0\\0\\1\end{bmatrix}, \begin{bmatrix}1\\2\\3\end{bmatrix}\right\}.$$

12. Use the Gram–Schmidt process to construct an orthonormal basis for the subspace W of the Euclidean space R_3 *spanned* by

$\{[1 \quad -1 \quad 1], [-2 \quad 2 \quad -2], [2 \quad -1 \quad 2], [0 \quad 0 \quad 0]\}$.

13. Prove Theorem 3.5.

14. Let $S = \{[1 \quad -1 \quad 0], [1 \quad 0 \quad -1]\}$ be a basis for a subspace W of the Euclidean space R_3.

(a) Use the Gram–Schmidt process to obtain an orthonormal basis for W.

(b) Using Theorem 3.5 write $\mathbf{u} = [5 \quad -2 \quad -3]$ as a linear combination of the vectors obtained in part (a).

15. Prove that if T is an orthonormal basis for a Euclidean space and

$$[\mathbf{v}]_T = \begin{bmatrix} a_1 \\ a_2 \\ \vdots \\ a_n \end{bmatrix},$$

then $\|\mathbf{v}\| = \sqrt{a_1^2 + a_2^2 + \cdots + a_n^2}$.

16. Let W be the subspace of the Euclidean space R^3 with basis

$$S = \left\{ \begin{bmatrix} 1 \\ 0 \\ -2 \end{bmatrix}, \begin{bmatrix} -3 \\ 2 \\ 1 \end{bmatrix} \right\}. \text{ Let } \mathbf{v} = \begin{bmatrix} -1 \\ 2 \\ -3 \end{bmatrix} \text{ be in } W.$$

(a) Find the length of $\mathbf{v}$ directly.

(b) Using the Gram–Schmidt process, transform S into an orthonormal basis T for W.

(c) Find the length of $\mathbf{v}$ using the coordinate vector of $\mathbf{v}$ with respect to T.

17. (a) Verify that $S = \left\{ \begin{bmatrix} \frac{1}{3} \\ \frac{2}{3} \\ \frac{2}{3} \end{bmatrix}, \begin{bmatrix} \frac{2}{3} \\ \frac{1}{3} \\ -\frac{2}{3} \end{bmatrix}, \begin{bmatrix} \frac{2}{3} \\ -\frac{2}{3} \\ \frac{1}{3} \end{bmatrix} \right\}$ is an orthonormal basis for the Euclidean space R^3.

(b) Use Theorem 3.5 to find the coordinate vector of $\mathbf{v} = \begin{bmatrix} 15 \\ 3 \\ 3 \end{bmatrix}$ with respect to S.

(c) Find the length of $\mathbf{v}$ directly and also using the coordinate vector found in part (b).

18. Apply the Gram–Schmidt process to the basis $\{1, t, t^2\}$ for the Euclidean space P_2 and obtain an orthonormal basis for P_2.

19. Let V be the Euclidean space of all 2×2 matrices with inner product defined by $(A, B) = \text{Tr}\,(B^T A)$. Prove that

$$S = \left\{ \begin{bmatrix} 1 & 0 \\ 0 & 0 \end{bmatrix}, \begin{bmatrix} 0 & 1 \\ 0 & 0 \end{bmatrix}, \begin{bmatrix} 0 & 0 \\ 1 & 0 \end{bmatrix}, \begin{bmatrix} 0 & 0 \\ 0 & 1 \end{bmatrix} \right\}$$

is an orthonormal basis for V.

20. Consider the orthonormal basis

$$S = \left\{ \begin{bmatrix} \frac{1}{\sqrt{5}} \\ 0 \\ \frac{2}{\sqrt{5}} \end{bmatrix}, \begin{bmatrix} \frac{-2}{\sqrt{5}} \\ 0 \\ \frac{1}{\sqrt{5}} \end{bmatrix}, \begin{bmatrix} 0 \\ 1 \\ 0 \end{bmatrix} \right\}$$

for R^3. Using Theorem 3.5 write the vector $\begin{bmatrix} 2 \\ -3 \\ 1 \end{bmatrix}$ as a linear combination of the vectors in S.

21. Let W be the subspace of R^3 with orthonormal basis $\{\mathbf{u}_1, \mathbf{u}_2\}$, where

$$\mathbf{u}_1 = \begin{bmatrix} 0 \\ 1 \\ 0 \end{bmatrix} \quad \text{and} \quad \mathbf{u}_2 = \begin{bmatrix} \frac{1}{\sqrt{5}} \\ 0 \\ \frac{2}{\sqrt{5}} \end{bmatrix}.$$

Write the vector $\mathbf{u} = \begin{bmatrix} 1 \\ 2 \\ -1 \end{bmatrix}$ as $\mathbf{v} + \mathbf{w}$, with $\mathbf{v}$ and $\mathbf{w}$ as in Theorem 3.8.

22. Let W be the subspace of R^4 with orthonormal basis $\{\mathbf{u}_1, \mathbf{u}_2, \mathbf{u}_3\}$, where

$$\mathbf{u}_1 = \begin{bmatrix} \frac{1}{\sqrt{2}} \\ 0 \\ 0 \\ \frac{-1}{\sqrt{2}} \end{bmatrix}, \quad \mathbf{u}_2 = \begin{bmatrix} 0 \\ 0 \\ 1 \\ 0 \end{bmatrix}, \quad \text{and } \mathbf{u}_3 = \begin{bmatrix} \frac{1}{\sqrt{2}} \\ 0 \\ 0 \\ \frac{1}{\sqrt{2}} \end{bmatrix}.$$

Write the vector $\mathbf{u} = \begin{bmatrix} 1 \\ 0 \\ 2 \\ 3 \end{bmatrix}$ as $\mathbf{v} + \mathbf{w}$ with $\mathbf{v}$ and $\mathbf{w}$ as in Theorem 3.8.

23. Let W be the subspace of continuous functions on $[-\pi, \pi]$ defined in Example 6. In each of the following cases, find $\text{proj}_W \mathbf{u}$.
(a) $\mathbf{u} = t$. (b) $\mathbf{u} = t^2$. (c) $\mathbf{u} = e^t$.

24. Let W be the subspace of R^3 defined in Exercise 21, and let $\mathbf{u} = \begin{bmatrix} -1 \\ 0 \\ 1 \end{bmatrix}$. Find the distance from $\mathbf{u}$ to W.

25. Let W be the subspace of R^4 defined in Exercise 22, and let $\mathbf{u} = \begin{bmatrix} 1 \\ 2 \\ -1 \\ 0 \end{bmatrix}$. Find the distance from $\mathbf{u}$ to W.

26. Let $\mathbf{u}, \mathbf{u}_1, \mathbf{u}_2, \ldots, \mathbf{u}_n$ be vectors in R^n. Show that if $\mathbf{u}$ is orthogonal to $\mathbf{u}_1, \mathbf{u}_2, \ldots, \mathbf{u}_n$, then $\mathbf{u}$ is orthogonal to every vector in

$$\text{span } \{\mathbf{u}_1, \mathbf{u}_2, \ldots, \mathbf{u}_n\}.$$

27. Let $\mathbf{u}$ be a fixed vector in R^n. Prove that the set of all vectors in R^n that are orthogonal to $\mathbf{u}$ is a subspace of R^n.

28. Prove that the vectors $\mathbf{v}$ and $\mathbf{w}$ in Theorem 3.8 are unique.

29. Let $S = \{\mathbf{v}_1, \mathbf{v}_2, \ldots, \mathbf{v}_k\}$ be an orthonormal basis for the Euclidean space V and $\{a_1, \ldots, a_k\}$ be any set of scalars none of which is zero. Prove that $T = \{a_1\mathbf{v}_1, a_2\mathbf{v}_2, \ldots, a_k\mathbf{v}_k\}$ is an orthogonal basis for V. How should the scalars $a_1, \ldots, a_k$ be chosen so that T is an orthonormal basis for V?

30. Let $S = \{\mathbf{u}_1, \mathbf{u}_2, \ldots, \mathbf{u}_k\}$ be an orthonormal basis for a subspace W of Euclidean space V which has dimension $n > k$. Discuss how to construct an orthonormal basis for V that includes S.

31. Determine if the software that you use has a command to obtain an orthonormal set of vectors from a linearly independent set of vectors in R^n. (Assume that the standard inner product is used.) If it does, compare the output from your command with the results in Example 2. To experiment further, use Exercises 2, 7, 8, 11, and 12.

3.5. *Least Squares (Optional)*

From Chapter 1 we recall that an $m \times n$ linear system $A\mathbf{x} = \mathbf{b}$ is inconsistent if it has no solution. In the discussion following Example 7 in Section 2.6 we show that $A\mathbf{x} = \mathbf{b}$ is consistent if and only if $\mathbf{b}$ belongs to the column space of A. Equivalently, $A\mathbf{x} = \mathbf{b}$ is inconsistent if and only if $\mathbf{b}$ is *not* in the column space of A. Inconsistent systems do indeed arise in many situations and we must determine how to deal with them. Our approach is to change the problem so that we do not require that the matrix equation $A\mathbf{x} = \mathbf{b}$ be satisfied. Instead, we seek a vector $\hat{\mathbf{x}}$ in R^n such that $A\hat{\mathbf{x}}$ is as close to $\mathbf{b}$ as possible. That is, find the vector in the column space of A that is closest to $\mathbf{b}$. From Theorem 3.9, if W is the column space of A, we determine $\hat{\mathbf{x}}$ such that $A\hat{\mathbf{x}} = \text{proj}_W \mathbf{b}$. As shown in the proof of Theorem 3.9, $\mathbf{b} - \text{proj}_W \mathbf{b} = \mathbf{b} - A\hat{\mathbf{x}}$ is orthogonal to every vector in W, the column space of A. It follows that $\mathbf{b} - A\hat{\mathbf{x}}$ is orthogonal to each column of A.

Expressing this in terms of a matrix equation, we have

$$A^T(A\hat{\mathbf{x}} - \mathbf{b}) = \mathbf{0}$$

or, equivalently,

$$A^TA\hat{\mathbf{x}} = A^T\mathbf{b}. \tag{1}$$

Any solution to (1) is called a **least squares solution** to the linear system $A\mathbf{x} = \mathbf{b}$. (**Warning:** In general $A\hat{\mathbf{x}} \neq \mathbf{b}$.) Equation (1) is called the **normal system** of equations associated with $A\mathbf{x} = \mathbf{b}$, or just the normal system. Observe that if A is nonsingular, a least squares solution to $A\mathbf{x} = \mathbf{b}$ is just the usual solution $\mathbf{x} = A^{-1}\mathbf{b}$ (see Exercise 1).

To compute a least squares solution $\hat{\mathbf{x}}$ to the linear system $A\mathbf{x} = \mathbf{b}$, we can proceed as follows. Compute $\text{proj}_W \mathbf{b}$ using Equation (1) in Section 3.4 and then solve $A\hat{\mathbf{x}} = \text{proj}_W \mathbf{b}$. To compute $\text{proj}_W \mathbf{b}$ requires that we have an orthonormal basis for W, the column space of A. We could first find a basis for W by determining the reduced row echelon form of A^T and taking the transposes of the nonzero rows. Next apply the Gram–Schmidt process to the basis to find an orthonormal basis for W. The procedure just outlined is theoretically valid when we assume that exact arithmetic is used. However, even small numerical errors, due to, say, roundoff, may adversely affect the results. Thus more sophisticated algorithms are required for numerical applications. [See D. Hill, *Experiments in Computational Matrix Algebra* (New York: Random House, Inc., 1988).] We shall not pursue the general case here, but turn our attention to an important special case.

Theorem 3.10. *If A is $m \times n$ with rank $A = n$, then A^TA is nonsingular and the linear system $A\mathbf{x} = \mathbf{b}$ has a unique least squares solution given by $\hat{\mathbf{x}} = (A^TA)^{-1}A^T\mathbf{b}$.*

Proof: If A has rank n, then the columns of A are linearly independent. Matrix A^TA is nonsingular provided that the linear system $A^TA\mathbf{x} = \mathbf{0}$ has only the zero solution. Multiplying both sides of $A^TA\mathbf{x} = \mathbf{0}$ by x^T gives

$$\mathbf{0} = \mathbf{x}^TA^TA\mathbf{x} = (A\mathbf{x})^T(A\mathbf{x}) = (A\mathbf{x}, A\mathbf{x})$$

using the standard inner product on R^m. It follows from Definition 3.2(a) that $A\mathbf{x} = \mathbf{0}$. But this implies that we have a linear combination of the linearly independent columns of A that is zero, hence $\mathbf{x} = \mathbf{0}$. Thus A^TA is nonsingular and Equation (1) has the unique solution $\hat{\mathbf{x}} = (A^TA)^{-1}A^T\mathbf{b}$. ■

When rank $A = n$, it is computationally more efficient to solve Equation (1) by Gaussian elimination than to determine $(A^TA)^{-1}$ and then form the product $(A^TA)^{-1}A^T\mathbf{b}$.

Example 1. Determine a least squares solution to $A\mathbf{x} = \mathbf{b}$, where

$$A = \begin{bmatrix} 1 & 2 & -1 & 3 \\ 2 & 1 & 1 & 2 \\ -2 & 3 & 4 & 1 \\ 4 & 2 & 1 & 0 \\ 0 & 2 & 1 & 3 \\ 1 & -1 & 2 & 0 \end{bmatrix}, \qquad \mathbf{b} = \begin{bmatrix} 1 \\ 5 \\ -2 \\ 1 \\ 3 \\ 5 \end{bmatrix}.$$

Solution: Using row reduction, we can show that rank $A = 4$ (verify). Then using Theorem 3.10, we form the normal system $A^TA\mathbf{x} = A^T\mathbf{b}$,

$$\begin{bmatrix} 26 & 5 & -1 & 5 \\ 5 & 23 & 13 & 17 \\ -1 & 13 & 24 & 6 \\ 5 & 17 & 6 & 23 \end{bmatrix}\mathbf{x} = \begin{bmatrix} 24 \\ 4 \\ 10 \\ 20 \end{bmatrix}.$$

Applying Gaussian elimination, we have the unique least squares solution (verify)

$$\hat{\mathbf{x}} \approx \begin{bmatrix} 0.9990 \\ -2.0643 \\ 1.1039 \\ 1.8902 \end{bmatrix}.$$

If W is the column space of A, then (verify)

$$\text{proj}_W \mathbf{b} = A\hat{\mathbf{x}} \approx \begin{bmatrix} 1.4371 \\ 4.8181 \\ -1.8852 \\ 0.9713 \\ 2.6459 \\ 5.2712 \end{bmatrix},$$

which is the vector in W such that $\|\mathbf{b} - \mathbf{y}\|$, $\mathbf{y}$ in W, is minimized. That is,

$$\min_{\mathbf{y} \text{ in } W} \|\mathbf{b} - \mathbf{y}\| = \|\mathbf{b} - A\hat{\mathbf{x}}\|. \qquad \blacksquare$$

Least squares problems often arise when one tries to construct a mathematical model of the form

$$y(t) = x_1 f_1(t) + x_2 f_2(t) + \cdots + x_n f_n(t) \tag{2}$$

to a data set $D = \{(t_i, y_i)\ i = 1, 2, \ldots, m\}$, where $m > n$. Ideally, we would like to determine $x_1, x_2, \ldots, x_n$ such that

$$y_i = x_1 f_1(t_i) + x_2 f_2(t_i) + \cdots + x_n f_n(t_i)$$

for each data point t_i, $i = 1, 2, \ldots, m$. In matrix form we have the linear system $A\mathbf{x} = \mathbf{b}$, where

$$A = \begin{bmatrix} f_1(t_1) & f_2(t_1) & \cdots & f_n(t_1) \\ f_1(t_2) & f_2(t_2) & \cdots & f_n(t_2) \\ \vdots & \vdots & \vdots\ \vdots\ \vdots & \vdots \\ f_1(t_m) & f_2(t_m) & \cdots & f_n(t_m) \end{bmatrix}, \tag{3}$$

$$\mathbf{x} = \begin{bmatrix} x_1 & x_2 & \cdots & x_n \end{bmatrix}^T, \quad \text{and} \quad \mathbf{b} = \begin{bmatrix} y_1 & y_2 & \cdots & y_m \end{bmatrix}^T.$$

As is often the case, the system $A\mathbf{x} = \mathbf{b}$ is inconsistent, so we determine a least squares solution $\hat{\mathbf{x}}$ of $A\mathbf{x} = \mathbf{b}$. If we set $x_i = \hat{x}_i$ in the model equation (2), we say that

$$\hat{y}(t) = \hat{x}_1 f_1(t) + \hat{x}_2 f_2(t) + \cdots + \hat{x}_n f_n(t)$$

gives a least squares model for data set D. In general, $\hat{y}(t_i) \neq y_i$, $i = 1, 2, \ldots, m$ (they may be equal, but there is no guarantee). Let $e_i = y_i - \hat{y}(t_i)$, $i = 1, 2, \ldots, m$, which represents the error incurred at t_i when $\hat{y}(t)$ is used as a mathematical model for data set D. If

$$\mathbf{e} = \begin{bmatrix} e_1 & e_2 & \cdots & e_m \end{bmatrix}^T,$$

then

$$\mathbf{e} = \mathbf{b} - A\hat{\mathbf{x}}$$

and Theorem 3.9 guarantees that $\|\mathbf{e}\| = \|\mathbf{b} - A\hat{\mathbf{x}}\|$ is as small as possible. That is,

$$\|\mathbf{e}\|^2 = (\mathbf{e}, \mathbf{e}) = (\mathbf{b} - A\hat{\mathbf{x}}, \mathbf{b} - A\hat{\mathbf{x}}) = \sum_{i=1}^{m} \left[y_i - \sum_{j=1}^{n} \hat{x}_j f_j(t_i) \right]^2$$

is minimized. We say that $\hat{\mathbf{x}}$, the least squares solution, minimizes the sum of the squares of the deviations between the observations y_i and the values $\hat{y}(t_i)$ predicted by the model equation.

Example 2. The following data shows atmospheric pollutants y_i (relative to an EPA standard) at half-hour intervals t_i.

t_i	1	1.5	2	2.5	3	3.5	4	4.5	5
y_i	−0.15	0.24	0.68	1.04	1.21	1.15	0.86	0.41	−0.08

A plot of these data as shown in Figure 3.23 suggests that a quadratic polynomial

$$y(t) = x_1 + x_2 t + x_3 t^2$$

may produce a good model for these data. With $f_1(t) = 1$, $f_2(t) = t$, and $f_3(t) = t^2$, (3) gives

Figure 3.23

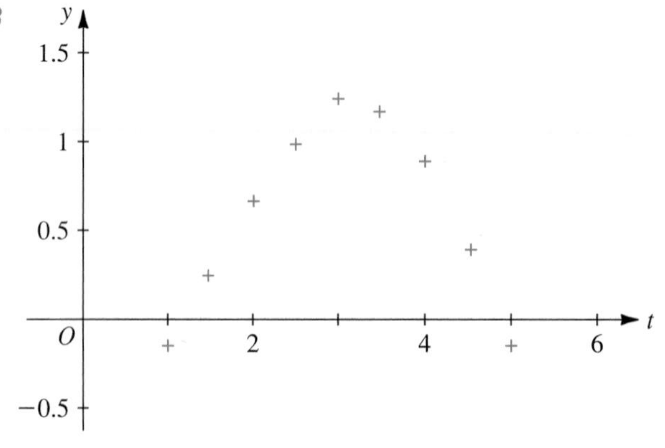

$$A = \begin{bmatrix} 1 & 1 & 1 \\ 1 & 1.5 & 2.25 \\ 1 & 2 & 4 \\ 1 & 2.5 & 6.25 \\ 1 & 3 & 9 \\ 1 & 3.5 & 12.25 \\ 1 & 4 & 16 \\ 1 & 4.5 & 20.25 \\ 1 & 5 & 25 \end{bmatrix}, \quad \mathbf{x} = \begin{bmatrix} x_1 \\ x_2 \\ x_3 \end{bmatrix}, \quad \mathbf{b} = \begin{bmatrix} -0.15 \\ 0.24 \\ 0.68 \\ 1.04 \\ 1.21 \\ 1.15 \\ 0.86 \\ 0.41 \\ -0.05 \end{bmatrix}.$$

The rank of A is 3 (verify), and the normal system is

$$\begin{bmatrix} 9 & 27 & 96 \\ 27 & 96 & 378 \\ 96 & 378 & 1583.25 \end{bmatrix} \begin{bmatrix} x_1 \\ x_2 \\ x_3 \end{bmatrix} = \begin{bmatrix} 5.36 \\ 16.71 \\ 54.65 \end{bmatrix}.$$

Applying Gaussian elimination gives (verify)

$$\hat{\mathbf{x}} \approx \begin{bmatrix} -1.9317 \\ 2.0067 \\ -0.3274 \end{bmatrix},$$

so we obtain the quadratic polynomial model

$$y(t) = -1.9317 + 2.0067t - 0.3274t^2.$$

Figure 3.24 shows the data set indicated with + and a graph of $y(t)$. We see that $y(t)$ is close to each data point but is not required to go through the data. ■

Figure 3.24

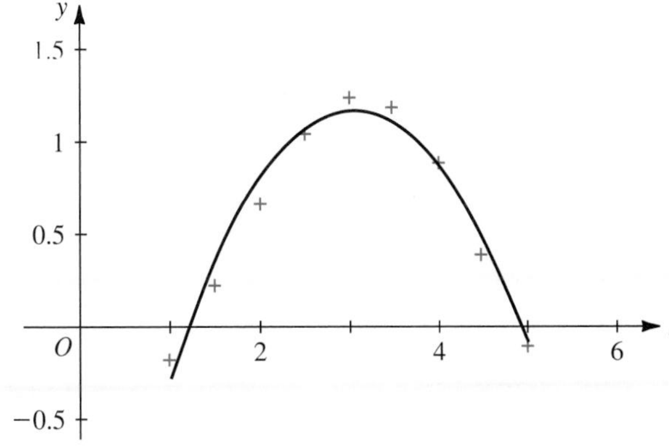

3.5. EXERCISES

1. Let A be $n \times n$ and nonsingular. From the normal system of equations in (1), show that the least squares solution to $A\mathbf{x} = \mathbf{b}$ is $\hat{\mathbf{x}} = A^{-1}\mathbf{b}$.

2. Determine the least squares solution to $A\mathbf{x} = \mathbf{b}$, where

$$A = \begin{bmatrix} 2 & 1 \\ 1 & 0 \\ 0 & -1 \\ -1 & 1 \end{bmatrix} \quad \text{and} \quad \mathbf{b} = \begin{bmatrix} 3 \\ 1 \\ 2 \\ -1 \end{bmatrix}.$$

3. Determine the least squares solution to $A\mathbf{x} = \mathbf{b}$, where

$$A = \begin{bmatrix} 1 & 2 & 1 \\ 1 & 3 & 2 \\ 2 & 5 & 3 \\ 2 & 0 & 1 \\ 3 & 1 & 1 \end{bmatrix} \quad \text{and} \quad \mathbf{b} = \begin{bmatrix} -1 \\ 2 \\ 0 \\ 1 \\ -2 \end{bmatrix}.$$

4. In the manufacture of product Z, the amount of compound A present in the product is controlled by the amount of ingredient B used in the refining process. In manufacturing a liter of Z, the amount of B used and the amount of A present are recorded. The following data were obtained:

B used (grams/liter)	2	4	6	8	10
A present (grams/liter)	3.5	8.2	10.5	12.9	14.6

Determine the least squares line to the data. [In Equation (2), use $f_1(t) = 1$, $f_2(t) = t$.] Also compute $\|\mathbf{e}\|$.

5. In Exercise 4, the least squares line to the data set $D = \{(t_i, y_i)\ i = 1, 2, \ldots, m\}$ is the line $y = \hat{x}_1 + \hat{x}_2 t$, which minimizes

$$E_1 = \sum_{i=1}^{m} [y_i - (x_1 + x_2 t_i)]^2.$$

Similarly, the least squares quadratic (see Example 2) to the data set D is the parabola $y = \hat{x}_1 + \hat{x}_2 t + \hat{x}_3 t^2$, which minimizes

$$E_2 = \sum_{i=1}^{m} [y_i - (x_1 + x_2 t_i + x_3 t_i^2)]^2.$$

Give a vector space argument to show that $E_2 \le E_1$.

6. The following table is a sample set of seasonal farm employment data (t_i, y_i) over about a two-year period, where t_i represents months and y_i represents millions of people. A plot of the data is given in Figure 3.25. It is decided to develop a least squares mathematical model of the form

$$y(t) = x_1 + x_2 t + x_3 \cos t.$$

t_i	y_i
3.1	3.7
4.3	4.4
5.6	5.3
6.8	5.2
8.1	4.0
9.3	3.6

t_i	y_i
10.6	3.6
11.8	5.0
13.1	5.0
14.3	3.8
15.6	2.8
16.8	3.3
18.1	4.5

Figure 3.25

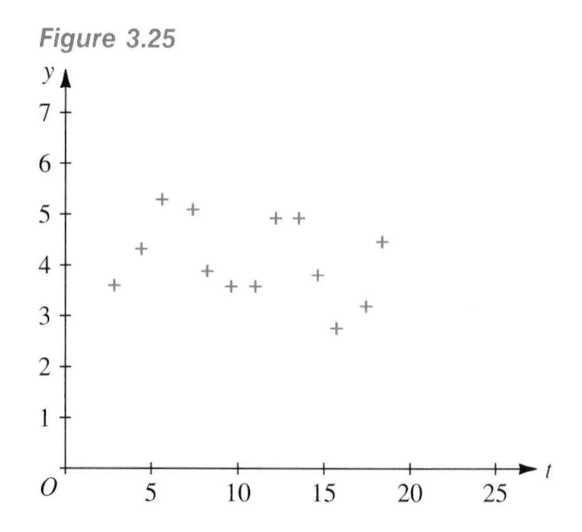

Determine the least squares model. Plot the resulting function $y(t)$ and the data set on the same coordinate system.

7. Given $A\mathbf{x} = \mathbf{b}$, where

$$A = \begin{bmatrix} 1 & 3 & -3 \\ 2 & 4 & -2 \\ 0 & -1 & 2 \\ 1 & 2 & -1 \end{bmatrix} \quad \text{and} \quad \mathbf{b} = \begin{bmatrix} 1 \\ 0 \\ 0 \\ 1 \end{bmatrix}.$$

(a) Show that rank $A = 2$.

(b) Since rank $A \neq$ number of columns, Theorem 3.10 cannot be used to determine a least squares solution $\hat{\mathbf{x}}$. Follow the general procedure as discussed prior to Theorem 3.10 to find a least squares solution. Is the solution unique?

C 8. In some software, the command for solving a linear system produces a least squares solution when the coefficient matrix is not square or nonsingular. Determine if this is the case in your software. If it is, compare your software's output with the solution given in Example 1. To experiment further, use Exercise 7.

Supplementary Exercises

1. Exercise 27 of Section 3.4 proves that the set of all vectors in R^n that are orthogonal to a fixed vector $\mathbf{u}$ forms a subspace of R^n. For $\mathbf{u} = \begin{bmatrix} 1 \\ -2 \\ 1 \end{bmatrix}$, find an orthogonal basis for the subspace of vectors in R^3 that are orthogonal to $\mathbf{u}$. [*Hint:* Solve the linear system $(\mathbf{u}, \mathbf{v}) = 0$, when $\mathbf{v} = \begin{bmatrix} x_1 \\ x_2 \\ x_3 \end{bmatrix}$.]

2. Use the Gram–Schmidt process to find an orthonormal basis for the subspace of R^4 with basis

$$\left\{ \begin{bmatrix} 1 \\ 0 \\ 0 \\ -1 \end{bmatrix}, \begin{bmatrix} 1 \\ -1 \\ 0 \\ 0 \end{bmatrix}, \begin{bmatrix} 0 \\ 1 \\ 0 \\ 1 \end{bmatrix} \right\}.$$

3. Given the orthonormal basis

$$S = \left\{ \begin{bmatrix} \frac{1}{\sqrt{2}} \\ 0 \\ \frac{-1}{\sqrt{2}} \end{bmatrix}, \begin{bmatrix} 0 \\ 1 \\ 0 \end{bmatrix}, \begin{bmatrix} \frac{1}{\sqrt{2}} \\ 0 \\ \frac{1}{\sqrt{2}} \end{bmatrix} \right\}$$

for R^3, write the vector

$$\mathbf{v} = \begin{bmatrix} 1 \\ 2 \\ 3 \end{bmatrix}$$

as a linear combination of the vectors in S.

4. Use the Gram–Schmidt process to find an orthonormal basis for the subspace of R^4 with basis

$$\left\{ \begin{bmatrix} 1 \\ 0 \\ -1 \\ 0 \end{bmatrix}, \begin{bmatrix} 1 \\ -1 \\ 0 \\ 0 \end{bmatrix}, \begin{bmatrix} 2 \\ 1 \\ 0 \\ 0 \end{bmatrix} \right\}.$$

5. Given vector $\mathbf{u} = \mathbf{i} + 2\mathbf{j} + \mathbf{k}$ and the plane P determined by the vectors

$$\mathbf{v} = \mathbf{i} - 3\mathbf{j} - 2\mathbf{k} \quad \text{and} \quad \mathbf{w} = 3\mathbf{i} - \mathbf{j} - 3\mathbf{k},$$

find the vector in P closest to $\mathbf{u}$ and the distance from $\mathbf{u}$ to P.

6. Find the distance from the point $(2, 3, -1)$ to the plane $3x - 2y + z = 0$. (*Hint:* First find an orthonormal basis for the plane.)

7. Consider the vector space of continuous functions on $[0, \pi]$ with an inner product defined by $(f, g) = \int_0^{\pi} f(t)g(t)\,dt$. Show that the collection of functions $\sin nt$, for $n = 1, 2, \ldots,$ is an orthogonal set.

8. Let V be the inner product space defined in Example 5 of Section 3.3. In each of the following let W be the subspace spanned by the given orthonormal vectors $\mathbf{u}_1, \mathbf{u}_2, \ldots, \mathbf{u}_n$. Find $\text{proj}_W \mathbf{u}$, for the vector $\mathbf{u}$ in V.

(a) $\mathbf{u} = t + t^2, \mathbf{u}_1 = \frac{1}{\sqrt{\pi}} \cos t, \mathbf{u}_2 = \frac{1}{\sqrt{\pi}} \sin 2t.$

(b) $\mathbf{u} = \sin \frac{1}{2}t, \mathbf{u}_1 = \frac{1}{\sqrt{2\pi}}, \mathbf{u}_2 = \frac{1}{\sqrt{\pi}} \sin t.$

(c) $\mathbf{u} = \cos^2 t, \quad \mathbf{u}_1 = \frac{1}{\sqrt{2\pi}}, \quad \mathbf{u}_2 = \frac{1}{\sqrt{\pi}} \cos t,$

$\mathbf{u}_3 = \frac{1}{\sqrt{\pi}} \cos 2t.$

9. Find an orthonormal basis for the set of vectors $\mathbf{x}$ such that $A\mathbf{x} = \mathbf{0}$ when:

(a) $A = \begin{bmatrix} 1 & 0 & 5 & -2 \\ 0 & 1 & -2 & -5 \end{bmatrix}.$

(b) $A = \begin{bmatrix} 1 & 0 & 5 & -2 \\ 0 & 1 & -2 & 4 \end{bmatrix}.$

10. If W is a subspace of the inner product space V, then the set of all vectors in V that are orthogonal to every vector in W is called the **orthogonal complement** of W.

(a) Prove that the orthogonal complement of W is a subspace of V.

(b) Prove that every vector $\mathbf{u}$ in V is uniquely expressible in the form $\mathbf{v} + \mathbf{w}$, where $\mathbf{v}$ is in W and $\mathbf{w}$ is in the orthogonal complement of W.

11. Let $W = \text{span}\left\{\begin{bmatrix} 1 \\ 0 \\ 1 \end{bmatrix}, \begin{bmatrix} 0 \\ 1 \\ 0 \end{bmatrix}\right\}$ in R^3.

(a) Find a basis for the orthogonal complement of W.

(b) Show that vectors $\begin{bmatrix} 1 \\ 0 \\ 1 \end{bmatrix}, \begin{bmatrix} 0 \\ 1 \\ 0 \end{bmatrix}$, and the basis for the orthogonal complement of W from part (a) form a basis for R^3.

(c) Find vectors $\mathbf{v}$ and $\mathbf{w}$ as defined in Exercise 10, so that $\mathbf{u} = \mathbf{v} + \mathbf{w}$ for each of the following vectors.

(i) $\mathbf{u} = \begin{bmatrix} 1 \\ 0 \\ 0 \end{bmatrix}.$ (ii) $\mathbf{u} = \begin{bmatrix} 1 \\ 2 \\ 3 \end{bmatrix}.$

12. Let $3x + 2y - z = 0$ be the equation of a plane π in 3-space.

(a) Show that π is a subspace of R^3.

(b) Describe the orthogonal complement of π.

13. Use the Gram–Schmidt process to find an orthonormal basis for P_2 with respect to the inner product $(f, g) = \int_{-1}^{1} f(t)g(t)\,dt$, starting with the standard basis $\{1, t, t^2\}$ for P_2. The polynomials thus obtained are called the **Legendre polynomials**.

14. Using the Legendre polynomials from Exercise 13 as an orthonormal basis for P_2 considered as a subspace of the vector space of continuous functions on $[-1, 1]$, find $\text{proj}_{P_2} \mathbf{u}$ for each of the following.

(a) $\mathbf{u} = t^3$. (b) $\mathbf{u} = \sin \pi t$. (c) $\mathbf{u} = \cos \pi t$.

15. Let A be an $n \times n$ symmetric matrix, and suppose that R^n has the standard inner product. Prove that if $(\mathbf{u}, A\mathbf{u}) = (\mathbf{u}, \mathbf{u})$ for all $\mathbf{u}$ in R^n, then $A = I_n$.

16. An $n \times n$ symmetric matrix A is **positive semidefinite** if $\mathbf{x}^T A \mathbf{x} \geq 0$ for all $\mathbf{x}$ in R^n. Prove:

(a) Every positive definite matrix is positive semidefinite.

(b) If A is singular and positive semidefinite, then A is not positive definite.

(c) A diagonal matrix A is positive semidefinite if and only if $a_{ii} \geq 0$ for $i = 1, 2, \ldots, n$.

17. In Chapter 6 the notion of an orthogonal matrix is discussed. It is shown that an $n \times n$ matrix P is orthogonal if and only if the columns of P, denoted $\mathbf{p}_1, \mathbf{p}_2, \ldots, \mathbf{p}_n$, form an orthonormal set in R^n using the standard inner product.

(a) For $\mathbf{x}$ in R^n, prove that $\|P\mathbf{x}\| = \|\mathbf{x}\|$, using Theorem 3.7.

(b) For $\mathbf{x}$ and $\mathbf{y}$ in R^n, prove that the angle between $P\mathbf{x}$ and $P\mathbf{y}$ is the same as that between $\mathbf{x}$ and $\mathbf{y}$.

18. Let A be an $n \times n$ skew-symmetric matrix. Prove that $\mathbf{x}^T A\mathbf{x} = 0$ for all $\mathbf{x}$ in R^n.

19. Let B be an $m \times n$ matrix with orthonormal columns $\mathbf{b}_1, \mathbf{b}_2, \ldots, \mathbf{b}_n$.

(a) Prove that $m \geq n$.

(b) Prove that $B^T B = I_n$.

20. Let $\{\mathbf{u}_1, \ldots, \mathbf{u}_k, \mathbf{u}_{k+1}, \ldots, \mathbf{u}_n\}$ be an orthonormal basis for Euclidean space V, $S = \text{span } \{\mathbf{u}_1, \ldots, \mathbf{u}_k\}$, and $T = \text{span } \{\mathbf{u}_{k+1}, \ldots, \mathbf{u}_n\}$. For any $\mathbf{x}$ in S and any $\mathbf{y}$ in T, show that $(\mathbf{x}, \mathbf{y}) = 0$.

4 Linear Transformations and Matrices

4.1. Definition and Examples

As we have noted earlier, much of calculus deals with the study of properties of functions. Indeed, properties of functions are of great importance in every branch of mathematics, and linear algebra is no exception. We have already encountered functions mapping one vector space into another vector space; these were the isomorphisms. If we drop some of the conditions that need to be satisfied by a function on a vector space to be an isomorphism, we get another very useful type of function, called a linear transformation. Linear transformations play an important role in many areas of mathematics, the physical and social sciences, and economics. All vector spaces considered henceforth are finite-dimensional.

Definition 4.1. Let V and W be vector spaces. A function $L: V \to W$ is called a **linear transformation** of V into W if
(a) $L(\mathbf{u} + \mathbf{v}) = L(\mathbf{u}) + L(\mathbf{v})$ for $\mathbf{u}$ and $\mathbf{v}$ in V.
(b) $L(c\mathbf{u}) = cL(\mathbf{u})$ for $\mathbf{u}$ in V, and c a real number.
If $V = W$, the linear transformation $L: V \to W$ is also called a **linear operator** on V. ■

In Definition 4.1, observe that in (a) the + in $\mathbf{u} + \mathbf{v}$ refers to the addition operation in V, whereas the + in $L(\mathbf{u}) + L(\mathbf{v})$ refers to the addition operation in

W. Similarly, in (b) the scalar product $c\mathbf{u}$ is in V, while the scalar product $cL(\mathbf{u})$ is in W.

We have already pointed out in Section 2.5 that an isomorphism is a linear transformation that is one-to-one and onto. Linear transformations occur very frequently, and we now look at some examples. (At this point it might be profitable to review the material of Section A.2.) It can be shown that $L: V \to W$ is a linear transformation if and only if $L(a\mathbf{u} + b\mathbf{v}) = aL(\mathbf{u}) + bL(\mathbf{v})$ for any real numbers a, b and any vectors $\mathbf{u}$, $\mathbf{v}$ in V (see Exercise 4).

Example 1. Let $L: R^3 \to R^2$ be defined by

$$L\left(\begin{bmatrix} a_1 \\ a_2 \\ a_3 \end{bmatrix}\right) = \begin{bmatrix} a_1 \\ a_2 \end{bmatrix}.$$

To show that L is a linear transformation, we let

$$\mathbf{u} = \begin{bmatrix} a_1 \\ a_2 \\ a_3 \end{bmatrix} \quad \text{and} \quad \mathbf{v} = \begin{bmatrix} b_1 \\ b_2 \\ b_3 \end{bmatrix}.$$

Then

$$L(\mathbf{u} + \mathbf{v}) = L\left(\begin{bmatrix} a_1 + b_1 \\ a_2 + b_2 \\ a_3 + b_3 \end{bmatrix}\right) = \begin{bmatrix} a_1 + b_1 \\ a_2 + b_2 \end{bmatrix} = \begin{bmatrix} a_1 \\ a_2 \end{bmatrix} + \begin{bmatrix} b_1 \\ b_2 \end{bmatrix} = L(\mathbf{u}) + L(\mathbf{v}).$$

Also, if c is a real number, then

$$L(c\mathbf{u}) = L\left(\begin{bmatrix} ca_1 \\ ca_2 \\ ca_3 \end{bmatrix}\right) = \begin{bmatrix} ca_1 \\ ca_2 \end{bmatrix} = c\begin{bmatrix} a_1 \\ a_2 \end{bmatrix} = cL(\mathbf{u}).$$

Hence L is a linear transformation, which is called a **projection**. It is simple and helpful to describe geometrically the effect of L. The image under L of a vector in R^3 with head $P(a_1, a_2, a_3)$ is found by drawing a line through P perpendicular to R^2, the (x, y)-plane. We obtain the point $Q(a_1, a_2)$ of intersection of this line with the (x, y)-plane. The vector in R^2 with head Q is the image of $\mathbf{u}$ under L (Figure 4.1). Referring to the optional material in Section 3.4, we see that $L(\mathbf{u})$ is the orthogonal projection of the vector $\mathbf{u}$ in R^3 onto the (x, y)-plane. ■

Figure 4.1 Projection.

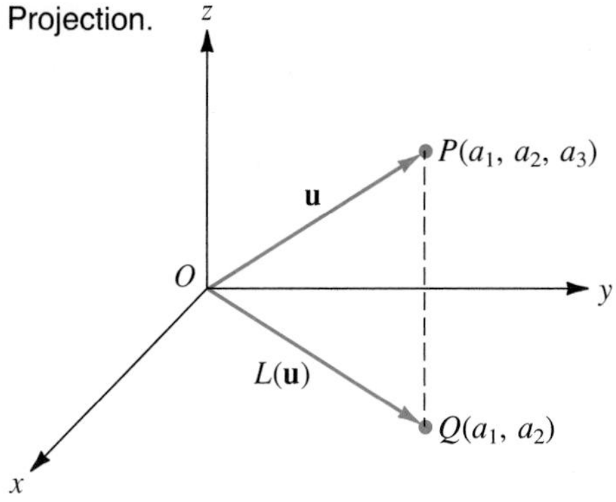

Example 2. We consider the mapping L from the vector space P_2 into the vector space P_1 defined by $L(at^2 + bt + c) = 2at + b$. Then L is ordinary differentiation and it is not difficult to show that L is a linear transformation (verify). ■

Example 3. Let $L: R^3 \to R^3$ be defined by

$$L\left(\begin{bmatrix} a_1 \\ a_2 \\ a_3 \end{bmatrix}\right) = r\begin{bmatrix} a_1 \\ a_2 \\ a_3 \end{bmatrix},$$

where r is a real number. Then L is a linear operator on R^3 (verify). If $r > 1$, L is called a **dilation**; if $0 < r < 1$, L is called a **contraction**. Thus dilation stretches a vector, while contraction shrinks it (Figure 4.2). ■

Figure 4.2

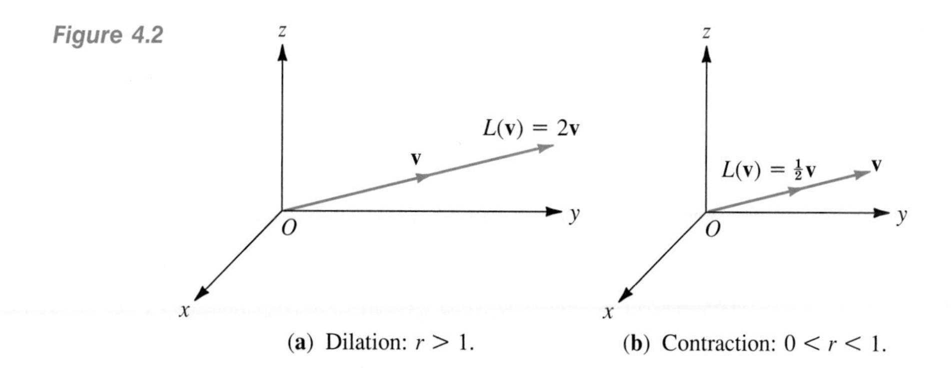

(**a**) Dilation: $r > 1$. (**b**) Contraction: $0 < r < 1$.

Example 4. Let $L: P_2 \to R$ (R = the vector space of real numbers under the usual operations of addition and scalar multiplication) be defined by

$$L(at^2 + bt + c) = \int_0^1 (at^2 + bt + c)\, dt.$$

Then L is a linear transformation (verify). ■

Example 5. Let $L: R^3 \to R^2$ be defined by

$$L\left(\begin{bmatrix} a_1 \\ a_2 \\ a_3 \end{bmatrix}\right) = \begin{bmatrix} 1 & 0 & 1 \\ 0 & 1 & -1 \end{bmatrix} \begin{bmatrix} a_1 \\ a_2 \\ a_3 \end{bmatrix}.$$

Then L is a linear transformation (verify). ■

Example 6. Let A be an $m \times n$ matrix. Define $L: R^n \to R^m$ by $L(\mathbf{x}) = A\mathbf{x}$ for $\mathbf{x}$ in R^n. Then L is a linear transformation (Exercise 19). ■

Example 7. Let $L: R^2 \to R^2$ be defined by

$$L\left(\begin{bmatrix} a_1 \\ a_2 \end{bmatrix}\right) = \begin{bmatrix} a_1 \\ -a_2 \end{bmatrix}.$$

Then L is a linear operator on R^2 (verify). Geometrically, the action of L is shown in Figure 4.3. Thus L is a **reflection** with respect to the x-axis. In a similar way we can also consider reflection with respect to the y-axis. ■

Figure 4.3 Reflection.

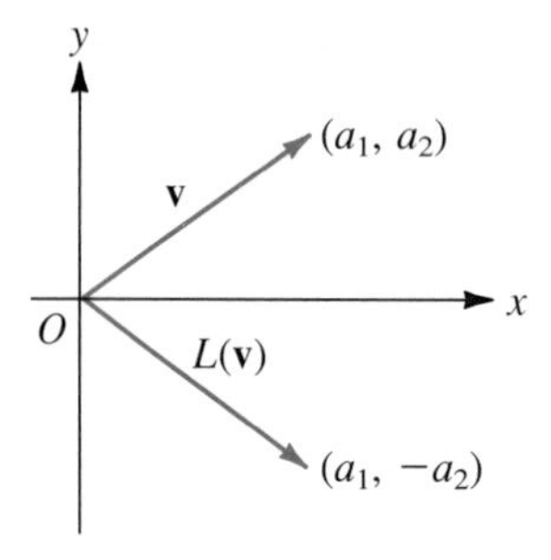

Example 8. Suppose that we rotate every point in the (x, y)-plane counterclockwise through an angle ϕ about the origin of a Cartesian coordinate system. Thus, if P has coordinates (x, y), then after rotating, we get P' with coordinates (x', y'), where

$$x' = x \cos \phi - y \sin \phi \qquad \text{and} \qquad y' = x \sin \phi + y \cos \phi,$$

or

$$x = x' \cos \phi + y' \sin \phi \qquad \text{and} \qquad y = -x' \sin \phi + y' \cos \phi.$$

Recall that we did this in calculus to simplify the general equation of second degree $ax^2 + bxy + cy^2 + dx + ey + f = 0$. Substituting for x and y in terms of x' and y', we obtain $a'x'^2 + b'x'y' + c'y'^2 + d'x' + e'y' + f' = 0$. The key point is to choose ϕ so that $b' = 0$. Once this is done (we might now have to perform a translation of coordinates), we identify the general equation of second degree as a circle, ellipse, hyperbola, parabola, or a degenerate form of these. This topic will be treated from a linear algebra point of view in Section 6.4. Note that we may perform this change of coordinates by considering the function L: $R^2 \to R^2$ defined by

$$L\left(\begin{bmatrix} x \\ y \end{bmatrix}\right) = \begin{bmatrix} \cos \phi & -\sin \phi \\ \sin \phi & \cos \phi \end{bmatrix} \begin{bmatrix} x \\ y \end{bmatrix}.$$

If $\mathbf{v} = \begin{bmatrix} x \\ y \end{bmatrix}$ is the vector from the origin to $P(x, y)$, then $L(\mathbf{v})$, the vector obtained by rotating $\mathbf{v}$ counterclockwise through the angle ϕ (Figure 4.4), is the vector from the origin to $P'(x', y')$. Then L is a linear transformation (verify) called a **rotation**. ■

Figure 4.4 Rotation.

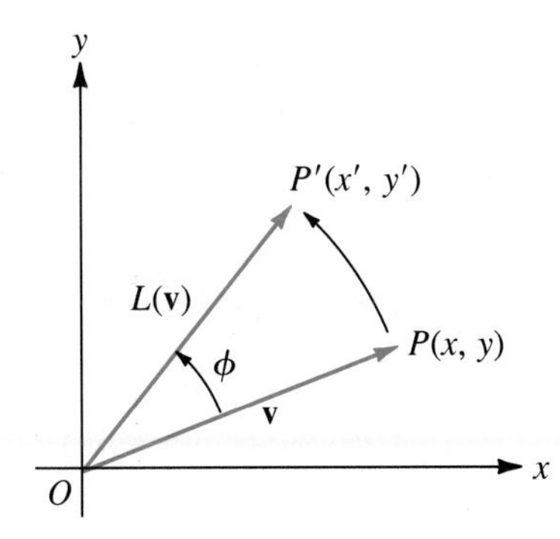

Example 9. Let $L: R^3 \to R^3$ be defined by

$$L\left(\begin{bmatrix} a_1 \\ a_2 \\ a_3 \end{bmatrix}\right) = \begin{bmatrix} a_1 + 1 \\ 2a_2 \\ a_3 \end{bmatrix}.$$

To determine whether L is a linear transformation, let

$$\mathbf{u} = \begin{bmatrix} a_1 \\ a_2 \\ a_3 \end{bmatrix} \quad \text{and} \quad \mathbf{v} = \begin{bmatrix} b_1 \\ b_2 \\ b_3 \end{bmatrix}.$$

Then

$$L(\mathbf{u} + \mathbf{v}) = L\left(\begin{bmatrix} a_1 \\ a_2 \\ a_3 \end{bmatrix} + \begin{bmatrix} b_1 \\ b_2 \\ b_3 \end{bmatrix}\right) = L\left(\begin{bmatrix} a_1 + b_1 \\ a_2 + b_2 \\ a_3 + b_3 \end{bmatrix}\right)$$
$$= \begin{bmatrix} (a_1 + b_1) + 1 \\ 2(a_2 + b_2) \\ a_3 + b_3 \end{bmatrix}.$$

On the other hand,

$$L(\mathbf{u}) + L(\mathbf{v}) = \begin{bmatrix} a_1 + 1 \\ 2a_2 \\ a_3 \end{bmatrix} + \begin{bmatrix} b_1 + 1 \\ 2b_2 \\ b_3 \end{bmatrix} = \begin{bmatrix} (a_1 + b_1) + 2 \\ 2(a_2 + b_2) \\ a_3 + b_3 \end{bmatrix}.$$

Since $L(\mathbf{u} + \mathbf{v}) \neq L(\mathbf{u}) + L(\mathbf{v})$, we conclude that the function L is not a linear transformation. ■

Example 10. Let $L: R_2 \to R_2$ be defined by

$$L([a_1 \quad a_2]) = [a_1^2 \quad 2a_2].$$

Is L a linear transformation?

Solution: Let

$$\mathbf{u} = [a_1 \quad a_2] \qquad \text{and} \qquad \mathbf{v} = [b_1 \quad b_2].$$

Then

$$\begin{aligned} L(\mathbf{u} + \mathbf{v}) &= L([a_1 \quad a_2] + [b_1 \quad b_2]) \\ &= L([a_1 + b_1 \quad a_2 + b_2]) \\ &= [(a_1 + b_1)^2 \quad 2(a_2 + b_2)]. \end{aligned}$$

On the other hand,

$$\begin{aligned} L(\mathbf{u}) + L(\mathbf{v}) &= [a_1^2 \quad 2a_2] + [b_1^2 \quad 2b_2] \\ &= [a_1^2 + b_1^2 \quad 2(a_2 + b_2)]. \end{aligned}$$

Since $L(\mathbf{u} + \mathbf{v}) \neq L(\mathbf{u}) + L(\mathbf{v})$ for all choices of a_1 and b_1, we conclude that L is not a linear transformation. ■

We know, from calculus, that a function can be specified by a formula that assigns to every member of the domain a unique element of the range. On the other hand, we can also specify a function by listing next to each member of the domain its assigned element of the range. An example of this would be provided by listing the names of all charge account customers of a department store along with their charge account number. At first glance it appears impossible to describe a linear transformation $L: V \to W$ of a vector space $V \neq \{\mathbf{0}\}$ into a vector space W in this latter manner, for V has infinitely many members in it. However, the following very useful theorem tells us that *once we know what a linear transformation L does to a basis for V, then we have completely specified L.* Thus, since in this book we only deal with finite-dimensional vector spaces, it is possible to describe L by giving only the images of a finite number of vectors in the domain V.

Theorem 4.1. *Let $L: V \to W$ be a linear transformation of an n-dimensional vector space V into a vector space W. Let $S = \{\mathbf{v}_1, \mathbf{v}_2, \ldots, \mathbf{v}_n\}$ be a basis for V. If $\mathbf{v}$ is any vector in V, then $L(\mathbf{v})$ is completely determined by $\{L(\mathbf{v}_1), L(\mathbf{v}_2), \ldots, L(\mathbf{v}_n)\}$.*

Proof: Since $\mathbf{v}$ is in V, we can write $\mathbf{v} = a_1\mathbf{v}_1 + a_2\mathbf{v}_2 + \cdots + a_n\mathbf{v}_n$, where $a_1, a_2, \ldots, a_n$ are uniquely determined real numbers. Then

$$L(\mathbf{v}) = L(a_1\mathbf{v}_1 + a_2\mathbf{v}_2 + \cdots + a_n\mathbf{v}_n)$$

$$= L(a_1\mathbf{v}_1) + L(a_2\mathbf{v}_2) + \cdots + L(a_n\mathbf{v}_n)$$
$$= a_1L(\mathbf{v}_1) + a_2L(\mathbf{v}_2) + \cdots + a_nL(\mathbf{v}_n).$$

Thus $L(\mathbf{v})$ has been completely determined by the vectors $L(\mathbf{v}_1), L(\mathbf{v}_2), \ldots, L(\mathbf{v}_n)$. ■

Theorem 4.1 can also be stated in the following useful form: Let $L: V \to W$ and $T: V \to W$ be linear transformations of the n-dimensional vector space V into a vector space W. Let $S = \{\mathbf{v}_1, \mathbf{v}_2, \ldots, \mathbf{v}_n\}$ be a basis for V. If $T(\mathbf{v}_i) = L(\mathbf{v}_i)$ for $i = 1, 2, \ldots, n$, then $T(\mathbf{v}) = L(\mathbf{v})$ for every $\mathbf{v}$ in V; that is, *if L and T agree on a basis for V, then L and T are identical linear transformations.*

Example 11. Let $L: R_4 \to R_2$ be a linear transformation and let $S = \{\mathbf{v}_1, \mathbf{v}_2, \mathbf{v}_3, \mathbf{v}_4\}$ be a basis for R_4 (see Example 3 in Section 2.4), where $\mathbf{v}_1 = [1 \quad 0 \quad 1 \quad 0]$, $\mathbf{v}_2 = [0 \quad 1 \quad -1 \quad 2]$, $\mathbf{v}_3 = [0 \quad 2 \quad 2 \quad 1]$, and $\mathbf{v}_4 = [1 \quad 0 \quad 0 \quad 1]$. Suppose that

$$L(\mathbf{v}_1) = [1 \quad 2],\ L(\mathbf{v}_2) = [0 \quad 3],\ L(\mathbf{v}_3) = [0 \quad 0],\text{ and } L(\mathbf{v}_4) = [2 \quad 0].$$

Then for

$$\mathbf{v} = [3 \quad -5 \quad -5 \quad 0] = 2\mathbf{v}_1 + \mathbf{v}_2 - 3\mathbf{v}_3 + \mathbf{v}_4$$

it follows by Theorem 4.1 that

$$L(\mathbf{v}) = 2L(\mathbf{v}_1) + L(\mathbf{v}_2) - 3L(\mathbf{v}_3) + L(\mathbf{v}_4) = [4 \quad 7].$$ ■

Theorem 4.2. *Let $L: V \to W$ be a linear transformation. Then*
(a) $L(\mathbf{0}_V) = \mathbf{0}_W$.
(b) $L(\mathbf{u} - \mathbf{v}) = L(\mathbf{u}) - L(\mathbf{v})$, *for* $\mathbf{u}, \mathbf{v}$ *in* V.

Proof: (a) We have

$$\mathbf{0}_V = \mathbf{0}_V + \mathbf{0}_V,$$

so

$$L(\mathbf{0}_V) = L(\mathbf{0}_V + \mathbf{0}_V)$$
$$L(\mathbf{0}_V) = L(\mathbf{0}_V) + L(\mathbf{0}_V)$$

Subtracting $L(\mathbf{0}_V)$ from both sides, we obtain

$$L(\mathbf{0}_V) = \mathbf{0}_W.$$

$$\text{(b) } L(\mathbf{u} - \mathbf{v}) = L(\mathbf{u} + (-1)(\mathbf{v})) = L(\mathbf{u}) + L((-1)\mathbf{v}) = L(\mathbf{u}) + (-1)L(\mathbf{v}) = L(\mathbf{u}) - L(\mathbf{v}).$$ ■

4.1. EXERCISES

1. Which of the following functions are linear transformations?
(a) $L: R_2 \rightarrow R_3$ defined by
$L([a_1 \quad a_2]) = [a_1 + 1 \quad a_2 \quad a_1 + a_2]$.
(b) $L: R_2 \rightarrow R_3$ defined by
$L([a_1 \quad a_2]) = [a_1 + a_2 \quad a_2 \quad a_1 - a_2]$.

2. Which of the following functions are linear transformations?
(a) $L: R_3 \rightarrow R_3$ defined by
$L([a_1 \quad a_2 \quad a_3]) = [a_1 \quad a_2^2 + a_3^2 \quad a_3^2]$.
(b) $L: R_3 \rightarrow R_3$ defined by
$L([a_1 \quad a_2 \quad a_3]) = [1 \quad a_3 \quad a_2]$.
(c) $L: R_3 \rightarrow R_3$ defined by
$L([a_1 \quad a_2 \quad a_3)] = [0 \quad a_3 \quad a_2]$.

3. Which of the following functions are linear transformations? [Here $p'(t)$ denotes the derivative of $p(t)$ with respect to t.]
(a) $L: P_2 \rightarrow P_3$ defined by
$L(p(t)) = t^3p'(0) + t^2p(0)$.
(b) $L: P_1 \rightarrow P_2$ defined by
$L(p(t)) = tp(t) + p(0)$.
(c) $L: P_1 \rightarrow P_2$ defined by
$L(p(t)) = tp(t) + 1$.

4. Let $L: V \rightarrow W$ be a mapping of a vector space V into a vector space W. Prove that L is a linear transformation if and only if $L(a\mathbf{u} + b\mathbf{v}) = aL(\mathbf{u}) + bL(\mathbf{v})$ for any real numbers a, b and any vectors $\mathbf{u}$, $\mathbf{v}$ in V.

5. Prove that the functions in Examples 2, 3, and 4 are linear transformations.

6. Prove that the functions in Examples 5 and 8 are linear transformations.

7. Consider the function $L: {}_3R_4 \rightarrow {}_2R_4$ defined by $L(A) = \begin{bmatrix} 2 & 3 & 1 \\ 1 & 2 & -3 \end{bmatrix} A$ for A in ${}_3R_4$.
(a) Find $L\left(\begin{bmatrix} 1 & 2 & 0 & -1 \\ 3 & 0 & 2 & 3 \\ 4 & 1 & -2 & 1 \end{bmatrix}\right)$.
(b) Show that L is a linear transformation.

8. In Section 4.3 we show that every linear transformation between finite-dimensional vector spaces can be represented as a matrix multiplication. For each of the following conjecture a matrix A such that $L(\mathbf{v}) = A\mathbf{v}$ and then verify your conjecture.
(a) The projection from R^3 to R^2 in Example 1.
(b) The dilation from R^3 to R^3 in Example 3.
(c) The reflection from R^2 to R^2 in Example 7.

9. Let $L: R^3 \rightarrow R^2$ be a linear transformation for which we know that

$$L\left(\begin{bmatrix} 1 \\ 0 \\ 0 \end{bmatrix}\right) = \begin{bmatrix} 2 \\ -4 \end{bmatrix},$$

$$L\left(\begin{bmatrix} 0 \\ 1 \\ 0 \end{bmatrix}\right) = \begin{bmatrix} 3 \\ -5 \end{bmatrix}, \quad L\left(\begin{bmatrix} 0 \\ 0 \\ 1 \end{bmatrix}\right) = \begin{bmatrix} 2 \\ 3 \end{bmatrix}.$$

(a) What is $L\left(\begin{bmatrix} 1 \\ -2 \\ 3 \end{bmatrix}\right)$?

(b) What is $L\left(\begin{bmatrix} a_1 \\ a_2 \\ a_3 \end{bmatrix}\right)$?

10. Let $L: R_2 \to R_2$ be a linear transformation for which we know that

$$L([1 \quad 1]) = [1 \quad -2], \qquad L([-1 \quad 1]) = [2 \quad 3].$$

(a) What is $L([-1 \quad 5])$?
(b) What is $L([a_1 \quad a_2])$?

11. Let $L: P_2 \to P_3$ be a linear transformation for which we know that $L(1) = 1$, $L(t) = t^2$, $L(t^2) = t^3 + t$.
(a) Find $L(2t^2 - 5t + 3)$.
(b) Find $L(at^2 + bt + c)$.

12. Let A be a fixed 3×3 matrix; also let L: ${}_3R_3 \to {}_3R_3$ be defined by $L(X) = AX - XA$, for X in ${}_3R_3$. Show that L is a linear transformation.

13. Let $L: R \to R$ be defined by $L(\mathbf{v}) = a\mathbf{v} + b$, where a and b are real numbers (of course, $\mathbf{v}$ is a vector in R, which in this case means that $\mathbf{v}$ is also a real number). Find all values of a and b such that L is a linear transformation.

14. Let V be an inner product space, and let $\mathbf{w}$ be a fixed vector in V. Let $L: V \to R$ be defined by $L(\mathbf{v}) = (\mathbf{v}, \mathbf{w})$ for $\mathbf{v}$ in V. Show that L is a linear transformation.

15. Describe the following linear transformations geometrically.

(a) $L\left(\begin{bmatrix} a_1 \\ a_2 \end{bmatrix}\right) = \begin{bmatrix} -a_1 \\ a_2 \end{bmatrix}$.

(b) $L\left(\begin{bmatrix} a_1 \\ a_2 \end{bmatrix}\right) = \begin{bmatrix} -a_1 \\ -a_2 \end{bmatrix}$.

(c) $L\left(\begin{bmatrix} a_1 \\ a_2 \end{bmatrix}\right) = \begin{bmatrix} -a_2 \\ a_1 \end{bmatrix}$.

16. Let V be a vector space and r a fixed scalar. Prove that the function $L: V \to V$ defined by $L(\mathbf{v}) = r\mathbf{v}$ is a linear operator on V.

17. Let V and W be vector spaces. Prove that the function $O: V \to W$ defined by $O(\mathbf{v}) = \mathbf{0}_W$ is a linear transformation, which is called the **zero linear transformation**.

18. Let $I: V \to V$ be defined by $I(\mathbf{v}) = \mathbf{v}$, for $\mathbf{v}$ in V. Show that I is a linear transformation, which is called the **identity operator** on V.

19. Let A be an $m \times n$ matrix, and suppose that L: $R^n \to R^m$ is defined by $L(\mathbf{x}) = A\mathbf{x}$ for $\mathbf{x}$ in R^n. Show that L is a linear transformation.

20. Let V be the vector space of all real-valued differentiable functions, and let W be the vector space of all real-valued functions. Define $L: V \to W$ by $L(f) = f'$, where f' is the derivative of f. Prove or disprove that L is a linear transformation.

21. Let V be the vector space of all real-valued functions that are integrable over the interval $[a, b]$; let $W = R^1$. Define $L: V \to W$ by $L(f) = \int_a^b f(x)\,dx$. Prove or disprove that L is a linear transformation.

22. Let A be an $n \times n$ matrix and suppose that L: ${}_nR_n \to {}_nR_n$ is defined by $L(X) = AX$, for X in ${}_nR_n$. Show that L is a linear transformation.

23. Let V be an n-dimensional vector space with ordered basis $S = \{\mathbf{v}_1, \mathbf{v}_2, \ldots, \mathbf{v}_n\}$ and let $T = \{\mathbf{w}_1, \mathbf{w}_2, \ldots, \mathbf{w}_n\}$ be an ordered set of vectors in V. Prove that there is a unique linear transformation $L: V \to V$ such that $L(\mathbf{v}_i) = \mathbf{w}_i$ for $i = 1, 2, \ldots, n$. [*Hint:* Let L be a mapping from V into V such that $L(\mathbf{v}_i) = \mathbf{w}_i$, then show how to extend L to be a linear transformation defined on all of V.]

24. If your software supports graphics (see Exercises 24 through 26 in Section 2.1), then dilations, contractions, reflections, and rotations of vectors in R^2 (and possibly R^3) can be displayed for visual inspection. Experiment with these linear transformations in your software.

25. Assuming that you have experimented as suggested in Exercise 24, try the following experiment. Let L_1 be a dilation with $r = 2$, L_2 a reflection with respect to the x-axis, and L_3 a rotation through $\phi = 30°$. For $\mathbf{v} = \begin{bmatrix} 3 \\ 1 \end{bmatrix}$, graphically determine if the following results involving successive linear transformations are true or false.

(a) $L_1(L_2(\mathbf{v})) = L_2(L_1(\mathbf{v}))$.

(b) $L_1(L_3(\mathbf{v})) = L_3(L_1(\mathbf{v}))$.

(c) $L_2(L_3(\mathbf{v})) = L_3(L_2(\mathbf{v}))$.

4.2. *The Kernel and Range of a Linear Transformation*

In this section we study special types of linear transformations; we formulate the notions of one-to-one linear transformations and onto linear transformations. We also develop methods for determining when a linear transformation is one-to-one or onto, and examine some applications of these notions.

Definition 4.2. A linear transformation $L: V \to W$ is called **one-to-one** if it is a one-to-one function; that is, if $\mathbf{v}_1 \neq \mathbf{v}_2$ implies that $L(\mathbf{v}_1) \neq L(\mathbf{v}_2)$. An equivalent statement is that L is one-to-one if $L(\mathbf{v}_1) = L(\mathbf{v}_2)$ implies that $\mathbf{v}_1 = \mathbf{v}_2$ (see Figure A.2 in Appendix A). ■

Example 1. Let $L: R^2 \to R^2$ be defined by

$$L\left(\begin{bmatrix} a_1 \\ a_2 \end{bmatrix}\right) = \begin{bmatrix} a_1 + a_2 \\ a_1 - a_2 \end{bmatrix}.$$

To determine whether L is one-to-one, we let

$$\mathbf{v}_1 = \begin{bmatrix} a_1 \\ a_2 \end{bmatrix} \quad \text{and} \quad \mathbf{v}_2 = \begin{bmatrix} b_1 \\ b_2 \end{bmatrix}.$$

Then if $L(\mathbf{v}_1) = L(\mathbf{v}_2)$, we have

$$\begin{aligned} a_1 + a_2 &= b_1 + b_2 \\ a_1 - a_2 &= b_1 - b_2. \end{aligned}$$

Adding these equations, we obtain $2a_1 = 2b_1$, or $a_1 = b_1$, which implies that $a_2 = b_2$. Hence $\mathbf{v}_1 = \mathbf{v}_2$ and L is one-to-one. ■

Example 2. Let $L: R^3 \to R^2$ be the linear transformation defined in Example 1 of Section 4.1 (a projection) by

$$L\left(\begin{bmatrix} a_1 \\ a_2 \\ a_3 \end{bmatrix}\right) = \begin{bmatrix} a_1 \\ a_2 \end{bmatrix}.$$

Since $L\left(\begin{bmatrix} 1 \\ 3 \\ 3 \end{bmatrix}\right) = L\left(\begin{bmatrix} 1 \\ 3 \\ -2 \end{bmatrix}\right)$, we conclude that L is not one-to-one. ■

We shall now develop some more efficient ways of determining whether or not a linear transformation is one-to-one.

Definition 4.3. Let $L: V \to W$ be a linear transformation of a vector space V into a vector space W. The **kernel** of L, ker L, is the subset of V consisting of all elements $\mathbf{v}$ of V such that $L(\mathbf{v}) = \mathbf{0}_W$. ■

We observe that Theorem 4.2 assures us that ker L is never an empty set, because if $L: V \to W$ is a linear transformation, then $\mathbf{0}_V$ is in ker L.

Example 3. Let $L: R^3 \to R^2$ be as defined in Example 2. The vector $\begin{bmatrix} 0 \\ 0 \\ 2 \end{bmatrix}$ is in ker L, since $L\left(\begin{bmatrix} 0 \\ 0 \\ 2 \end{bmatrix}\right) = \begin{bmatrix} 0 \\ 0 \end{bmatrix}$. However, the vector $\begin{bmatrix} 2 \\ -3 \\ 4 \end{bmatrix}$ is not in ker L, since $L\left(\begin{bmatrix} 2 \\ -3 \\ 4 \end{bmatrix}\right) = \begin{bmatrix} 2 \\ -3 \end{bmatrix}$. To find ker L, we must determine all $\mathbf{v}$ in R^3 so that $L(\mathbf{v}) = \mathbf{0}_{R^2}$. That is, we seek $\mathbf{v} = \begin{bmatrix} a_1 \\ a_2 \\ a_3 \end{bmatrix}$ so that $L(\mathbf{v}) = L\left(\begin{bmatrix} a_1 \\ a_2 \\ a_3 \end{bmatrix}\right) = \begin{bmatrix} 0 \\ 0 \end{bmatrix} = \mathbf{0}_{R^2}$. However, $L(\mathbf{v}) = \begin{bmatrix} a_1 \\ a_2 \end{bmatrix}$. Thus $\begin{bmatrix} a_1 \\ a_2 \end{bmatrix} = \begin{bmatrix} 0 \\ 0 \end{bmatrix}$, so $a_1 = 0$, $a_2 = 0$, and

a_3 can be any real number. Hence ker L consists of all vectors in R^3 of the form $\begin{bmatrix} 0 \\ 0 \\ a \end{bmatrix}$, where a is any real number. It is clear that ker L consists of the z-axis in (x, y, z) three-dimensional space R^3. ■

An examination of the elements in ker L allows us to decide whether L is or is not one-to-one.

Theorem 4.3. *Let $L: V \to W$ be a linear transformation of a vector space V into a vector space W. Then*
(a) ker L *is a subspace of* V.
(b) L *is one-to-one if and only if* ker $L = \{\mathbf{0}_V\}$.

Proof: (a) We show that if $\mathbf{v}$ and $\mathbf{w}$ are in ker L, then so are $\mathbf{v} + \mathbf{w}$ and $c\mathbf{v}$ for any real number c. If $\mathbf{v}$ and $\mathbf{w}$ are in ker L, then $L(\mathbf{v}) = \mathbf{0}_W$, and $L(\mathbf{w}) = \mathbf{0}_W$. Then since L is a linear transformation,

$$L(\mathbf{v} + \mathbf{w}) = L(\mathbf{v}) + L(\mathbf{w}) = \mathbf{0}_W + \mathbf{0}_W = \mathbf{0}_W.$$

Thus $\mathbf{v} + \mathbf{w}$ is in ker L. Also,

$$L(c\mathbf{v}) = cL(\mathbf{v}) = c\mathbf{0}_W = \mathbf{0}_W,$$

so $c\mathbf{v}$ is in ker L. Hence ker L is a subspace of V.

(b) Let L be one-to-one. We show that ker $L = \{\mathbf{0}_V\}$. Let $\mathbf{v}$ be in ker L. Then $L(\mathbf{v}) = \mathbf{0}_W$. Also, we already know that $L(\mathbf{0}_V) = \mathbf{0}_W$. Then $L(\mathbf{v}) = L(\mathbf{0}_V)$. Since L is one-to-one, we conclude that $\mathbf{v} = \mathbf{0}_V$. Hence ker $L = \{\mathbf{0}_V\}$.

Conversely, suppose that ker $L = \{\mathbf{0}_V\}$. We wish to show that L is one-to-one. Let $L(\mathbf{v}_1) = L(\mathbf{v}_2)$ for $\mathbf{v}_1$ and $\mathbf{v}_2$ in V. Then

$$L(\mathbf{v}_1) - L(\mathbf{v}_2) = \mathbf{0}_W,$$

so that $L(\mathbf{v}_1 - \mathbf{v}_2) = \mathbf{0}_W$. This means that $\mathbf{v}_1 - \mathbf{v}_2$ is in ker L, so $\mathbf{v}_1 - \mathbf{v}_2 = \mathbf{0}_V$. Hence $\mathbf{v}_1 = \mathbf{v}_2$, and L is one-to-one. ■

Note that we can also state Theorem 4.3(b) as: *L is one-to-one if and only if dim ker $L = 0$.*

Example 4. Let $L: P_2 \rightarrow R$ be as defined in Example 4 of Section 4.1:

$$L(at^2 + bt + c) = \int_0^1 (at^2 + bt + c)\, dt.$$

(a) Find ker L.
(b) Find dim ker L.
(c) Is L one-to-one?

Solution: (a) To find ker L, we seek an element $\mathbf{v} = at^2 + bt + c$ in P_2 such that $L(\mathbf{v}) = L(at^2 + bt + c) = \mathbf{0}_R = 0$. Now

$$L(\mathbf{v}) = \frac{at^3}{3} + \frac{bt^2}{2} + ct \Big|_0^1 = \frac{a}{3} + \frac{b}{2} + c.$$

Thus $c = -a/3 - b/2$. Then ker L consists of all polynomials in P_2 of the form $at^2 + bt + (-a/3 - b/2)$, for a and b any real numbers.

(b) To find the dimension of ker L, we obtain a basis for ker L. Any vector in ker L can be written as

$$at^2 + bt + \left(-\frac{a}{3} - \frac{b}{2}\right) = a\left(t^2 - \frac{1}{3}\right) + b\left(t - \frac{1}{2}\right).$$

Thus the elements $(t^2 - \frac{1}{3})$ and $(t - \frac{1}{2})$ in P_2 span ker L. Now, these elements are also linearly independent, since they are not constant multiples of each other. Thus $\{t^2 - \frac{1}{3}, t - \frac{1}{2}\}$ is a basis for ker L, and dim ker $L = 2$.

(c) Since dim ker $L = 2$, L is not one-to-one. ■

Definition 4.4. If $L: V \rightarrow W$ is a linear transformation of a vector space V into a vector space W, then the **range** of L or **image** of V under L, denoted by range L, consists of all those vectors in W that are images under L of vectors in V. Thus $\mathbf{w}$ is in range L if we can find some vector $\mathbf{v}$ in V such that $L(\mathbf{v}) = \mathbf{w}$. The linear transformation L is called **onto** if range $L = W$. ■

Theorem 4.4. If $L: V \rightarrow W$ is a linear transformation of a vector space V into a vector space W, then range L is a subspace of W.

Proof: Let $\mathbf{w}_1$ and $\mathbf{w}_2$ be in range L. Then $\mathbf{w}_1 = L(\mathbf{v}_1)$ and $\mathbf{w}_2 = L(\mathbf{v}_2)$ for some $\mathbf{v}_1$ and $\mathbf{v}_2$ in V. Now

$$\mathbf{w}_1 + \mathbf{w}_2 = L(\mathbf{v}_1) + L(\mathbf{v}_2) = L(\mathbf{v}_1 + \mathbf{v}_2),$$

which implies that $\mathbf{w}_1 + \mathbf{w}_2$ is in range L. Also, if $\mathbf{w}$ is in range L, then $\mathbf{w} = L(\mathbf{v})$ for some $\mathbf{v}$ in V. Then $c\mathbf{w} = cL(\mathbf{v}) = L(c\mathbf{v})$ where c is a scalar, so that $c\mathbf{w}$ is in range L. Hence range L is a subspace of W. ■

Example 5. Consider Example 2 of this section again. Is the projection L onto?

Solution: We choose any vector $\mathbf{w} = \begin{bmatrix} c \\ d \end{bmatrix}$ in R^2 and seek a vector $\mathbf{v} = \begin{bmatrix} a_1 \\ a_2 \\ a_3 \end{bmatrix}$ in V such that $L(\mathbf{v}) = \mathbf{w}$. Now $L(\mathbf{v}) = \begin{bmatrix} a_1 \\ a_2 \end{bmatrix}$, so if $a_1 = c$ and $a_2 = d$, then $L(\mathbf{v}) = \mathbf{w}$. Therefore, L is onto and dim range $L = 2$. ■

Example 6. Consider Example 4 of this section; is L onto?

Solution: Given a vector $\mathbf{w}$ in R, $\mathbf{w} = r$, a real number, can we find a vector $\mathbf{v} = at^2 + bt + c$ in P_2 so that $L(\mathbf{v}) = \mathbf{w} = r$?

Now $L(\mathbf{v}) = \int_0^1 (at^2 + bt + c)\, dt = a/3 + b/2 + c$. We can let $a = b = 0$ and $c = r$. Hence L is onto. Moreover, dim range $L = 1$. ■

Example 7. Let $L\colon R^3 \to R^3$ be defined by

$$L\left(\begin{bmatrix} a_1 \\ a_2 \\ a_3 \end{bmatrix}\right) = \begin{bmatrix} 1 & 0 & 1 \\ 1 & 1 & 2 \\ 2 & 1 & 3 \end{bmatrix} \begin{bmatrix} a_1 \\ a_2 \\ a_3 \end{bmatrix}.$$

(a) Is L onto?
(b) Find a basis for range L.
(c) Find ker L.
(d) Is L one-to-one?

Solution: (a) Given any $\mathbf{w} = \begin{bmatrix} a \\ b \\ c \end{bmatrix}$ in R^3, where a, b, and c are any real numbers, can we find $\mathbf{v} = \begin{bmatrix} a_1 \\ a_2 \\ a_3 \end{bmatrix}$ so that $L(\mathbf{v}) = \mathbf{w}$? We seek a solution to the linear system

$$\begin{bmatrix} 1 & 0 & 1 \\ 1 & 1 & 2 \\ 2 & 1 & 3 \end{bmatrix} \begin{bmatrix} a_1 \\ a_2 \\ a_3 \end{bmatrix} = \begin{bmatrix} a \\ b \\ c \end{bmatrix}$$

and we find the reduced row echelon form of the augmented matrix to be (verify)

$$\left[\begin{array}{ccc:c} 1 & 0 & 1 & a \\ 0 & 1 & 1 & b-a \\ 0 & 0 & 0 & c-b-a \end{array}\right].$$

Thus a solution exists only for $c - b - a = 0$, so L is not onto.

(b) To find a basis for range L, we note that

$$L\left(\begin{bmatrix} a_1 \\ a_2 \\ a_3 \end{bmatrix}\right) = \begin{bmatrix} 1 & 0 & 1 \\ 1 & 1 & 2 \\ 2 & 1 & 3 \end{bmatrix} \begin{bmatrix} a_1 \\ a_2 \\ a_3 \end{bmatrix} = \begin{bmatrix} a_1 + a_3 \\ a_1 + a_2 + 2a_3 \\ 2a_1 + a_2 + 3a_3 \end{bmatrix}$$

$$= a_1 \begin{bmatrix} 1 \\ 1 \\ 2 \end{bmatrix} + a_2 \begin{bmatrix} 0 \\ 1 \\ 1 \end{bmatrix} + a_3 \begin{bmatrix} 1 \\ 2 \\ 3 \end{bmatrix}.$$

This means that $\left\{\begin{bmatrix} 1 \\ 1 \\ 2 \end{bmatrix}, \begin{bmatrix} 0 \\ 1 \\ 1 \end{bmatrix}, \begin{bmatrix} 1 \\ 2 \\ 3 \end{bmatrix}\right\}$ spans range L. That is, range L is the subspace of R^3 spanned by the columns of the matrix defining L.

The first two vectors in this set are linearly independent, since they are not constant multiples of each other. The third vector is the sum of the first two. Therefore, the first two vectors form a basis for range L, and dim range $L = 2$.

(c) To find ker L, we wish to find all $\mathbf{v}$ in R^3 so that $L(\mathbf{v}) = \mathbf{0}_{R^3}$. Solving the resulting homogeneous system, we find (verify) that $a_1 = -a_3$ and $a_2 = -a_3$. Thus ker L consists of all vectors of the form $\begin{bmatrix} -a \\ -a \\ a \end{bmatrix} = a \begin{bmatrix} -1 \\ -1 \\ 1 \end{bmatrix}$, where a is any real number.

(d) Since ker $L \neq \{\mathbf{0}_{R^3}\}$, it follows from Theorem 4.3(b) that L is not one-to-one. ■

The problem of finding a basis for ker L always reduces to the problem of finding a basis for the solution space of a homogeneous system; this latter problem has been solved in Example 10 of Section 2.4.

A basis for range L can be obtained by the method discussed in Theorem 2.7 or by the procedure given in Section 2.6. Both approaches are illustrated in the next example.

Example 8. Let $L: R_4 \rightarrow R_3$ be defined by

$$L([a_1 \quad a_2 \quad a_3 \quad a_4]) = [a_1 + a_2 \quad a_3 + a_4 \quad a_1 + a_3].$$

Find a basis for range L.

Solution: We have

$$L([a_1 \quad a_2 \quad a_3 \quad a_4]) = a_1[1 \quad 0 \quad 1] + a_2[1 \quad 0 \quad 0] + a_3[0 \quad 1 \quad 1] + a_4[0 \quad 1 \quad 0].$$

Thus

$$S = \{[1 \quad 0 \quad 1], [1 \quad 0 \quad 0], [0 \quad 1 \quad 1], [0 \quad 1 \quad 0]\}$$

spans range L. To find a subset of S that is a basis for range L, we proceed as in Theorem 2.7 by first writing

$$a_1[1 \quad 0 \quad 1] + a_2[1 \quad 0 \quad 0] + a_3[0 \quad 1 \quad 1] + a_4[0 \quad 1 \quad 0] = [0 \quad 0 \quad 0].$$

The reduced row echelon form of the augmented matrix of this homogeneous system is (verify)

$$\left[\begin{array}{cccc|c} 1 & 0 & 0 & -1 & 0 \\ 0 & 1 & 0 & 1 & 0 \\ 0 & 0 & 1 & 1 & 0 \end{array}\right].$$

Since the leading 1's appear in columns 1, 2, and 3, we conclude that the first three vectors in S form a basis for range L. Thus

$$\{[1 \quad 0 \quad 1], [1 \quad 0 \quad 1], [0 \quad 1 \quad 1]\}$$

is a basis for range L.

Alternatively, we may proceed as in Section 2.6 to form the matrix whose rows are the given vectors

$$\begin{bmatrix} 1 & 0 & 1 \\ 1 & 0 & 0 \\ 0 & 1 & 1 \\ 0 & 1 & 0 \end{bmatrix}.$$

Transforming this matrix to reduced row echelon form, we obtain (verify)

$$\begin{bmatrix} 1 & 0 & 0 \\ 0 & 1 & 0 \\ 0 & 0 & 1 \\ 0 & 0 & 0 \end{bmatrix}.$$

Hence $\{[1 \quad 0 \quad 0], [0 \quad 1 \quad 0], [0 \quad 0 \quad 1]\}$ is a basis for range L. ■

To determine if a linear transformation is one-to-one or onto, we must solve a linear system. This is one further demonstration of the frequency with which linear systems must be solved to answer many questions in linear algebra. Finally, from Example 7 we saw that

$$\dim \ker L + \dim \text{range } L = \dim \text{domain } L.$$

This very important result is always true and we now prove it in the following theorem.

Theorem 4.5. *If $L: V \to W$ is a linear transformation of an n-dimensional vector space V into a vector space W, then*

$$\dim \ker L + \dim \text{range } L = \dim V.$$

Proof: Let $k = \dim \ker L$. If $k = n$, then $\ker L = V$ (Exercise 36, Section 2.4), which implies that $L(\mathbf{v}) = \mathbf{0}_W$ for every $\mathbf{v}$ in V. Hence range $L = \{\mathbf{0}_W\}$, dim range $L = 0$, and the conclusion holds. Next, suppose that $1 \le k < n$. We shall prove that dim range $L = n - k$. Let $\{\mathbf{v}_1, \mathbf{v}_2, \ldots, \mathbf{v}_k\}$ be a basis for ker L. By Theorem 2.9 we can extend this basis to a basis

$$S = \{\mathbf{v}_1, \mathbf{v}_2, \ldots, \mathbf{v}_k, \mathbf{v}_{k+1}, \ldots, \mathbf{v}_n\}$$

for V. We prove that the set $T = \{L(\mathbf{v}_{k+1}), L(\mathbf{v}_{k+2}), \ldots, L(\mathbf{v}_n)\}$ is a basis for range L.

First, we show that T spans range L. Let $\mathbf{w}$ be any vector in range L. Then $\mathbf{w} = L(\mathbf{v})$ for some $\mathbf{v}$ in V. Since S is a basis for V, we can find a unique set of real numbers $a_1, a_2, \ldots, a_n$ such that $\mathbf{v} = a_1\mathbf{v}_1 + a_2\mathbf{v}_2 + \cdots + a_n\mathbf{v}_n$. Then

$$\begin{aligned}\mathbf{w} = L(\mathbf{v}) &= L(a_1\mathbf{v}_1 + a_2\mathbf{v}_2 + \cdots + a_k\mathbf{v}_k + a_{k+1}\mathbf{v}_{k+1} + \cdots + a_n\mathbf{v}_n)\\ &= a_1L(\mathbf{v}_1) + a_2L(\mathbf{v}_2) + \cdots + a_kL(\mathbf{v}_k) + a_{k+1}L(\mathbf{v}_{k+1}) + \cdots + a_nL(\mathbf{v}_n)\\ &= a_{k+1}L(\mathbf{v}_{k+1}) + \cdots + a_nL(\mathbf{v}_n)\end{aligned}$$

because $\mathbf{v}_1, \mathbf{v}_2, \ldots, \mathbf{v}_k$ are in ker L. Hence T spans range L.

Now we show that T is linearly independent. Suppose that

$$a_{k+1}L(\mathbf{v}_{k+1}) + a_{k+2}L(\mathbf{v}_{k+2}) + \cdots + a_nL(\mathbf{v}_n) = \mathbf{0}_W.$$

Then

$$L(a_{k+1}\mathbf{v}_{k+1} + a_{k+2}\mathbf{v}_{k+2} + \cdots + a_n\mathbf{v}_n) = \mathbf{0}_W.$$

Hence the vector $a_{k+1}\mathbf{v}_{k+1} + a_{k+2}\mathbf{v}_{k+2} + \cdots + a_n\mathbf{v}_n$ is in ker L, and we can write

$$a_{k+1}\mathbf{v}_{k+1} + a_{k+2}\mathbf{v}_{k+2} + \cdots + a_n\mathbf{v}_n = b_1\mathbf{v}_1 + b_2\mathbf{v}_2 + \cdots + b_k\mathbf{v}_k,$$

where $b_1, b_2, \ldots, b_k$ are uniquely determined real numbers. We then have

$$b_1\mathbf{v}_1 + b_2\mathbf{v}_2 + \cdots + b_k\mathbf{v}_k - a_{k+1}\mathbf{v}_{k+1} - a_{k+2}\mathbf{v}_{k+2} - \cdots - a_n\mathbf{v}_n = \mathbf{0}_V.$$

Since S is linearly independent, we find that $b_1 = b_2 = \cdots = b_k = a_{k+1} = \cdots = a_n = 0$. Hence T is linearly independent and forms a basis for range L.

If $k = 0$, then ker L has no basis; we let $\{\mathbf{v}_1, \mathbf{v}_2, \ldots, \mathbf{v}_n\}$ be a basis for V. The proof now proceeds as above. ■

The dimension of ker L is also called the **nullity** of L.

We have seen that a linear transformation may be one-to-one and not onto or onto and not one-to-one. However, the following corollary shows that each of these properties implies the other if the vector spaces V and W have the same dimensions.

Corollary 4.1. *If $L: V \to W$ is a linear transformation of a vector space V into a vector space W and* dim V = dim W, *then:*

(a) *If L is one-to-one, then it is onto.*
(b) *If L is onto, then it is one-to-one.*

Proof: Exercise. ■

If $A = [a_{ij}]$ is an $m \times n$ matrix, then we can define a function $L\colon R^n \to R^m$ by $L(\mathbf{x}) = A\mathbf{x}$ for $\mathbf{x}$ a vector in R^n. It is easy to verify that L is a linear transformation (Exercise 19 of Section 4.1). Now if $A\mathbf{x} = \mathbf{b}$ is a system of m linear equations in n unknowns, then finding a solution to $A\mathbf{x} = \mathbf{b}$ is equivalent to obtaining, for a given vector $\mathbf{b}$ in R^m, a vector $\mathbf{x}$ in R^n such that $L(\mathbf{x}) = \mathbf{b}$. Then ker L consists of all vectors $\mathbf{x}$ in R^n such that $L(\mathbf{x}) = A\mathbf{x} = \mathbf{0}$, that is, of all solutions to the homogeneous system $A\mathbf{x} = \mathbf{0}$. Moreover, it is easy to see that range L is the subspace of R^m spanned by the columns of A; that is, range L = the column space of A (Exercise 11). This means that dim range L = rank A. It then follows from Theorem 4.5 that dim ker $L = n -$ rank A. Thus we have established the result that the solution space of a homogeneous system $A\mathbf{x} = \mathbf{0}$, where A is an $m \times n$ matrix, has dimension $n - r$, where r = rank A, or the dimension of the solution space is the number of columns minus rank A.

A linear transformation $L\colon V \to W$ of a vector space V into a vector space W is called **invertible** if it is an invertible function, that is, if there exists a unique function $L^{-1}\colon W \to V$ such that $L \circ L^{-1} = I_W$ and $L^{-1} \circ L = I_V$, where I_V = identity linear transformation on V and I_W = identity linear transformation on W. We now prove the following theorem.

Theorem 4.6. *A linear transformation $L\colon V \to W$ is invertible if and only if L is one-to-one and onto. Moreover, L^{-1} is a linear transformation and $(L^{-1})^{-1} = L$.*

Proof: Let L be one-to-one and onto. We define a function $H\colon W \to V$ as follows. If $\mathbf{w}$ is in W, then since L is onto, $\mathbf{w} = L(\mathbf{v})$ for some $\mathbf{v}$ in V, and since L is one-to-one, $\mathbf{v}$ is unique. Let $H(\mathbf{w}) = \mathbf{v}$; H is a function and $L(H(\mathbf{w})) = L(\mathbf{v}) = \mathbf{w}$, so that $L \circ H = I_W$. Also, $H(L(\mathbf{v})) = H(\mathbf{w}) = \mathbf{v}$, so $H \circ L = I_V$. Thus H is an inverse of L. Now H is unique, for if $H_1\colon W \to V$ is a function such that $L \circ H_1 = I_W$ and $H_1 \circ L = I_V$, then $L(H(\mathbf{w})) = \mathbf{w} = L(H_1(\mathbf{w}))$ for any $\mathbf{w}$ in W. Since L is one-to-one, we conclude that $H(\mathbf{w}) = H_1(\mathbf{w})$. Hence $H = H_1$. Thus $H = L^{-1}$ and L is invertible.

Conversely, let L be invertible; that is, $L \circ L^{-1} = I_W$ and $L^{-1} \circ L = I_V$. We show that L is one-to-one and onto. Suppose that $L(\mathbf{v}_1) = L(\mathbf{v}_2)$ for $\mathbf{v}_1$, $\mathbf{v}_2$ in V. Then $L^{-1}(L(\mathbf{v}_1)) = L^{-1}(L(\mathbf{v}_2))$, so $\mathbf{v}_1 = \mathbf{v}_2$, which means that L is one-to-one. Also, if $\mathbf{w}$ is a vector in W, then $L(L^{-1}(\mathbf{w})) = \mathbf{w}$, so if we let $L^{-1}(\mathbf{w}) = \mathbf{v}$, then $L(\mathbf{v}) = \mathbf{w}$. Thus L is onto.

We now show that L^{-1} is a linear transformation. Let $\mathbf{w}_1$, $\mathbf{w}_2$ be in W, where $L(\mathbf{v}_1) = \mathbf{w}_1$ and $L(\mathbf{v}_2) = \mathbf{w}_2$ for $\mathbf{v}_1$, $\mathbf{v}_2$ in V. Then since

$$L(a\mathbf{v}_1 + b\mathbf{v}_2) = aL(\mathbf{v}_1) + bL(\mathbf{v}_2) = a\mathbf{w}_1 + b\mathbf{w}_2 \qquad \text{for } a,\ b \text{ real numbers,}$$

we have

$$L^{-1}(a\mathbf{v}_1 + b\mathbf{v}_2) = a\mathbf{v}_1 + b\mathbf{v}_2 = aL^{-1}(\mathbf{w}_1) + bL^{-1}(\mathbf{w}_2).$$

which implies that L^{-1} is a linear transformation.

Finally, since $L \circ L^{-1} = I_W$, $L^{-1} \circ L = I_V$, and inverses are unique, we conclude that $(L^{-1})^{-1} = L$. ■

Example 9. Consider the linear transformation $L: R^3 \to R^3$ defined by

$$L\left(\begin{bmatrix} a_1 \\ a_2 \\ a_3 \end{bmatrix}\right) = \begin{bmatrix} 1 & 1 & 1 \\ 2 & 2 & 1 \\ 0 & 1 & 1 \end{bmatrix} \begin{bmatrix} a_1 \\ a_2 \\ a_3 \end{bmatrix}.$$

Since $\ker L = \{\mathbf{0}\}$ (verify), L is one-to-one and by Corollary 4.1 it is also onto, so it is invertible. To obtain L^{-1}, we let

$$L(\mathbf{v}) = L\left(\begin{bmatrix} a_1 \\ a_2 \\ a_3 \end{bmatrix}\right) = \mathbf{w} = \begin{bmatrix} b_1 \\ b_2 \\ b_3 \end{bmatrix}.$$

We are then solving the linear system

$$\begin{aligned} a_1 + a_2 + a_3 &= b_1 \\ 2a_1 + 2a_2 + a_3 &= b_2 \\ a_2 + a_3 &= b_3 \end{aligned}$$

for a_1, a_2, and a_3. We find that

$$L^{-1}\left(\begin{bmatrix} b_1 \\ b_2 \\ b_3 \end{bmatrix}\right) = \begin{bmatrix} 1 & 1 & 1 \\ 2 & 2 & 1 \\ 0 & 1 & 1 \end{bmatrix}^{-1} \begin{bmatrix} b_1 \\ b_2 \\ b_3 \end{bmatrix} = \begin{bmatrix} 1 & 0 & -1 \\ -2 & 1 & 1 \\ 2 & -1 & 0 \end{bmatrix} \begin{bmatrix} b_1 \\ b_2 \\ b_3 \end{bmatrix}$$

$$= \begin{bmatrix} b_1 - b_3 \\ -2b_1 + b_2 + b_3 \\ 2b_1 - b_2 \end{bmatrix}.$$

■

The following useful theorem shows that one-to-one linear transformations preserve linear independence of a set of vectors. Moreover, if this property holds, then L is one-to-one.

Theorem 4.7. *A linear transformation $L: V \to W$ is one-to-one if and only if the image of every linearly independent set of vectors in V is a linearly independent set of vectors in W.*

Proof: Let $S = \{\mathbf{v}_1, \mathbf{v}_2, \ldots, \mathbf{v}_k\}$ be a linearly independent set of vectors in V and let $T = \{L(\mathbf{v}_1), L(\mathbf{v}_2), \ldots, L(\mathbf{v}_k)\}$. Suppose that L is one-to-one; we show that T is linearly independent. Let

$$a_1L(\mathbf{v}_1) + a_2L(\mathbf{v}_2) + \cdots + a_kL(\mathbf{v}_k) = \mathbf{0}_W,$$

where $a_1, a_2, \ldots, a_k$ are real numbers. Then

$$L(a_1\mathbf{v}_1 + a_2\mathbf{v}_2 + \cdots + a_k\mathbf{v}_k) = \mathbf{0}_W = L(\mathbf{0}_V).$$

Since L is one-to-one, we conclude that

$$a_1\mathbf{v}_1 + a_2\mathbf{v}_2 + \cdots + a_k\mathbf{v}_k = \mathbf{0}_V.$$

Now S is linearly independent, so $a_1 = a_2 = \cdots = a_k = 0$. Hence T is linearly independent.

Conversely, suppose that the image of any linearly independent set of vectors in V is a linearly independent set of vectors in W. Now $\{\mathbf{v}\}$, where $\mathbf{v} \neq \mathbf{0}_V$, is a linearly independent set in V. Since the set $\{L(\mathbf{v})\}$ is linearly independent, $L(\mathbf{v}) \neq \mathbf{0}_W$, so $\ker L = \{\mathbf{0}_V\}$, which means that L is one-to-one. ■

It follows from this theorem that *if $L: V \to W$ is a linear transformation and* $\dim V = \dim W$, *then L is one-to-one, and thus invertible, if and only if the image of a basis for V under L is a basis for W* (see Exercise 14).

We now make one final remark for a linear system $A\mathbf{x} = \mathbf{b}$, where A is $n \times n$. We again consider the linear transformation $L: R^n \to R^n$ defined by $L(\mathbf{x}) = A\mathbf{x}$, for $\mathbf{x}$ in R^n. If A is a nonsingular matrix, then $\dim \text{range } L = \text{rank } A = n$, so $\dim \ker L = 0$. Thus L is one-to-one and hence onto. This means that the given linear system has a unique solution (of course, we already knew this result from other considerations). However, if A is singular, then $\text{rank } A < n$. This means that $\dim \ker L = n - \text{rank } A > 0$, so L is not one-to-one and not onto. Therefore, there exists a vector $\mathbf{b}$ in R^n, for which the system $A\mathbf{x} = \mathbf{b}$ has no solution. Moreover, since A is singular, $A\mathbf{x} = \mathbf{0}$ has a nontrivial solution $\mathbf{x}_0$. If $A\mathbf{x} = \mathbf{b}$ has a solution $\mathbf{y}$, then $\mathbf{x}_0 + \mathbf{y}$ is a solution to $A\mathbf{x} = \mathbf{b}$ (verify). Thus, for A singular, if a solution to $A\mathbf{x} = \mathbf{b}$ exists, then it is not unique.

The following statements are then equivalent:

1. A is nonsingular.
2. $A\mathbf{x} = \mathbf{0}$ has only the trivial solution.
3. A is row (column) equivalent to I_n.
4. The linear system $A\mathbf{x} = \mathbf{b}$ has a unique solution for every vector $\mathbf{b}$ in R^n.
5. A is a product of elementary matrices.
6. A has rank n.
7. The rows (columns) of A form a linearly independent set of vectors in R^n.
8. The dimension of the solution space of $A\mathbf{x} = \mathbf{0}$ is zero.
9. The linear transformation $L: R^n \to R^n$ defined by $L(\mathbf{x}) = A\mathbf{x}$, for $\mathbf{x}$ in R^n, is one-to-one and onto.

We can summarize the conditions under which a linear transformation L of an n-dimensional vector space V into itself (or more generally to an n-dimensional vector space W) is invertible by the following equivalent statements:

1. L is invertible.
2. L is one-to-one.
3. L is onto.

4.2. EXERCISES

1. Let $L: R^2 \to R^2$ be the linear transformation defined by $L\left(\begin{bmatrix} a_1 \\ a_2 \end{bmatrix}\right) = \begin{bmatrix} a_1 \\ 0 \end{bmatrix}$.

(a) Is $\begin{bmatrix} 0 \\ 2 \end{bmatrix}$ in ker L?

(b) Is $\begin{bmatrix} 2 \\ 2 \end{bmatrix}$ in ker L?

(c) Is $\begin{bmatrix} 3 \\ 0 \end{bmatrix}$ in range L?

(d) Is $\begin{bmatrix} 3 \\ 2 \end{bmatrix}$ in range L?

(e) Find ker L.

(f) Find range L.

2. Let $L: R^2 \to R^2$ be the linear transformation defined by $L\left(\begin{bmatrix} a_1 \\ a_2 \end{bmatrix}\right) = \begin{bmatrix} 1 & 2 \\ 2 & 4 \end{bmatrix}\begin{bmatrix} a_1 \\ a_2 \end{bmatrix}$.

(a) Is $\begin{bmatrix} 1 \\ 2 \end{bmatrix}$ in ker L?

(b) Is $\begin{bmatrix} 2 \\ -1 \end{bmatrix}$ in ker L?

(c) Is $\begin{bmatrix} 3 \\ 6 \end{bmatrix}$ in range L?

(d) Is $\begin{bmatrix} 2 \\ 3 \end{bmatrix}$ in range L?

(e) Find ker L.

(f) Find a set of vectors spanning range L.

3. Let $L: R_4 \to R_2$ be the linear transformation defined by

$$L([a_1 \quad a_2 \quad a_3 \quad a_4]) = [a_1 + a_3 \quad a_2 + a_4].$$

(a) Is $[2 \quad 3 \quad -2 \quad 3]$ in ker L?
(b) Is $[4 \quad -2 \quad -4 \quad 2]$ in ker L?
(c) Is $[1 \quad 2]$ in range L?
(d) Is $[0 \quad 0]$ in range L?
(e) Find ker L.
(f) Find a set of vectors spanning range L.

4. Let $L: R_2 \to R_3$ be the linear transformation defined by $L([a_1 \quad a_2]) = [a_1 \quad a_1 + a_2 \quad a_2]$.
(a) Find ker L. (b) Is L one-to-one?
(c) Is L onto?

5. Let $L: R_4 \to R_3$ be the linear transformation defined by

$$L([a_1 \quad a_2 \quad a_3 \quad a_4]) = [a_1 + a_2 \quad a_3 + a_4 \quad a_1 + a_3].$$

(a) Find a basis for ker L.
(b) What is dim ker L?
(c) Find a basis for range L.
(d) What is dim range L?

6. Let $L: P_2 \to P_3$ be the linear transformation defined by $L(p(t)) = t^2p'(t)$.
(a) Find a basis for and the dimension of ker L.
(b) Find a basis for and the dimension of range L.

7. Let $L: {}_2R_3 \to {}_3R_3$ be the linear transformation defined by

$$L(A) = \begin{bmatrix} 2 & -1 \\ 1 & 2 \\ 3 & 1 \end{bmatrix} A \qquad \text{for } A \text{ in } {}_2R_3.$$

(a) Find the dimension of ker L.
(b) Find the dimension of range L.

8. Let $L: V \to W$ be a linear transformation.
(a) Show that dim range $L \le$ dim V.
(b) Prove that if L is onto, then dim $W \le$ dim V.

9. Verify Theorem 4.5 for the following linear transformations.
(a) $L: P_2 \to P_2$ defined by $L(p(t)) = tp'(t)$.
(b) $L: R_3 \to R_2$ defined by $L([a_1 \quad a_2 \quad a_3]) = [a_1 + a_2 \quad a_1 + a_3]$.
(c) $L: {}_2R_2 \to {}_2R_3$ defined by

$$L(A) = A\begin{bmatrix} 1 & 2 & 3 \\ 2 & 1 & 3 \end{bmatrix}, \text{ for } A \text{ in } {}_2R_2.$$

10. Prove Corollary 4.1.

11. Let A be an $m \times n$ matrix, and consider the linear transformation $L: R^n \to R^m$ defined by $L(\mathbf{x}) = A\mathbf{x}$, for $\mathbf{x}$ in R^n. Show that

$$\text{range } L = \text{column space of } A.$$

12. Let $L: R^5 \to R^4$ be the linear transformation defined by

$$L\left(\begin{bmatrix} a_1 \\ a_2 \\ a_3 \\ a_4 \\ a_5 \end{bmatrix}\right) = \begin{bmatrix} 1 & 0 & -1 & 3 & -1 \\ 1 & 0 & 0 & 2 & -1 \\ 2 & 0 & -1 & 5 & -1 \\ 0 & 0 & -1 & 1 & 0 \end{bmatrix}\begin{bmatrix} a_1 \\ a_2 \\ a_3 \\ a_4 \\ a_5 \end{bmatrix}.$$

(a) Find a basis for and the dimension of ker L.
(b) Find a basis for and the dimension of range L.

13. Let $L: R_3 \to R_3$ be the linear transformation defined by $L(\mathbf{e}_1^T) = L([1 \quad 0 \quad 0]) = [3 \quad 0 \quad 0]$, $L(\mathbf{e}_2^T) = L([0 \quad 1 \quad 0]) = [1 \quad 1 \quad 1]$, and $L(\mathbf{e}_3^T) =$

$L([0\ \ 0\ \ 1]) = [2\ \ 1\ \ 1]$. Is the set $\{L(\mathbf{e}_1^T), L(\mathbf{e}_2^T), L(\mathbf{e}_3^T)\} = \{[3\ \ 0\ \ 0], [1\ \ 1\ \ 1], [2\ \ 1\ \ 1]\}$ a basis for R_3?

14. Let $L: V \to W$ be a linear transformation, and let $\dim V = \dim W$. Prove that L is invertible if and only if the image of a basis for V under L is a basis for W.

15. Let $L: R^3 \to R^3$ be defined by

$$L\left(\begin{bmatrix}1\\0\\0\end{bmatrix}\right) = \begin{bmatrix}1\\2\\3\end{bmatrix}, \quad L\left(\begin{bmatrix}0\\1\\0\end{bmatrix}\right) = \begin{bmatrix}0\\1\\1\end{bmatrix},$$

$$L\left(\begin{bmatrix}0\\0\\1\end{bmatrix}\right) = \begin{bmatrix}1\\1\\0\end{bmatrix}.$$

(a) Prove that L is invertible.

(b) Find $L^{-1}\left(\begin{bmatrix}2\\3\\4\end{bmatrix}\right)$.

16. Let $L: V \to W$ be a linear transformation, and let $S = \{\mathbf{v}_1, \mathbf{v}_2, \ldots, \mathbf{v}_n\}$ be a set of vectors in V. Prove that if $T = \{L(\mathbf{v}_1), L(\mathbf{v}_2), \ldots, L(\mathbf{v}_n)\}$ is linearly independent, then so is S. What can we say about the converse?

17. Find the dimension of the solution space for the following homogeneous system:

$$\begin{bmatrix}1 & 2 & 1 & 3\\2 & 1 & -1 & 2\\1 & 0 & 0 & -1\\4 & 1 & -1 & 0\end{bmatrix}\begin{bmatrix}x_1\\x_2\\x_3\\x_4\end{bmatrix} = \begin{bmatrix}0\\0\\0\\0\end{bmatrix}.$$

18. Find a linear transformation $L: R_2 \to R_3$ such that $S = \{[1\ \ -1\ \ 2], [3\ \ 1\ \ -1]\}$ is a basis for range L.

19. Let $L: R^3 \to R^3$ be the linear transformation defined by

$$L\left(\begin{bmatrix}a_1\\a_2\\a_3\end{bmatrix}\right) = \begin{bmatrix}1 & 1 & 1\\0 & 1 & 2\\1 & 2 & 2\end{bmatrix}\begin{bmatrix}a_1\\a_2\\a_3\end{bmatrix}.$$

(a) Prove that L is invertible.

(b) Find $L^{-1}\left(\begin{bmatrix}a_1\\a_2\\a_3\end{bmatrix}\right)$.

20. Let $L: V \to W$ be a linear transformation. Prove that L is one-to-one if and only if dim range $L = \dim V$.

21. Let $L: R^4 \to R^6$ be a linear transformation.
(a) If dim ker $L = 2$, what is dim range L?
(b) If dim range $L = 3$, what is dim ker L?

22. Let $L: V \to R^5$ be a linear transformation.
(a) If L is onto and dim ker $L = 2$, what is $\dim V$?
(b) If L is one-to-one and onto, what is $\dim V$?

23. Let L be the linear transformation defined in Exercise 20, Section 4.1. Prove or disprove:
(a) L is one-to-one.
(b) L is onto.

24. Let L be the linear transformation defined in Exercise 21, Section 4.1. Prove or disprove:
(a) L is one-to-one.
(b) L is onto.

4.3. *The Matrix of a Linear Transformation*

In Section 4.2 we saw that if A is an $m \times n$ matrix, then we can define a linear transformation $L: R^n \to R^m$ by $L(\mathbf{x}) = A\mathbf{x}$ for $\mathbf{x}$ in R^n. In Section 4.4 we shall prove that if A is an $m \times n$ matrix, V is an n-dimensional vector space with ordered basis S and W is an m-dimensional vector space with ordered basis T, then we can define a linear transformation $L: V \to W$ associated with A and the two bases. We shall now develop the converse notion; if $L: V \to W$ is a linear transformation of an n-dimensional vector space V into an m-dimensional vector space W, and if we choose ordered bases for V and W, then we can associate a unique $m \times n$ matrix A with L that will enable us to find $L(\mathbf{x})$ for $\mathbf{x}$ in V by merely performing matrix multiplication.

Theorem 4.8. *Let $L: V \to W$ be a linear transformation of an n-dimensional vector space V into an m-dimensional vector space W ($n \neq 0$, $m \neq 0$) and let $S = \{\mathbf{v}_1, \mathbf{v}_2, \ldots, \mathbf{v}_n\}$ and $T = \{\mathbf{w}_1, \mathbf{w}_2, \ldots, \mathbf{w}_m\}$ be ordered bases for V and W, respectively. Then the $m \times n$ matrix A whose jth column is the coordinate vector $[L(\mathbf{v}_j)]_T$ of $L(\mathbf{v}_j)$ with respect to T has the following property: If $\mathbf{y} = L(\mathbf{x})$ for some $\mathbf{x}$ in V, then*

$$[\mathbf{y}]_T = A[\mathbf{x}]_S, \tag{1}$$

where $[\mathbf{x}]_S$ and $[\mathbf{y}]_T$ are the coordinate vectors of $\mathbf{x}$ and $\mathbf{y}$ with respect to S and T, respectively. Moreover, A is the only matrix with this property.

Proof: We show how to construct the matrix A. Consider the vector $\mathbf{v}_j$ in V for $j = 1, 2, \ldots, n$. Then $L(\mathbf{v}_j)$ is a vector in W, and since T is an ordered basis for W, we can then express this vector as a linear combination of the vectors in T in a unique manner. Thus

$$L(\mathbf{v}_j) = c_{1j}\mathbf{w}_1 + c_{2j}\mathbf{w}_2 + \cdots + c_{mj}\mathbf{w}_m. \tag{2}$$

This means that the coordinate vector of $L(\mathbf{v}_j)$ with respect to T is

$$[L(\mathbf{v}_j)]_T = \begin{bmatrix} c_{1j} \\ c_{2j} \\ \vdots \\ c_{mj} \end{bmatrix}.$$

Recall from Section 2.5 that to find the coordinate vector $[L(\mathbf{v}_j)]_T$ we must solve a linear system. We now define an $m \times n$ matrix A by choosing $[L(\mathbf{v}_j)]_T$ as the jth column of A and show that this matrix satisfies the properties stated in the theorem.

Let $\mathbf{x}$ be any vector in V, and let $L[\mathbf{x}] = \mathbf{y}$, where $\mathbf{y}$ is in W. Now let

$$[\mathbf{x}]_S = \begin{bmatrix} a_1 \\ a_2 \\ \vdots \\ a_n \end{bmatrix} \quad \text{and} \quad [\mathbf{y}]_T = \begin{bmatrix} b_1 \\ b_2 \\ \vdots \\ b_m \end{bmatrix}.$$

This means that $\mathbf{x} = a_1\mathbf{v}_1 + a_2\mathbf{v}_2 + \cdots + a_n\mathbf{v}_n$. Then

$$\begin{aligned} L(\mathbf{x}) &= a_1L(\mathbf{v}_1) + a_2L(\mathbf{v}_2) + \cdots + a_nL(\mathbf{v}_n) \\ &= a_1(c_{11}\mathbf{w}_1 + c_{21}\mathbf{w}_2 + \cdots + c_{m1}\mathbf{w}_m) \\ &\quad + a_2(c_{12}\mathbf{w}_1 + c_{22}\mathbf{w}_2 + \cdots + c_{m2}\mathbf{w}_m) \\ &\quad + \cdots + a_n(c_{1n}\mathbf{w}_1 + c_{2n}\mathbf{w}_2 + \cdots + c_{mn}\mathbf{w}_m) \\ &= (c_{11}a_1 + c_{12}a_2 + \cdots + c_{1n}a_n)\mathbf{w}_1 + (c_{21}a_1 + c_{22}a_2 + \cdots + c_{2n}a_n)\mathbf{w}_2 \\ &\quad + \cdots + (c_{m1}a_1 + c_{m2}a_2 + \cdots + c_{mn}a_n)\mathbf{w}_m. \end{aligned}$$

Now $L(\mathbf{x}) = \mathbf{y} = b_1\mathbf{w}_1 + b_2\mathbf{w}_2 + \cdots + b_m\mathbf{w}_m$. Hence

$$b_i = c_{i1}a_1 + c_{i2}a_2 + \cdots + c_{in}a_n \qquad \text{for } i = 1, 2, \ldots, m.$$

Next, we verify Equation (1). We have

$$\begin{aligned} A[\mathbf{x}]_S &= \begin{bmatrix} c_{11} & c_{12} & \cdots & c_{1n} \\ c_{21} & c_{22} & \cdots & c_{2n} \\ \vdots & \vdots & & \vdots \\ c_{m1} & c_{m2} & \cdots & c_{mn} \end{bmatrix} \begin{bmatrix} a_1 \\ a_2 \\ \vdots \\ a_n \end{bmatrix} \\ &= \begin{bmatrix} c_{11}a_1 + c_{12}a_2 + \cdots + c_{1n}a_n \\ c_{21}a_1 + c_{22}a_2 + \cdots + c_{2n}a_n \\ \vdots \\ c_{m1}a_1 + c_{m2}a_2 + \cdots + c_{mn}a_n \end{bmatrix} = \begin{bmatrix} b_1 \\ b_2 \\ \vdots \\ b_m \end{bmatrix} = [\mathbf{y}]_T. \end{aligned}$$

Finally, we show that $A = [c_{ij}]$ is the only matrix with this property. Suppose that we have another matrix $A^* = [c^*_{ij}]$ with the same properties as A, and that $A^* \neq A$. All the elements of A and A^* cannot be equal, so say that the kth columns of these matrices are unequal. Now the coordinate vector of $\mathbf{v}_k$ with respect to the basis S is

$$[\mathbf{v}_k]_S = \begin{bmatrix} 0 \\ 0 \\ \vdots \\ 1 \\ 0 \\ \vdots \\ 0 \end{bmatrix} \leftarrow k\text{th row.}$$

Then

$$[L(\mathbf{v}_k)]_T = A \begin{bmatrix} 0 \\ 0 \\ \vdots \\ 1 \\ 0 \\ \vdots \\ 0 \end{bmatrix} = \begin{bmatrix} a_{1k} \\ a_{2k} \\ \vdots \\ a_{mk} \end{bmatrix} = \text{the } k\text{th column of } A$$

and

$$[L(\mathbf{v}_k)]_T = A^* \begin{bmatrix} 0 \\ 0 \\ \vdots \\ 1 \\ 0 \\ \vdots \\ 0 \end{bmatrix} = \begin{bmatrix} a^*_{1k} \\ a^*_{2k} \\ \vdots \\ a^*_{mk} \end{bmatrix} = \text{the } k\text{th column of } A^*.$$

This means that $L(\mathbf{v}_k)$ has two different coordinate vectors with respect to the same ordered basis, which is impossible. Hence the matrix A is unique. ■

We now summarize the procedure given in Theorem 4.8 for computing the matrix of a linear transformation $L: V \to W$ with respect to the ordered bases $S = \{\mathbf{v}_1, \mathbf{v}_2, \ldots, \mathbf{v}_n\}$ and $T = \{\mathbf{w}_1, \mathbf{w}_2, \ldots, \mathbf{w}_m\}$ for V and W, respectively.

STEP 1. Compute $L(\mathbf{v}_j)$ for $j = 1, 2, \ldots, n$.

STEP 2. Find the coordinate vector $[L(\mathbf{v}_j)]_T$ of $L(\mathbf{v}_j)$ with respect to T. This means that we have to express $L(\mathbf{v}_j)$ as a linear combination of the vectors in T [see Equation (2)] and this requires the solution of a linear system.

STEP 3. The matrix A of the linear transformation L with respect to the ordered bases S and T is formed by choosing for each j from 1 to n, $[L(\mathbf{v}_j)]_T$ as the jth column of A.

Example 1. Let $L: P_2 \rightarrow P_1$ be defined by $L(p(t)) = p'(t)$ and consider the ordered bases $S = \{t^2, t, 1\}$ and $T = \{t, 1\}$ for P_2 and P_1, respectively.
(a) Find the matrix A associated with L.
(b) If $p(t) = 5t^2 - 3t + 2$, compute $L(p(t))$ directly and using A.

Solution: (a) We have

$$L(t^2) = 2t = 2t + 0(1), \qquad \text{so } [L(t^2)]_T = \begin{bmatrix} 2 \\ 0 \end{bmatrix}.$$

$$L(t) = 1 = 0(t) + 1(1), \qquad \text{so } [L(t)]_T = \begin{bmatrix} 0 \\ 1 \end{bmatrix}.$$

$$L(1) = 0 = 0(t) + 0(1), \qquad \text{so } [L(1)]_T = \begin{bmatrix} 0 \\ 0 \end{bmatrix}.$$

In this case, the coordinates of $L(t^2)$, $L(t)$, and $L(1)$ with respect to the T-basis are obtained by observation since the T-basis is quite simple. Thus $A = \begin{bmatrix} 2 & 0 & 0 \\ 0 & 1 & 0 \end{bmatrix}$.

(b) Since $p(t) = 5t^2 - 3t + 2$, then $L(p(t)) = 10t - 3$. However, we can find $L(p(t))$ using the matrix A as follows. Since

$$[p(t)]_S = \begin{bmatrix} 5 \\ -3 \\ 2 \end{bmatrix},$$

then

$$[L(p(t))]_T = \begin{bmatrix} 2 & 0 & 0 \\ 0 & 1 & 0 \end{bmatrix} \begin{bmatrix} 5 \\ -3 \\ 2 \end{bmatrix} = \begin{bmatrix} 10 \\ -3 \end{bmatrix},$$

which means that $L(p(t)) = 10t - 3$. ■

Example 2. Let $L: P_2 \rightarrow P_1$ be defined as in Example 1 and consider the ordered bases $S = \{1, t, t^2\}$ and $T = \{t, 1\}$ for P_2 and P_1, respectively. We then find that the matrix A associated with L is $\begin{bmatrix} 0 & 0 & 2 \\ 0 & 1 & 0 \end{bmatrix}$ (verify). Notice that if we change the order of the vectors in S or T, the matrix may change. ■

Example 3. Let $L\colon P_2 \to P_1$ be defined as in Example 1 and consider the ordered bases $S = \{t^2, t, 1\}$ and $T = \{t+1, t-1\}$ for P_2 and P_1, respectively.
(a) Find the matrix A associated with L.
(b) If $p(t) = 5t^2 - 3t + 2$, compute $L(p(t))$.

Solution: (a) We have

$$L(t^2) = 2t.$$

To find the coordinates of $L(t^2)$ with respect to the T-basis, we form

$$L(t^2) = 2t = a_1(t+1) + a_2(t-1),$$

which leads to the linear system

$$\begin{aligned} a_1 + a_2 &= 2 \\ a_1 - a_2 &= 0, \end{aligned}$$

whose solution is $a_1 = 1$, $a_2 = 1$ (verify). Hence

$$[L(t^2)]_T = \begin{bmatrix} 1 \\ 1 \end{bmatrix}.$$

Similarly,

$$L(t) = 1 = \tfrac{1}{2}(t+1) - \tfrac{1}{2}(t-1), \qquad \text{so } [L(t)]_T = \begin{bmatrix} \tfrac{1}{2} \\ -\tfrac{1}{2} \end{bmatrix}.$$

$$L(1) = 0 = 0(t+1) + 0(t-1), \qquad \text{so } [L(1)]_T = \begin{bmatrix} 0 \\ 0 \end{bmatrix}.$$

Hence $A = \begin{bmatrix} 1 & \tfrac{1}{2} & 0 \\ 1 & -\tfrac{1}{2} & 0 \end{bmatrix}$.

(b) We have

$$[L(p(t))]_T = \begin{bmatrix} 1 & \frac{1}{2} & 0 \\ 1 & -\frac{1}{2} & 0 \end{bmatrix} \begin{bmatrix} 5 \\ -3 \\ 2 \end{bmatrix} = \begin{bmatrix} \frac{7}{2} \\ \frac{13}{2} \end{bmatrix},$$

so $L(p(t)) = \frac{7}{2}(t + 1) + \frac{13}{2}(t - 1) = 10t - 3$, which agrees with the result obtained in Example 1. ■

Notice that the matrices obtained in Examples 1, 2, and 3 are different even though L is the same in all three examples. In Section 4.5 we discuss the relationship between any two of these three matrices.

The matrix A is called the **representation of L with respect to the ordered bases S and T**. We also say that **A represents L with respect to S and T**. Having A enables us to replace L by A, $\mathbf{x}$ by $[\mathbf{x}]_S$, $\mathbf{y}$ by $[\mathbf{y}]_T$ in the expression $L(\mathbf{x}) = \mathbf{y}$ to get $A[\mathbf{x}]_S = [\mathbf{y}]_T$. Thus the result of applying L to $\mathbf{x}$ in V to obtain $\mathbf{y}$ in W can be obtained by multiplying the matrix A by the matrix $[\mathbf{x}]_S$. That is, we can work with matrices rather than with linear transformations. Physicists and others who deal at great length with linear transformations perform most of their computations with the matrix representations of the linear transformations. Of course, it is easier to work with matrices on a computer than with our abstract definition of a linear transformation. The relationship between linear transformations and matrices is a much stronger one than mere computational convenience. In the next section we shall show that the set of all linear transformations from an n-dimensional vector space V to an m-dimensional vector space W is a vector space that is isomorphic to the vector space of all $m \times n$ matrices.

We might also mention that if $L\colon R^n \to R^m$ is a linear transformation, then one often uses the natural bases for R^n and R^m, which simplifies the task of obtaining a representation of L.

Example 4. Let $L\colon R^3 \to R^2$ be defined by

$$L\left(\begin{bmatrix} x_1 \\ x_2 \\ x_3 \end{bmatrix}\right) = \begin{bmatrix} 1 & 1 & 1 \\ 1 & 2 & 3 \end{bmatrix} \begin{bmatrix} x_1 \\ x_2 \\ x_3 \end{bmatrix}.$$

Let

$$\mathbf{e}_1 = \begin{bmatrix} 1 \\ 0 \\ 0 \end{bmatrix}, \quad \mathbf{e}_2 = \begin{bmatrix} 0 \\ 1 \\ 0 \end{bmatrix}, \quad \mathbf{e}_3 = \begin{bmatrix} 0 \\ 0 \\ 1 \end{bmatrix},$$

$$\overline{\mathbf{e}}_1 = \begin{bmatrix} 1 \\ 0 \end{bmatrix}, \quad \text{and} \quad \overline{\mathbf{e}}_2 = \begin{bmatrix} 0 \\ 1 \end{bmatrix}.$$

Then $S = \{\mathbf{e}_1, \mathbf{e}_2, \mathbf{e}_3\}$ and $T = \{\overline{\mathbf{e}}_1, \overline{\mathbf{e}}_2\}$ are the natural bases for R^3 and R^2, respectively.

Now

$$L(\mathbf{e}_1) = \begin{bmatrix} 1 & 1 & 1 \\ 1 & 2 & 3 \end{bmatrix} \begin{bmatrix} 1 \\ 0 \\ 0 \end{bmatrix} = \begin{bmatrix} 1 \\ 1 \end{bmatrix} = 1\overline{\mathbf{e}}_1 + 1\overline{\mathbf{e}}_2, \qquad \text{so } [L(\mathbf{e}_1)]_T = \begin{bmatrix} 1 \\ 1 \end{bmatrix},$$

$$L(\mathbf{e}_2) = \begin{bmatrix} 1 & 1 & 1 \\ 1 & 2 & 3 \end{bmatrix} \begin{bmatrix} 0 \\ 1 \\ 0 \end{bmatrix} = \begin{bmatrix} 1 \\ 2 \end{bmatrix} = 1\overline{\mathbf{e}}_1 + 2\overline{\mathbf{e}}_2, \qquad \text{so } [L(\mathbf{e}_2)]_T = \begin{bmatrix} 1 \\ 2 \end{bmatrix},$$

$$L(\mathbf{e}_3) = \begin{bmatrix} 1 & 1 & 1 \\ 1 & 2 & 3 \end{bmatrix} \begin{bmatrix} 0 \\ 0 \\ 1 \end{bmatrix} = \begin{bmatrix} 1 \\ 3 \end{bmatrix} = 1\overline{\mathbf{e}}_1 + 3\overline{\mathbf{e}}_2, \qquad \text{so } [L(\mathbf{e}_3)_T = \begin{bmatrix} 1 \\ 3 \end{bmatrix}.$$

In this case, the coordinate vectors of $L(\mathbf{e}_1)$, $L(\mathbf{e}_2)$, and $L(\mathbf{e}_3)$ with respect to the T-basis are easily obtained because T is the natural basis for R^2. Then the representation of L with respect to S and T is

$$A = \begin{bmatrix} 1 & 1 & 1 \\ 1 & 2 & 3 \end{bmatrix}.$$

The reason that A is the same matrix as the one involved in the definition of L is that the natural bases are being used for R^3 and R^2. ■

Example 5. Let L: $R^3 \to R^2$ be defined as in Example 4 and consider the ordered bases

$$S = \left\{ \begin{bmatrix} 1 \\ 1 \\ 0 \end{bmatrix}, \begin{bmatrix} 0 \\ 1 \\ 1 \end{bmatrix}, \begin{bmatrix} 0 \\ 0 \\ 1 \end{bmatrix} \right\} \quad \text{and} \quad T = \left\{ \begin{bmatrix} 1 \\ 2 \end{bmatrix}, \begin{bmatrix} 1 \\ 3 \end{bmatrix} \right\}$$

for R^3 and R^2, respectively. Then

$$L\left(\begin{bmatrix} 1 \\ 1 \\ 0 \end{bmatrix}\right) = \begin{bmatrix} 1 & 1 & 1 \\ 1 & 2 & 3 \end{bmatrix} \begin{bmatrix} 1 \\ 1 \\ 0 \end{bmatrix} = \begin{bmatrix} 2 \\ 3 \end{bmatrix}.$$

Similarly,

$$L\left(\begin{bmatrix} 0 \\ 1 \\ 1 \end{bmatrix}\right) = \begin{bmatrix} 2 \\ 5 \end{bmatrix} \quad \text{and} \quad L\left(\begin{bmatrix} 0 \\ 0 \\ 1 \end{bmatrix}\right) = \begin{bmatrix} 1 \\ 3 \end{bmatrix}.$$

To determine the coordinates of the images of the S-basis, we must solve the three linear systems

$$a_1\begin{bmatrix} 1 \\ 2 \end{bmatrix} + a_2\begin{bmatrix} 1 \\ 3 \end{bmatrix} = \mathbf{b},$$

where $\mathbf{b} = \begin{bmatrix} 2 \\ 3 \end{bmatrix}, \begin{bmatrix} 2 \\ 5 \end{bmatrix}$, and $\begin{bmatrix} 1 \\ 3 \end{bmatrix}$. This can be done simultaneously as in Section 2.5, by transforming the partitioned matrix

$$\left[\begin{array}{cc:c:c:c} 1 & 1 & 2 & 2 & 1 \\ 2 & 3 & 3 & 5 & 3 \end{array}\right]$$

to reduced row echelon form, obtaining (verify)

$$\left[\begin{array}{cc:c:c:c} 1 & 0 & 3 & 1 & 0 \\ 0 & 1 & -1 & 1 & 1 \end{array}\right].$$

The last three columns of this matrix are the desired coordinate vectors of the S-basis with respect to the T-basis. That is, the last three columns form the matrix A representing L with respect to S and T. Thus

$$A = \begin{bmatrix} 3 & 1 & 0 \\ -1 & 1 & 1 \end{bmatrix}.$$

This matrix, of course, differs from the one that defined L. Thus, although a matrix A may be involved in the definition of a linear transformation L, we cannot conclude that it is necessarily the representation of L that we seek. ■

From Example 5 we see that if $L: R^n \to R^m$ is a linear transformation, then a computationally efficient way to obtain a matrix representation A of L with respect to the ordered bases $S = \{\mathbf{v}_1, \mathbf{v}_2, \ldots, \mathbf{v}_n\}$ for R^n and $T = \{\mathbf{w}_1, \mathbf{w}_2, \ldots, \mathbf{w}_m\}$ for R^m is to proceed as follows: Transform the partitioned matrix

$$[\mathbf{w}_1 \quad \mathbf{w}_2 \quad \cdots \quad \mathbf{w}_m \mid L(\mathbf{v}_1) \mid L(\mathbf{v}_2) \mid \cdots \mid L(\mathbf{v}_n)]$$

to reduced row echelon form. The matrix A consists of the last n columns of this last matrix.

If $L: V \to V$ is a linear operator on an n-dimensional space V, then to obtain a representation of L we fix ordered bases S and T for V, and obtain a matrix A representing L with respect to S and T. However, it is often convenient in this case to choose $S = T$. To avoid verbosity in this case, we refer to A as the **representation of L with respect to S.**

It is, of course, clear that the matrix of the identity operator (see Exercise 18 in Section 4.1) on an n-dimensional space, with respect to any basis, is I_n.

If $L: V \to V$ is an invertible linear operator and if A is the representation of L with respect to an ordered basis S for V, then A^{-1} is the representation of L^{-1} with respect to S. This fact, which can be proved directly at this point, will follow almost trivially in Section 4.4.

Suppose that $L: V \to W$ is a linear transformation and that A is the matrix representing L with respect to ordered bases for V and W. Then the problem of finding ker L reduces to the problem of finding the solution space of $A\mathbf{x} = \mathbf{0}$. Moreover, the problem of finding range L reduces to the problem of finding the column space of A.

We can summarize the conditions under which a linear transformation L of an n-dimensional vector space V into itself (or more generally into an n-dimensional vector space W) is invertible by the following equivalent statements:

1. L is invertible.
2. L is one-to-one.
3. L is onto.
4. The matrix A representing L with respect to ordered bases S and T for V and W is nonsingular.

4.3. EXERCISES

1. Let $L: R^2 \rightarrow R^2$ be defined by

$$L\left(\begin{bmatrix} x_1 \\ x_2 \end{bmatrix}\right) = \begin{bmatrix} x_1 + 2x_2 \\ 2x_1 - x_2 \end{bmatrix}.$$

Let S be the natural basis for R^2 and let $T = \left\{\begin{bmatrix} -1 \\ 2 \end{bmatrix}, \begin{bmatrix} 2 \\ 0 \end{bmatrix}\right\}$. Find the representation of L with respect to:

(a) S. (b) S and T. (c) T and S. (d) T.

(e) Find $L\left(\begin{bmatrix} 1 \\ 2 \end{bmatrix}\right)$ using the definition of L and also using the matrices obtained in parts (a) through (d).

2. Let $L: R_4 \rightarrow R_3$ be defined by

$$L([x_1 \quad x_2 \quad x_3 \quad x_4]) = [x_1 \quad x_2 + x_3 \quad x_3 + x_4].$$

Let S and T be the natural bases for R_4 and R_3, respectively. Let $S' = \{[1 \quad 0 \quad 0 \quad 1], [0 \quad 0 \quad 0 \quad 1], [1 \quad 1 \quad 0 \quad 0], [0 \quad 1 \quad 1 \quad 0]\}$ and $T' = \{[1 \quad 1 \quad 0], [0 \quad 1 \quad 0], [1 \quad 0 \quad 1]\}$.

(a) Find the representation of L with respect to S and T.

(b) Find the representation of L with respect to S' and T'.

(c) Find $L([2 \quad 1 \quad -1 \quad 3])$ using the matrices obtained in parts (a) and (b) and compare this answer with that obtained from the definition for L.

3. Let $L: R^4 \rightarrow R^3$ be defined by

$$L\left(\begin{bmatrix} x_1 \\ x_2 \\ x_3 \\ x_4 \end{bmatrix}\right) = \begin{bmatrix} 1 & 0 & 1 & 1 \\ 0 & 1 & 2 & 1 \\ -1 & -2 & 1 & 0 \end{bmatrix} \begin{bmatrix} x_1 \\ x_2 \\ x_3 \\ x_4 \end{bmatrix}.$$

Let S and T be the natural bases for R^4 and R^3, respectively, and consider the ordered bases

$$S' = \left\{\begin{bmatrix} 1 \\ 1 \\ 0 \\ 0 \end{bmatrix}, \begin{bmatrix} 0 \\ 1 \\ 0 \\ 0 \end{bmatrix}, \begin{bmatrix} 0 \\ 0 \\ 1 \\ 1 \end{bmatrix}, \begin{bmatrix} 0 \\ 1 \\ 1 \\ 0 \end{bmatrix}\right\} \quad \text{and}$$

$$T' = \left\{\begin{bmatrix} 1 \\ 0 \\ 1 \end{bmatrix}, \begin{bmatrix} 0 \\ 1 \\ 1 \end{bmatrix}, \begin{bmatrix} 0 \\ 0 \\ 1 \end{bmatrix}\right\}$$

for R^4 and R^3, respectively. Find the representation of L with respect to (a) S and T; (b) S' and T'.

4. Let $L: R^2 \rightarrow R^2$ be the linear transformation rotating R^2 counterclockwise through an angle ϕ. Find the representation of L with respect to the natural basis for R^2.

5. Let $L: R^3 \rightarrow R^3$ be defined by

$$L\left(\begin{bmatrix} 1 \\ 0 \\ 0 \end{bmatrix}\right) = \begin{bmatrix} 1 \\ 1 \\ 0 \end{bmatrix}, \quad L\left(\begin{bmatrix} 0 \\ 1 \\ 0 \end{bmatrix}\right) = \begin{bmatrix} 2 \\ 0 \\ 1 \end{bmatrix},$$

$$L\left(\begin{bmatrix} 0 \\ 0 \\ 1 \end{bmatrix}\right) = \begin{bmatrix} 1 \\ 0 \\ 1 \end{bmatrix}.$$

(a) Find the representation of L with respect to the natural basis S for R^3.

(b) Find $L\left(\begin{bmatrix} 1 \\ 2 \\ 3 \end{bmatrix}\right)$ using the definition of L and also using the matrix obtained in part (a).

6. Let $L: R^3 \rightarrow R^3$ be defined as in Exercise 5. Let $T = \{L(\mathbf{e}_1), L(\mathbf{e}_2), L(\mathbf{e}_3)\}$ be an ordered basis for R^3, and let S be the natural basis for R^3.

(a) Find the representation of L with respect to S and T.

(b) Find $L\left(\begin{bmatrix}1\\2\\3\end{bmatrix}\right)$ using the matrix obtained in part (a).

7. Let $L: R^3 \to R^3$ be the linear transformation represented by the matrix $\begin{bmatrix}1 & 3 & 1\\1 & 2 & 0\\0 & 1 & 1\end{bmatrix}$ with respect to the natural basis for R^3. Find:

(a) $L\left(\begin{bmatrix}1\\2\\3\end{bmatrix}\right)$. (b) $L\left(\begin{bmatrix}0\\1\\1\end{bmatrix}\right)$.

8. Let $L: {}_2R_2 \to {}_2R_2$ be defined by $L(A) = \begin{bmatrix}1 & 2\\3 & 4\end{bmatrix}A$ for A in ${}_2R_2$. Consider the ordered bases

$$S = \left\{\begin{bmatrix}1 & 0\\0 & 0\end{bmatrix}, \begin{bmatrix}0 & 1\\0 & 0\end{bmatrix}, \begin{bmatrix}0 & 0\\1 & 0\end{bmatrix}, \begin{bmatrix}0 & 0\\0 & 1\end{bmatrix}\right\}$$

and

$$T = \left\{\begin{bmatrix}1 & 0\\0 & 1\end{bmatrix}, \begin{bmatrix}1 & 1\\0 & 0\end{bmatrix}, \begin{bmatrix}1 & 0\\1 & 0\end{bmatrix}, \begin{bmatrix}0 & 1\\0 & 0\end{bmatrix}\right\}$$

for ${}_2R_2$. Find the representation of L with respect to:

(a) S. (b) T. (c) S and T. (d) T and S.

9. Let V be the vector space with basis $S = \{1, t, e^t, te^t\}$, and let $L: V \to V$ be a linear operator defined by $L(f) = f' = df/dt$. Find the representation of L with respect to S.

10. Let $L: P_1 \to P_2$ be defined by $L(p(t)) = tp(t) + p(0)$. Consider the ordered bases $S = \{t, 1\}$ and $S' = \{t + 1, t - 1\}$ for P_1, and $T = \{t^2, t, 1\}$ and $T' = \{t^2 + 1, t - 1, t + 1\}$ for P_2. Find the representation of L with respect to:

(a) S and T. (b) S' and T'.

(c) Find $L(-3t - 3)$ using the definition of L and the matrices obtained in parts (a) and (b).

11. Let $A = \begin{bmatrix}1 & 2\\3 & 4\end{bmatrix}$, and let $L: {}_2R_2 \to {}_2R_2$ be the linear transformation defined by $L(X) = AX - XA$ for X in ${}_2R_2$. Let S and T be the ordered bases for ${}_2R_2$ defined in Exercise 8. Find the representation of L with respect to:

(a) S. (b) T. (c) S and T. (d) T and S.

12. Let $L: V \to V$ be a linear operator. A nontrivial subspace U of V is called **invariant** under L if $L(U)$ is contained in U. Let L be a linear operator with invariant subspace U. Show that if $\dim U = m$, and $\dim V = n$, then L has a representation with respect to a basis S for V of the form

$\begin{bmatrix}A & B\\O & C\end{bmatrix}$, where A is $m \times m$, B is $m \times (n - m)$, O is the zero $(n - m) \times m$ matrix, and C is $(n - m) \times (n - m)$.

13. Let $L: R^2 \to R^2$ be defined by

$$L\left(\begin{bmatrix}x_1\\x_2\end{bmatrix}\right) = \begin{bmatrix}x_1\\-x_2\end{bmatrix},$$

reflection about the x-axis. Consider the natural basis S and the ordered basis

$T = \left\{\begin{bmatrix}1\\1\end{bmatrix}, \begin{bmatrix}-1\\1\end{bmatrix}\right\}$, for R^2. Find the representation of L with respect to:

(a) S. (b) T. (c) S and T. (d) T and S.

14. If $L: R_3 \to R_2$ is the linear transformation whose representation with respect to the natural bases for R_3 and R_2 is $\begin{bmatrix}1 & -1 & 2\\2 & 1 & 3\end{bmatrix}$, find:

(a) $L([1 \;\; 2 \;\; 3])$. (b) $L([-1 \;\; 2 \;\; -1])$.
(c) $L([0 \;\; 1 \;\; 2])$. (d) $L([0 \;\; 1 \;\; 0])$.
(e) $L([0 \;\; 0 \;\; 1])$.

15. If $O: V \to W$ is the zero linear transformation, show that the representation of L with respect to any ordered bases for V and W is the $m \times n$ zero matrix, where $n = \dim V$, $m = \dim W$.

16. If $I: V \to V$ is the identity linear operator on V defined by $I(\mathbf{v}) = \mathbf{v}$ for $\mathbf{v}$ in V prove that the matrix of I with respect to any ordered basis S for V is I_n, where $\dim V = n$.

17. Let $I: R_2 \to R_2$ be the identity linear operator on R_2. Let $S = \{[1 \quad 0], [0 \quad 1]\}$ and $T = \{[1 \quad -1], [2 \quad 3]\}$ be ordered bases for R_2. Find the representation of I with respect to:
(a) S. (b) T. (c) S and T. (d) T and S.

18. Let V be the vector space of continuous functions with basis $S = \{e^t, e^{-t}\}$. Find the representation of the linear operator $L: V \to V$ defined by $L(f) = f'$ with respect to S.

19. Let V be the vector space of continuous functions with ordered basis $S = \{\sin t, \cos t\}$. Find the representation of the linear operator $L: V \to V$ defined by $L(f) = f'$ with respect to S.

20. Let V be the vector space of continuous functions with ordered basis $S = \{\sin t, \cos t\}$ and consider $T = \{\sin t - \cos t, \sin t + \cos t\}$, another ordered basis for V. Find the representation of the linear operator $L: V \to V$ defined by $L(f) = f'$ with respect to:
(a) S. (b) T. (c) S and T. (d) T and S.

21. Let $L: V \to V$ be a linear operator defined by $L(\mathbf{v}) = c\mathbf{v}$, where c is a fixed constant. Prove that the representation of L with respect to any ordered basis for V is a scalar matrix (see Section 1.4).

22. Let the representation of $L: R^3 \to R^2$ with respect to the ordered bases $S = \{\mathbf{v}_1, \mathbf{v}_2, \mathbf{v}_3\}$ and $T = \{\mathbf{w}_1, \mathbf{w}_2\}$ be

$$A = \begin{bmatrix} 1 & 2 & 1 \\ -1 & 1 & 0 \end{bmatrix},$$

where

$$\mathbf{v}_1 = \begin{bmatrix} -1 \\ 1 \\ 0 \end{bmatrix}, \quad \mathbf{v}_2 = \begin{bmatrix} 0 \\ 1 \\ 1 \end{bmatrix}, \quad \mathbf{v}_3 = \begin{bmatrix} 1 \\ 0 \\ 0 \end{bmatrix},$$

$$\mathbf{w}_1 = \begin{bmatrix} 1 \\ 2 \end{bmatrix}, \quad \text{and} \quad \mathbf{w}_2 = \begin{bmatrix} 1 \\ -1 \end{bmatrix}.$$

(a) Compute $[L(\mathbf{v}_1)]_T$, $[L(\mathbf{v}_2)]_T$, and $[L(\mathbf{v}_3)]_T$.
(b) Compute $L(\mathbf{v}_1)$, $L(\mathbf{v}_2)$, and $L(\mathbf{v}_3)$.
(c) Compute $L\left(\begin{bmatrix} 2 \\ 1 \\ -1 \end{bmatrix}\right)$.

23. The results in this section have implications for computer graphics (see also Exercises 24 and 25 in Section 4.1): namely, that elementary graphics operations such as dilations, contractions, reflections, and rotations can be performed by matrix multiplication using appropriate matrices. For R^2 determine the matrices for these linear transformations and experiment with them using the graphics in your software.

24. For Exercise 25 in Section 4.1, rephrase parts (a), (b), and (c) in terms of matrix questions. Use your software to determine if the answers to these questions are true or false.

4.4. *The Vector Space of Matrices and the Vector Space of Linear Transformations (Optional)*

We have already seen in Section 2.2 that the set ${}_mR_n$ of all $m \times n$ matrices is a vector space under the operations of matrix addition and scalar multiplication.

We now show in this section that the set of all linear transformations of an n-dimensional vector space V into an m-dimensional vector space W is also a vector space U under two suitably defined operations, and we shall examine the relation between U and ${}_mR_n$.

Definition 4.5. Let V and W be two vector spaces of dimensions n and m, respectively. Also let $L_1: V \to W$ and $L_2: V \to W$ be linear transformations. We define a mapping $L: V \to W$ by $L(\mathbf{x}) = L_1(\mathbf{x}) + L_2(\mathbf{x})$, for $\mathbf{x}$ in V. Of course, the $+$ here is vector addition in W. We shall denote L by $L_1 \boxplus L_2$ and call it the **sum** of L_1 and L_2. Also, if $L: V \to W$ is a linear transformation and c is a real number, we define a mapping $H: V \to W$ by $H(\mathbf{x}) = cL(\mathbf{x})$ for $\mathbf{x}$ in V. Of course, the operation on the right-hand side is scalar multiplication in W. We denote H by $c \boxdot L$ and call it the **scalar multiple** of L by c. ■

Example 1. Let $V = R_3$ and $W = R_2$. Let $L_1: R_3 \to R_2$ and $L_2: R_3 \to R_2$ be defined by

$$L_1(\mathbf{x}) = L_1([a_1 \quad a_2 \quad a_3]) = [a_1 + a_2 \quad a_2 + a_3]$$

and

$$L_2(\mathbf{x}) = L_2([a_1 \quad a_2 \quad a_3]) = [a_1 + a_3 \quad a_2].$$

Then

$$(L_1 \boxplus L_2)(\mathbf{x}) = [2a_1 + a_2 + a_3 \quad 2a_2 + a_3]$$

and

$$(3 \boxdot L_1)(\mathbf{x}) = [3a_1 + 3a_2 \quad 3a_2 + 3a_3].$$ ■

We leave it to the reader (see the exercises in this section) to verify that if L, L_1, and L_2 are linear transformations of V into W and if c is a real number, then $L_1 \boxplus L_2$ and $c \boxdot L$ are linear transformations. We also let the reader show that the set U of all linear transformations of V into W is a vector space under the operations $\boxplus$ and $\boxdot$. The linear transformation $O: V \to W$ defined by $O(\mathbf{x}) = \mathbf{0}_W$ for $\mathbf{x}$ in V is the zero vector in U. That is, $L \boxplus O = O \boxplus L = L$ for any L in U. Also, if L is in U, then $L \boxplus (-1 \boxdot L) = O$, so we may write $(-1) \boxdot L$ as $-L$. Of course, to say that $S = \{L_1, L_2, \ldots, L_k\}$ is a linearly dependent set in U merely means that there exist k scalars $a_1, a_2, \ldots, a_k$ not all zero such that

$$(a_1 \boxdot L_1) \boxplus (a_2 \boxdot L_2) \boxplus \cdots \boxplus (a_k \boxdot L_k) = O,$$

where O is the zero linear transformation.

Example 2. Let L_1: $R_2 \rightarrow R_3$, L_2: $R_2 \rightarrow R_3$, and L_3: $R_2 \rightarrow R_3$ be defined by

$$\begin{aligned} L_1([a_1 \quad a_2]) &= [a_1 + a_2 \quad 2a_1 \quad a_2], \\ L_2([a_1 \quad a_2]) &= [a_2 - a_1 \quad 2a_1 + a_2 \quad a_1], \\ L_3([a_1 \quad a_2]) &= [3a_1 \quad -2a_2 \quad a_1 + 2a_2]. \end{aligned}$$

Determine whether $S = \{L_1, L_2, L_3\}$ is linearly independent.

Solution: Suppose that

$$(a_1 \boxdot L_1) \boxplus (a_2 \boxdot L_2) \boxplus (a_3 \boxdot L_3) = O,$$

where a_1, a_2, and a_3 are real numbers. Then for $\mathbf{e}_1^T = [1 \quad 0]$, we have

$$(a_1 \boxdot L_1) \boxplus (a_2 \boxdot L_2) \boxplus (a_3 \boxdot L_3)(\mathbf{e}_1^T) = O(\mathbf{e}_1^T) = [0 \quad 0 \quad 0],$$

so

$$\begin{aligned} a_1L_1(\mathbf{e}_1^T) + a_2L_2(\mathbf{e}_1^T) + a_3L_3(\mathbf{e}_1^T) &= a_1[1 \quad 2 \quad 0] + a_2[-1 \quad 2 \quad 1] \\ &\quad + a_3[3 \quad 0 \quad 1] = [0 \quad 0 \quad 0]. \end{aligned}$$

Thus we must solve the homogeneous system

$$\begin{bmatrix} 1 & -1 & 3 \\ 2 & 2 & 0 \\ 0 & 1 & 1 \end{bmatrix} \begin{bmatrix} a_1 \\ a_2 \\ a_3 \end{bmatrix} = \begin{bmatrix} 0 \\ 0 \\ 0 \end{bmatrix},$$

obtaining $a_1 = a_2 = a_3 = 0$ (verify). Hence S is linearly independent. ■

Theorem 4.9. *Let U be the vector space of all linear transformations of an n-dimensional vector space V into an m-dimensional vector space W, $n \neq 0$ and $m \neq 0$, under the operations $\boxplus$ and $\boxdot$. Then U is isomorphic to the vector space ${}_mR_n$ of all $m \times n$ matrices.*

Proof: Let $S = \{\mathbf{v}_1, \mathbf{v}_2, \ldots, \mathbf{v}_n\}$ and $T = \{\mathbf{w}_1, \mathbf{w}_2, \ldots, \mathbf{w}_m\}$ be ordered bases for V and W, respectively. We define a function $M: U \to {}_mR_n$ by letting $M(L)$ = the matrix representing L with respect to the bases S and T. We now show that M is an isomorphism.

First, M is one-to-one, for if L_1 and L_2 are two different elements in U, then $L_1(\mathbf{v}_j) \neq L_2(\mathbf{v}_j)$ for some $j = 1, 2, \ldots, n$. This means that the jth columns of $M(L_1)$ and $M(L_2)$, which are the coordinate vectors of $L_1(\mathbf{v}_j)$ and $L_2(\mathbf{v}_j)$, respectively, with respect to T, are different, so $M(L_1) \neq M(L_2)$. Hence M is one-to-one.

Next, M is onto. Let $A = [a_{ij}]$ be a given $m \times n$ matrix; that is, A is an element of ${}_mR_n$. Then we define a function $L: V \to W$ by

$$L(\mathbf{v}_i) = \sum_{k=1}^{m} a_{ki}\mathbf{w}_k, \qquad i = 1, 2, \ldots, n,$$

and if $\mathbf{x} = c_1\mathbf{v}_1 + c_2\mathbf{v}_2 + \cdots + c_n\mathbf{v}_n$, we define $L(\mathbf{x})$ by

$$L(\mathbf{x}) = \sum_{i=1}^{n} c_i L(\mathbf{v}_i).$$

It is easy to show that L is a linear transformation; moreover, the matrix representing L with respect to S and T is $A = [a_{ij}]$ (verify). Thus $M(L) = A$, and so M is onto.

Now let $M(L_1) = A = [a_{ij}]$ and $M(L_2) = B = [b_{ij}]$. We show that $M(L_1 \boxplus L_2) = A + B$. First, note that the jth column of $M(L_1 \boxplus L_2)$ is

$$[(L_1 \boxplus L_2)(\mathbf{v}_j)]_T = [L_1(\mathbf{v}_j) + L_2(\mathbf{v}_j)]_T = [L_1(\mathbf{v}_j)]_T + [L_2(\mathbf{v}_j)]_T.$$

Thus the jth column of $M(L_1 \boxplus L_2)$ is the sum of the jth columns of $M(L_1) = A$ and $M(L_2) = B$. Hence $M(L_1 \boxplus L_2) = A + B$.

Finally, let $M(L) = A$, and c = a real number. Following the idea in the above paragraph, we can show that $M(c \boxdot L) = cA$ (verify). Hence U and ${}_nR_m$ are isomorphic. ■

This theorem implies that the dimension of U is mn, for dim ${}_mR_n = mn$. Also, it means that when dealing with finite-dimensional vector spaces, we can always replace all linear transformations by their matrix representations and work only with the matrices. Moreover, it should be noted again that matrices lend themselves much more readily than linear transformations to computer implementations.

Example 3. Let $A = \begin{bmatrix} 1 & 2 & -1 \\ 2 & -1 & 3 \end{bmatrix}$, and $S = \{\mathbf{e}_1, \mathbf{e}_2, \mathbf{e}_3\}$ and $T = \{\overline{\mathbf{e}}_1, \overline{\mathbf{e}}_2\}$ be the natural bases for R^3 and R^2, respectively.

(a) Find the unique linear transformation $L: R^3 \to R^2$ whose representation with respect to S and T is A.

(b) Let

$$S' = \left\{ \begin{bmatrix} 1 \\ 0 \\ 1 \end{bmatrix}, \begin{bmatrix} 1 \\ 1 \\ 0 \end{bmatrix}, \begin{bmatrix} 0 \\ 1 \\ 1 \end{bmatrix} \right\} \quad \text{and} \quad T' = \left\{ \begin{bmatrix} 1 \\ 3 \end{bmatrix}, \begin{bmatrix} 2 \\ -1 \end{bmatrix} \right\}$$

be ordered bases for R^3 and R^2, respectively. Determine the linear transformation $L: R^3 \to R^2$, whose representation with respect to S' and T' is A.

(c) Compute $L\left(\begin{bmatrix} 1 \\ 2 \\ 3 \end{bmatrix}\right)$ using L as determined in part (b).

Solution: (a) Let

$$L(\mathbf{e}_1) = 1\overline{\mathbf{e}}_1 + 2\overline{\mathbf{e}}_2 = \begin{bmatrix} 1 \\ 2 \end{bmatrix},$$

$$L(\mathbf{e}_2) = 2\overline{\mathbf{e}}_1 - 1\overline{\mathbf{e}}_2 = \begin{bmatrix} 2 \\ -1 \end{bmatrix},$$

$$L(\mathbf{e}_3) = -\overline{\mathbf{e}}_1 + 3\overline{\mathbf{e}}_2 = \begin{bmatrix} -1 \\ 3 \end{bmatrix}.$$

Now if $\mathbf{x} = \begin{bmatrix} a_1 \\ a_2 \\ a_3 \end{bmatrix}$ is in R^3, we define $L(\mathbf{x})$ by

$$\begin{aligned} L(\mathbf{x}) &= L(a_1\mathbf{e}_1 + a_2\mathbf{e}_2 + a_3\mathbf{e}_3) \\ &= a_1L(\mathbf{e}_1) + a_2L(\mathbf{e}_2) + a_3L(\mathbf{e}_3) \\ &= a_1\begin{bmatrix} 1 \\ 2 \end{bmatrix} + a_2\begin{bmatrix} 2 \\ -1 \end{bmatrix} + a_3\begin{bmatrix} -1 \\ 3 \end{bmatrix} \\ &= \begin{bmatrix} a_1 + 2a_2 - a_3 \\ 2a_1 - a_2 + 3a_3 \end{bmatrix}. \end{aligned}$$

Note that $L(\mathbf{x}) = \begin{bmatrix} 1 & 2 & -1 \\ 2 & -1 & 3 \end{bmatrix} \begin{bmatrix} a_1 \\ a_2 \\ a_3 \end{bmatrix}$, so we could have defined L by $L(\mathbf{x}) = A\mathbf{x}$ for $\mathbf{x}$ in R^3. We can do this when the bases S and T are the natural bases.

(b) Let

$$L\left(\begin{bmatrix} 1 \\ 0 \\ 1 \end{bmatrix}\right) = 1\begin{bmatrix} 1 \\ 3 \end{bmatrix} + 2\begin{bmatrix} 2 \\ -1 \end{bmatrix} = \begin{bmatrix} 5 \\ 1 \end{bmatrix},$$

$$L\left(\begin{bmatrix} 1 \\ 1 \\ 0 \end{bmatrix}\right) = 2\begin{bmatrix} 1 \\ 3 \end{bmatrix} - 1\begin{bmatrix} 2 \\ -1 \end{bmatrix} = \begin{bmatrix} 0 \\ 7 \end{bmatrix},$$

$$L\left(\begin{bmatrix} 0 \\ 1 \\ 1 \end{bmatrix}\right) = -1\begin{bmatrix} 1 \\ 3 \end{bmatrix} + 3\begin{bmatrix} 2 \\ -1 \end{bmatrix} = \begin{bmatrix} 5 \\ -6 \end{bmatrix}.$$

Then if $\mathbf{x} = \begin{bmatrix} a_1 \\ a_2 \\ a_3 \end{bmatrix}$, we express $\mathbf{x}$ in terms of the basis S' as

$$\mathbf{x} = b_1\begin{bmatrix} 1 \\ 0 \\ 1 \end{bmatrix} + b_2\begin{bmatrix} 1 \\ 1 \\ 0 \end{bmatrix} + b_3\begin{bmatrix} 0 \\ 1 \\ 1 \end{bmatrix}, \text{ hence } [\mathbf{x}]_{S'} = \begin{bmatrix} b_1 \\ b_2 \\ b_3 \end{bmatrix}.$$

Define $L(\mathbf{x})$ by

$$\begin{aligned} L(\mathbf{x}) &= b_1 L\left(\begin{bmatrix} 1 \\ 0 \\ 1 \end{bmatrix}\right) + b_2 L\left(\begin{bmatrix} 1 \\ 1 \\ 0 \end{bmatrix}\right) + b_3 L\left(\begin{bmatrix} 0 \\ 1 \\ 1 \end{bmatrix}\right) \\ &= b_1\begin{bmatrix} 5 \\ 1 \end{bmatrix} + b_2\begin{bmatrix} 0 \\ 7 \end{bmatrix} + b_3\begin{bmatrix} 5 \\ -6 \end{bmatrix}, \end{aligned}$$

so

$$L(\mathbf{x}) = \begin{bmatrix} 5b_1 + 5b_3 \\ b_1 + 7b_2 - 6b_3 \end{bmatrix}. \tag{1}$$

(c) To find $L\left(\begin{bmatrix}1\\2\\3\end{bmatrix}\right)$, we first have (verify)

$$\begin{bmatrix}1\\2\\3\end{bmatrix} = 1\begin{bmatrix}1\\0\\1\end{bmatrix} + 0\begin{bmatrix}1\\1\\0\end{bmatrix} + 2\begin{bmatrix}0\\1\\1\end{bmatrix}, \text{ hence } \left[\begin{bmatrix}1\\2\\3\end{bmatrix}\right]_{S'} = \begin{bmatrix}1\\0\\2\end{bmatrix}.$$

Then using $b_1 = 1$, $b_2 = 0$, and $b_3 = 2$ in Equation (1), we obtain

$$L\left(\begin{bmatrix}1\\2\\3\end{bmatrix}\right) = \begin{bmatrix}15\\-11\end{bmatrix}.$$

■

The linear transformations obtained in Example 3 depend on the ordered bases for R^2 and R^3. Thus if L is as in part (a), then

$$L\left(\begin{bmatrix}1\\2\\3\end{bmatrix}\right) = \begin{bmatrix}2\\9\end{bmatrix},$$

which differs from the answer obtained in part (c), since the linear transformation in part (b) differs from that in part (a).

Now let V_1 be an n-dimensional vector space, V_2 an m-dimensional vector space, and V_3 a p-dimensional vector space. Let $L_1: V_1 \rightarrow V_2$ and $L_2: V_2 \rightarrow V_3$ be linear transformations. We can define the composite function

$$L_2 \circ L_1: V_1 \rightarrow V_3 \qquad \text{by} \qquad (L_2 \circ L_1)(\mathbf{x}) = L_2(L_1(\mathbf{x}))$$

for $\mathbf{x}$ in V_1. Then it is easy to see that $L_2 \circ L_1$ is a linear transformation. If $L: V \rightarrow V$, then $L \circ L$ is written as L^2.

Example 4. Let $L_1: R^2 \rightarrow R^2$ and $L_2: R^2 \rightarrow R^2$ be defined by

$$L_1\left(\begin{bmatrix}a_1\\a_2\end{bmatrix}\right) = \begin{bmatrix}a_1\\-a_2\end{bmatrix} \quad \text{and} \quad L_2\left(\begin{bmatrix}a_1\\a_2\end{bmatrix}\right) = \begin{bmatrix}a_2\\a_1\end{bmatrix}.$$

Then

$$(L_1 \circ L_2)\left(\begin{bmatrix} a_1 \\ a_2 \end{bmatrix}\right) = L_1\left(\begin{bmatrix} a_2 \\ a_1 \end{bmatrix}\right) = \begin{bmatrix} a_2 \\ -a_1 \end{bmatrix},$$

while

$$(L_2 \circ L_1)\left(\begin{bmatrix} a_1 \\ a_2 \end{bmatrix}\right) = L_2\left(\begin{bmatrix} a_1 \\ -a_2 \end{bmatrix}\right) = \begin{bmatrix} -a_2 \\ a_1 \end{bmatrix}.$$

Thus $L_1 \circ L_2 \neq L_2 \circ L_1$. ■

Theorem 4.10. *Let V_1 be an n-dimensional vector space, V_2 an m-dimensional vector space, and V_3 a p-dimensional vector space with linear transformations L_1 and L_2 such that L_1: $V_1 \rightarrow V_2$ and L_2: $V_2 \rightarrow V_3$. If ordered bases P, S, and T are chosen for V_1, V_2, and V_3, respectively, then $M(L_2 \circ L_1) = M(L_2)M(L_1)$.*

Proof: Let $M(L_1) = A$, with respect to the P and S ordered bases for V_1 and V_2, respectively, and let $M(L_2) = B$, with respect to the S and T ordered bases for V_2 and V_3, respectively. For any vector $\mathbf{x}$ in V_1, $[L_1(\mathbf{x})]_S = A[\mathbf{x}]_P$ and for any vector $\mathbf{y}$ in V_2, $[L_2(\mathbf{y})]_T = B[\mathbf{y}]_S$. Then it follows that

$$\begin{aligned} [(L_2 \circ L_1)(\mathbf{x})]_T &= [L_2(L_1(\mathbf{x}))]_T = B[L_1(\mathbf{x})]_S \\ &= B(A[\mathbf{x}]_P) = (BA)[\mathbf{x}]_P. \end{aligned}$$

Since a linear transformation has a unique representation with respect to ordered bases (see Theorem 4.8), we have $M(L_2 \circ L_1) = BA$ and it follows that $M(L_2 \circ L_1) = M(L_2)M(L_1)$. ■

Since AB need not equal BA for matrices A and B, it is thus not surprising that $L_1 \circ L_2$ need not be the same linear transformation as $L_2 \circ L_1$.

If L: $V \rightarrow V$ is an invertible linear operator and if A is a representation of L with respect to an ordered basis S for V, then the representation of the identity operator $L \circ L^{-1}$ is the identity matrix I_n. Thus $M(L)M(L^{-1}) = I_n$, which means that A^{-1} is the representation of L^{-1} with respect to S.

4.4. EXERCISES

1. Let L_1, L_2, and L be linear transformations of V into W. Prove:

(a) $L_1 \boxplus L_2$ is a linear transformation of V into W.

(b) If c is a real number, then $c \boxdot L$ is a linear transformation of V into W.

(c) If A represents L with respect to the ordered bases S and T for V and W, respectively, then cA represents $c \boxdot L$ with respect to S and T, where c = a real number.

2. Let U be the set of all linear transformations of V into W, and let $O: V \to W$ be the zero linear transformation defined by $O(\mathbf{x}) = \mathbf{0}_W$ for $\mathbf{x}$ in V.

(a) Prove that $O \boxplus L = L \boxplus O = L$ for any L in U.

(b) Show that if L is in U, then

$$L \boxplus ((-1) \boxdot L) = O.$$

3. Let $L_1: R_3 \to R_3$ and $L_2: R_3 \to R_3$ be linear transformations such that $L_1(\mathbf{e}_1^T) = [-1 \quad 2 \quad 1]$, $L_1(\mathbf{e}_2^T) = [0 \quad 1 \quad 2]$, $L_1(\mathbf{e}_3^T) = [-1 \quad 1 \quad 3]$, and $L_2(\mathbf{e}_1^T) = [0 \quad 1 \quad 3]$, $L_2(\mathbf{e}_2^T) = [4 \quad -2 \quad 1]$, $L_2(\mathbf{e}_3^T) = [0 \quad 2 \quad 2]$, where $S = \{\mathbf{e}_1^T, \mathbf{e}_2^T, \mathbf{e}_3^T\}$ is the natural basis for R_3. Find:

(a) $(L_1 \boxplus L_2)([a_1 \quad a_2 \quad a_3])$.

(b) $(L_1 \boxplus L_2)([2 \quad 1 \quad -3])$.

(c) The representation of $L_1 \boxplus L_2$ with respect to S.

(d) $(-2 \boxdot L_1)([a_1 \quad a_2 \quad a_3])$.

(e) $(-2 \boxdot L_1) \boxplus (4 \boxdot L_2)([2 \quad 1 \quad -3])$.

(f) The representation of $(-2 \boxdot L_1) \boxplus (4 \boxdot L_2)$ with respect to S.

4. Verify that the set U of all linear transformations of V into W is a vector space under the operations $\boxplus$ and $\boxdot$.

5. Let $L_i: R_3 \to R_3$ be a linear transformation defined by

$$L_i(\mathbf{x}) = A_i\mathbf{x}, \qquad i = 1, 2,$$

where

$$A_1 = \begin{bmatrix} 1 & 1 & 1 \\ 2 & 2 & 2 \\ 3 & 3 & 3 \end{bmatrix} \quad \text{and} \quad A_2 = \begin{bmatrix} 1 & 1 & 2 \\ 2 & 2 & 4 \\ -2 & -2 & -4 \end{bmatrix}.$$

(a) For $\mathbf{x} = [3 \quad -2 \quad 1]^T$, find $(L_1 \boxdot L_2)(\mathbf{x})$.

(b) Determine $\ker L_1$, $\ker L_2$, and $\ker L_1 \cap \ker L_2$ (intersection of kernels).

(c) Determine $\ker (L_1 \boxdot L_2)$.

(d) What is the relationship between $\ker L_1 \cap \ker L_2$ and $\ker (L_1 \boxdot L_2)$?

6. Let V_1, V_2, and V_3 be vector spaces of dimensions n, m, and p, respectively. Also let $L_1: V_1 \to V_2$ and $L_2: V_2 \to V_3$ be linear transformations. Prove that $L_2 \circ L_1: V_1 \to V_3$ is a linear transformation.

7. Let $L_1: P_1 \to P_2$ be the linear transformation defined by $L_1(p(t)) = tp(t)$ and let $L_2: P_2 \to P_3$ be the linear transformation defined by $L_2(p(t)) = t^2p'(t)$. Let $P = \{t + 1, t - 1\}$, $S = \{t^2, t - 1, t + 2\}$, and $T = \{t^3, t^2 - 1, t, t + 1\}$ be ordered bases for P_1, P_2, and P_3, respectively.

(a) Find the representation C of $L_2 \circ L_1$ with respect to P and T.

(b) Compute the representation A of L_1 with respect to P and S and the representation B of L_2 with respect to S and T. Verify that BA is the matrix C obtained in part (a).

8. Let L_1, L_2, and S be as in Exercise 3. Find:

(a) $(L_1 \circ L_2)([a_1 \quad a_2 \quad a_3])$,

(b) $(L_2 \circ L_1)([a_1 \quad a_2 \quad a_3])$.

(c) The representation of $L_1 \circ L_2$ with respect to S.

(d) The representation of $L_2 \circ L_1$ with respect to S.

9. If $\begin{bmatrix} 1 & 4 & -1 \\ 2 & 1 & 3 \\ 1 & -1 & 2 \end{bmatrix}$ is the representation of a linear operator $L: R^3 \to R^3$ with respect to ordered bases S and T for R^3, find the representation with respect to S and T of:
(a) $2 \boxdot L$.
(b) $2 \boxdot L \boxplus L \circ L$.

10. Let L_1, L_2, and L_3 be linear transformations of R_3 into R_2 defined by

$$L_1([a_1 \quad a_2 \quad a_3]) = [a_1 + a_2 \quad a_1 - a_3],$$
$$L_2([a_1 \quad a_2 \quad a_3]) = [a_1 - a_2 \quad a_3],$$
and $$L_3([a_1 \quad a_2 \quad a_3]) = [a_1 \quad a_2 + a_3].$$

Prove that $S = \{L_1, L_2, L_3\}$ is a linearly independent set in the vector space U of all linear transformations of R_3 into R_2.

11. Find the dimension of the vector space U of all linear transformations of V into W for each of the following.
(a) $V = R^2$, $W = R^3$.
(b) $V = P_2$, $W = P_1$.
(c) $V = {}_2R_1$, $W = {}_3R_2$.
(d) $V = R_3$, $W = R_4$.

12. Repeat Exercise 11 for each of the following.
(a) $V = W$ is the vector space with basis $\{\sin t, \cos t\}$.
(b) $V = W$ is the vector space with basis $\{1, t, e^t, te^t\}$.
(c) V is the vector space *spanned* by $\{1, t, 2t\}$, and W is the vector space with basis $\{t^2, t, 1\}$.

13. Let $A = [a_{ij}]$ be a given $m \times n$ matrix, and let V and W be given vector spaces of dimensions n and m, respectively. Let $S = \{\mathbf{v}_1, \mathbf{v}_2, \ldots, \mathbf{v}_n\}$ be an ordered basis for V and $T = \{\mathbf{w}_1, \mathbf{w}_2, \ldots, \mathbf{w}_m\}$ be an ordered basis for W. Define a function L: $V \to W$ by $L(\mathbf{v}_i) = \sum_{k=1}^{m} a_{ki}\mathbf{w}_k$, $i = 1, 2, \ldots, n$, and if $\mathbf{x} = c_1\mathbf{v}_1 + c_2\mathbf{v}_2 + \cdots + c_n\mathbf{v}_n$, we define $L(\mathbf{x})$ by $L(\mathbf{x}) = \sum_{i=1}^{n} c_i L(\mathbf{v}_i)$.
(a) Show that L is a linear transformation.
(b) Show that A represents L with respect to S and T.

14. Let $A = \begin{bmatrix} 1 & 2 & -2 \\ 3 & 4 & -1 \end{bmatrix}$. Let S be the natural basis for R^3 and T the natural basis for R^2.
(a) Find the linear transformation $L: R^3 \to R^2$ determined by A.
(b) Find $L\left(\begin{bmatrix} a_1 \\ a_2 \\ a_3 \end{bmatrix}\right)$. (c) Find $L\left(\begin{bmatrix} 1 \\ 2 \\ 3 \end{bmatrix}\right)$.

15. Let A be as in Exercise 14. Consider the ordered bases $S = \{t^2, t, 1\}$ and $T = \{t, 1\}$ for P_2 and P_1, respectively.
(a) Find the linear transformation $L: P_2 \to P_1$ determined by A.
(b) Find $L(at^2 + bt + c)$.
(c) Find $L(2t^2 - 5t + 4)$.

16. Find two linear transformations L_1: $R^2 \to R^2$ and L_2: $R^2 \to R^2$ such that $L_2 \circ L_1 \neq L_1 \circ L_2$.

17. Find a linear transformation L: $R^2 \to R^2$, $L \neq I$, the identity operator, such that $L^2 = L \circ L = I$.

18. Find a linear transformation L: $R^2 \to R^2$, $L \neq O$, the zero transformation, such that $L^2 = O$.

19. Find a linear transformation L: $R^2 \to R^2$, $L \neq I$, $L \neq O$, such that $L^2 = L$.

20. Let L: $R^3 \to R^3$ be the linear transformation defined in Exercise 15 of Section 4.2. Find the matrix representing L^{-1} with respect to the natural basis for R^3.

21. Let L: $R^3 \to R^3$ be the linear transformation defined in Exercise 19 of Section 4.2. Find the matrix representing L^{-1} with respect to the natural basis for R^3.

22. Let $L: R^3 \to R^3$ be the invertible linear transformation represented by $A = \begin{bmatrix} 2 & 0 & 4 \\ -1 & 1 & -2 \\ 2 & 3 & 3 \end{bmatrix}$ with respect to an ordered basis S for R^3. Find the representation of L^{-1} with respect to S.

23. Let $L: V \to V$ be a linear transformation represented by a matrix A with respect to an ordered basis S for V. Show that A^2 represents $L^2 = L \circ L$ with respect to S. Moreover, show that if k is a positive integer, then A^k represents $L^k = L \circ L \circ \cdots \circ L$ (k times) with respect to S.

24. Let $L: P_1 \to P_1$ be the invertible linear transformation represented by $A = \begin{bmatrix} 4 & 2 \\ 3 & -1 \end{bmatrix}$ with respect to an ordered basis S for P_1. Find the representation of L^{-1} with respect to S.

4.5. *Similarity*

In Section 4.3 we saw how the matrix representing a linear transformation of an n-dimensional vector space V into an m-dimensional vector space W depends upon the ordered bases we choose for V and W. We now see how this matrix changes when the bases for V and W are changed.

Theorem 4.11. *Let $L: V \to W$ be a linear transformation of an n-dimensional vector space V into an m-dimensional vector space W. Let $S = \{\mathbf{v}_1, \mathbf{v}_2, \ldots, \mathbf{v}_n\}$ and $S' = \{\mathbf{v}'_1, \mathbf{v}'_2, \ldots, \mathbf{v}'_n\}$ be ordered bases for V, with transition matrix P from S' to S; let $T = \{\mathbf{w}_1, \mathbf{w}_2, \ldots, \mathbf{w}_m\}$ and $T' = \{\mathbf{w}'_1, \mathbf{w}'_2, \ldots, \mathbf{w}'_m\}$ be ordered bases for W with transition matrix Q from T' to T. If A is the representation of L with respect to S and T, then $Q^{-1}AP$ is the representation of L with respect to S' and T'.*

Proof: Recalling Section 2.5, where the transition matrix was first introduced, if P is the transition matrix from S' to S, and $\mathbf{x}$ is a vector in V, then

$$[\mathbf{x}]_S = P[\mathbf{x}]_{S'}, \tag{1}$$

where the jth column of P is the coordinate vector $[\mathbf{v}'_j]_S$ of $\mathbf{v}'_j$ with respect to S. Similarly, if Q is the transition matrix from T' to T and $\mathbf{y}$ is a vector in W, then

$$[\mathbf{y}]_T = Q[\mathbf{y}]_{T'}, \tag{2}$$

where the jth column of Q is the coordinate vector $[\mathbf{w}'_j]_T$ of $\mathbf{w}'_j$ with respect to T. Now if A is the representation of L with respect to S and T, then

$$[L(\mathbf{x})]_T = A[\mathbf{x}]_S \tag{3}$$

for $\mathbf{x}$ in V. Substituting $\mathbf{y} = L(\mathbf{x})$ in (2), we have $[L(\mathbf{x})]_T = Q[L(\mathbf{x})]_{T'}$. Now using first (3) and then (1) in this last equation, we obtain

$$Q[L(\mathbf{x})]_{T'} = AP[\mathbf{x}]_{S'},$$

so

$$[L(\mathbf{x})]_{T'} = Q^{-1}AP[\mathbf{x}]_{S'}.$$

This means that $Q^{-1}AP$ is the representation of L with respect to S' and T'. ■

Theorem 4.11 can be illustrated by the diagram shown in Figure 4.5, where Q^{-1} is the transition matrix from the T-basis to the T'-basis. This figure shows that there are two ways of going from $\mathbf{x}$ in V to $L(\mathbf{x})$ in W: directly using matrix B, or indirectly using the matrices P, A, and Q^{-1}.

Figure 4.5 $B = Q^{-1}AP$

$$\begin{array}{ccccc} S'\text{-basis} & V & \xrightarrow[B \text{ represents } L]{L} & W & T'\text{-basis} \\ & P \downarrow & & \uparrow Q^{-1} & \\ S\text{-basis} & V & \xrightarrow[A \text{ represents } L]{L} & W & T\text{-basis} \end{array}$$

From Section 1.7 we see that *two representations of a linear transformation with respect to different pairs of bases are equivalent.*

Example 1. Let $L: R^3 \to R^2$ be defined by $L\left(\begin{bmatrix} a_1 \\ a_2 \\ a_3 \end{bmatrix}\right) = \begin{bmatrix} a_1 + a_3 \\ a_2 - a_3 \end{bmatrix}$. Consider the ordered bases

$$S = \left\{\begin{bmatrix} 1 \\ 0 \\ 0 \end{bmatrix}, \begin{bmatrix} 0 \\ 1 \\ 0 \end{bmatrix}, \begin{bmatrix} 0 \\ 0 \\ 1 \end{bmatrix}\right\} \quad \text{and} \quad S' = \left\{\begin{bmatrix} 1 \\ 1 \\ 0 \end{bmatrix}, \begin{bmatrix} 0 \\ 1 \\ 1 \end{bmatrix}, \begin{bmatrix} 0 \\ 0 \\ 1 \end{bmatrix}\right\}$$

for R^3, and

$$T = \left\{\begin{bmatrix} 1 \\ 0 \end{bmatrix}, \begin{bmatrix} 0 \\ 1 \end{bmatrix}\right\} \quad \text{and} \quad T' = \left\{\begin{bmatrix} 1 \\ 1 \end{bmatrix}, \begin{bmatrix} 1 \\ 3 \end{bmatrix}\right\}$$

for R^2. We can easily establish (verify) that $A = \begin{bmatrix} 1 & 0 & 1 \\ 0 & 1 & -1 \end{bmatrix}$ is the representation of L with respect to S and T.

The transition matrix P from S' to S is the matrix whose jth column is the coordinate vector of the jth vector in the basis S' with respect to S. Thus $P = \begin{bmatrix} 1 & 0 & 0 \\ 1 & 1 & 0 \\ 0 & 1 & 1 \end{bmatrix}$ and the transition matrix Q from T' to T is $Q = \begin{bmatrix} 1 & 1 \\ 1 & 3 \end{bmatrix}$. Now $Q^{-1} = \begin{bmatrix} \frac{3}{2} & -\frac{1}{2} \\ -\frac{1}{2} & \frac{1}{2} \end{bmatrix}$. (We could also obtain Q^{-1} as the transition matrix from T to T'.) Then the representation of L with respect to S' and T' is

$$B = Q^{-1}AP = \begin{bmatrix} 1 & \frac{3}{2} & 2 \\ 0 & -\frac{1}{2} & -1 \end{bmatrix}.$$

On the other hand, we can compute the representation of L with respect to S' and T' directly. We have

$$L\left(\begin{bmatrix} 1 \\ 1 \\ 0 \end{bmatrix}\right) = \begin{bmatrix} 1 \\ 1 \end{bmatrix} = 1\begin{bmatrix} 1 \\ 1 \end{bmatrix} + 0\begin{bmatrix} 1 \\ 3 \end{bmatrix}, \qquad \text{so} \left[L\left(\begin{bmatrix} 1 \\ 1 \\ 0 \end{bmatrix}\right)\right]_{T'} = \begin{bmatrix} 1 \\ 0 \end{bmatrix}.$$

$$L\left(\begin{bmatrix} 0 \\ 1 \\ 1 \end{bmatrix}\right) = \begin{bmatrix} 1 \\ 0 \end{bmatrix} = \frac{3}{2}\begin{bmatrix} 1 \\ 1 \end{bmatrix} - \frac{1}{2}\begin{bmatrix} 1 \\ 3 \end{bmatrix}, \qquad \text{so} \left[L\left(\begin{bmatrix} 0 \\ 1 \\ 1 \end{bmatrix}\right)\right]_{T'} = \begin{bmatrix} \frac{3}{2} \\ -\frac{1}{2} \end{bmatrix}.$$

$$L\left(\begin{bmatrix} 0 \\ 0 \\ 1 \end{bmatrix}\right) = \begin{bmatrix} 1 \\ -1 \end{bmatrix} = 2\begin{bmatrix} 1 \\ 1 \end{bmatrix} - 1\begin{bmatrix} 1 \\ 3 \end{bmatrix}, \qquad \text{so} \left[L\left(\begin{bmatrix} 0 \\ 0 \\ 1 \end{bmatrix}\right)\right]_{T'} = \begin{bmatrix} 2 \\ -1 \end{bmatrix}.$$

Then the representation of L with respect to S' and T' is

$$\begin{bmatrix} 1 & \frac{3}{2} & 2 \\ 0 & -\frac{1}{2} & -1 \end{bmatrix},$$

which agrees with our earlier result. ■

Taking $V = W$ in Theorem 4.10, we obtain an important result, which we state as Corollary 4.2.

Corollary 4.2. *Let $L: V \rightarrow V$ be a linear operator on an n-dimensional vector space V. Let*

$S = \{\mathbf{v}_1, \mathbf{v}_2, \ldots, \mathbf{v}_n\}$ and $S' = \{\mathbf{v}'_1, \mathbf{v}'_2, \ldots, \mathbf{v}'_n\}$ be ordered bases for V with transition matrix P from S' to S. If A is the representation of L with respect to S, then $P^{-1}AP$ is the representation of L with respect to S'. ■

We may define the **rank** of a linear transformation $L: V \rightarrow W$, rank L, as the rank of any matrix representing L. This definition makes sense, since if A and B represent L, then A and B are equivalent; by Section 1.7 we know that equivalent matrices have the same rank.

We can now restate Theorem 4.5 as follows. If $L: V \rightarrow W$ is a linear transformation, then

$$\text{nullity } L + \text{rank } L = \dim V.$$

We also speak of the **nullity** of an $m \times n$ matrix A. This is the nullity of the linear transformation $L: R^n \rightarrow R^m$ defined by $L(\mathbf{x}) = A\mathbf{x}$, for $\mathbf{x}$ in R^n (see Section 4.2). That is, nullity $A = \dim \ker L$.

Theorem 4.12. *Let $L: V \rightarrow W$ be a linear transformation. Then* rank $L = \dim$ range L.

Proof: Let $n = \dim V$, $m = \dim W$, and $r = \dim$ range L. Then, from Theorem 4.5, $\dim \ker L = n - r$. Let $\mathbf{v}_{r+1}, \mathbf{v}_{r+2}, \ldots, \mathbf{v}_n$ be a basis for ker L. By Theorem 2.9, there exist vectors $\mathbf{v}_1, \mathbf{v}_2, \ldots, \mathbf{v}_r$ in V such that $S = \{\mathbf{v}_1, \mathbf{v}_2, \ldots, \mathbf{v}_r, \mathbf{v}_{r+1}, \ldots, \mathbf{v}_n\}$ is a basis for V. The vectors $\mathbf{w}_1 = L(\mathbf{v}_1)$, $\mathbf{w}_2 = L(\mathbf{v}_2), \ldots, \mathbf{w}_r = L(\mathbf{v}_r)$ form a basis for range L (they clearly span range L and there are r of them, so Theorem 2.10 applies). Again by Theorem 2.9, there exist vectors $\mathbf{w}_{r+1}, \mathbf{w}_{r+2}, \ldots, \mathbf{w}_m$ in W such that $T = \{\mathbf{w}_1, \mathbf{w}_2, \ldots, \mathbf{w}_r, \mathbf{w}_{r+1}, \mathbf{w}_{r+2}, \ldots, \mathbf{w}_m\}$ is a basis for W. Let A denote the $m \times n$ matrix which represents L with respect to S and T. The columns of A are (Theorem 4.8)

$$[L(\mathbf{v}_i)]_T = [\mathbf{w}_i]_T = \mathbf{e}_i \qquad i = 1, 2, \ldots, r$$

and

$$[L(\mathbf{v}_j)]_T = [\mathbf{0}_W]_T = \mathbf{0}_{R^m} \qquad j = r + 1, r + 2, \ldots, n.$$

Hence

$$A = \begin{bmatrix} I_r & O \\ O & O \end{bmatrix}.$$

Therefore,

$$\text{rank } L = \text{rank } A = r = \dim \text{ range } L. \qquad \blacksquare$$

Definition 4.6. If A and B are $n \times n$ matrices, we say that B is **similar** to A if there is a nonsingular matrix P such that $B = P^{-1}AP$. ■

It is easy to show (Exercise 1) that B is similar to A if and only if A is similar to B. Thus we replace the statements "A is similar to B" and "B is similar to A" by "A and B are similar."

We then see that any two representations of a linear transformation $L: V \to V$ are similar. Conversely, let $A = [a_{ij}]$ and $B = [b_{ij}]$ be similar $n \times n$ matrices and let V be an n-dimensional vector space (we may take V as R^n). We wish to show that A and B represent the same linear transformation $L: V \to V$ with respect to different bases. Since A and B are similar, $B = P^{-1}AP$ for some nonsingular matrix $P = [p_{ij}]$. Let $S = \{\mathbf{v}_1, \mathbf{v}_2, \ldots, \mathbf{v}_n\}$ be an ordered basis for V; from the proof of Theorem 4.9, we know that there exists a linear transformation $L: V \to V$, which is represented by A with respect to S. Then

$$[L(\mathbf{x})]_S = A[\mathbf{x}]_S. \tag{4}$$

We wish to prove that B also represents L with respect to some basis for V. Let

$$\mathbf{w}_j = \sum_{i=1}^{n} p_{ij}\mathbf{v}_i. \tag{5}$$

We first show that $T = \{\mathbf{w}_1, \mathbf{w}_2, \ldots, \mathbf{w}_n\}$ is also a basis for V. Suppose that

$$a_1\mathbf{w}_1 + a_2\mathbf{w}_2 + \cdots + a_n\mathbf{w}_n = \mathbf{0}.$$

Then, from (5), we have

$$a_1\left(\sum_{i=1}^{n} p_{i1}\mathbf{v}_i\right) + a_2\left(\sum_{i=1}^{n} p_{i2}\mathbf{v}_i\right) + \cdots + a_n\left(\sum_{i=1}^{n} p_{in}\mathbf{v}_i\right) = \mathbf{0},$$

which can be rewritten as

$$\left(\sum_{j=1}^{n} p_{1j}a_j\right)\mathbf{v}_1 + \left(\sum_{j=1}^{n} p_{2j}a_j\right)\mathbf{v}_2 + \cdots + \left(\sum_{j=1}^{n} p_{nj}a_j\right)\mathbf{v}_n = \mathbf{0}. \tag{6}$$

Since S is linearly independent, each of the coefficients in (6) is zero. Thus

$$\sum_{j=1}^{n} p_{ij}a_j = 0 \qquad i = 1, 2, \ldots, n,$$

or equivalently,

$$P\mathbf{a} = 0,$$

where $\mathbf{a} = [a_1 \quad a_2 \quad \cdots \quad a_n]^T$. This is a homogeneous system of n equations in the n unknowns $a_1, a_2, \ldots, a_n$, whose coefficient matrix is P. Since P is nonsingular, the only solution is the trivial one. Hence $a_1 = a_2 = \cdots = a_n = 0$ and T is linearly independent. Moreover, Equation (5) implies that P is the transition matrix from T to S (see Section 2.5). Thus

$$[\mathbf{x}]_S = P[\mathbf{x}]_T. \tag{7}$$

Using (7), Equation (4), the matrix equation of L, becomes

$$P[L(\mathbf{x})]_T = AP[\mathbf{x}]_T \qquad \text{or} \qquad [L(\mathbf{x})]_T = P^{-1}AP[\mathbf{x}]_T,$$

which means that the representation of L with respect to T is $B = P^{-1}AP$. We can summarize these results in the following theorem.

Theorem 4.13. *Let V be any n-dimensional vector space and let A and B be any $n \times n$ matrices. Then A and B are similar if and only if A and B represent the same linear transformation $L\colon V \to V$ with respect to two ordered bases for V.* ■

Example 2. Let $L\colon R_3 \to R_3$ be defined by

$$L([a_1 \quad a_2 \quad a_3]) = [2a_1 - a_3 \qquad a_1 + a_2 - a_3 \qquad a_3].$$

Let $S = \{[1 \quad 0 \quad 0], [0 \quad 1 \quad 0], [0 \quad 0 \quad 1]\}$ be the natural basis for R_3. The representation of L with respect to S is $A = \begin{bmatrix} 2 & 0 & -1 \\ 1 & 1 & -1 \\ 0 & 0 & 1 \end{bmatrix}$. Now consider the ordered basis

$$S' = \{[1 \quad 0 \quad 1], [0 \quad 1 \quad 0], [1 \quad 1 \quad 0]\}$$

for R_3. The transition matrix P from S' to S is $P = \begin{bmatrix} 1 & 0 & 1 \\ 0 & 1 & 1 \\ 1 & 0 & 0 \end{bmatrix}$; moreover,

$$P^{-1} = \begin{bmatrix} 0 & 0 & 1 \\ -1 & 1 & 1 \\ 1 & 0 & -1 \end{bmatrix}.$$

Then the representation of L with respect to S' is

$$B = P^{-1}AP = \begin{bmatrix} 1 & 0 & 0 \\ 0 & 1 & 0 \\ 0 & 0 & 2 \end{bmatrix}.$$

The same result can be obtained directly (verify). The matrices A and B are similar. ■

Observe that the matrix B obtained in Example 2 is diagonal. We can now ask a number of related questions. First, given $L: V \rightarrow V$, when can we choose a basis S for V such that the representation of L with respect to S is diagonal? How do we choose such a basis? In Example 2 we apparently pulled our basis S' "out of the air." If we cannot choose a basis giving a representation of L that is diagonal, can we choose a basis giving a matrix that is close in appearance to a diagonal matrix? What do we gain from having such simple representations? First, we already know from Section 4.4 that if A represents $L: V \rightarrow V$ with respect to some ordered basis S for V, then A^k represents $L \circ L \circ \cdots \circ L = L^k$ with respect to S; now if A is similar to B, then B^k also represents L^k. Of course, if B is diagonal, then it is a trivial matter to compute B^k: the diagonal elements of B^k are those of B raised to the k power. We shall also find that if A is similar to a diagonal matrix, then we can easily solve a homogeneous linear system of differential equations with constant coefficients. The answers to these questions will be taken up in detail in Chapter 6.

Similar matrices enjoy some other nice properties. For example, if A and B are similar, then Tr (A) = Tr (B) (see Exercise 23 in Section 1.2 for a definition of trace). Also, if A and B are similar, then A^k and B^k are similar, if k is a positive integer. Proofs of these results are not difficult and are left as exercises.

We obtain one final result on similar matrices.

Theorem 4.14. *If A and B are similar $n \times n$ matrices, then* rank A = rank B.

Proof: We know from Theorem 4.13 that A and B represent the same linear transformation $L: R^n \to R^n$ with respect to different bases. Hence rank A = rank L = rank B. ■

4.5. EXERCISES

1. Let A, B, and C be square matrices. Show that:
(a) A is similar to A.
(b) If A is similar to B, then B is similar to A.
(c) If A is similar to B and B is similar to C, then A is similar to C.

2. Let L be the linear transformation defined in Exercise 2, Section 4.3.
(a) Find the transition matrix P from S' to S.
(b) Find the transition matrix from S to S' and verify that it is P^{-1}.
(c) Find the transition matrix Q from T' to T.
(d) Find the representation of L with respect to S' and T'.
(e) What is the dimension of range L?

3. Do Exercise 1(d) of Section 4.3 using transition matrices.

4. Do Exercise 8(b) of Section 4.3 using transition matrices.

5. Let $L: R^2 \to R^2$ be defined by

$$L\left(\begin{bmatrix} a_1 \\ a_2 \end{bmatrix}\right) = \begin{bmatrix} a_1 \\ -a_2 \end{bmatrix}.$$

(a) Find the representation of L with respect to the natural basis S for R^2.
(b) Find the representation of L with respect to the ordered basis

$$T = \left\{\begin{bmatrix} 1 \\ -1 \end{bmatrix}, \begin{bmatrix} 1 \\ 2 \end{bmatrix}\right\}.$$

(c) Verify that the matrices obtained in parts (a) and (b) are similar.
(d) Verify that the ranks of the matrices obtained in parts (a) and (b) are equal.

6. Show that if A and B are similar matrices, then A^k and B^k are similar for any positive integer k. (*Hint:* If $B = P^{-1}AP$, find $B^2 = BB$, and so on.)

7. Show that if A and B are similar, then A^T and B^T are similar.

8. Prove that if A and B are similar, then Tr (A) = Tr (B). (*Hint:* See Exercise 23 in Section 1.2 for a definition of trace.)

9. Let $L: R_3 \to R_2$ be the linear transformation whose representation is $A = \begin{bmatrix} 2 & -1 & 3 \\ 3 & 1 & 0 \end{bmatrix}$ with respect to the ordered bases

$$S = \{[1 \quad 0 \quad -1], [0 \quad 2 \quad 0], [1 \quad 2 \quad 3]\}$$

and

$$T = \{[1 \quad -1], [2 \quad 0]\}.$$

Find the representation of L with respect to the natural bases for R_3 and R_2.

10. Let $L: R^3 \to R^3$ be the linear transformation whose representation with respect to the natural basis for R^3 is $A = [a_{ij}]$. Let $P = \begin{bmatrix} 0 & 1 & 1 \\ 1 & 0 & 1 \\ 1 & 1 & 0 \end{bmatrix}$. Find a basis T for R^3 with respect to which $B = P^{-1}AP$ represents L. (*Hint:* See the solution of Example 2.)

11. Let A and B be similar. Prove:
(a) If A is nonsingular, then B is nonsingular.
(b) If A is nonsingular, then A^{-1} and B^{-1} are similar.

12. Do Exercise 13(b) of Section 4.3 using transition matrices.

13. Do Exercise 17(b) of Section 4.3 using transition matrices.

14. Do Exercise 10(b) of Section 4.3 using transition matrices.

15. Do Exercise 20(b) of Section 4.3 using transition matrices.

16. Prove that A and O_n are similar if and only if $A = O_n$.

Supplementary Exercises

1. For an $n \times n$ matrix A, the trace of A, $\text{Tr}(A)$, is defined as the sum of the diagonal entries of A (see Exercise 23 in Section 1.2). Prove that the trace defines a linear transformation from ${}_nR_n$ to the vector space of all real numbers.

2. Let L: ${}_nR_m \to {}_mR_n$ be the function defined by $L(A) = A^T$ (the transpose of A), for A in V. Is L a linear transformation? Justify your answer.

3. Let V be the vector space of all $n \times n$ matrices and let L: $V \to V$ be the function defined by

$$L(A) = \begin{cases} A^{-1} & \text{if } A \text{ is nonsingular} \\ O & \text{if } A \text{ is singular} \end{cases}$$

for A in V. Is L a linear transformation? Justify your answer.

4. Let L: $R_3 \to R_3$ be a linear transformation for which we know that

$$L([1 \quad 0 \quad 1]) = [1 \quad 2 \quad 3],$$
$$L([0 \quad 1 \quad 2]) = [1 \quad 0 \quad 0],$$
$$L([1 \quad 1 \quad 0]) = [1 \quad 0 \quad 1].$$

(a) What is $L([4 \quad 1 \quad 0])$?
(b) What is $L([0 \quad 0 \quad 0])$?
(c) What is $L([a_1 \quad a_2 \quad a_3])$?

5. Let L: $P_1 \to P_1$ be a linear transformation defined by

$$L(t - 1) = t + 2 \qquad \text{and} \qquad L(t + 1) = 2t + 1.$$

(a) What is $L(5t + 1)$?
(b) What is $L(at + b)$?

6. Let L: $P_2 \to P_2$ be the linear transformation defined by

$$L(at^2 + bt + c) = (a + c)t^2 + (b + c)t.$$

(a) Is $t^2 - t - 1$ in ker L?
(b) Is $t^2 + t - 1$ in ker L?
(c) Is $2t^2 - t$ in range L?
(d) Is $t^2 - t + 2$ in range L?
(e) Find a basis for ker L.
(f) Find a basis for range L.

7. Let L: $P_3 \to P_3$ be the linear transformation defined by

$$L(at^3 + bt^2 + ct + d) = (a - b)t^3 + (c - d)t.$$

(a) Is $t^3 + t^2 + t - 1$ in ker L?
(b) Is $t^3 - t^2 + t - 1$ in ker L?
(c) Is $3t^3 + t$ in range L?
(d) Is $3t^3 - t^2$ in range L?
(e) Find a basis for ker L.
(f) Find a basis for range L.

8. For vector space V and linear transformation L of Exercise 20 in Section 4.1, find a basis for ker L.

9. Let V be the vector space of continuous functions on $[0, 1]$ and let $L: V \to R$ be given by $L(f) = f(0)$, for f in V.
(a) Show that L is a linear transformation.
(b) Describe the kernel of L and give examples of polynomials, quotients of polynomials, and trigonometric functions that belong to ker L.
(c) If we redefine L by $L(f) = f(\frac{1}{2})$, is it still a linear transformation? Explain.

10. Let $L: P_1 \to R$ be the linear transformation defined by

$$L(p(t)) = \int_0^1 p(t)\, dt.$$

(a) Find a basis for ker L.
(b) Find a basis for range L.
(c) Verify Theorem 4.5 for L.

11. Let $L: P_2 \to P_2$ be the linear transformation defined by

$$L(at^2 + bt + c) = (a + 2c)t^2 + (b - c)t + (a - c).$$

Let $S = \{1, t, t^2\}$ and $T = \{t^2 - 1, t, t - 1\}$ be ordered bases for P_2.
(a) Find the matrix of L with respect to S and T.
(b) If $p(t) = 2t^2 - 3t + 1$, compute $L(p(t))$ using the matrix obtained in part (a).

12. Let $L: P_1 \to P_1$ be a linear transformation which is represented by the matrix

$$A = \begin{bmatrix} 2 & -3 \\ 1 & 2 \end{bmatrix}$$

with respect to the basis $S = \{p_1(t), p_2(t)\}$, where

$$p_1(t) = t - 2 \quad \text{and} \quad p_2(t) = t + 1.$$

(a) Compute $L(p_1(t))$ and $L(p_2(t))$.
(b) Compute $[L(p_1(t))]_S$ and $[L(p_2(t))]_S$.
(c) Compute $L(t + 2)$.

13. Let $L: P_3 \to P_3$ be defined by

$$L(at^3 + bt^2 + ct + d) = 3at^2 + 2bt + c.$$

Find the matrix of L with respect to the basis $S = \{t^3, t^2, t, 1\}$ for P_3.

14. Consider R^n as an inner product space with the standard inner product and let $L: R^n \to R^n$ be a linear transformation. Prove that for any vector $\mathbf{u}$ in R^n, $\|L(\mathbf{u})\| = \|\mathbf{u}\|$ if and only if $(L(\mathbf{u}), L(\mathbf{v})) = (\mathbf{u}, \mathbf{v})$ for any vectors $\mathbf{u}$ and $\mathbf{v}$ in R^n. Such a linear transformation is said to **preserve inner products**.

15. Let $L_1: V \to V$ and $L_2: V \to V$ be linear transformations on a vector space V. Prove that

$$(L_1 + L_2)^2 = L_1^2 + 2L_1 \circ L_2 + L_2^2$$

if and only if $L_1 \circ L_2 = L_2 \circ L_1$ (see Exercise 23 in Section 4.4).

16. Let $\mathbf{u}$ and $\mathbf{v}$ be nonzero vectors in R^n. In Section 3.3 we defined the angle between $\mathbf{u}$ and $\mathbf{v}$ to be the angle θ such that

$$\cos\theta = \frac{(\mathbf{u}, \mathbf{v})}{\|\mathbf{u}\|\,\|\mathbf{v}\|} \qquad 0 \le \theta \le \pi.$$

A linear transformation $L: R^n \to R^n$ is called **angle preserving** if the angle between $\mathbf{u}$ and $\mathbf{v}$ is the same as that between $L(\mathbf{u})$ and $L(\mathbf{v})$. Prove that if L is inner product preserving (see Exercise 14), then it is angle preserving.

17. Let $L: R^n \to R^n$ be a linear transformation that preserves inner products (see Exercise 14), and let the $n \times n$ matrix A represent L with respect to some ordered basis S for R^n.
(a) Prove that ker $L = \{\mathbf{0}\}$.
(b) Prove that $A^TA = I_n$. (*Hint:* Use Supplementary Exercise 15 in Chapter 3.)

5 Determinants

5.1. Definition

In Exercise 23, Section 1.2, we defined the trace of a square ($n \times n$) matrix $A = [a_{ij}]$ by $\text{Tr}(A) = \sum_{i=1}^{n} a_{ii}$. Another very important number associated with a square matrix A is the determinant of A, which we now define. Determinants first arose in the solution of linear systems. Although the methods given in Chapter 1 for solving such systems are more efficient than those involving determinants, determinants will be vital for our further study of a linear transformation $L: V \to V$. First, we deal briefly with permutations, which are used in our definition of determinant. Throughout this chapter, when we use the term "matrix" we mean "square matrix."

Definition 5.1. Let $S = \{1, 2, \ldots, n\}$ be the set of integers from 1 to n, arranged in ascending order. A rearrangement $j_1 j_2 \cdots j_n$ of the elements of S is called a **permutation** of S. Thus a permutation of S is a one-to-one mapping of S onto itself. ■

We can put any one of the n elements of S in first position, any one of the remaining $n - 1$ elements in second position, any one of the remaining $n - 2$ elements in third position, and so on until the nth position can only be filled by the last remaining element. Thus there are $n(n-1)(n-2) \cdots 2 \cdot 1 = n!$ permutations of S; we denote the set of all permutations of S by S_n.

Example 1. Let $S = \{1, 2, 3\}$. The set S_3 of all permutations of S consists of the $3! = 6$ permutations 123, 132, 213, 231, 312, and 321. ■

A permutation $j_1 j_2 \cdots j_n$ of S is said to have an **inversion** if a larger integer, j_r, precedes a smaller one, j_s. A permutation is called **even** if the total number of inversions in it is even, or **odd** if the total number of inversions in it is odd. If $n \geq 2$, there are $n!/2$ even and $n!/2$ odd permutations in S_n.

Example 2. S_1 has only $1! = 1$ permutation: 1, which is even because there are no inversions. ■

Example 3. S_2 has $2! = 2$ permutations: 12, which is even (no inversions), and 21, which is odd (one inversion). ■

Example 4. In the permutation 4312 in S_4, 4 precedes 3, 4 precedes 1, 4 precedes 2, 3 precedes 1, and 3 precedes 2. Thus the total number of inversions in this permutation is 5, and 4312 is odd. ■

Example 5. S_3 has $3! = 3 \cdot 2 \cdot 1 = 6$ permutations: 123, 231, and 312, which are even, and 132, 213, and 321, which are odd. ■

Definition 5.2. Let $A = [a_{ij}]$ be an $n \times n$ matrix. The determinant function, denoted by **det**, is defined by

$$\det(A) = \sum (\pm) a_{1j_1} a_{2j_2} \cdots a_{nj_n},$$

where the summation is over all permutations $j_1 j_2 \cdots j_n$ of the set $S = \{1, 2, \ldots, n\}$. The sign is taken as $+$ or $-$ according to whether the permutation $j_1 j_2 \cdots j_n$ is even or odd. ■

In each term $(\pm) a_{1j_1} a_{2j_2} \cdots a_{nj_n}$ of $\det(A)$, the row subscripts are in natural order while the column subscripts are in the order $j_1 j_2 \cdots j_n$. Thus each term in $\det(A)$, with its appropriate sign, is a product of n entries of A, with exactly one entry from each row and exactly one entry from each column. Since we sum over all permutations of S, $\det(A)$ has $n!$ terms in the sum.

Another notation for $\det(A)$ is $|A|$. We shall use both $\det(A)$ and $|A|$.

Example 6. If $A = [a_{11}]$ is a 1×1 matrix, then $\det(A) = a_{11}$. ■

Example 7. If $A = \begin{bmatrix} a_{11} & a_{12} \\ a_{21} & a_{22} \end{bmatrix}$, then to obtain det (A) we write down the terms $a_{1-}a_{2-}$, and replace the dashes with all possible elements of S_2: These are 12 and 21. Now 12 is an even permutation and 21 is an odd permutation. Thus

$$|A| = a_{11}a_{22} - a_{12}a_{21}.$$

Hence we see that det (A) can be obtained by forming the product of the entries on the line from left to right and subtracting from this number the product of the entries on the line from right to left:

$$\begin{matrix} a_{11} & a_{12}. \\ a_{21} & a_{22}. \end{matrix}$$

Thus, if $A = \begin{bmatrix} 2 & -3 \\ 4 & 5 \end{bmatrix}$, then $|A| = (2)(5) - (-3)(4) = 22$. ■

Example 8. If $A = \begin{bmatrix} a_{11} & a_{12} & a_{13} \\ a_{21} & a_{22} & a_{23} \\ a_{31} & a_{32} & a_{33} \end{bmatrix}$, then to compute $|A|$ we write down the six terms $a_{1-}a_{2-}a_{3-}$, $a_{1-}a_{2-}a_{3-}$, $a_{1-}a_{2-}a_{3-}$, $a_{1-}a_{2-}a_{3-}$, $a_{1-}a_{2-}a_{3-}$, $a_{1-}a_{2-}a_{3-}$. All the elements of S_3 are used to replace the dashes, and if we prefix each term by + or − according to whether the permutation is even or odd, we find that

$$\begin{aligned} |A| = {} & a_{11}a_{22}a_{33} + a_{12}a_{23}a_{31} + a_{13}a_{21}a_{32} - a_{11}a_{23}a_{32} \\ & -a_{12}a_{21}a_{33} - a_{13}a_{22}a_{31}. \end{aligned} \tag{1}$$

We can also obtain $|A|$ as follows. Repeat the first and second columns of A as shown below. Form the sum of the products of the entries on the lines from left to right, and subtract from this number the products of the entries on the lines from right to left (verify):

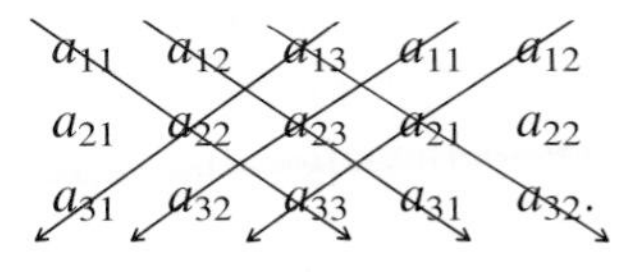

It should be emphasized that for $n \geq 4$, *there is no "easy" method for evaluating* det (A) *as in Examples* 7 *and* 8. ■

Example 9. Let

$$A = \begin{bmatrix} 1 & 2 & 3 \\ 2 & 1 & 3 \\ 3 & 1 & 2 \end{bmatrix}.$$

Evaluate $|A|$.

Solution: Substituting in (1), we find that

$$\begin{vmatrix} 1 & 2 & 3 \\ 2 & 1 & 3 \\ 3 & 1 & 2 \end{vmatrix} = (1)(1)(2) + (2)(3)(3) + (3)(2)(1) - (1)(3)(1) - (2)(2)(2) - (3)(1)(3) = 6.$$

We could obtain the same result by using the easy method illustrated above, as follows:

$$\begin{array}{ccccc} 1 & 2 & 3 & 1 & 2 \\ 2 & 1 & 3 & 2 & 1 \\ 3 & 1 & 2 & 3 & 1 \end{array}$$

$$|A| = (1)(1)(2) + (2)(3)(3) + (3)(2)(1) - (3)(1)(3) - (1)(3)(1) - (2)(2)(2) = 6.$$

■

It may already have struck the reader that Definition 5.2 is an extremely tedious way of computing determinants for a sizable value of n. In fact, $10! = 3.6288 \times 10^6$ and $20! = 2.4329 \times 10^{18}$, each an enormous number. In Section 5.2 we develop properties of determinants that will greatly reduce the computational effort.

Permutations are studied to some depth in abstract algebra courses and in courses dealing with group theory. As we just noted, we shall develop methods for evaluating determinants other than those involving permutations. However, we do require the following important property of permutations. If we interchange two numbers in the permutation $j_1j_2 \cdots j_n$, then the number of inversions is either increased or decreased by an odd number.

A proof of this fact can be given by first noting that if two adjacent numbers in the permutation $j_1j_2 \cdots j_n$ are interchanged, then the number of inversions is either increased or decreased by 1. Thus consider the permutations $j_1j_2 \cdots j_ej_f \cdots j_n$ and $j_1j_2 \cdots j_fj_e \cdots j_n$. If j_ej_f is an inversion, then j_fj_e is not an inversion and the second permutation has one fewer inversion than the first one; if j_ej_f is not an inversion, then j_fj_e is, and so the second permutation has one more inversion than the first. Now an interchange of any two numbers in a permutation $j_1 j_2 \cdots j_n$ can always be achieved by an odd number of successive interchanges of adjacent numbers. Thus, if we wish to interchange j_c and $j_k (c < k)$ and there are s numbers between j_c and j_k, we move j_c to the right, by interchanging adjacent numbers, until j_c follows j_k. This requires $s + 1$ steps. Next, we move j_k to the left, by interchanging adjacent numbers until it is where j_c was. This requires s steps. Thus the total number of adjacent interchanges required is $(s + 1) + s = 2s + 1$, which is always odd. Since each adjacent interchange changes the number of inversions by 1 or -1, and since a sum of an odd number of numbers each of which is 1 or -1 is always odd, we conclude that the number of inversions is changed by an odd number. Thus the number of inversions in 54132 is 8 and the number of inversions in 52134 (obtained by interchanging 2 and 4) is 5.

5.1. EXERCISES

1. Find the number of inversions in each of the following permutations of $S = \{1, 2, 3, 4, 5\}$.
(a) 52134. (b) 45213. (c) 42135.

2. Find the number of inversions in each of the following permutations of $S = \{1, 2, 3, 4, 5\}$.
(a) 13542. (b) 35241. (c) 12345.

3. Determine whether each of the following permutations of $S = \{1, 2, 3, 4\}$ is even or odd.
(a) 4213. (b) 1243. (c) 1234.

4. Determine whether each of the following permutations of $S = \{1, 2, 3, 4\}$ is even or odd.
(a) 3214. (b) 1423. (c) 2143.

5. Determine the sign associated with each of the following permutations of the column indices in the expansion of a 5×5 matrix.
(a) 25431. (b) 31245. (c) 21345.

6. Determine the sign associated with each of the following permutations of the column indices in the expansion of a 5×5 matrix.
(a) 52341. (b) 34125. (c) 14523.

7. (a) Find the number of inversions in the permutation 436215.

(b) Verify that the number of inversions in the permutation 416235, obtained from that in part (a) by interchanging two numbers, differs from the answer in part (a) by an odd number.

8. Evaluate:

(a) $\begin{vmatrix} 2 & -1 \\ 3 & 2 \end{vmatrix}$. (b) $\begin{vmatrix} 2 & 1 \\ 4 & 3 \end{vmatrix}$.

9. Evaluate:

(a) $\begin{vmatrix} 1 & 2 \\ 2 & 4 \end{vmatrix}$. (b) $\begin{vmatrix} 3 & 1 \\ -3 & -1 \end{vmatrix}$.

10. Let $A = [a_{ij}]$ be a 4×4 matrix. Develop the general expression for $\det(A)$.

11. Evaluate:

(a) $\begin{vmatrix} 2 & 1 & 3 \\ 3 & 2 & 1 \\ 0 & 1 & 2 \end{vmatrix}$. (b) $\begin{vmatrix} 2 & 1 & 3 \\ -3 & 2 & 1 \\ -1 & 3 & 4 \end{vmatrix}$.

(c) $\begin{vmatrix} 0 & 0 & 0 & 3 \\ 0 & 0 & 4 & 0 \\ 0 & 2 & 0 & 0 \\ 6 & 0 & 0 & 0 \end{vmatrix}$.

12. Evaluate:

(a) $\begin{vmatrix} 2 & 0 & 0 \\ 0 & -3 & 0 \\ 0 & 0 & 4 \end{vmatrix}$. (b) $\begin{vmatrix} 2 & 4 & 5 \\ 0 & -6 & 2 \\ 0 & 0 & 3 \end{vmatrix}$.

(c) $\begin{vmatrix} 0 & 0 & 2 & 0 \\ 0 & 3 & 0 & 0 \\ 6 & 0 & 0 & 0 \\ 0 & 0 & 0 & 5 \end{vmatrix}$.

13. Evaluate:

(a) $\begin{vmatrix} t-1 & 2 \\ 3 & t-2 \end{vmatrix}$. (b) $\begin{vmatrix} t-1 & -1 & -2 \\ 0 & t & 2 \\ 0 & 0 & t-3 \end{vmatrix}$.

14. Evaluate:

(a) $\begin{vmatrix} t & 4 \\ 5 & t-8 \end{vmatrix}$. (b) $\begin{vmatrix} t-1 & 0 & 1 \\ -2 & t & -1 \\ 0 & 0 & t+1 \end{vmatrix}$.

15. For each of the matrices in Exercise 13, find values of t for which the determinant is 0.

16. For each of the matrices in Exercise 14, find values of t for which the determinant is 0.

5.2. *Properties of Determinants*

In this section we examine properties of determinants that simplify their computation.

Theorem 5.1. If A is a matrix, then det $(A) = $ det (A^T).

Proof: Let $A = [a_{ij}]$ and $A^T = [b_{ij}]$, where $b_{ij} = a_{ji}$. We have

$$\det(A^T) = \sum (\pm) b_{1j_1} b_{2j_2} \cdots b_{nj_n} = \sum (\pm) a_{j_1 1} a_{j_2 2} \cdots a_{j_n n}.$$

We can then write $b_{1j_1} b_{2j_2} \cdots b_{nj_n} = a_{j_1 1} a_{j_2 2} \cdots a_{j_n n} = a_{1k_1} a_{2k_2} \cdots a_{nk_n}$, which is a term of det (A). Thus the terms in det (A^T) and det (A) are identical. We must now check that the signs of corresponding terms are also identical. It can be shown that the number of inversions in the permutation $k_1 k_2 \cdots k_n$, which determines the sign associated with the term $a_{1k_1} a_{2k_2} \cdots a_{nk_n}$, is the same as the number of inversions in the permutation $j_1 j_2 \cdots j_n$, which determines the sign associated with the term $b_{1j_1} b_{2j_2} \cdots b_{nj_n}$. As an example,

$$b_{13} b_{24} b_{35} b_{41} b_{52} = a_{31} a_{42} a_{53} a_{14} a_{25} = a_{14} a_{25} a_{31} a_{42} a_{53};$$

the number of inversions in the permutation 45123 is 6 and the number of inversions in the permutation 34512 is also 6. Since the signs of corresponding terms are identical, we conclude that det $(A^T) = $ det (A). ■

Example 1. Let A be the matrix in Example 9 of Section 5.1. Then

$$A^T = \begin{bmatrix} 1 & 2 & 3 \\ 2 & 1 & 1 \\ 3 & 3 & 2 \end{bmatrix}.$$

Substituting in (1) of Section 5.1 (or using the method of lines given in Example 8 of Section 5.1), we find that

$$\begin{aligned} |A^T| &= (1)(1)(2) + (2)(1)(3) + (3)(2)(3) \\ &\quad -(1)(1)(3) - (2)(2)(2) - (3)(1)(3) = 6 = |A|. \end{aligned}$$ ■

Theorem 5.1 will enable us to replace ''row'' by ''column'' in many of the additional properties of determinants; we see how to do this in the following theorem.

Theorem 5.2. If matrix B results from matrix A by interchanging two rows (columns) of A, then $\det(B) = -\det(A)$.

Proof: Suppose that B arises from A by interchanging rows r and s of A, say $r < s$. Then we have $b_{rj} = a_{sj}$, $b_{sj} = a_{rj}$, and $b_{ij} = a_{ij}$ for $i \neq r$, $i \neq s$. Now

$$\begin{aligned} \det(B) &= \sum (\pm) b_{1j_1} b_{2j_2} \cdots b_{rj_r} \cdots b_{sj_s} \cdots b_{nj_n} \\ &= \sum (\pm) a_{1j_1} a_{2j_2} \cdots a_{sj_r} \cdots a_{rj_s} \cdots a_{nj_n} \\ &= \sum (\pm) a_{1j_1} a_{2j_2} \cdots a_{rj_s} \cdots a_{sj_r} \cdots a_{nj_n}. \end{aligned}$$

The permutation $j_1 j_2 \cdots j_s \cdots j_r \cdots j_n$ results from the permutation $j_1 j_2 \cdots j_r \cdots j_s \cdots j_n$ by an interchange of two numbers and the number of inversions in the former differs by an odd number from the number of inversions in the latter. This means that the sign of each term in $\det(B)$ is the negative of the sign of the corresponding term in $\det(A)$. Hence $\det(B) = -\det(A)$.

Now let B arise from A by interchanging two columns of A. Then B^T arises from A^T by interchanging two rows of A^T. So $\det(B^T) = -\det(A^T)$, but $\det(B^T) = \det(B)$ and $\det(A^T) = \det(A)$. Hence $\det(B) = -\det(A)$. ■

In the results to follow, proofs will be given only for the rows of A; the proofs for the corresponding column cases proceed as at the end of the proof of Theorem 5.2.

Example 2. We have $\begin{vmatrix} 2 & -1 \\ 3 & 2 \end{vmatrix} = -\begin{vmatrix} 3 & 2 \\ 2 & -1 \end{vmatrix} = \begin{vmatrix} 2 & 3 \\ -1 & 2 \end{vmatrix} = 7.$ ■

Theorem 5.3. If two rows (columns) of A are equal, then $\det(A) = 0$.

Proof: Suppose that rows r and s of A are equal. Interchange rows r and s of A to obtain a matrix B. Then $\det(B) = -\det(A)$. On the other hand, $B = A$, so $\det(B) = \det(A)$. Thus $\det(A) = -\det(A)$, and so $\det(A) = 0$. ■

Example 3. We have $\begin{vmatrix} 1 & 2 & 3 \\ -1 & 0 & 7 \\ 1 & 2 & 3 \end{vmatrix} = 0$ (verify by the use of Definition 5.2). ■

Theorem 5.4. If a row (column) of A consists entirely of zeros, then $\det(A) = 0$.

Proof: Let the ith row of A consist entirely of zeros. Since each term in Definition 5.2 for the determinant of A contains a factor from the ith row, each term in $\det(A)$ is zero. Hence $\det(A) = 0$. ■

Example 4. We have $\begin{vmatrix} 1 & 2 & 3 \\ 4 & 5 & 6 \\ 0 & 0 & 0 \end{vmatrix} = 0$ (verify by the use of Definition 5.2). ■

Theorem 5.5. If B is obtained from A by multiplying a row (column) of A by a real number c, then $\det(B) = c\det(A)$.

Proof: Suppose that the rth row of $A = [a_{ij}]$ is multiplied by c to obtain $B = [b_{ij}]$. Then $b_{ij} = a_{ij}$ if $i \neq r$ and $b_{rj} = ca_{rj}$. Using Definition 5.2, we obtain $\det(B)$ as

$$\begin{aligned} \det(B) &= \sum (\pm) b_{1j_1} b_{2j_2} \cdots b_{rj_r} \cdots b_{nj_n} \\ &= \sum (\pm) a_{1j_1} a_{2j_2} \cdots (ca_{rj_r}) \cdots a_{nj_n} \\ &= c\left(\sum (\pm) a_{1j_1} a_{2j_2} \cdots a_{rj_r} \cdots a_{nj_n}\right) = c\det(A). \end{aligned}$$ ■

Example 5. We have $\begin{vmatrix} 2 & 6 \\ 1 & 12 \end{vmatrix} = 2\begin{vmatrix} 1 & 3 \\ 1 & 12 \end{vmatrix} = (2)(3)\begin{vmatrix} 1 & 1 \\ 1 & 4 \end{vmatrix} = 6(4-1) = 18.$ ■

We can use Theorem 5.5 to simplify the computation of $\det(A)$ by factoring out common factors from rows and columns of A.

Example 6. We have

$$\begin{vmatrix} 1 & 2 & 3 \\ 1 & 5 & 3 \\ 2 & 8 & 6 \end{vmatrix} = 2\begin{vmatrix} 1 & 2 & 3 \\ 1 & 5 & 3 \\ 1 & 4 & 3 \end{vmatrix} = (2)(3)\begin{vmatrix} 1 & 2 & 1 \\ 1 & 5 & 1 \\ 1 & 4 & 1 \end{vmatrix} = (2)(3)(0) = 0.$$

Here we have factored out 2 from the third row and 3 from the third column, and then used Theorem 5.3, since the first and third columns are equal. ■

Theorem 5.6. *If $B = [b_{ij}]$ is obtained from $A = [a_{ij}]$ by adding to each element of the* rth *row (column) of A c times the corresponding element of the* sth *row (column), $r \neq s$, of A, then* det (B) = det (A).

Proof: We prove the theorem for rows. We have $b_{ij} = a_{ij}$ for $i \neq r$, and $b_{rj} = a_{rj} + ca_{sj}$, $r \neq s$, say $r < s$. Then

$$\begin{aligned} \det(B) &= \sum (\pm) b_{1j_1} b_{2j_2} \cdots b_{rj_r} \cdots b_{nj_n} \\ &= \sum (\pm) a_{1j_1} a_{2j_2} \cdots (a_{rj_r} + ca_{sj_r}) \cdots a_{nj_n} \\ &= \sum (\pm) a_{1j_1} a_{2j_2} \cdots a_{rj_r} \cdots a_{sj_s} \cdots a_{nj_n} \\ &\quad + \sum (\pm) a_{1j_1} a_{2j_2} \cdots (ca_{sj_r}) \cdots a_{sj_s} \cdots a_{nj_n}. \end{aligned}$$

Now the first term in this last expression is det (A) while the second is $c\left[\sum (\pm) a_{1j_1} a_{2j_2} \cdots a_{sj_r} \cdots a_{sj_s} \cdots a_{nj_n}\right]$. Note that

$$\sum (\pm) a_{1j_1} a_{2j_2} \cdots a_{sj_r} \cdots a_{sj_s} \cdots a_{nj_n} = \begin{vmatrix} a_{11} & a_{12} & \cdots & a_{1n} \\ a_{21} & a_{22} & \cdots & a_{2n} \\ & \vdots & & \\ a_{s1} & a_{s2} & \cdots & a_{sn} \\ & \vdots & & \\ a_{s1} & a_{s2} & \cdots & a_{sn} \\ & \vdots & & \\ a_{n1} & a_{n2} & \cdots & a_{nn} \end{vmatrix} \begin{matrix} \\ \\ \\ \leftarrow r\text{th row} \\ \\ \leftarrow s\text{th row} \\ \\ \\ \end{matrix}$$

$$= 0$$

because this matrix has two equal rows. Hence det (B) = det $(A) + 0$ = det (A). ■

Example 7. We have $\begin{vmatrix} 1 & 2 & 3 \\ 2 & -1 & 3 \\ 1 & 0 & 1 \end{vmatrix} = \begin{vmatrix} 5 & 0 & 9 \\ 2 & -1 & 3 \\ 1 & 0 & 1 \end{vmatrix}$, obtained by adding twice the second row to the first row. By applying the definition of determinant to the second determinant, both are seen to have the value 4. ■

Theorem 5.7. *If a matrix $A = [a_{ij}]$ is upper (lower) triangular, then* $\det(A) = a_{11}a_{22}\cdots a_{nn}$; *that is, the determinant of a triangular matrix is the product of the elements on the main diagonal.*

Proof: Let $A = [a_{ij}]$ be upper triangular (that is, $a_{ij} = 0$ for $i > j$). Then a term $a_{1j_1}a_{2j_2}\cdots a_{nj_n}$ in the expression for $|A|$ can be nonzero only for $1 \leq j_1$, $2 \leq j_2, \ldots, n \leq j_n$. Now $j_1j_2\cdots j_n$ must be a permutation, or rearrangement, of $\{1, 2, \ldots, n\}$. Hence we must have $j_1 = 1, j_2 = 2, \ldots, j_n = n$. Thus the only term of $\det(A)$ that can be nonzero is the product of the elements on the main diagonal of A. Hence $\det(A) = a_{11}a_{22}\cdots a_{nn}$.

We leave the proof of the lower triangular case to the reader. ■

Theorems 5.2, 5.5, and 5.6 are very useful in evaluating determinants. What we do is transform A by means of our elementary row or column operations to a triangular matrix. Of course, we must keep track of how the determinant of the resulting matrices changes as we perform the elementary row or column operations.

Example 8. We have

$$\begin{vmatrix} 4 & 3 & 2 \\ 3 & -2 & 5 \\ 2 & 4 & 6 \end{vmatrix} \underset{\frac{1}{2}\mathbf{r}_3 \to \mathbf{r}_3}{=} 2\begin{vmatrix} 4 & 3 & 2 \\ 3 & -2 & 5 \\ 1 & 2 & 3 \end{vmatrix} \underset{\mathbf{r}_1 \leftrightarrow \mathbf{r}_3}{=}$$

$$-2\begin{vmatrix} 1 & 2 & 3 \\ 3 & -2 & 5 \\ 4 & 3 & 2 \end{vmatrix} \underset{\mathbf{r}_2 - 3\mathbf{r}_1 \to \mathbf{r}_2}{=} -2\begin{vmatrix} 1 & 2 & 3 \\ 0 & -8 & -4 \\ 4 & 3 & 2 \end{vmatrix} \underset{\mathbf{r}_3 - 4\mathbf{r}_1 \to \mathbf{r}_3}{=}$$

$$-2\begin{vmatrix} 1 & 2 & 3 \\ 0 & -8 & -4 \\ 0 & -5 & -10 \end{vmatrix} \underset{\frac{1}{4}\mathbf{r}_2 \to \mathbf{r}_2}{=} (-2)(4)\begin{vmatrix} 1 & 2 & 3 \\ 0 & -2 & -1 \\ 0 & -5 & -10 \end{vmatrix} \underset{\frac{1}{5}\mathbf{r}_3 \to \mathbf{r}_3}{=}$$

$$(-2)(4)(5)\begin{vmatrix} 1 & 2 & 3 \\ 0 & -2 & -1 \\ 0 & -1 & -2 \end{vmatrix} \underset{\mathbf{r}_3 - \frac{1}{2}\mathbf{r}_2 \to \mathbf{r}_3}{=} (-2)(4)(5)\begin{vmatrix} 1 & 2 & 3 \\ 0 & -2 & -1 \\ 0 & 0 & -\frac{3}{2} \end{vmatrix}$$

$$= (-2)(4)(5)(1)(-2)(-\tfrac{3}{2}) = -120.$$

Here $\frac{1}{2}\mathbf{r}_3 \to \mathbf{r}_3$ means that $\frac{1}{2}$ times the third row $\mathbf{r}_3$ replaces the third row; $\mathbf{r}_1 \leftrightarrow \mathbf{r}_3$ means that we interchange the first and third rows; $\mathbf{r}_2 - 3\mathbf{r}_1 \to \mathbf{r}_2$ means that the second row minus three times the first row replaces the second row. ■

We can now compute the determinant of the identity matrix I_n: $\det(I_n) = 1$. We can also compute the determinants of the elementary matrices discussed in Section 1.6, as follows.

Let E_1 be an elementary matrix of type I; that is, E_1 is obtained from I_n by interchanging, say, the ith and jth rows of I_n. By Theorem 5.2 we have that $\det(E_1) = -\det(I_n) = -1$. Now let E_2 be an elementary matrix of type II; that is, E_2 is obtained from I_n by multiplying, say, the ith row of I_n by $c \neq 0$. By Theorem 5.5 we have that $\det(E_2) = c \det(I_n) = c$. Finally, let E_3 be an elementary matrix of type III; that is, E_3 is obtained from I_n by adding c times the sth row of I_n to the rth row of I_n $(r \neq s)$. By Theorem 5.6 we have that $\det(E_3) = \det(I_n) = 1$. Thus the determinant of an elementary matrix is never zero.

Next, we prove that the determinant of a product of two matrices is the product of their determinants and that A is nonsingular if and only if $\det(A) \neq 0$.

Lemma 5.1. *If E is an elementary matrix, then* $\det(EA) = \det(E)\det(A)$, *and* $\det(AE) = \det(A)\det(E)$.

Proof: If E is an elementary matrix of type I, then EA is obtained from A by interchanging two rows of A, so $\det(EA) = -\det(A)$. Also $\det(E) = -1$. Thus $\det(EA) = \det(E)\det(A)$.

If E is an elementary matrix of type II, then EA is obtained from A by multiplying a given row of A by $c \neq 0$. Then $\det(EA) = c \det(A)$ and $\det(E) = c$, so $\det(EA) = \det(E)\det(A)$.

Finally, if E is an elementary matrix of type III, then EA is obtained from A by adding a multiple of a row of A to a different row of A. Then $\det(EA) = \det(A)$ and $\det(E) = 1$, so $\det(EA) = \det(E)\det(A)$.

Thus, in all cases, $\det(EA) = \det(E)\det(A)$. By a similar proof, we can show that $\det(AE) = \det(A)\det(E)$. ■

It also follows from Lemma 5.1 that if $B = E_r E_{r-1} \cdots E_2 E_1 A$, then

$$\begin{aligned} \det(B) &= \det(E_r(E_{r-1} \cdots E_2 E_1 A)) \\ &= \det(E_r) \det(E_{r-1} E_{r-2} \cdots E_2 E_1 A) \\ &\vdots \\ &= \det(E_r) \det(E_{r-1}) \cdots \det(E_2) \det(E_1) \det(A). \end{aligned}$$

Theorem 5.8. If A is an $n \times n$ matrix, then A is nonsingular if and only if $\det(A) \neq 0$.

Proof: If A is nonsingular, then A is a product of elementary matrices (Theorem 1.16). Thus let $A = E_1 E_2 \cdots E_k$. Then $\det(A) = \det(E_1 E_2 \cdots E_k) = \det(E_1) \det(E_2) \cdots \det(E_k) \neq 0$.

If A is singular, then A is row equivalent to a matrix B that has a row of zeros (Theorem 1.18). Then $A = E_1 E_2 \cdots E_r B$, where $E_1, E_2, \ldots, E_r$ are elementary matrices. It then follows by the observation following Lemma 5.1 that $\det(A) = \det(E_1 E_2 \cdots E_r B) = \det(E_1) \det(E_2) \cdots \det(E_r) \det(B) = 0$, since $\det(B) = 0$. ■

Corollary 5.1. If A is an $n \times n$ matrix, then rank $A = n$ if and only if $\det(A) \neq 0$.

Proof: By Corollary 2.7 we know that A is nonsingular if and only if rank $A = n$. ■

Corollary 5.2. If A is an $n \times n$ matrix, then $A\mathbf{x} = \mathbf{0}$ has a nontrivial solution if and only if $\det(A) = 0$.

Proof: Exercise. ■

Theorem 5.9. If A and B are $n \times n$ matrices, then $\det(AB) = \det(A) \det(B)$.

Proof: If A is nonsingular, then A is row equivalent to I_n. Thus $A = E_k E_{k-1} \cdots E_2 E_1 I_n = E_k E_{k-1} \cdots E_2 E_1$, where $E_1, E_2, \ldots, E_k$ are elementary matrices. Then

$$\det(A) = \det(E_k E_{k-1} \cdots E_2 E_1) = \det(E_k) \det(E_{k-1}) \cdots \det(E_2) \det(E_1).$$

Now

$$
\begin{aligned}
\det(AB) &= \det(E_k E_{k-1} \cdots E_2 E_1 B) \\
&= \det(E_k) \det(E_{k-1}) \cdots \det(E_2) \det(E_1) \det(B) = \det(A) \det(B).
\end{aligned}
$$

If A is singular, then $\det(A) = 0$ by Theorem 5.8. Moreover, if A is singular, then A is row equivalent to a matrix C that has a row consisting entirely of zeros (Theorem 1.18). Thus $C = E_k E_{k-1} \cdots E_2 E_1 A$, so

$$CB = E_k E_{k-1} \cdots E_2 E_1 AB.$$

This means that AB is row equivalent to CB, and since CB has a row consisting entirely of zeros, it follows that AB is singular. Hence $\det(AB) = 0$ and in this case we also have $\det(AB) = \det(A) \det(B)$. ■

Example 9. Let

$$A = \begin{bmatrix} 1 & 2 \\ 3 & 4 \end{bmatrix} \quad \text{and} \quad B = \begin{bmatrix} 2 & -1 \\ 1 & 2 \end{bmatrix}.$$

Then

$$|A| = -2 \quad \text{and} \quad |B| = 5.$$

On the other hand, $AB = \begin{bmatrix} 4 & 3 \\ 10 & 5 \end{bmatrix}$ and $|AB| = -10 = |A|\,|B|$. ■

Corollary 5.3. *If A is nonsingular, then* $\det(A^{-1}) = 1/\det A$.

Proof: Exercise. ■

Corollary 5.4. *If A and B are similar matrices, then* $\det(A) = \det(B)$.

Proof: Exercise. ■

The determinant of a sum of two $n \times n$ matrices A and B is, in general, not the sum of the determinants of A and B. The best result we can give along these lines is that if A, B, and C are $n \times n$ matrices all of whose entries are equal except for the kth row (column), and the kth row (column) of C is the sum of the kth rows (columns) of A and B, then $\det(C) = \det(A) + \det(B)$. We shall not prove this result, but will consider an example.

Example 10. Let

$$A = \begin{bmatrix} 2 & 2 & 3 \\ 0 & 3 & 4 \\ 0 & 2 & 4 \end{bmatrix}, \qquad B = \begin{bmatrix} 2 & 2 & 3 \\ 0 & 3 & 4 \\ 1 & -2 & -4 \end{bmatrix},$$

and

$$C = \begin{bmatrix} 2 & 2 & 3 \\ 0 & 3 & 4 \\ 1 & 0 & 0 \end{bmatrix}.$$

Then $|A| = 8$, $|B| = -9$, and $|C| = -1$, so $|C| = |A| + |B|$. ■

5.2. EXERCISES

1. Using the technique in Example 8 or by citing a particular theorem or corollary, evaluate:

(a) $\begin{vmatrix} 3 & 0 \\ 2 & 1 \end{vmatrix}$. (b) $\begin{vmatrix} 2 & 1 \\ 4 & 3 \end{vmatrix}$.

(c) $\begin{vmatrix} 4 & 0 & 0 \\ 0 & 2 & 0 \\ 0 & 0 & 3 \end{vmatrix}$. (d) $\begin{vmatrix} 4 & 1 & 3 \\ 2 & 3 & 0 \\ 1 & 3 & 2 \end{vmatrix}$.

(e) $\begin{vmatrix} 4 & 2 & 2 & 0 \\ 2 & 0 & 0 & 0 \\ 3 & 0 & 0 & 1 \\ 0 & 0 & 1 & 0 \end{vmatrix}$. (f) $\begin{vmatrix} 4 & 2 & 3 & -4 \\ 3 & -2 & 1 & 5 \\ -2 & 0 & 1 & -3 \\ 8 & -2 & 6 & 4 \end{vmatrix}$.

2. Using the technique in Example 8 or by citing a particular theorem or corollary, evaluate:

(a) $\begin{vmatrix} 2 & -2 \\ 3 & -1 \end{vmatrix}$. (b) $\begin{vmatrix} 4 & 2 & 0 \\ 0 & -2 & 5 \\ 0 & 0 & 3 \end{vmatrix}$.

(c) $\begin{vmatrix} 3 & 4 & 2 \\ 2 & 5 & 0 \\ 3 & 0 & 0 \end{vmatrix}$. (d) $\begin{vmatrix} 4 & -3 & 5 \\ 5 & 2 & 0 \\ 2 & 0 & 4 \end{vmatrix}$.

(e) $\begin{vmatrix} 4 & 0 & 0 & 0 \\ -1 & 2 & 0 & 0 \\ 1 & 2 & -3 & 0 \\ 1 & 5 & 3 & 5 \end{vmatrix}$. (f) $\begin{vmatrix} 2 & 0 & 1 & \\ 3 & 2 & -4 & - \\ 2 & 3 & -1 & \\ 11 & 8 & -4 & \end{vmatrix}$

3. If $\begin{vmatrix} a_1 & a_2 & a_3 \\ b_1 & b_2 & b_3 \\ c_1 & c_2 & c_3 \end{vmatrix} = 3$, find

$$\begin{vmatrix} a_1 + 2b_1 - 3c_1 & a_2 + 2b_2 - 3c_2 & a_3 + 2b_3 - 3c_3 \\ b_1 & b_2 & b_3 \\ c_1 & c_2 & c_3 \end{vmatrix}.$$

4. Verify that $\det(AB) = \det(A)\det(B)$ for the following.

(a) $A = \begin{bmatrix} 1 & -2 & 3 \\ -2 & 3 & 1 \\ 0 & 1 & 0 \end{bmatrix}$,

$B = \begin{bmatrix} 1 & 0 & 2 \\ 3 & -2 & 5 \\ 2 & 1 & 3 \end{bmatrix}$.

(b) $A = \begin{bmatrix} 2 & 3 & 6 \\ 0 & 3 & 2 \\ 0 & 0 & -4 \end{bmatrix}$,

$B = \begin{bmatrix} 3 & 0 & 0 \\ 4 & 5 & 0 \\ 2 & 1 & -2 \end{bmatrix}$.

5. Evaluate:

(a) $\begin{vmatrix} -4 & 2 & 0 & 0 \\ 2 & 3 & 1 & 0 \\ 3 & 1 & 0 & 2 \\ 1 & 3 & 0 & 3 \end{vmatrix}$.

(b) $\begin{vmatrix} 2 & 0 & 0 & 0 \\ -5 & 3 & 0 & 0 \\ 3 & 2 & 4 & 0 \\ 4 & 2 & 1 & -5 \end{vmatrix}$.

(c) $\begin{vmatrix} t-1 & -1 & -2 \\ 0 & t-2 & 2 \\ 0 & 0 & t-3 \end{vmatrix}$.

(d) $\begin{vmatrix} t+1 & 4 \\ 2 & t-3 \end{vmatrix}$.

6. Is $\det(AB) = \det(BA)$? Justify your answer.

7. If $\det(AB) = 0$, is $\det(A) = 0$ or $\det(B) = 0$? Give reasons for your answer.

8. Show that if c is a scalar and A is $n \times n$, then $\det(cA) = c^n \det(A)$.

9. Show that if A is $n \times n$ with n odd and skew symmetric, then $\det(A) = 0$.

10. Show that if A is a matrix such that in each row and in each column one and only one element is $\neq 0$, then $\det(A) \neq 0$.

11. Prove Corollary 5.2.

12. Show that if $AB = I_n$, then $\det(A) \neq 0$ and $\det(B) \neq 0$.

13. (a) Show that if $A = A^{-1}$, then $\det(A) = \pm 1$.
(b) If $A^T = A^{-1}$, what is $\det(A)$?

14. Show that if A and B are square matrices, then
$\begin{vmatrix} A & O \\ O & B \end{vmatrix} = |A|\,|B|$.

15. If A is a nonsingular matrix such that $A^2 = A$, what is $\det(A)$?

16. Prove Corollary 5.3.

17. Show that if A, B, and C are square matrices, then
$\begin{vmatrix} A & O \\ C & B \end{vmatrix} = |A|\,|B|$.

18. Show that if A and B are both $n \times n$, then:
(a) $\det(A^T B^T) = \det(A)\det(B^T)$.
(b) $\det(A^T B^T) = \det(A^T)\det(B)$.

19. Verify the result in Exercise 14 for $A = \begin{bmatrix} 1 & 2 \\ 3 & 4 \end{bmatrix}$ and $B = \begin{bmatrix} 2 & 1 \\ -3 & 2 \end{bmatrix}$.

20. Use the properties of Section 5.2 to prove that

$$\begin{vmatrix} 1 & a & a^2 \\ 1 & b & b^2 \\ 1 & c & c^2 \end{vmatrix} = (b-a)(c-a)(c-b).$$

(*Hint:* Use factorization.) This determinant is called a **Vandermonde determinant.**

21. If $\det(A) = 2$, find $\det(A^5)$.

22. Use Theorem 5.8 to determine which of the following matrices are nonsingular.

(a) $\begin{bmatrix} 1 & 2 & 3 \\ 0 & 1 & 2 \\ 2 & -3 & 1 \end{bmatrix}$. (b) $\begin{bmatrix} 1 & 2 \\ 3 & 4 \end{bmatrix}$.

23. Use Theorem 5.8 to determine which of the following matrices are nonsingular.

(a) $\begin{bmatrix} 1 & 3 & 2 \\ 2 & 1 & 4 \\ 1 & -7 & 2 \end{bmatrix}$.

(b) $\begin{bmatrix} 1 & 2 & 0 & 5 \\ 3 & 4 & 1 & 7 \\ -2 & 5 & 2 & 0 \\ 0 & 1 & 2 & -7 \end{bmatrix}$.

24. Use Corollary 5.1 to find out whether rank $A = 3$ for the following matrices.

(a) $A = \begin{bmatrix} 1 & 2 & 3 \\ 2 & 1 & 0 \\ -3 & 1 & 2 \end{bmatrix}$.

(b) $A = \begin{bmatrix} 1 & 3 & -4 \\ -2 & 1 & 2 \\ -9 & 15 & 0 \end{bmatrix}$.

(c) $A = \begin{bmatrix} 1 & 0 & 1 \\ 1 & 1 & 0 \\ 2 & 1 & 0 \end{bmatrix}$.

25. Use Corollary 5.2 to find out whether the following homogeneous system has a nontrivial solution (do *not* solve).

$$\begin{aligned} x_1 - 2x_2 + x_3 &= 0 \\ 2x_1 + 3x_2 + x_3 &= 0 \\ 3x_1 + x_2 + 2x_3 &= 0. \end{aligned}$$

26. Repeat Exercise 25 for the following homogeneous system:

$$\begin{bmatrix} 1 & 2 & 0 & 1 \\ 0 & 1 & 2 & 3 \\ 0 & 0 & 1 & 2 \\ 0 & 1 & 2 & -1 \end{bmatrix} \begin{bmatrix} x_1 \\ x_2 \\ x_3 \\ x_4 \end{bmatrix} = \begin{bmatrix} 0 \\ 0 \\ 0 \\ 0 \end{bmatrix}.$$

27. Let $A = [a_{ij}]$ be an upper triangular matrix. Prove that A is nonsingular if and only if $a_{ii} \neq 0$, for $i = 1, 2, \ldots, n$.

28. Let $A^2 = A$. Prove that either A is singular or $\det(A) = 1$.

29. Prove Corollary 5.4.

30. Let $AB = AC$. Prove that if $\det(A) \neq 0$, then $B = C$.

C 31. Determine if the software you are using has a command for computing the determinant of a matrix. If it does, verify the computations in Examples 8, 9, and 10. Experiment further using the matrices in Exercises 1 and 2.

C 32. Assuming that your software has a command to compute the determinant of a matrix, read the accompanying software documentation to determine the method used. Is the description closest to that in Section 5.1, Example 8 in Section 5.2, or the material in Section 1.8?

C 33. **Warning:** Theorem 5.8 assumes that all calculations for $\det(A)$ are done using exact arithmetic. As noted previously, this is usually not the case in software. Hence, computationally the determinant may not be a valid test for nonsingularity. Perform the following experiment. Let $A = \begin{bmatrix} 1 & 2 & 3 \\ 4 & 5 & 6 \\ 7 & 8 & 9 \end{bmatrix}$. Show that $\det(A)$ is 0 either by hand or using your software. Next show by hand computation that

$$\det(B) = -3\epsilon, \text{ where } B = \begin{bmatrix} 1 & 2 & 3 \\ 4 & 5 & 6 \\ 7 & 8 & 9+\epsilon \end{bmatrix}.$$

Hence, theoretically, for any $\epsilon \neq 0$ matrix B is nonsingular. Let your software compute det (B) for $\epsilon = \pm 10^{-k}$, $k = 5, 6, \ldots, 20$. Do the computational results match the theoretical result? If not, formulate a conjecture to explain why not.

5.3. *Cofactor Expansion*

So far we have evaluated determinants by using Definition 5.2 and the properties established in Section 5.2. We now develop a method for evaluating the determinant of an $n \times n$ matrix which reduces the problem to the evaluation of determinants of matrices of order $n - 1$. We can then repeat the process for these $(n-1) \times (n-1)$ matrices until we get to 2×2 matrices.

Definition 5.3. Let $A = [a_{ij}]$ be an $n \times n$ matrix. Let M_{ij} be the $(n-1) \times (n-1)$ submatrix of A obtained by deleting the ith row and jth column of A. The determinant det (M_{ij}) is called the **minor** of a_{ij}. ■

Definition 5.4. Let $A = [a_{ij}]$ be an $n \times n$ matrix. The **cofactor** A_{ij} of a_{ij} is defined as $A_{ij} = (-1)^{i+j} \det(M_{ij})$. ■

Example 1. Let $A = \begin{bmatrix} 3 & -1 & 2 \\ 4 & 5 & 6 \\ 7 & 1 & 2 \end{bmatrix}$. Then

$$|M_{12}| = \begin{vmatrix} 4 & 6 \\ 7 & 2 \end{vmatrix} = 8 - 42 = -34, \qquad |M_{23}| = \begin{vmatrix} 3 & -1 \\ 7 & 1 \end{vmatrix} = 3 + 7 = 10,$$

and

$$|M_{31}| = \begin{vmatrix} -1 & 2 \\ 5 & 6 \end{vmatrix} = -6 - 10 = -16.$$

Also, $A_{12} = (-1)^{1+2}|M_{12}| = (-1)(-34) = 34$,
$A_{23} = (-1)^{2+3}|M_{23}| = (-1)(10) = -10$,
and $A_{31} = (-1)^{1+3}|M_{31}| = (1)(-16) = -16$. ■

If we think of the sign $(-1)^{i+j}$ as being located in position (i, j) of an $n \times n$ matrix, then the signs form a checkerboard pattern that has a + in the (1, 1) position. The patterns for $n = 3$ and $n = 4$ are as follows:

$$\begin{bmatrix} + & - & + \\ - & + & - \\ + & - & + \end{bmatrix} \quad \begin{bmatrix} + & - & + & - \\ - & + & - & + \\ + & - & + & - \\ - & + & - & + \end{bmatrix}$$

$n = 3$ $\qquad$ $n = 4$

Theorem 5.10. Let $A = [a_{ij}]$, *be an* $n \times n$ *matrix. Then*

$$\det(A) = a_{i1}A_{i1} + a_{i2}A_{i2} + \cdots + a_{in}A_{in}$$
(expansion of $\det(A)$ *about the* ith *row)*

and

$$\det(A) = a_{1j}A_{1j} + a_{2j}A_{2j} + \cdots + a_{nj}A_{nj}.$$
(expansion of $\det(A)$ *about the* jth *column).*

Proof: The first formula follows from the second by Theorem 5.1, that is, from the fact that $\det(A^T) = \det(A)$. We omit the general proof and consider the 3×3 matrix $A = [a_{ij}]$. From (1) in Section 5.1,

$$\begin{aligned} \det(A) = {} & a_{11}a_{22}a_{33} + a_{12}a_{23}a_{31} + a_{13}a_{21}a_{32} \\ & -a_{11}a_{23}a_{32} - a_{12}a_{21}a_{33} - a_{13}a_{22}a_{31}. \end{aligned} \tag{1}$$

We can write this expression as

$$\begin{aligned} \det(A) = {} & a_{11}(a_{22}a_{33} - a_{23}a_{32}) + a_{12}(a_{23}a_{31} - a_{21}a_{33}) \\ & +a_{13}(a_{21}a_{32} - a_{22}a_{31}). \end{aligned}$$

Now,

$$A_{11} = (-1)^{1+1}\begin{vmatrix} a_{22} & a_{23} \\ a_{32} & a_{33} \end{vmatrix} = (a_{22}a_{33} - a_{23}a_{32}),$$

$$A_{12} = (-1)^{1+2}\begin{vmatrix} a_{21} & a_{23} \\ a_{31} & a_{33} \end{vmatrix} = (a_{23}a_{31} - a_{21}a_{33}),$$

$$A_{13} = (-1)^{1+3}\begin{vmatrix} a_{21} & a_{22} \\ a_{31} & a_{32} \end{vmatrix} = (a_{21}a_{32} - a_{22}a_{31}).$$

Hence

$$\det(A) = a_{11}A_{11} + a_{12}A_{12} + a_{13}A_{13},$$

which is the expansion of $|A|$ about the first row.

If we now write (1) as

$$\begin{aligned}\det(A) = \ & a_{13}(a_{21}a_{32} - a_{22}a_{31}) + a_{23}(a_{12}a_{31} - a_{11}a_{32}) \\ & + a_{33}(a_{11}a_{22} - a_{12}a_{21}),\end{aligned}$$

we can easily verify that

$$\det(A) = a_{13}A_{13} + a_{23}A_{23} + a_{33}A_{33},$$

which is the expansion of $|A|$ about the third column. ■

Example 2. To evaluate

$$\begin{vmatrix} 1 & 2 & -3 & 4 \\ -4 & 2 & 1 & 3 \\ 3 & 0 & 0 & -3 \\ 2 & 0 & -2 & 3 \end{vmatrix},$$

it is best to expand about either the second column or the third row because they each have two zeros. Obviously, the optimal course of action is to expand about the row or column that has the largest number of zeros, because in that case the cofactors A_{ij} of those a_{ij} which are zero do not have to be evaluated since $a_{ij}A_{ij} = (0)(A_{ij}) = 0$. Thus, expanding about the third row, we have

$$\begin{vmatrix} 1 & 2 & -3 & 4 \\ -4 & 2 & 1 & 3 \\ 3 & 0 & 0 & -3 \\ 2 & 0 & -2 & 3 \end{vmatrix}$$

$$= (-1)^{3+1}(3)\begin{vmatrix} 2 & -3 & 4 \\ 2 & 1 & 3 \\ 0 & -2 & 3 \end{vmatrix} + (-1)^{3+2}(0)\begin{vmatrix} 1 & -3 & 4 \\ -4 & 1 & 3 \\ 2 & -2 & 3 \end{vmatrix}$$

$$+ (-1)^{3+3}(0)\begin{vmatrix} 1 & 2 & 4 \\ -4 & 2 & 3 \\ 2 & 0 & 3 \end{vmatrix} + (-1)^{3+4}(-3)\begin{vmatrix} 1 & 2 & -3 \\ -4 & 2 & 1 \\ 2 & 0 & -2 \end{vmatrix}$$

$$= (+1)(3)(20) + 0 + 0 + (-1)(-3)(-4) = 48. \quad \blacksquare$$

We can use the properties of Section 5.2 to introduce many zeros in a given row or column and then expand about that row or column. Consider the following example.

Example 3. We have

$$\begin{vmatrix} 1 & 2 & -3 & 4 \\ -4 & 2 & 1 & 3 \\ 1 & 0 & 0 & -3 \\ 2 & 0 & -2 & 3 \end{vmatrix} \underset{\mathbf{c}_4 + 3\mathbf{c}_1 \to \mathbf{c}_4}{=} \begin{vmatrix} 1 & 2 & -3 & 7 \\ -4 & 2 & 1 & -9 \\ 1 & 0 & 0 & 0 \\ 2 & 0 & -2 & 9 \end{vmatrix}$$

$$= (-1)^{3+1}(1)\begin{vmatrix} 2 & -3 & 7 \\ 2 & 1 & -9 \\ 0 & -2 & 9 \end{vmatrix} \underset{\mathbf{r}_1 - \mathbf{r}_2 \to \mathbf{r}_1}{=} (-1)^4(1)\begin{vmatrix} 0 & -4 & 16 \\ 2 & 1 & -9 \\ 0 & -2 & 9 \end{vmatrix}$$

$$= (-1)^4(1)(8) = 8.$$

Here $\mathbf{c}_4 + 3\mathbf{c}_1 \to \mathbf{c}_4$ means that the fourth column $\mathbf{c}_4$ plus three times the first column $\mathbf{c}_1$ replaces the fourth column. $\blacksquare$

5.3. EXERCISES

1. Let $A = \begin{bmatrix} 1 & 0 & -2 \\ 3 & 1 & 4 \\ 5 & 2 & -3 \end{bmatrix}$.

Find the following minors.
(a) $|M_{13}|$. (b) $|M_{22}|$.
(c) $|M_{31}|$. (d) $|M_{32}|$.

2. Let $A = \begin{bmatrix} 2 & -1 & 0 & 3 \\ 1 & 2 & -2 & 4 \\ -1 & 1 & -3 & -2 \\ 0 & 2 & -1 & 5 \end{bmatrix}$.

Find the following minors.
(a) $|M_{12}|$. (b) $|M_{23}|$.
(c) $|M_{33}|$. (d) $|M_{41}|$.

3. Let $A = \begin{bmatrix} -1 & 2 & 3 \\ -2 & 5 & 4 \\ 0 & 1 & -3 \end{bmatrix}$.

Find the following cofactors.
(a) A_{13}. (b) A_{21}. (c) A_{32}. (d) A_{33}.

4. Let $A = \begin{bmatrix} 1 & 0 & 3 & 0 \\ 2 & 1 & -4 & -1 \\ 3 & 2 & 4 & 0 \\ 0 & 3 & -1 & 0 \end{bmatrix}$.

Find the following cofactors.
(a) A_{12}. (b) A_{23}. (c) A_{33}. (d) A_{41}.

5. Use Theorem 5.10 to evaluate the determinants in Exercise 1(a), (d), and (e) of Section 5.2.

6. Use Theorem 5.10 to evaluate the determinants in Exercise 1(b), (c), and (f) of Section 5.2.

7. Use Theorem 5.10 to evaluate the determinants in Exercise 2(a), (c), and (f) of Section 5.2.

8. Use Theorem 5.10 to evaluate the determinants in Exercise 2(b), (d), and (e) of Section 5.2.

9. Show by a column (row) expansion that if $A = [a_{ij}]$ is upper (lower) triangular, then $\det(A) = a_{11}a_{22}\cdots a_{nn}$.

10. If $A = [a_{ij}]$ is a 3×3 matrix, develop the general expression for $\det(A)$ by expanding:
(a) About the second column.
(b) About the third row.
Compare these answers with those obtained for Example 8 in Section 5.1.

11. Find all values of t for which:
(a) $\begin{vmatrix} t-2 & 2 \\ 3 & t-3 \end{vmatrix} = 0.$
(b) $\begin{vmatrix} t-1 & -4 \\ 0 & t-4 \end{vmatrix} = 0.$

12. Find all values of t for which

$$\begin{vmatrix} t-1 & 0 & 1 \\ -2 & t+2 & -1 \\ 0 & 0 & t+1 \end{vmatrix} = 0.$$

13. Let A be an $n \times n$ matrix.
(a) Show that $f(t) = \det(tI_n - A)$ is a polynomial in t of degree n.
(b) What is the coefficient of t^n in $f(t)$?
(c) What is the constant term in $f(t)$?

14. Verify your answers to Exercise 13 with the following matrices.
(a) $\begin{bmatrix} 1 & 2 \\ 3 & 4 \end{bmatrix}$. (b) $\begin{bmatrix} 1 & 3 & 2 \\ 2 & -1 & 3 \\ 3 & 0 & 1 \end{bmatrix}$.
(c) $\begin{bmatrix} 1 & 1 \\ 1 & 1 \end{bmatrix}$.

5.4. *The Inverse of a Matrix*

We saw in Section 5.3 that Theorem 5.10 provides formulas for expanding $|A|$ about either a row or a column of A. Thus $\det(A) = a_{i1}A_{i1} + a_{i2}A_{i2} + \cdots + a_{in}A_{in}$ is the expansion of $\det(A)$ about the ith row. It is interesting to ask what $a_{i1}A_{k1} + a_{i2}A_{k2} + \cdots + a_{in}A_{kn}$ is for $i \neq k$, because as soon as we answer this question we shall obtain another method for finding the inverse of a nonsingular matrix.

Theorem 5.11. *If $A = [a_{ij}]$ is an $n \times n$ matrix, then*

$$a_{i1}A_{k1} + a_{i2}A_{k2} + \cdots + a_{in}A_{kn} = 0 \qquad \text{for } i \neq k;$$
$$a_{1j}A_{1k} + a_{2j}A_{2k} + \cdots + a_{nj}A_{nk} = 0 \qquad \text{for } j \neq k.$$

Proof: We prove only the first formula. The second follows from the first by Theorem 5.1.

Consider the matrix B obtained from A by replacing the kth row of A by the ith row of A. Thus B is a matrix with two identical rows—the ith and kth, so $\det(B) = 0$. Now expand $\det(B)$ about the kth row. The elements of the kth row of B are $a_{i1}, a_{i2}, \ldots, a_{in}$. The cofactors of the kth row are $A_{k1}, A_{k2}, \ldots, A_{kn}$. Thus

$$0 = \det(B) = a_{i1}A_{k1} + a_{i2}A_{k2} + \cdots + a_{in}A_{kn},$$

as we wanted to show. ■

This theorem says that if we sum the products of the elements of any row (column) times the corresponding cofactors of any other row (column), then we obtain zero.

Example 1. Let $A = \begin{bmatrix} 1 & 2 & 3 \\ -2 & 3 & 1 \\ 4 & 5 & -2 \end{bmatrix}$. Then

$$A_{21} = (-1)^{2+1}\begin{vmatrix} 2 & 3 \\ 5 & -2 \end{vmatrix} = 19,$$

$$A_{22} = (-1)^{2+2}\begin{vmatrix} 1 & 3 \\ 4 & -2 \end{vmatrix} = -14, \quad \text{and} \quad A_{23} = (-1)^{2+3}\begin{vmatrix} 1 & 2 \\ 4 & 5 \end{vmatrix} = 3.$$

Now

$$a_{31}A_{21} + a_{32}A_{22} + a_{33}A_{23} = (4)(19) + (5)(-14) + (-2)(3) = 0,$$

and

$$a_{11}A_{21} + a_{12}A_{22} + a_{13}A_{23} = (1)(19) + 2(-14) + (3)(3) = 0. \quad ■$$

We may summarize our expansion results by writing

$$\begin{aligned} a_{i1}A_{k1} + a_{i2}A_{k2} + \cdots + a_{in}A_{kn} &= \det(A) \qquad \text{if } i = k \\ &= 0 \qquad \text{if } i \neq k \end{aligned}$$

and

$$a_{1j}A_{1k} + a_{2j}A_{2k} + \cdots + a_{nj}A_{nk} = \det(A) \qquad \text{if } j = k$$
$$= 0 \qquad \text{if } j \neq k.$$

Definition 5.5. Let $A = [a_{ij}]$ be an $n \times n$ matrix. The $n \times n$ matrix adj A, called the **adjoint** of A, is the matrix whose (i, j) entry is the cofactor A_{ji} of a_{ji}. Thus

$$\text{adj } A = \begin{bmatrix} A_{11} & A_{21} & \cdots & A_{n1} \\ A_{12} & A_{22} & \cdots & A_{n2} \\ \vdots & \vdots & & \vdots \\ A_{1n} & A_{2n} & \cdots & A_{nn} \end{bmatrix}.$$ ■

REMARK. It should be noted that the term *adjoint* has other meanings in linear algebra in addition to its use in Definition 5.5.

Example 2. Let $A = \begin{bmatrix} 3 & -2 & 1 \\ 5 & 6 & 2 \\ 1 & 0 & -3 \end{bmatrix}$. Compute adj A.

Solution: We first compute the cofactors of A. We have

$$A_{11} = (-1)^{1+1}\begin{vmatrix} 6 & 2 \\ 0 & -3 \end{vmatrix} = -18,$$

$$A_{12} = (-1)^{1+2}\begin{vmatrix} 5 & 2 \\ 1 & -3 \end{vmatrix} = 17, \quad \text{and} \quad A_{13} = (-1)^{1+3}\begin{vmatrix} 5 & 6 \\ 1 & 0 \end{vmatrix} = -6;$$

$$A_{21} = (-1)^{2+1}\begin{vmatrix} -2 & 1 \\ 0 & -3 \end{vmatrix} = -6,$$

$$A_{22} = (-1)^{2+2}\begin{vmatrix} 3 & 1 \\ 1 & -3 \end{vmatrix} = -10, \quad \text{and} \quad A_{23} = (-1)^{2+3}\begin{vmatrix} 3 & -2 \\ 1 & 0 \end{vmatrix} = -2;$$

$$A_{31} = (-1)^{3+1}\begin{vmatrix} -2 & 1 \\ 6 & 2 \end{vmatrix} = -10,$$

$$A_{32} = (-1)^{3+2}\begin{vmatrix} 3 & 1 \\ 5 & 2 \end{vmatrix} = -1, \quad \text{and} \quad A_{33} = (-1)^{3+3}\begin{vmatrix} 3 & -2 \\ 5 & 6 \end{vmatrix} = 28.$$

Hence

$$\text{adj } A = \begin{bmatrix} -18 & -6 & -10 \\ 17 & -10 & -1 \\ -6 & -2 & 28 \end{bmatrix}.$$ ■

Theorem 5.12. If $A = [a_{ij}]$ *is an* $n \times n$ *matrix, then* $A(\text{adj } A) = (\text{adj } A)A = \det(A)I_n$.

Proof: We have

$$A(\text{adj } A) = \begin{bmatrix} a_{11} & a_{12} & \cdots & a_{1n} \\ a_{21} & a_{22} & \cdots & a_{2n} \\ \vdots & \vdots & & \vdots \\ a_{i1} & a_{i2} & \cdots & a_{in} \\ \vdots & \vdots & & \vdots \\ a_{n1} & a_{n2} & \cdots & a_{nn} \end{bmatrix} \begin{bmatrix} A_{11} & A_{21} & \cdots & A_{j1} & \cdots & A_{n1} \\ A_{12} & A_{22} & \cdots & A_{j2} & \cdots & A_{n2} \\ \vdots & \vdots & & \vdots & & \vdots \\ A_{1n} & A_{2n} & \cdots & A_{jn} & \cdots & A_{nn} \end{bmatrix}.$$

The (i, j) entry in the product matrix $A(\text{adj } A)$ is

$$\begin{aligned} a_{i1}A_{j1} + a_{i2}A_{j2} + \cdots + a_{in}A_{jn} &= \det(A) && \text{if } i = j \\ &= 0 && \text{if } i \neq j. \end{aligned}$$

This means that

$$A(\text{adj } A) = \begin{bmatrix} \det(A) & 0 & \cdots & 0 \\ 0 & \det(A) & & \vdots \\ \vdots & & & 0 \\ 0 & \cdots & 0 & \det(A) \end{bmatrix} = \det(A)I_n.$$

The (i, j) entry in the product matrix $(\text{adj } A)\,A$ is

$$\begin{aligned} A_{1i}a_{1j} + A_{2i}a_{2j} + \cdots + A_{ni}a_{nj} &= \det(A) && \text{if } i = j \\ &= 0 && \text{if } i \neq j. \end{aligned}$$

Thus $(\text{adj } A)A = \det(A)I_n$. ■

Example 3. Consider the matrix of Example 2. Then

$$\begin{bmatrix} 3 & -2 & 1 \\ 5 & 6 & 2 \\ 1 & 0 & -3 \end{bmatrix} \begin{bmatrix} -18 & -6 & -10 \\ 17 & -10 & -1 \\ -6 & -2 & 28 \end{bmatrix} = \begin{bmatrix} -94 & 0 & 0 \\ 0 & -94 & 0 \\ 0 & 0 & -94 \end{bmatrix} = -94 \begin{bmatrix} 1 & 0 & 0 \\ 0 & 1 & 0 \\ 0 & 0 & 1 \end{bmatrix}$$

and

$$\begin{bmatrix} -18 & -6 & -10 \\ 17 & -10 & -1 \\ -6 & -2 & 28 \end{bmatrix}\begin{bmatrix} 3 & -2 & 1 \\ 5 & 6 & 2 \\ 1 & 0 & -3 \end{bmatrix} = -94\begin{bmatrix} 1 & 0 & 0 \\ 0 & 1 & 0 \\ 0 & 0 & 1 \end{bmatrix}. \qquad \blacksquare$$

We now have a new method for finding the inverse of a nonsingular matrix, and we state this result as a corollary.

Corollary 5.5. If A is an $n \times n$ *matrix and* $\det(A) \neq 0$, *then*

$$A^{-1} = \frac{1}{\det(A)}(\text{adj } A) = \begin{bmatrix} \frac{A_{11}}{\det(A)} & \frac{A_{21}}{\det(A)} & \cdots & \frac{A_{n1}}{\det(A)} \\ \frac{A_{12}}{\det(A)} & \frac{A_{22}}{\det(A)} & \cdots & \frac{A_{n2}}{\det(A)} \\ \vdots & \vdots & & \vdots \\ \frac{A_{1n}}{\det(A)} & \frac{A_{2n}}{\det(A)} & \cdots & \frac{A_{nn}}{\det(A)} \end{bmatrix}.$$

Proof: By Theorem 5.12, $A(\text{adj } A) = \det(A)I_n$, so if $\det(A) \neq 0$, then

$$A\left(\frac{1}{\det(A)}(\text{adj } A)\right) = \frac{1}{\det(A)}(A(\text{adj } A)) = \frac{1}{\det(A)}(\det(A)I_n) = I_n.$$

Hence $A^{-1} = \frac{1}{\det(A)}(\text{adj } A)$. ■

Example 4. Again consider the matrix of Example 2. Then $|A| = -94$, and

$$A^{-1} = \frac{1}{\det(A)}(\text{adj } A) = \begin{bmatrix} \frac{18}{94} & \frac{6}{94} & \frac{10}{94} \\ -\frac{17}{94} & \frac{10}{94} & \frac{1}{94} \\ \frac{6}{94} & \frac{2}{94} & -\frac{28}{94} \end{bmatrix}. \qquad \blacksquare$$

We might note that the method of inverting a nonsingular matrix given in Corollary 5.5 is much less efficient than the methods given in Chapter 1. In fact, the computation of A^{-1} using determinants, as given in Corollary 5.5, becomes too expensive from a computing point of view for $n > 4$. We discuss these matters in Section 5.6, where we deal with determinants from a computational point of view. However, Corollary 5.5 is still a useful result on other grounds.

5.4. EXERCISES

1. Verify Theorem 5.11 for the matrix

$$A = \begin{bmatrix} -2 & 3 & 0 \\ 4 & 1 & -3 \\ 2 & 0 & 1 \end{bmatrix}$$

by computing $a_{11}A_{12} + a_{21}A_{22} + a_{31}A_{32}$.

2. Let $A = \begin{bmatrix} 2 & 1 & 3 \\ -1 & 2 & 0 \\ 3 & -2 & 1 \end{bmatrix}$.

(a) Find adj A.
(b) Compute det (A).
(c) Verify Theorem 5.12; that is, show that $A(\text{adj } A) = (\text{adj } A)A = \det(A)I_3$.

3. Let $A = \begin{bmatrix} 6 & 2 & 8 \\ -3 & 4 & 1 \\ 4 & -4 & 5 \end{bmatrix}$.

(a) Find adj A.
(b) Compute det (A).
(c) Verify Theorem 5.12; that is, show that $A(\text{adj } A) = (\text{adj } A)A = \det(A)I_3$.

4. Find the inverse of the matrix in Exercise 2 by the method given in Corollary 5.5.

5. Repeat Exercise 9 of Section 1.6 by the method given in Corollary 5.5. Compare your results with those obtained earlier.

6. Prove that if A is a symmetric matrix, then adj A is symmetric.

7. Use the method given in Corollary 5.5 to find the inverse, if it exists, of

(a) $\begin{bmatrix} 0 & 2 & 1 & 3 \\ 2 & -1 & 3 & 4 \\ -2 & 1 & 5 & 2 \\ 0 & 1 & 0 & 2 \end{bmatrix}$.

(b) $\begin{bmatrix} 4 & 2 & 2 \\ 0 & 1 & 2 \\ 1 & 0 & 3 \end{bmatrix}$. (c) $\begin{bmatrix} 3 & 2 \\ -3 & 4 \end{bmatrix}$.

8. Prove that if A is a nonsingular upper triangular matrix, then A^{-1} is upper triangular.

9. Use the method given in Corollary 5.5 to find the inverse of

$$A = \begin{bmatrix} a & b \\ c & d \end{bmatrix} \quad \text{if } ad - bc \neq 0.$$

10. Use the method given in Corollary 5.5 to find the inverse of

$$A = \begin{bmatrix} 1 & a & a^2 \\ 1 & b & b^2 \\ 1 & c & c^2 \end{bmatrix}.$$

[*Hint:* See Exercise 20 in Section 5.2, where det (A) was computed.]

11. Use the method given in Corollary 5.5 to find the inverse of

$$A = \begin{bmatrix} 4 & 0 & 0 \\ 0 & -3 & 0 \\ 0 & 0 & 2 \end{bmatrix}.$$

12. Use the method given in Corollary 5.5 to find the inverse of

$$A = \begin{bmatrix} 4 & 1 & 2 \\ 0 & -3 & 3 \\ 0 & 0 & 2 \end{bmatrix}.$$

13. Prove that if A is singular, then adj A is singular. [*Hint:* First show that if A is singular, then $A(\text{adj } A) = O$.]

14. Prove that if A is an $n \times n$ matrix, then $\det(\text{adj}(A)) = [\det(A)]^{n-1}$.

15. Assuming that your software has a command for computing the inverse of a matrix (see Exercise 41 in Section 1.4), read the accompanying software documentation to determine the method used. Is the description closer to that in Section 1.6 or Corollary 5.5? See also the comments in Section 5.6.

5.5. *Other Applications of Determinants*

We can use the results developed in Theorem 5.12 to obtain another method for solving a linear system of n equations in n unknowns. This method is known as **Cramer's rule**.

Theorem 5.13. (Cramer's† Rule) *Let*

$$\begin{aligned} a_{11}x_1 + a_{12}x_2 + \cdots + a_{1n}x_n &= b_1 \\ a_{21}x_1 + a_{22}x_2 + \cdots + a_{2n}x_n &= b_2 \\ &\vdots \\ a_{n1}x_1 + a_{n2}x_2 + \cdots + a_{nn}x_n &= b_n \end{aligned}$$

be a linear system of n equations in n unknowns and let $A = [a_{ij}]$ be the coefficient matrix, so that we can write the given system as $A\mathbf{x} = \mathbf{b}$, where

† Gabriel Cramer (1704–1752) was born in Geneva, Switzerland, and lived there all his life. Remaining single, he traveled extensively, taught at the Académie de Calvin, and participated actively in civic affairs.

The rule for solving systems of linear equations appeared in an appendix to his 1750 book, *Introduction à l'analyse des lignes courbes algébriques*. It was known previously by other mathematicians but was not widely known or very clearly explained until its appearance in Cramer's influential work.

$\mathbf{b} = \begin{bmatrix} b_1 \\ b_2 \\ \vdots \\ b_n \end{bmatrix}$. *If* det $(A) \neq 0$, *then the system has the unique solution*

$$x_1 = \frac{\det(A_1)}{\det(A)}, \; x_2 = \frac{\det(A_2)}{\det(A)}, \; . \; . \; . \; , \; x_n = \frac{\det(A_n)}{\det(A)},$$

where A_i *is the matrix obtained from A by replacing its* ith *column by* **b**.

Proof: If det $(A) \neq 0$, then, by Theorem 5.8, A is nonsingular. Hence

$$\mathbf{x} = \begin{bmatrix} x_1 \\ x_2 \\ \vdots \\ x_n \end{bmatrix} = A^{-1}\mathbf{b} = \begin{bmatrix} \frac{A_{11}}{\det(A)} & \frac{A_{21}}{\det(A)} & \cdots & \frac{A_{n1}}{\det(A)} \\ \frac{A_{12}}{\det(A)} & \frac{A_{22}}{\det(A)} & \cdots & \frac{A_{n2}}{\det(A)} \\ \vdots & \vdots & & \vdots \\ \frac{A_{1i}}{\det(A)} & \frac{A_{2i}}{\det(A)} & \cdots & \frac{A_{ni}}{\det(A)} \\ \vdots & \vdots & & \vdots \\ \frac{A_{1n}}{\det(A)} & \frac{A_{2n}}{\det(A)} & \cdots & \frac{A_{nn}}{\det(A)} \end{bmatrix} \begin{bmatrix} b_1 \\ b_2 \\ \vdots \\ b_n \end{bmatrix}.$$

This means that

$$x_i = \frac{A_{1i}}{\det(A)}b_1 + \frac{A_{2i}}{\det(A)}b_2 + \cdots + \frac{A_{ni}}{\det(A)}b_n \quad \text{for } i = 1, 2, \ldots, n.$$

Now let

$$A_i = \begin{bmatrix} a_{11} & a_{12} & \cdots & a_{1i-1} & b_1 & a_{1i+1} & \cdots & a_{1n} \\ a_{21} & a_{22} & \cdots & a_{2i-1} & b_2 & a_{2i+1} & \cdots & a_{2n} \\ \vdots & \vdots & & \vdots & \vdots & \vdots & & \vdots \\ a_{n1} & a_{n2} & \cdots & a_{ni-1} & b_n & a_{ni+1} & \cdots & a_{nn} \end{bmatrix}.$$

If we evaluate det (A_i) by expanding about the cofactors of the ith column, we find that

$$\det(A_i) = A_{1i}b_1 + A_{2i}b_2 + \cdots + A_{ni}b_n.$$

Hence

$$x_i = \frac{\det(A_i)}{\det(A)} \qquad \text{for } i = 1, 2, \ldots, n. \tag{1}$$

■

In the expression for x_i given in Equation (1), the determinant, $\det(A_i)$, of A_i can be calculated by any method desired. It was only in the *derivation* of the expression for x_i that we had to evaluate $\det(A_i)$ by expanding about the ith column.

Example 1. Consider the following linear system:

$$\begin{aligned} -2x_1 + 3x_2 - x_3 &= 1 \\ x_1 + 2x_2 - x_3 &= 4 \\ -2x_1 - x_2 + x_3 &= -3. \end{aligned}$$

We have $|A| = \begin{vmatrix} -2 & 3 & -1 \\ 1 & 2 & -1 \\ -2 & -1 & 1 \end{vmatrix} = -2$. Then

$$x_1 = \frac{\begin{vmatrix} 1 & 3 & -1 \\ 4 & 2 & -1 \\ -3 & -1 & 1 \end{vmatrix}}{|A|} = \frac{-4}{-2} = 2,$$

$$x_2 = \frac{\begin{vmatrix} -2 & 1 & -1 \\ 1 & 4 & -1 \\ -2 & -3 & 1 \end{vmatrix}}{|A|} = \frac{-6}{-2} = 3,$$

and

$$x_3 = \frac{\begin{vmatrix} -2 & 3 & 1 \\ 1 & 2 & 4 \\ -2 & -1 & -3 \end{vmatrix}}{|A|} = \frac{-8}{-2} = 4.$$

■

We note that Cramer's rule is applicable only to the case in which we have n equations in n unknowns and the coefficient matrix A is nonsingular. Cramer's rule becomes computationally inefficient if $n \geq 4$ and it is then better to use the

Gaussian elimination or Gauss–Jordan reduction methods discussed in Section 1.5.

Our next application of determinants will enable us to tell whether a set of n vectors in R^n or R_n is linearly independent.

Theorem 5.14. *Let $S = \{\mathbf{v}_1, \mathbf{v}_2, \ldots, \mathbf{v}_n\}$ be a set of n vectors in R^n (R_n). Let A be the matrix whose columns (rows) are the elements of S. Then S is linearly independent if and only if* $\det(A) \neq 0$.

Proof: We shall prove the result for columns only; the proof for rows is analogous.

If S is linearly independent, then the dimension of the column space of A = rank $A = n$, and from Corollary 5.1 it follows that $\det(A) \neq 0$. Conversely, if $\det(A) \neq 0$, by Corollary 5.1 we know that rank $A = n$-dim column space of A. Hence S is linearly independent. ■

Example 2. Is $S = \{[1 \quad 2 \quad 3], [0 \quad 1 \quad 2], [3 \quad 0 \quad -1]\}$ a linearly independent set of vectors in R^3?

Solution: We form the matrix A whose rows are the vectors in S:

$$A = \begin{bmatrix} 1 & 2 & 3 \\ 0 & 1 & 2 \\ 3 & 0 & -1 \end{bmatrix}.$$

Since $|A| = 2$ (verify), we conclude that S is linearly independent. ■

Example 3. To find out if $S = \{t^2 + t, t + 1, t - 1\}$ is a basis for P_2, we note that P_2 is a three-dimensional vector space isomorphic to R^3 under the mapping $L: P_2 \to R^3$ defined by $L(at^2 + bt + c) = \begin{bmatrix} a \\ b \\ c \end{bmatrix}$. Therefore, S is a basis for P_2 if and only if $T = \{L(t^2 + t), L(t + 1), L(t - 1)\}$ is a basis for R^3. To decide whether this is so, we apply Theorem 5.14. Thus let A be the matrix whose columns are $L(t^2 + t)$, $L(t + 1)$, $L(t - 1)$, respectively. Now

$$L(t^2 + t) = \begin{bmatrix} 1 \\ 1 \\ 0 \end{bmatrix}, \qquad L(t + 1) = \begin{bmatrix} 0 \\ 1 \\ 1 \end{bmatrix}, \qquad \text{and} \qquad L(t - 1) = \begin{bmatrix} 0 \\ 1 \\ -1 \end{bmatrix},$$

so

$$A = \begin{bmatrix} 1 & 0 & 0 \\ 1 & 1 & 1 \\ 0 & 1 & -1 \end{bmatrix}.$$

Since $|A| = -2$ (verify), we conclude that T is linearly independent. Hence S is linearly independent, and since $\dim P_2 = 3$, S is a basis for P_2. ■

We now recall that if a set S of n vectors in R^n (R_n) is linearly independent, then S spans R^n (R_n), and conversely, if S spans R^n (R_n), then S is linearly independent (Theorem 2.10 in Section 2.4). Thus the condition in Theorem 5.14 (that $|A| \neq 0$) is also necessary and sufficient for S to span R^n (R_n).

We may summarize our results on the application of determinants by noting that the following statements are equivalent for an $n \times n$ matrix A:

1. $\det(A) \neq 0$.
2. A is nonsingular.
3. The rows (columns) of A are linearly independent.
4. $A\mathbf{x} = \mathbf{0}$ has only the trivial solution.
5. $\text{rank}\ A = n$.

Determinants and Cross Product (Optional)

Our final application of determinants is to cross product. Recall the definition given in Section 3.2 for the cross product $\mathbf{u} \times \mathbf{v}$ of the vectors $\mathbf{u} = u_1\mathbf{i} + u_2\mathbf{j} + u_3\mathbf{k}$ and $\mathbf{v} = v_1\mathbf{i} + v_2\mathbf{j} + v_3\mathbf{k}$ in R^3:

$$\mathbf{u} \times \mathbf{v} = (u_2v_3 - u_3v_2)\mathbf{i} + (u_3v_1 - u_1v_3)\mathbf{j} + (u_1v_2 - u_2v_1)\mathbf{k}.$$

If we formally write the matrix

$$C = \begin{bmatrix} \mathbf{i} & \mathbf{j} & \mathbf{k} \\ u_1 & u_2 & u_3 \\ v_1 & v_2 & v_3 \end{bmatrix},$$

then the determinant of C, obtained by expanding about the cofactors of the first row, is $\mathbf{u} \times \mathbf{v}$, that is,

$$\mathbf{u} \times \mathbf{v} = |C| = \begin{vmatrix} u_2 & u_3 \\ v_2 & v_3 \end{vmatrix}\mathbf{i} - \begin{vmatrix} u_1 & u_3 \\ v_1 & v_3 \end{vmatrix}\mathbf{j} + \begin{vmatrix} u_1 & u_2 \\ v_1 & v_2 \end{vmatrix}\mathbf{k}.$$

Of course, C is not really a matrix and $|C|$ is not really a determinant, but it is convenient to think of the computation in this way.

Example 4. If $\mathbf{u} = 2\mathbf{i} + \mathbf{j} + 2\mathbf{k}$ and $\mathbf{v} = 3\mathbf{i} - \mathbf{j} - 3\mathbf{k}$, as in Example 1 of Section 3.2, then

$$C = \begin{bmatrix} \mathbf{i} & \mathbf{j} & \mathbf{k} \\ 2 & 1 & 2 \\ 3 & -1 & -3 \end{bmatrix}$$

and $|C| = \mathbf{u} \times \mathbf{v} = -\mathbf{i} + 12\mathbf{j} - 5\mathbf{k}$, when expanded about its first row. ■

5.5. EXERCISES

1. If possible, solve the following linear system by Cramer's rule:

$$\begin{aligned} 2x_1 + 4x_2 + 6x_3 &= 2 \\ x_1 \qquad + 2x_3 &= 0 \\ 2x_1 + 3x_2 - x_3 &= -5. \end{aligned}$$

2. Repeat Exercise 1 for the linear system

$$\begin{bmatrix} 1 & 1 & 1 & -2 \\ 0 & 2 & 1 & 3 \\ 2 & 1 & -1 & 2 \\ 1 & -1 & 0 & 1 \end{bmatrix}\begin{bmatrix} x_1 \\ x_2 \\ x_3 \\ x_4 \end{bmatrix} = \begin{bmatrix} -4 \\ 4 \\ 5 \\ 4 \end{bmatrix}.$$

3. Is $S = \left\{ \begin{bmatrix} 2 \\ 2 \\ 3 \end{bmatrix}, \begin{bmatrix} 1 \\ 0 \\ 2 \end{bmatrix}, \begin{bmatrix} 0 \\ 1 \\ 3 \end{bmatrix} \right\}$ a linearly independent set of vectors in R^3? Use Theorem 5.14.

4. Solve the following linear system for x_3, by Cramer's rule:

$$\begin{aligned} 2x_1 + x_2 + x_3 &= 6 \\ 3x_1 + 2x_2 - 2x_3 &= -2 \\ x_1 + x_2 + 2x_3 &= -4. \end{aligned}$$

5. Repeat Exercise 5 of Section 1.5; use Cramer's rule.

6. Is $S = \{[0 \;\; -1 \;\; 4], [4 \;\; 1 \;\; 2], [2 \;\; 0 \;\; 3]\}$, a basis for R_3? Use Theorem 5.14.

7. Repeat Exercise 1 for the following linear system:

$$\begin{aligned} 2x_1 - x_2 + 3x_3 &= 0 \\ x_1 + 2x_2 - 3x_3 &= 0 \\ 4x_1 + 2x_2 + x_3 &= 0. \end{aligned}$$

8. Does the set $S = \left\{ \begin{bmatrix} 3 \\ 1 \\ 0 \end{bmatrix}, \begin{bmatrix} 0 \\ -1 \\ 1 \end{bmatrix}, \begin{bmatrix} 1 \\ 2 \\ -1 \end{bmatrix} \right\}$ span R^3? Use Theorem 5.14.

9. Repeat Exercise 6(b) of Section 1.5; use Cramer's rule.

10. Repeat Exercise 1 for the following linear system:

$$\begin{aligned} 2x_1 + 3x_2 + 7x_3 &= 0 \\ -2x_1 \qquad - 4x_3 &= 0 \\ x_1 + 2x_2 + 4x_3 &= 0. \end{aligned}$$

11. Is $S = \{t^3 + t + 1, 2t^2 + 3, t - 1, 2t^3 - 2t^2\}$ a basis for P_3? Use Theorem 5.14.

12. Is

$$S = \left\{ \begin{bmatrix} 1 & 1 \\ 0 & 1 \end{bmatrix}, \begin{bmatrix} 0 & 2 \\ 1 & 3 \end{bmatrix}, \begin{bmatrix} 0 & 0 \\ -1 & 2 \end{bmatrix}, \begin{bmatrix} 0 & -2 \\ 0 & 3 \end{bmatrix} \right\}$$

a basis for ${}_2R_2$? Use Theorem 5.14.

13. For what values of c is the set $\{t + 3, 2t + c^2 + 2\}$ linearly independent?

14. Let $\mathbf{u} = u_1\mathbf{i} + u_2\mathbf{j} + u_3\mathbf{k}$, $\mathbf{v} = v_1\mathbf{i} + v_2\mathbf{j} + v_3\mathbf{k}$, and $\mathbf{w} = w_1\mathbf{i} + w_2\mathbf{j} + w_3\mathbf{k}$ be vectors in R^3. Show that

$$(\mathbf{u} \times \mathbf{v}) \cdot \mathbf{w} = \begin{vmatrix} u_1 & u_2 & u_3 \\ v_1 & v_2 & v_3 \\ w_1 & w_2 & w_3 \end{vmatrix}.$$

15. Compute $\mathbf{u} \times \mathbf{v}$ by the method of Example 4.

(a) $\mathbf{u} = 2\mathbf{i} + 3\mathbf{j} + 4\mathbf{k}$, $\mathbf{v} = -\mathbf{i} + 3\mathbf{j} - \mathbf{k}$.
(b) $\mathbf{u} = \mathbf{i} + \mathbf{k}$, $\mathbf{v} = 2\mathbf{i} + 3\mathbf{j} - \mathbf{k}$.
(c) $\mathbf{u} = \mathbf{i} - \mathbf{j} + 2\mathbf{k}$, $\mathbf{v} = 3\mathbf{i} - 4\mathbf{j} + \mathbf{k}$.
(d) $\mathbf{u} = 2\mathbf{i} + \mathbf{j} - 2\mathbf{k}$, $\mathbf{v} = \mathbf{i} + 3\mathbf{k}$.

5.6. *Determinants from a Computational Point of View*

So far we have developed two methods for solving a linear system: Gaussian elimination (or Gauss–Jordan reduction) and Cramer's rule. We also have two methods for inverting a nonsingular matrix: the method involving determinants and the method developed in Section 1.6, which uses elementary matrices. We must then develop some criteria for choosing one or the other method depending upon our needs.

In general, if we are seeking numerical answers, then any method involving determinants can be used for $n \leq 4$. We shall compare Gaussian elimination and Cramer's rule for solving the linear system $A\mathbf{x} = \mathbf{b}$, when A is 25×25.

If we find $\mathbf{x}$ by Cramer's rule, then we must first obtain $\det(A)$. We can find $\det(A)$ by cofactor expansion, say $\det(A) = a_{11}A_{11} + a_{21}A_{21} + \cdots + a_{n1}A_{n1}$, where we have expanded about the first column. Note that if each cofactor is available, we require 25 multiplications. Now each cofactor A_{ij} is the determinant of a 24×24 matrix, and it can be expanded about a given row or column, requiring 24 multiplications. Thus the computation of $\det(A)$ requires more than $25 \times 24 \times 23 \times \cdots \times 2 \times 1 = 25!$ multiplications. If we were to use a futuristic computer capable of performing *100 billion* multiplications *per second,* it would take about *4 million years* to evaluate $\det(A)$. However, Gaussian elimination takes about 25^3 multiplications, and we would find the solution in less than *1 second*. Of course, if A is $n \times n$, we can compute $\det(A)$ in a much more efficient way, by using elementary row operations to reduce A to triangular form and then using Theorem 5.7 (see Example 8 in Section 5.2). When implemented this way, Cramer's rule will require approximately n^4 multiplications, compared to n^3 multiplications for Gaussian elimination.

The importance of determinants does not lie in their computational usage; determinants enable one to express the inverse of a matrix and the solution to a system of n linear equations in n unknowns by means of *expressions* or *formulas*. Gaussian elimination and the method for finding A^{-1} using elementary matrices have the property that we cannot write a *formula* for the answer; we must proceed numerically to obtain the answer. Sometimes we do not need a numerical answer but merely an expression for the answer, because we may wish to further manipulate the answer, for example, integrate it. However, the most important reason for studying determinants is that they play a key role in the further study of the properties of a linear transformation mapping a vector space V into itself. This study will be undertaken in Chapter 6.

Supplementary Exercises

1. Compute $|A|$ for each of the following.

(a) $A = \begin{bmatrix} 2 & 3 & 4 \\ 1 & 2 & 4 \\ 4 & 3 & 1 \end{bmatrix}$. (b) $A = \begin{bmatrix} 2 & 1 & 0 \\ 1 & 2 & 1 \\ 0 & 1 & 2 \end{bmatrix}$.

(c) $A = \begin{bmatrix} 2 & 1 & -1 & 2 \\ 2 & -3 & -1 & 4 \\ 1 & 3 & 2 & -3 \\ 1 & -2 & -1 & 1 \end{bmatrix}$.

(d) $A = \begin{bmatrix} 2 & 1 & 0 & 0 \\ 1 & 2 & 1 & 0 \\ 0 & 1 & 2 & 1 \\ 0 & 0 & 1 & 2 \end{bmatrix}$.

2. Find all values of t for which $|tI_3 - A| = 0$ for each of the following.

(a) $A = \begin{bmatrix} 1 & 4 & 0 \\ 0 & 4 & 0 \\ 0 & 0 & 1 \end{bmatrix}$. (b) $A = \begin{bmatrix} 2 & -2 & 0 \\ -3 & 1 & 0 \\ 0 & 0 & 3 \end{bmatrix}$.

(c) $A = \begin{bmatrix} 0 & 1 & 0 \\ 0 & 0 & 1 \\ 6 & -11 & 6 \end{bmatrix}$.

(d) $A = \begin{bmatrix} 0 & 1 & 0 \\ 0 & 0 & 1 \\ 3 & 1 & -3 \end{bmatrix}$.

3. Show that if $A^n = O$ for some positive integer n (i.e., if A is a nilpotent matrix), then $\det(A) = 0$.

4. Using only elementary row or elementary column operations and Theorems 5.2, 5.5, and 5.6 (do not expand the determinants), verify the following.

(a) $\begin{vmatrix} a-b & 1 & a \\ b-c & 1 & b \\ c-a & 1 & c \end{vmatrix} = \begin{vmatrix} a & 1 & b \\ b & 1 & c \\ c & 1 & a \end{vmatrix}$.

(b) $\begin{vmatrix} 1 & a & bc \\ 1 & b & ca \\ 1 & c & ab \end{vmatrix} = \begin{vmatrix} 1 & a & a^2 \\ 1 & b & b^2 \\ 1 & c & c^2 \end{vmatrix}$.

5. If (x_1, y_1) and (x_2, y_2) are distinct points in the plane, show that

$$\begin{vmatrix} x & y & 1 \\ x_1 & y_1 & 1 \\ x_2 & y_2 & 1 \end{vmatrix} = 0$$

is the equation of the line through (x_1, y_1) and (x_2, y_2). Use this result to develop a test for collinearity of three points.

6. Let $P_i(x_i, y_i, z_i)$, $i = 1, 2, 3$, be three points in 3-space. Show that

$$\begin{vmatrix} x & y & z & 1 \\ x_1 & y_1 & z_1 & 1 \\ x_2 & y_2 & z_2 & 1 \\ x_3 & y_3 & z_3 & 1 \end{vmatrix} = 0$$

is the equation of a plane (see Section 3.2) through points P_i, $i = 1, 2, 3$.

7. Show that if A is an $n \times n$ matrix, then $\det(AA^T) \geq 0$.

8. Prove or disprove that the determinant function is a linear transformation of ${}_nR_n$ into R^1.

9. Show that if A is a nonsingular matrix, then adj A is nonsingular and

$$(\text{adj } A)^{-1} = \frac{1}{\det(A)} A = \text{adj }(A^{-1}).$$

10. Prove that if two rows (columns) of the $n \times n$ matrix A are proportional, then $\det(A) = 0$.

11. Let Q be the $n \times n$ real matrix in which each entry is 1. Show that $\det(Q - nI_n) = 0$.

12. Let A be an $n \times n$ matrix with integer entries. Prove that A is nonsingular and A^{-1} has integer entries if and only if $\det(A) = \pm 1$.

6 Eigenvalues and Eigenvectors

6.1. *Diagonalization*

At this point we return to the question raised at the end of Section 4.5: If L: $V \to V$ is a linear transformation of an n-dimensional vector space V into itself (a linear operator on V), when and how can we find a basis S for V such that L is represented with respect to S by a diagonal matrix D? In this section we formulate this problem precisely; we also define some pertinent terminology. Later in the chapter we settle the question for symmetric matrices ($A = A^T$) and briefly discuss the situation in the general case. In this chapter every matrix is square.

Definition 6.1. Let L: $V \to V$ be a linear transformation of an n-dimensional vector space V into itself. We say that L is **diagonalizable** or can be **diagonalized** if there exists a basis S for V such that L is represented with respect to S by a diagonal matrix D. ■

Example 1. In Example 2 of Section 4.5 we considered the linear transformation L: $R_3 \to R_3$ defined by

$$L([a_1 \quad a_2 \quad a_3]) = [2a_1 - a_3 \quad a_1 + a_2 - a_3 \quad a_3].$$

In that example we used the basis

$$S' = \{[1 \quad 0 \quad 1], [0 \quad 1 \quad 0], [1 \quad 1 \quad 0]\}$$

for R_3 and showed that the representation of L with respect to S' is

$$B = \begin{bmatrix} 1 & 0 & 0 \\ 0 & 1 & 0 \\ 0 & 0 & 2 \end{bmatrix}.$$

Hence L is a diagonalizable linear transformation. ■

Let $L: V \rightarrow V$ be a diagonalizable linear transformation of an n-dimensional vector space V into itself and let $S = \{\mathbf{x}_1, \mathbf{x}_2, \ldots, \mathbf{x}_n\}$ be a basis for V such that L is represented with respect to S by a diagonal matrix

$$D = \begin{bmatrix} \lambda_1 & 0 & \cdots & & 0 \\ 0 & \lambda_2 & \cdots & & 0 \\ \vdots & \vdots & & & \vdots \\ & & & & 0 \\ 0 & 0 & \cdots & 0 & \lambda_n \end{bmatrix},$$

where $\lambda_1, \lambda_2, \ldots, \lambda_n$ are real scalars. Now recall that if D represents L with respect to S, then the jth column of D is the coordinate vector $[L(\mathbf{x}_j)]_S$ of $L(\mathbf{x}_j)$ with respect to S. Thus we have

$$[L(\mathbf{x}_j)]_S = \begin{bmatrix} 0 \\ 0 \\ \vdots \\ 0 \\ \lambda_j \\ 0 \\ \vdots \\ 0 \end{bmatrix} \leftarrow j\text{th row},$$

which means that

$$L(\mathbf{x}_j) = 0\mathbf{x}_1 + 0\mathbf{x}_2 + \cdots + 0\mathbf{x}_{j-1} + \lambda_j\mathbf{x}_j + 0\mathbf{x}_{j+1} + \cdots + 0\mathbf{x}_n = \lambda_j\mathbf{x}_j.$$

Conversely, let $S = \{\mathbf{x}_1, \mathbf{x}_2, \ldots, \mathbf{x}_n\}$ be a basis for V such that

$$\begin{aligned} L(\mathbf{x}_j) &= \lambda_j\mathbf{x}_j \\ &= 0\mathbf{x}_1 + 0\mathbf{x}_2 + \cdots + 0\mathbf{x}_{j-1} + \lambda_j\mathbf{x}_j \\ &\quad + 0\mathbf{x}_{j+1} + \cdots + 0\mathbf{x}_n \qquad \text{for } j = 1, 2, \ldots, n. \end{aligned}$$

We now determine the matrix representing L with respect to S. The jth column of this matrix is

$$[L(\mathbf{x}_j)]_S = \begin{bmatrix} 0 \\ 0 \\ \vdots \\ 0 \\ \lambda_j \\ 0 \\ \vdots \\ 0 \end{bmatrix}.$$

Hence

$$D = \begin{bmatrix} \lambda_1 & 0 & \cdots & & 0 \\ 0 & \lambda_2 & \cdots & & 0 \\ \vdots & \vdots & & & \vdots \\ & & & & 0 \\ 0 & 0 & \cdots & 0 & \lambda_n \end{bmatrix},$$

a diagonal matrix, represents L with respect to S, so L is diagonalizable.

The situation described in the preceding paragraph is extremely important, and we now formulate some terminology for it.

Definition 6.2. Let $L\colon V \to V$ be a linear transformation of an n-dimensional vector space V into itself (a linear operator on V). The real number λ is called an **eigenvalue** of L if there exists a nonzero vector $\mathbf{x}$ in V such that

$$L(\mathbf{x}) = \lambda\mathbf{x}. \tag{1}$$

Every nonzero vector $\mathbf{x}$ satisfying this equation is then called an **eigenvector** of L **associated with the eigenvalue** $\boldsymbol{\lambda}$. We might mention that the word "eigenvalue" is a hybrid ("eigen" in German means "proper"). Eigenvalues are also called **proper**, **characteristic**, or **latent values**, and eigenvectors are also called **proper**, **characteristic**, or **latent vectors**. ■

Note that $\mathbf{x} = \mathbf{0}_V$ always satisfies (1), but $\mathbf{0}_V$ is not an eigenvector, since we insist that an eigenvector be a nonzero vector.

In some applications one encounters matrices with complex entries and vector spaces with scalars that are complex numbers (see Sections B.1 and B.2, respectively). In such a setting the preceding definition of eigenvalue is modified so that an eigenvalue can be a real *or* a complex number. An introduction to this approach, a treatment usually presented in more advanced books, is given in Section B.2. Throughout the rest of this book we require that an eigenvalue be a real number. We can now state the following theorem, whose proof has been given above.

Theorem 6.1. *Let $L: V \to V$ be a linear transformation of an n-dimensional vector space V into itself. Then L is diagonalizable if and only if V has a basis S of eigenvectors of L. Moreover, if D is the diagonal matrix representing L with respect to S, then the entries on the main diagonal of D are the eigenvalues of L.* ■

Example 2. Let $L: V \to V$ be the linear operator defined by $L(\mathbf{x}) = 2\mathbf{x}$. We can see that the only eigenvalue of L is $\lambda = 2$ and that every nonzero vector in V is an eigenvector of L associated with the eigenvalue $\lambda = 2$. ■

Let $L: V \to V$ be a linear operator on V. If λ is an eigenvalue of L and $\mathbf{x}$ is an eigenvector associated with λ, then $\mathbf{x}$ has the property that L maps $\mathbf{x}$ into a scalar multiple of itself, the multiplier being the eigenvalue. We thus have a rather nice geometric interpretation of eigenvalues and eigenvectors.

In Figure 6.1 we show $\mathbf{x}$ and $L(\mathbf{x})$ for the cases $\lambda > 1$, $0 < \lambda < 1$, and $\lambda < 0$.

Figure 6.1

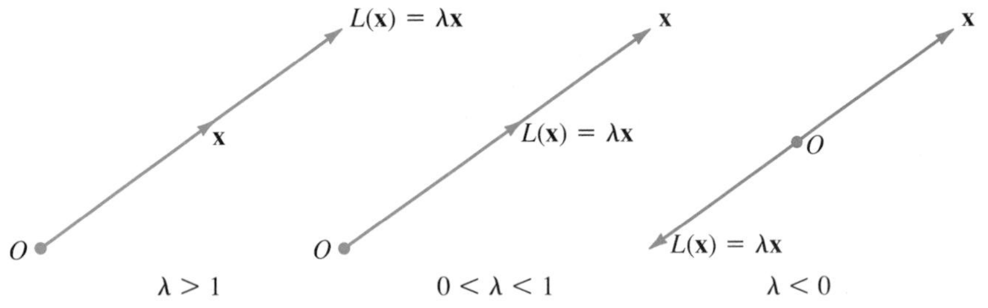

Example 3. Let $L: R^2 \to R^2$ be the linear transformation defined by $L\left(\begin{bmatrix} a_1 \\ a_2 \end{bmatrix}\right) = \begin{bmatrix} a_2 \\ a_1 \end{bmatrix}$. Then we can see that

$$L\left(\begin{bmatrix} a \\ a \end{bmatrix}\right) = 1\begin{bmatrix} a \\ a \end{bmatrix} \quad \text{and} \quad L\left(\begin{bmatrix} a \\ -a \end{bmatrix}\right) = -1\begin{bmatrix} a \\ -a \end{bmatrix}.$$

Thus any vector of the form $\begin{bmatrix} a \\ a \end{bmatrix}$, where a is any nonzero real number, for example, $\mathbf{x}_1 = \begin{bmatrix} 1 \\ 1 \end{bmatrix}$, is an eigenvector of L associated with the eigenvalue $\lambda = 1$; any vector of the form $\begin{bmatrix} a \\ -a \end{bmatrix}$, where a is any nonzero real number, such as $\mathbf{x}_2 = \begin{bmatrix} 1 \\ -1 \end{bmatrix}$, is an eigenvector of L associated with the eigenvalue $\lambda = -1$ (see Figure 6.2). ■

Figure 6.2

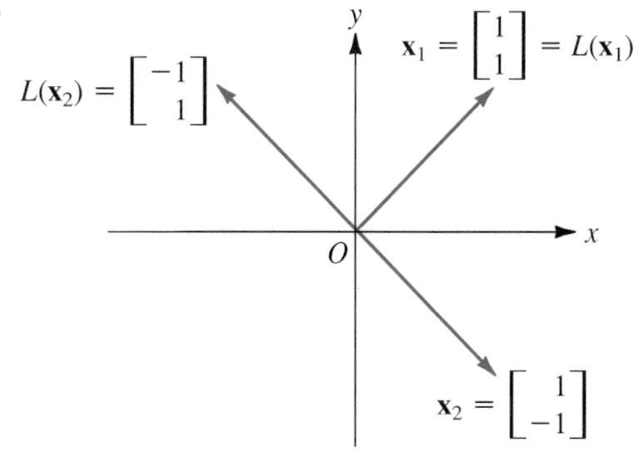

Example 2 shows that an eigenvalue λ can have associated with it many different eigenvectors. In fact, if $\mathbf{x}$ is an eigenvector of L associated with the eigenvalue λ [i.e., $L(\mathbf{x}) = \lambda\mathbf{x}$], then

$$L(r\mathbf{x}) = rL(\mathbf{x}) = r(\lambda\mathbf{x}) = \lambda(r\mathbf{x}),$$

for any real number r. Thus, if $r \neq 0$, then $r\mathbf{x}$ is also an eigenvector of L associated with λ so that eigenvectors are never unique.

Example 4. Let $L\colon R^2 \rightarrow R^2$ be the linear transformation performing a counterclockwise 90° rotation. Note that the only vector $\mathbf{x}$ mapped by L into a multiple of itself is the zero vector. Hence L has no eigenvalues and no eigenvectors. ■

Example 5. Let $L\colon R_2 \rightarrow R_2$ be defined by $L([a_1 \quad a_2]) = [0 \quad a_2]$. We can then see that

$$L([a \quad 0]) = 0[a \quad 0]$$

so that a vector of the form $[a \quad 0]$, where a is any nonzero real number (such as $[2 \quad 0]$), is an eigenvector of L associated with the eigenvalue $\lambda = 0$. Also,

$$L([0 \quad a]) = 1[0 \quad a]$$

so that a vector of the form $[0 \quad a]$ (such as $[0 \quad 1]$), where a is any nonzero real number, is an eigenvector of L associated with the eigenvalue $\lambda = 1$. ■

By definition the zero vector cannot be an eigenvector. However, Example 5 shows that the scalar zero can be an eigenvalue.

Example 6. Let V be the vector space of all real-valued differentiable functions of a single variable. Let $L: V \to V$ be the linear operator defined by

$$L(f) = f'.$$

Then the problem presented in Definition 6.2 can be stated as follows: Can we find a real number λ and a function $f \neq 0$ in V so that

$$L(f) = \lambda f? \tag{2}$$

If $y = f(x)$, then (2) can be written as

$$\frac{dy}{dx} = \lambda y. \tag{3}$$

Equation (3) states that the quantity y is one whose rate of change, with respect to x, is proportional to y itself. Examples of physical phenomena in which a quantity satisfies (3) include growth of human population, growth of bacteria and other organisms, investment problems, radioactive decay, carbon dating, and concentration of a drug in the body.

For each real number λ (an eigenvalue of L) we obtain, by using calculus, an associated eigenvector given by

$$f(x) = Ke^{\lambda x},$$

where K is an arbitrary nonzero constant. ■

Equation (3) is a simple example of a differential equation. The subject of differential equations is a major area in mathematics. In Chapter 7 we provide a brief introduction to homogeneous linear systems of differential equations.

Let L be a linear transformation of an n-dimensional vector space V into itself. If $S = \{\mathbf{x}_1, \mathbf{x}_2, \ldots, \mathbf{x}_n\}$ is a basis for V, then there is an $n \times n$ matrix A that represents L with respect to S (see Section 4.3). To determine an eigenvalue λ of L and an eigenvector $\mathbf{x}$ of L associated with the eigenvalue λ, we solve the equation

$$L(\mathbf{x}) = \lambda\mathbf{x}.$$

Using Theorem 4.8, we see that an equivalent matrix equation is

$$A[\mathbf{x}]_S = \lambda[\mathbf{x}]_S.$$

This formulation allows us to use techniques for solving linear systems in R^n to determine eigenvalue–eigenvector pairs of L.

Example 7. Let L: $P_2 \to P_2$ be a linear transformation defined by

$$L(at^2 + bt + c) = ct^2 + bt + a.$$

The eigen-problem for L can be formulated in terms of a matrix representing L with respect to a specific basis for P_2. Find the corresponding matrix eigen-problem for each of the bases $S = \{1 - t, 1 + t, t^2\}$ and $T = \{1, t, t^2\}$ for P_2.

Solution: To find the matrix A that represents L with respect to S, we compute (verify)

$$L(1 - t) = t^2 - t = \tfrac{1}{2}(1 - t) - \tfrac{1}{2}(1 + t) + t^2, \qquad \text{so } [L(1 - t)]_S = \begin{bmatrix} \frac{1}{2} \\ -\frac{1}{2} \\ 1 \end{bmatrix},$$

$$L(1 + t) = t^2 + t = -\tfrac{1}{2}(1 - t) + \tfrac{1}{2}(1 + t) + t^2, \qquad \text{so } [L(1 + t)]_S = \begin{bmatrix} -\frac{1}{2} \\ \frac{1}{2} \\ 1 \end{bmatrix},$$

$$L(t^2) = 1 = \tfrac{1}{2}(1 - t) + \tfrac{1}{2}(1 + t), \qquad \text{so } [L(t^2)]_S = \begin{bmatrix} \frac{1}{2} \\ \frac{1}{2} \\ 0 \end{bmatrix}.$$

Then

$$A = \begin{bmatrix} \frac{1}{2} & -\frac{1}{2} & \frac{1}{2} \\ -\frac{1}{2} & \frac{1}{2} & \frac{1}{2} \\ 1 & 1 & 0 \end{bmatrix}$$

and the matrix eigen-problem for L with respect to S is that of finding a real number λ and a nonzero vector $\mathbf{x}$ in R^3 so that

$$A\mathbf{x} = \lambda\mathbf{x}.$$

In a similar fashion we can show that the matrix B which represents L with respect to T is

$$B = \begin{bmatrix} 0 & 0 & 1 \\ 0 & 1 & 0 \\ 1 & 0 & 0 \end{bmatrix}$$

(verify) and the corresponding matrix eigen-problem for L with respect to T is

$$B\mathbf{x} = \lambda\mathbf{x}.$$

Thus the matrix eigen-problem for L depends upon the basis selected for V. ■

As we have seen in Example 7, the eigen-problem for a linear transformation can be expressed in terms of a matrix representing L. We now formulate the notions of eigenvalue and eigenvector for *any* square matrix. If A is an $n \times n$ matrix, we can consider, as in Section 4.1, the linear transformation $L: R^n \to R^n$ defined by $L(\mathbf{x}) = A\mathbf{x}$ for $\mathbf{x}$ in R^n. If λ is a scalar and $\mathbf{x} \neq \mathbf{0}$ a vector in R^n such that

$$A\mathbf{x} = \lambda\mathbf{x}, \tag{4}$$

then we shall say that λ is an **eigenvalue** of A and $\mathbf{x}$ is an **eigenvector** of A **associated with $\boldsymbol{\lambda}$**. That is, λ is an eigenvalue of L and $\mathbf{x}$ is an eigenvector of L associated with λ. We also say that A is **diagonalizable**, or can be **diagonalized**, if A is similar to a diagonal matrix D. Recall from Theorem 4.13 that if A and D are similar, then they both represent the same linear transformation $L: R^n \to R^n$, with respect to corresponding ordered bases S and T for R^n. Since L is represented by D with respect to T, we see that L is diagonalizable. Thus we can restate Theorem 6.1 as follows.

Theorem 6.2. *An $n \times n$ matrix A is similar to a diagonal matrix D if and only if R^n has a basis of eigenvectors of A. Moreover, the elements on the main diagonal of D are the eigenvalues of A.* ■

To use Theorem 6.2, we need only show that there is a set of n eigenvectors of A that are linearly independent, since n linearly independent vectors in R^n form a basis for R^n.

Example 8. Let $A = \begin{bmatrix} 1 & 1 \\ -2 & 4 \end{bmatrix}$. We wish to find the eigenvalues of A and their associated eigenvectors. Thus we wish to find all scalars λ and all nonzero vectors $\mathbf{x} = \begin{bmatrix} a_1 \\ a_2 \end{bmatrix}$ that satisfy Equation (4):

$$\begin{bmatrix} 1 & 1 \\ -2 & 4 \end{bmatrix}\begin{bmatrix} a_1 \\ a_2 \end{bmatrix} = \lambda\begin{bmatrix} a_1 \\ a_2 \end{bmatrix},$$

which yields

$$\begin{aligned} a_1 + a_2 &= \lambda a_1 \\ -2a_1 + 4a_2 &= \lambda a_2 \end{aligned} \quad \text{or} \quad \begin{aligned} (\lambda - 1)a_1 - a_2 &= 0 \\ 2a_1 + (\lambda - 4)a_2 &= 0. \end{aligned} \tag{5}$$

This homogeneous system of two equations in two unknowns has a nontrivial solution if and only if the determinant of the coefficient matrix is zero. Thus

$$\begin{vmatrix} \lambda - 1 & -1 \\ 2 & \lambda - 4 \end{vmatrix} = 0.$$

This means that

$$\lambda^2 - 5\lambda + 6 = 0 = (\lambda - 3)(\lambda - 2),$$

and so $\lambda = 2$ and $\lambda = 3$ are the eigenvalues of A. To find all eigenvectors of A associated with $\lambda = 2$, we substitute $\lambda = 2$ in Equation (4):

$$\begin{bmatrix} 1 & 1 \\ -2 & 4 \end{bmatrix}\begin{bmatrix} a_1 \\ a_2 \end{bmatrix} = 2\begin{bmatrix} a_1 \\ a_2 \end{bmatrix},$$

which yields

$$\begin{aligned} a_1 + a_2 &= 2a_1 \\ -2a_1 + 4a_2 &= 2a_2 \end{aligned} \quad \text{or} \quad \begin{aligned} (2 - 1)a_1 - a_2 &= 0 \\ 2a_1 + (2 - 4)a_2 &= 0 \end{aligned}$$

$$\text{or} \quad \begin{aligned} a_1 - a_2 &= 0 \\ 2a_1 - 2a_2 &= 0. \end{aligned}$$

Note that we could have obtained this last homogeneous system by merely substituting $\lambda = 2$ in (5). All solutions to this last system are given by

$$\begin{aligned} a_1 &= a_2 \\ a_2 &= \text{any real number } r. \end{aligned}$$

Hence all eigenvectors associated with the eigenvalue $\lambda_1 = 2$ are given by $\begin{bmatrix} r \\ r \end{bmatrix}$, r any nonzero real number. In particular, for $r = 1$, $\mathbf{x}_1 = \begin{bmatrix} 1 \\ 1 \end{bmatrix}$ is an eigenvector associated with the eigenvalue $\lambda_1 = 2$. Similarly, substituting $\lambda_2 = 3$ in Equation (5), we obtain

$$\begin{aligned} (3-1)a_1 \qquad - a_2 &= 0 \\ 2a_1 + (3-4)a_2 &= 0 \end{aligned} \qquad \text{or} \qquad \begin{aligned} 2a_1 - a_2 &= 0 \\ 2a_1 - a_2 &= 0. \end{aligned}$$

All solutions to this last system are given by

$$\begin{aligned} a_1 &= \tfrac{1}{2}a_2 \\ a_2 &= \text{any real number } r. \end{aligned}$$

Hence all eigenvectors associated with the eigenvalue $\lambda_2 = 3$ are given by $\begin{bmatrix} r/2 \\ r \end{bmatrix}$, r any nonzero real number. In particular, for $r = 2$, $\mathbf{x}_2 = \begin{bmatrix} 1 \\ 2 \end{bmatrix}$ is an eigenvector associated with the eigenvalue $\lambda_2 = 3$.

Since $S = \left\{ \begin{bmatrix} 1 \\ 1 \end{bmatrix}, \begin{bmatrix} 1 \\ 2 \end{bmatrix} \right\}$ is linearly independent (verify), R^2 has a basis of two eigenvectors of A and hence A can be diagonalized. From Theorem 6.2 we conclude that A is similar to $D = \begin{bmatrix} 2 & 0 \\ 0 & 3 \end{bmatrix}$. ■

Example 9. Let $A = \begin{bmatrix} 1 & 1 \\ 0 & 1 \end{bmatrix}$. To find out if A can be diagonalized, we form Equation (4) with $\mathbf{x} = \begin{bmatrix} a_1 \\ a_2 \end{bmatrix}$:

$$\begin{bmatrix} 1 & 1 \\ 0 & 1 \end{bmatrix} \begin{bmatrix} a_1 \\ a_2 \end{bmatrix} = \lambda \begin{bmatrix} a_1 \\ a_2 \end{bmatrix}.$$

This means that

$$a_1 + a_2 = \lambda a_1 \qquad \text{or} \qquad (\lambda - 1)a_1 \quad - a_2 = 0$$
$$a_2 = \lambda a_2 \qquad\qquad\qquad (\lambda - 1)a_2 = 0.$$

This homogeneous system has a nontrivial solution only for $\lambda = 1$. If $\lambda = 1$, then $a_2 = 0$. Thus $\begin{bmatrix} 1 \\ 0 \end{bmatrix}$ is an eigenvector of A associated with the eigenvalue $\lambda = 1$. In fact, every eigenvector associated with $\lambda = 1$ is of the form $\begin{bmatrix} r \\ 0 \end{bmatrix}$, with $r \neq 0$. Since every eigenvector of A associated with $\lambda = 1$ is a multiple of $\begin{bmatrix} 1 \\ 0 \end{bmatrix}$, A does not have two linearly independent eigenvectors. Hence A cannot be diagonalized. ■

If an $n \times n$ matrix A is similar to a diagonal matrix D, then $P^{-1}AP = D$ for some nonsingular matrix P. This means that $AP = PD$. Let

$$D = \begin{bmatrix} \lambda_1 & 0 & \cdots & & 0 \\ 0 & \lambda_2 & \cdots & & 0 \\ \vdots & 0 & \cdots & & \vdots \\ & \vdots & & & 0 \\ 0 & 0 & \cdots & 0 & \lambda_n \end{bmatrix},$$

and let $\mathbf{x}_j$, $j = 1, 2, \ldots, n$, be the jth column of P. Note that the jth column of AP is $A\mathbf{x}_j$, and the jth column of PD is $\lambda_j\mathbf{x}_j$ (see Exercise 26 in Section 1.2). Thus we have

$$A\mathbf{x}_j = \lambda_j\mathbf{x}_j,$$

which means that λ_j is an eigenvalue of A and $\mathbf{x}_j$ is an associated eigenvector.

Conversely, if $\lambda_1, \lambda_2, \ldots, \lambda_n$ are n eigenvalues of an $n \times n$ matrix A and $\mathbf{x}_1$, $\mathbf{x}_2, \ldots, \mathbf{x}_n$ are associated eigenvectors forming a linearly independent set in R^n, we let P be the matrix whose jth column is $\mathbf{x}_j$. From Theorem 5.14, P is nonsingular. Since $A\mathbf{x}_j = \lambda_j\mathbf{x}_j$, $j = 1, 2, \ldots, n$, we have $AP = PD$, or $P^{-1}AP = D$, which means that A is diagonalizable. Thus, if n eigenvectors $\mathbf{x}_1, \mathbf{x}_2, \ldots, \mathbf{x}_n$ of the $n \times n$ matrix A form a linearly independent set in R^n, we can diagonalize A by letting P be the matrix whose jth column is $\mathbf{x}_j$, and we find that $P^{-1}AP = D$, a diagonal matrix whose entries on the main diagonal are the associated eigenvalues of A. Of course, the order of the columns of P determines the order of the diagonal entries of D.

Example 10. Let A be as in Example 8. The eigenvalues of A are $\lambda_1 = 2$ and $\lambda_2 = 3$, and associated eigenvectors are $\mathbf{x}_1 = \begin{bmatrix} 1 \\ 1 \end{bmatrix}$ and $\mathbf{x}_2 = \begin{bmatrix} 1 \\ 2 \end{bmatrix}$, respectively. Thus $P = \begin{bmatrix} 1 & 1 \\ 1 & 2 \end{bmatrix}$ and $P^{-1} = \begin{bmatrix} 2 & -1 \\ -1 & 1 \end{bmatrix}$ (verify). Hence

$$P^{-1}AP = \begin{bmatrix} 2 & -1 \\ -1 & 1 \end{bmatrix} \begin{bmatrix} 1 & 1 \\ -2 & 4 \end{bmatrix} \begin{bmatrix} 1 & 1 \\ 1 & 2 \end{bmatrix} = \begin{bmatrix} 2 & 0 \\ 0 & 3 \end{bmatrix}.$$

On the other hand, if we let $\lambda_1 = 3$ and $\lambda_2 = 2$, then $\mathbf{x}_1 = \begin{bmatrix} 1 \\ 2 \end{bmatrix}$ and $\mathbf{x}_2 = \begin{bmatrix} 1 \\ 1 \end{bmatrix}$;

$P = \begin{bmatrix} 1 & 1 \\ 2 & 1 \end{bmatrix}$ and $P^{-1} = \begin{bmatrix} -1 & 1 \\ 2 & -1 \end{bmatrix}$, and

$$P^{-1}AP = \begin{bmatrix} -1 & 1 \\ 2 & -1 \end{bmatrix} \begin{bmatrix} 1 & 1 \\ -2 & 4 \end{bmatrix} \begin{bmatrix} 1 & 1 \\ 2 & 1 \end{bmatrix} = \begin{bmatrix} 3 & 0 \\ 0 & 2 \end{bmatrix}.$$

■

In Examples 2 through 5 we found eigenvalues and eigenvectors by inspection. In Example 8 we proceeded in a more systematic fashion. We can use the procedure of Example 8 as our standard method, as follows.

Definition 6.3. Let $A = \begin{bmatrix} a_{11} & a_{12} & \cdots & a_{1n} \\ a_{21} & a_{22} & \cdots & a_{2n} \\ \vdots & \vdots & & \vdots \\ a_{n1} & a_{n2} & \cdots & a_{nn} \end{bmatrix}$ be an $n \times n$ matrix. Then the determinant of the matrix

$$\lambda I_n - A = \begin{bmatrix} \lambda - a_{11} & -a_{12} & \cdots & -a_{1n} \\ -a_{21} & \lambda - a_{22} & \cdots & -a_{2n} \\ \vdots & \vdots & & \vdots \\ -a_{n1} & -a_{n2} & \cdots & \lambda - a_{nn} \end{bmatrix}$$

is called the **characteristic polynomial** of A. The equation

$$f(\lambda) = \det(\lambda I_n - A) = 0$$

is called the **characteristic equation** of A. ■

Recall from Chapter 5 that each term in the expansion of the determinant of an $n \times n$ matrix is a product of n entries of the matrix, containing exactly one entry from each row and exactly one entry from each column. Thus, if we expand $|\lambda I_n - A|$, we obtain a polynomial of degree n. The expression involving λ^n in the characteristic polynomial of A comes from the product

$$(\lambda - a_{11})(\lambda - a_{22}) \cdots (\lambda - a_{nn}),$$

and so the coefficient of λ^n is 1. We can then write

$$\det(\lambda I_n - A) = f(\lambda) = \lambda^n + a_1\lambda^{n-1} + a_2\lambda^{n-2} + \cdots + a_{n-1}\lambda + a_n.$$

Note that if we let $\lambda = 0$ in $\det(\lambda I_n - A)$ as well as in the expression on the right, then we get $\det(-A) = a_n$, and thus the constant term of the characteristic polynomial of A is $a_n = (-1)^n \det(A)$.

Example 11. Let $A = \begin{bmatrix} 1 & 2 & -1 \\ 1 & 0 & 1 \\ 4 & -4 & 5 \end{bmatrix}$. The characteristic polynomial of A is

$$f(\lambda) = \det(\lambda I_3 - A) = \begin{vmatrix} \lambda - 1 & -2 & 1 \\ -1 & \lambda & -1 \\ -4 & 4 & \lambda - 5 \end{vmatrix} = \lambda^3 - 6\lambda^2 + 11\lambda - 6$$

(verify). ■

We now connect the characteristic polynomial of a matrix with its eigenvalues in the following theorem.

Theorem 6.3. *Let A be an $n \times n$ matrix. The eigenvalues of A are the real roots of the characteristic polynomial of A.*

Proof: Let $\mathbf{x}$ in R^n be an eigenvector of A associated with the eigenvalue λ. Then

$$A\mathbf{x} = \lambda\mathbf{x}, \qquad \text{or} \qquad A\mathbf{x} = (\lambda I_n)\mathbf{x}, \qquad \text{or} \qquad (\lambda I_n - A)\mathbf{x} = \mathbf{0}.$$

This is a homogeneous system of n equations in n unknowns; a nontrivial solution exists if and only if $\det(\lambda I_n - A) = 0$. Hence λ is a real root of the characteristic polynomial of A.

Conversely, if λ is a real root of the characteristic polynomial of A, then $\det(\lambda I_n - A) = 0$, so the homogeneous system $(\lambda I_n - A)\mathbf{x} = \mathbf{0}$ has a nontrivial solution. Hence λ is an eigenvalue of A. ■

Thus, to find the eigenvalues of a given matrix A, we must find the real roots of the characteristic polynomial $f(\lambda)$. There are many methods for finding approximations to the roots of a polynomial, some of them more effective than others. Two results that are sometimes useful in this connection are: (1) the product of all the roots of the polynomial

$$f(\lambda) = \lambda^n + a_1\lambda^{n-1} + \cdots + a_{n-1}\lambda + a_n$$

is $(-1)^n a_n$, and (2) if $a_1, a_2, \ldots, a_n$ are integers, then $f(\lambda)$ cannot have a rational root that is not already an integer. Thus as possible rational roots of $f(\lambda)$ one need only try the integer factors of a_n. Of course, $f(\lambda)$ might well have irrational roots.

To minimize the computational effort and as a convenience to the reader, *all the characteristic polynomials to be solved in the rest of this chapter have only integer roots,* and each of these roots is a factor of the constant term of the characteristic polynomial of A. The corresponding eigenvectors are obtained by substituting for λ in the matrix equation

$$(\lambda I_n - A)\mathbf{x} = \mathbf{0} \tag{6}$$

and solving the resulting homogeneous system.

Example 12. Compute the eigenvalues and associated eigenvectors of the matrix A defined in Example 11.

Solution: In Example 11 we found the characteristic polynomial of A to be

$$f(\lambda) = \lambda^3 - 6\lambda^2 + 11\lambda - 6.$$

The possible integer roots of $f(\lambda)$ are ± 1, ± 2, ± 3, and ± 6. By substituting these values in $f(\lambda)$, we find that $f(1) = 0$, so that $\lambda = 1$ is a root of $f(\lambda)$. Hence $(\lambda - 1)$ is a factor of $f(\lambda)$. Dividing $f(\lambda)$ by $(\lambda - 1)$, we obtain

$$f(\lambda) = (\lambda - 1)(\lambda^2 - 5\lambda + 6) \text{ (verify)}.$$

Factoring $\lambda^2 - 5\lambda + 6$, we have

$$f(\lambda) = (\lambda - 1)(\lambda - 2)(\lambda - 3).$$

The eigenvalues of A are then $\lambda_1 = 1$, $\lambda_2 = 2$, and $\lambda_3 = 3$. To find an eigenvector $\mathbf{x}_1$ associated with $\lambda_1 = 1$, we substitute $\lambda = 1$ in (6), obtaining

$$\begin{bmatrix} 1-1 & -2 & 1 \\ -1 & 1 & -1 \\ -4 & 4 & 1-5 \end{bmatrix} \begin{bmatrix} a_1 \\ a_2 \\ a_3 \end{bmatrix} = \begin{bmatrix} 0 \\ 0 \\ 0 \end{bmatrix}$$

or

$$\begin{bmatrix} 0 & -2 & 1 \\ -1 & 1 & -1 \\ -4 & 4 & -4 \end{bmatrix} \begin{bmatrix} a_1 \\ a_2 \\ a_3 \end{bmatrix} = \begin{bmatrix} 0 \\ 0 \\ 0 \end{bmatrix}.$$

The vector $\begin{bmatrix} -\frac{r}{2} \\ \frac{r}{2} \\ r \end{bmatrix}$ is a solution for any real number r. Thus $\mathbf{x}_1 = \begin{bmatrix} -1 \\ 1 \\ 2 \end{bmatrix}$ is an eigenvector of A associated with $\lambda_1 = 1$ (r was taken as 2).

To find an eigenvector $\mathbf{x}_2$ associated with $\lambda_2 = 2$, we substitute $\lambda = 2$ in (6), obtaining

$$\begin{bmatrix} 2-1 & -2 & 1 \\ -1 & 2 & -1 \\ -4 & 4 & 2-5 \end{bmatrix} \begin{bmatrix} a_1 \\ a_2 \\ a_3 \end{bmatrix} = \begin{bmatrix} 0 \\ 0 \\ 0 \end{bmatrix}$$

or

$$\begin{bmatrix} 1 & -2 & 1 \\ -1 & 2 & -1 \\ -4 & 4 & -3 \end{bmatrix} \begin{bmatrix} a_1 \\ a_2 \\ a_3 \end{bmatrix} = \begin{bmatrix} 0 \\ 0 \\ 0 \end{bmatrix}.$$

The vector $\begin{bmatrix} -\dfrac{r}{2} \\ \dfrac{r}{4} \\ r \end{bmatrix}$ is a solution for any real number r. Thus $\mathbf{x}_2 = \begin{bmatrix} -2 \\ 1 \\ 4 \end{bmatrix}$ is an eigenvector of A associated with $\lambda_2 = 2$ (r was taken as 4).

To find an eigenvector $\mathbf{x}_3$ associated with $\lambda_3 = 3$, we substitute $\lambda = 3$ in (6), obtaining

$$\begin{bmatrix} 3-1 & -2 & 1 \\ -1 & 3 & -1 \\ -4 & 4 & 3-5 \end{bmatrix} \begin{bmatrix} a_1 \\ a_2 \\ a_3 \end{bmatrix} = \begin{bmatrix} 0 \\ 0 \\ 0 \end{bmatrix}$$

or

$$\begin{bmatrix} 2 & -2 & 1 \\ -1 & 3 & -1 \\ -4 & 4 & -2 \end{bmatrix} \begin{bmatrix} a_1 \\ a_2 \\ a_3 \end{bmatrix} = \begin{bmatrix} 0 \\ 0 \\ 0 \end{bmatrix}.$$

The vector $\begin{bmatrix} -\dfrac{r}{4} \\ \dfrac{r}{4} \\ r \end{bmatrix}$ is a solution for any real number r. Thus $\mathbf{x}_3 = \begin{bmatrix} -1 \\ 1 \\ 4 \end{bmatrix}$ is an eigenvector of A associated with $\lambda_3 = 3$ (r was taken as 4). ■

Of course, the characteristic polynomial of a matrix may have complex roots and it may even have no real roots. However, in the important case of symmetric matrices, all the roots of the characteristic polynomial are real. We shall prove this in Section 6.2 (Theorem 6.6).

Example 13. The characteristic polynomial of $A = \begin{bmatrix} 0 & 1 \\ -1 & 0 \end{bmatrix}$ is $f(\lambda) = \lambda^2 + 1$, which has no real roots. Thus A has no eigenvalues, according to our definition, and cannot be diagonalized. ■

Now that we know how to find the eigenvalues of A in a systematic fashion, the following is a useful theorem, because it identifies a large class of matrices that can be diagonalized.

Theorem 6.4. *An $n \times n$ matrix A is diagonalizable if all the roots of its characteristic polynomial are real and distinct.*

Proof: Let $\{\lambda_1, \lambda_2, \ldots, \lambda_n\}$ be the set of distinct eigenvalues of A, and let $S = \{\mathbf{x}_1, \mathbf{x}_2, \ldots, \mathbf{x}_n\}$ be a set of associated eigenvectors. We wish to prove that S is a basis for R^n, and it suffices to show that S is linearly independent.

Suppose that S is linearly dependent. Then Theorem 2.5 implies that some vector $\mathbf{x}_j$ is a linear combination of the preceding vectors in S. We can assume that $S_1 = \{\mathbf{x}_1, \mathbf{x}_2, \ldots, \mathbf{x}_{j-1}\}$ is linearly independent, for otherwise one of the vectors in S_1 is a linear combination of the preceding ones, and we can choose a new set S_2, and so on. We thus have that S_1 is linearly independent and that

$$\mathbf{x}_j = a_1\mathbf{x}_1 + a_2\mathbf{x}_2 + \cdots + a_{j-1}\mathbf{x}_{j-1}, \tag{7}$$

where $a_1, a_2, \ldots, a_{j-1}$ are real numbers. This means that

$$\begin{aligned} A\mathbf{x}_j &= A(a_1\mathbf{x}_1 + a_2\mathbf{x}_2 + \cdots + a_{j-1}\mathbf{x}_{j-1}) \\ &= a_1A\mathbf{x}_1 + a_2A\mathbf{x}_2 + \cdots + a_{j-1}A\mathbf{x}_{j-1}. \end{aligned} \tag{8}$$

Since $\lambda_1, \lambda_2, \ldots, \lambda_j$ are eigenvalues and $\mathbf{x}_1, \mathbf{x}_2, \ldots, \mathbf{x}_j$ are associated eigenvectors, we know that $A\mathbf{x}_i = \lambda_i\mathbf{x}_i$, for $i = 1, 2, \ldots, n$. Substituting in (8), we have

$$\lambda_j\mathbf{x}_j = a_1\lambda_1\mathbf{x}_1 + a_2\lambda_2\mathbf{x}_2 + \cdots + a_{j-1}\lambda_{j-1}\mathbf{x}_{j-1}. \tag{9}$$

Multiplying (7) by λ_j, we get

$$\lambda_j\mathbf{x}_j = \lambda_j a_1\mathbf{x}_1 + \lambda_j a_2\mathbf{x}_2 + \cdots + \lambda_j a_{j-1}\mathbf{x}_{j-1}. \tag{10}$$

Subtracting (10) from (9), we have

$$\mathbf{0} = \lambda_j\mathbf{x}_j - \lambda_j\mathbf{x}_j = a_1(\lambda_1 - \lambda_j)\mathbf{x}_1 + a_2(\lambda_2 - \lambda_j)\mathbf{x}_2 + \cdots + a_{j-1}(\lambda_{j-1} - \lambda_j)\mathbf{x}_{j-1}.$$

Since S_1 is linearly independent, we must have

$$a_1(\lambda_1 - \lambda_j) = 0, \quad a_2(\lambda_2 - \lambda_j) = 0, \quad \ldots, \quad a_{j-1}(\lambda_{j-1} - \lambda_j) = 0.$$

Now $(\lambda_1 - \lambda_j) \neq 0$, $(\lambda_2 - \lambda_j) \neq 0$, $\ldots$, $(\lambda_{j-1} - \lambda_j) \neq 0$, since the λ's are distinct, which implies that

$$a_1 = a_2 = \cdots = a_{j-1} = 0.$$

This means that $\mathbf{x}_j = \mathbf{0}$, which is impossible if $\mathbf{x}_j$ is an eigenvector. Hence S is linearly independent, so A is diagonalizable. ■

In the proof of Theorem 6.4 we have actually proved that eigenvectors associated with distinct eigenvalues form a linearly independent set (Exercise 30).

If all the roots of the characteristic polynomial of A are real and not all distinct, then A may or may not be diagonalizable. The characteristic polynomial of A can be written as the product of n factors, each of the form $\lambda - \lambda_0$, where λ_0 is a root of the characteristic polynomial. Now the eigenvalues of A are the real roots of the characteristic polynomial of A. Thus the characteristic polynomial can be written as

$$(\lambda - \lambda_1)^{k_1}(\lambda - \lambda_2)^{k_2} \cdots (\lambda - \lambda_r)^{k_r},$$

where $\lambda_1, \lambda_2, \ldots, \lambda_r$ are the distinct eigenvalues of A, and $k_1, k_2, \ldots, k_r$ are integers whose sum is n. The integer k_i is called the **multiplicity** of λ_i. Thus, in Example 9, $\lambda = 1$ is an eigenvalue of $A = \begin{bmatrix} 1 & 1 \\ 0 & 1 \end{bmatrix}$ of multiplicity 2. It can be shown that if the roots of the characteristic polynomial of A are all real, then A can be diagonalized if and only if for each eigenvalue λ of multiplicity k, we can find k linearly independent eigenvectors. This means that the solution space of the homogeneous system $(\lambda I_n - A)\mathbf{x} = \mathbf{0}$ has dimension k. It can also be shown that if λ is an eigenvalue of A of multiplicity k, then we can never find more than k linearly independent eigenvectors associated with λ.

Example 14. Let $A = \begin{bmatrix} 0 & 0 & 1 \\ 0 & 1 & 2 \\ 0 & 0 & 1 \end{bmatrix}$. The characteristic polynomial of A is $f(\lambda) = \lambda(\lambda - 1)^2$ (verify), so the eigenvalues of A are $\lambda_1 = 0$, $\lambda_2 = 1$, and $\lambda_3 = 1$; $\lambda_2 = 1$ is an eigenvalue of multiplicity 2. We now consider the eigenvectors associated with the eigenvalues $\lambda_2 = \lambda_3 = 1$. They are obtained by solving the homogeneous system (6):

$$\begin{bmatrix} 1 & 0 & -1 \\ 0 & 0 & -2 \\ 0 & 0 & 0 \end{bmatrix} \begin{bmatrix} a_1 \\ a_2 \\ a_3 \end{bmatrix} = \begin{bmatrix} 0 \\ 0 \\ 0 \end{bmatrix}.$$

The solutions are the vectors of the form $\begin{bmatrix} 0 \\ r \\ 0 \end{bmatrix}$, where r is any real number, so the dimension of the solution space of $(1I_3 - A)\mathbf{x} = \mathbf{0}$ is 1 (why?), and we cannot find two linearly independent eigenvectors. Thus A cannot be diagonalized. ■

Example 15. Let $A = \begin{bmatrix} 0 & 0 & 0 \\ 0 & 1 & 0 \\ 1 & 0 & 1 \end{bmatrix}$. The characteristic polynomial of A is $f(\lambda) = \lambda(\lambda - 1)^2$ (verify), so the eigenvalues of A are $\lambda_1 = 0$, $\lambda_2 = 1$, and $\lambda_3 = 1$; and $\lambda_2 = 1$ is again an eigenvalue of multiplicity 2. Now we consider the eigenvectors associated with the eigenvalues $\lambda_2 = \lambda_3 = 1$. They are obtained by solving the homogeneous system (6):

$$\begin{bmatrix} 1 & 0 & 0 \\ 0 & 0 & 0 \\ -1 & 0 & 0 \end{bmatrix} \begin{bmatrix} a_1 \\ a_2 \\ a_3 \end{bmatrix} = \begin{bmatrix} 0 \\ 0 \\ 0 \end{bmatrix}.$$

The solutions are the vectors of the form $\begin{bmatrix} 0 \\ r \\ s \end{bmatrix}$ for any real numbers r and s.

Thus $\mathbf{x}_2 = \begin{bmatrix} 0 \\ 1 \\ 0 \end{bmatrix}$ and $\mathbf{x}_3 = \begin{bmatrix} 0 \\ 0 \\ 1 \end{bmatrix}$ are eigenvectors.

Next we look for an eigenvector associated with $\lambda_1 = 0$. We have to solve (6):

$$\begin{bmatrix} 0 & 0 & 0 \\ 0 & -1 & 0 \\ -1 & 0 & -1 \end{bmatrix} \begin{bmatrix} a_1 \\ a_2 \\ a_3 \end{bmatrix} = \begin{bmatrix} 0 \\ 0 \\ 0 \end{bmatrix}.$$

The solutions are the vectors of the form $\begin{bmatrix} r \\ 0 \\ -r \end{bmatrix}$ for any real number r. Thus

$\mathbf{x}_1 = \begin{bmatrix} 1 \\ 0 \\ -1 \end{bmatrix}$ is an eigenvector associated with $\lambda_1 = 0$. Now $S = \{\mathbf{x}_1, \mathbf{x}_2, \mathbf{x}_3\}$ is linearly independent and so A can be diagonalized. ■

Thus a matrix may fail to be diagonalizable either because not all the roots of its characteristic polynomial are real numbers, or because its eigenvectors do not form a basis for R^n.

We next prove an important result as Theorem 6.5.

Theorem 6.5. Similar matrices have the same characteristic polynomial.

Proof: Let A and B be similar. Then $B = P^{-1}AP$, for some nonsingular matrix P. Let $f_A(\lambda)$ and $f_B(\lambda)$ be the characteristic polynomials of A and B, respectively. Then

$$\begin{aligned} f_B(\lambda) &= \det(\lambda I_n - B) = \det(\lambda I_n - P^{-1}AP) = \det(P^{-1}\lambda I_n P - P^{-1}AP) \\ &= \det(P^{-1}(\lambda I_n - A)P) = \det(P^{-1}) \det(\lambda I_n - A) \det(P) \\ &= \det(P^{-1}) \det(P) \det(\lambda I_n - A) = \det(\lambda I_n - A) = f_A(\lambda). \end{aligned}$$

■

Note that in the proof of Theorem 6.5 we have used the fact that the product of det (P^{-1}) and det (P) is 1, and that determinants are real numbers, so their order as factors in multiplication does not matter.

We now define the **characteristic polynomial** of a linear operator $L: V \to V$, as the characteristic polynomial of any matrix representing L; by Theorem 6.5 all representations of L will give the same characteristic polynomial. Of course, a scalar λ is an eigenvalue of L if and only if λ is a real root of the characteristic polynomial of L.

Example 16. In Example 7 we derived the matrix eigen-problem for the linear transformation $L: P_2 \to P_2$ defined by $L(at^2 + bt + c) = ct^2 + bt + a$. To compute the eigenvalues and corresponding eigenvectors of L, we could use the matrix eigenproblem for either basis S or T. Since it appears that the matrix

$$B = \begin{bmatrix} 0 & 0 & 1 \\ 0 & 1 & 0 \\ 1 & 0 & 0 \end{bmatrix},$$

the matrix representing L with respect to the basis $T = \{1, t, t^2\}$ for P_2 may be easier to use, we will find its eigenvalues and corresponding eigenvectors and transform them to the eigenvalues and eigenvectors of L.

The characteristic polynomial of B is $f(\lambda) = (\lambda + 1)(\lambda - 1)^2$ (verify), so the eigenvalues of B are $\lambda_1 = -1$, $\lambda_2 = 1$, and $\lambda_3 = 1$; λ_2 is an eigenvalue of multiplicity 2. An eigenvector associated with λ_1 is $\begin{bmatrix} 1 \\ 0 \\ -1 \end{bmatrix}$ and a pair of linearly independent eigenvectors associated with λ_2 are $\begin{bmatrix} 1 \\ 0 \\ 1 \end{bmatrix}$ and $\begin{bmatrix} 0 \\ 1 \\ 0 \end{bmatrix}$ (verify). We must recall that the eigenvectors of B represent the coordinates of a vector in P_2 with respect to the T-basis. Thus an eigenvector of L associated with λ_1 is $1(1) + 0(t) - 1(t^2) = -t^2 + 1$ and associated with λ_2, we have the linearly independent eigenvectors $1 + t^2$ and t (verify). ■

Eigenvalues and eigenvectors satisfy many important and interesting properties. For example, if A is an upper (lower) triangular matrix, then the eigenvalues of A are the elements on the main diagonal of A, since the determinant of such a matrix is the product of the elements on its main diagonal. Other properties are developed in the exercises for this section.

It must be pointed out that the method for finding the eigenvalues of a linear transformation or matrix by obtaining the real roots of the characteristic polynomial is not practical for $n > 4$, owing to the need for evaluating a determinant. Efficient numerical methods for finding eigenvalues are studied in numerical analysis courses.

6.1. EXERCISES

1. Let $L\colon R^2 \to R^2$ be counterclockwise rotation through an angle π. Find the eigenvalues and associated eigenvectors of L.

2. Find the characteristic polynomial of each of the following matrices.

(a) $\begin{bmatrix} 1 & 2 & 1 \\ 0 & 1 & 2 \\ -1 & 3 & 2 \end{bmatrix}$. (b) $\begin{bmatrix} 2 & 1 \\ -1 & 3 \end{bmatrix}$.

(c) $\begin{bmatrix} 4 & -1 & 3 \\ 0 & 2 & 1 \\ 0 & 0 & 3 \end{bmatrix}$.

3. Find the characteristic polynomial, the eigenvalues, and their associated eigenvectors for each of the following matrices.

(a) $\begin{bmatrix} 0 & 1 & 2 \\ 0 & 0 & 3 \\ 0 & 0 & 0 \end{bmatrix}$. (b) $\begin{bmatrix} 1 & 0 & 0 \\ -1 & 3 & 0 \\ 3 & 2 & -2 \end{bmatrix}$.

(c) $\begin{bmatrix} 1 & 1 \\ 1 & 1 \end{bmatrix}$.

4. Prove that if A is an upper (lower) triangular matrix, then the eigenvalues of A are the elements on the main diagonal of A.

5. Find the characteristic polynomial, the eigenvalues and their associated eigenvectors for each of the following matrices.

(a) $\begin{bmatrix} 1 & -1 \\ 2 & 4 \end{bmatrix}$. (b) $\begin{bmatrix} 2 & -2 & 3 \\ 0 & 3 & -2 \\ 0 & -1 & 2 \end{bmatrix}$.

(c) $\begin{bmatrix} 2 & 2 & 3 \\ 1 & 2 & 1 \\ 2 & -2 & 1 \end{bmatrix}$.

6. Prove that A and A^T have the same eigenvalues. What, if anything, can we say about the associated eigenvectors of A and A^T?

7. Let $L\colon P_2 \to P_2$ be the linear transformation defined by $L(p(t)) = p'(t)$ for $p(t)$ in P_2. Is L diagonalizable? If it is, find a basis S for P_2 with respect to which L is represented by a diagonal matrix.

8. Which of the following matrices are diagonalizable?

(a) $\begin{bmatrix} 1 & 4 \\ 1 & -2 \end{bmatrix}$. (b) $\begin{bmatrix} 1 & 0 \\ -2 & 1 \end{bmatrix}$.

(c) $\begin{bmatrix} 1 & 1 & -2 \\ 4 & 0 & 4 \\ 1 & -1 & 4 \end{bmatrix}$. (d) $\begin{bmatrix} 1 & 2 & 3 \\ 0 & -1 & 2 \\ 0 & 0 & 2 \end{bmatrix}$.

(e) $\begin{bmatrix} 3 & 1 & 0 \\ 0 & 3 & 1 \\ 0 & 0 & 3 \end{bmatrix}$.

9. Let $L: V \to V$ be a linear transformation, where V is an n-dimensional vector space. Let λ be an eigenvalue of L. Prove that the subset of V consisting of $\mathbf{0}_V$ and all eigenvectors of L associated with λ is a subspace of V. This subspace is called the **eigenspace** associated with λ.

10. Prove that if λ is an eigenvalue of a matrix A with associated eigenvector $\mathbf{x}$, and k is a positive integer, then λ^k is an eigenvalue of the matrix $A^k = A \cdot A \cdot \cdots \cdot A$ (k factors) with associated eigenvector $\mathbf{x}$.

11. Let $A = \begin{bmatrix} 1 & 4 \\ 1 & -2 \end{bmatrix}$ be the matrix of Exercise 8(a). Find the eigenvalues and eigenvectors of A^2 and verify Exercise 10.

12. Prove that if $A^k = O$ for some positive integer k (i.e., if A is a nilpotent matrix), then 0 is the only eigenvalue of A. (*Hint:* Use Exercise 10.)

13. Let V be the vector space of continuous functions with basis $\{\sin t, \cos t\}$, and let $L: V \to V$ be defined as $L(g(t)) = g'(t)$; that is, L is differentiation. Is L diagonalizable?

14. For each of the following matrices find, if possible, a nonsingular matrix P such that $P^{-1}AP$ is diagonal.

(a) $\begin{bmatrix} 4 & 2 & 3 \\ 2 & 1 & 2 \\ -1 & -2 & 0 \end{bmatrix}$. (b) $\begin{bmatrix} 1 & 1 & 2 \\ 0 & 1 & 0 \\ 0 & 1 & 3 \end{bmatrix}$.

(c) $\begin{bmatrix} 1 & 2 & 3 \\ 0 & 1 & 0 \\ 2 & 1 & 2 \end{bmatrix}$. (d) $\begin{bmatrix} 0 & -1 \\ 2 & 3 \end{bmatrix}$.

(e) $\begin{bmatrix} 3 & -2 & 1 \\ 0 & 2 & 0 \\ 0 & 0 & 0 \end{bmatrix}$.

15. Let λ be an eigenvalue of the $n \times n$ matrix A. Prove that the subset of R^n consisting of the zero vector and of all eigenvectors of A associated with λ is a subspace of R^n. This subspace is called the **eigenspace** associated with λ. (This result is a corollary to the result in Exercise 9.)

16. Let $L: R_3 \to R_3$ be defined by

$$L[a_1 \quad a_2 \quad a_3] = [2a_1 + 3a_2 \quad -a_2 + 4a_3 \quad 3a_3].$$

Find the characteristic polynomial and the eigenvalues and eigenvectors of L.

17. Let A be an $n \times n$ matrix.
(a) Show that det (A) is the product of all the roots of the characteristic polynomial of A.
(b) Show that A is singular if and only if 0 is an eigenvalue of A.
(c) Also prove the analogous statement for a linear transformation: If $L: V \to V$ is a linear transformation, show that L is not one-to-one if and only if 0 is an eigenvalue of L.
(d) Show that if A is nilpotent, then A is singular.

18. Let $L: V \to V$ be an invertible linear operator and let λ be an eigenvalue of L with associated eigenvector $\mathbf{x}$.
(a) Show that $1/\lambda$ is an eigenvalue of L^{-1} with associated eigenvector $\mathbf{x}$.
(b) State and prove the analogous statement for matrices.

19. Let $L: P_2 \rightarrow P_2$ be the linear transformation defined by

$$L(at^2 + bt + c) = (2a + b + c)t^2 + (2c - 3b)t + 4c.$$

Find the eigenvalues and eigenvectors of L. Is L diagonalizable?

20. Prove that if A and B are similar matrices, then A and B have the same eigenvalues.

21. Let $A = \begin{bmatrix} a & b \\ c & d \end{bmatrix}$. Find necessary and sufficient conditions for A to be diagonalizable.

22. Let

$$A = \begin{bmatrix} 1 & 2 & 3 & 4 \\ 0 & -1 & 3 & 2 \\ 0 & 0 & 3 & 3 \\ 0 & 0 & 0 & 2 \end{bmatrix}$$

represent the linear transformation $L: {}_2R_2 \rightarrow {}_2R_2$ with respect to the basis

$$S = \left\{ \begin{bmatrix} 1 & 0 \\ 0 & 0 \end{bmatrix}, \begin{bmatrix} 0 & 1 \\ 0 & 0 \end{bmatrix}, \begin{bmatrix} 0 & 0 \\ 1 & 0 \end{bmatrix}, \begin{bmatrix} 0 & 0 \\ 0 & 1 \end{bmatrix} \right\}.$$

Find the eigenvalues and eigenvectors of L.

23. Let A and B be nonsingular $n \times n$ matrices. Prove that AB and BA have the same eigenvalues.

24. Let $A = \begin{bmatrix} 2 & 2 & 3 & 4 \\ 0 & 2 & 3 & 2 \\ 0 & 0 & 1 & 1 \\ 0 & 0 & 0 & 1 \end{bmatrix}$.

(a) Find a basis for the eigenspace (see Exercise 15) associated with the eigenvalue $\lambda_1 = 1$.

(b) Find a basis for the eigenspace associated with the eigenvalue $\lambda_2 = 2$.

25. Which of the following matrices are similar to a diagonal matrix?

(a) $\begin{bmatrix} 2 & 3 & 0 \\ 0 & 1 & 0 \\ 0 & 0 & 2 \end{bmatrix}$. (b) $\begin{bmatrix} 2 & 3 & 1 \\ 0 & 1 & 0 \\ 0 & 0 & 2 \end{bmatrix}$.

(c) $\begin{bmatrix} -3 & 0 \\ 1 & 2 \end{bmatrix}$.

26. Let $D = \begin{bmatrix} 2 & 0 \\ 0 & -2 \end{bmatrix}$. Compute D^9.

27. Let $A = \begin{bmatrix} 3 & -5 \\ 1 & -3 \end{bmatrix}$. Compute A^9. (*Hint:* Find a matrix P such that $P^{-1}AP$ is a diagonal matrix D and show that $A^9 = PD^9P^{-1}$.)

28. Let V be the vector space of continuous functions with basis $\{e^t, e^{-t}\}$. Let $L: V \rightarrow V$ be defined by $L(g(t)) = g'(t)$ for $g(t)$ in V. Show that L is diagonalizable.

29. Prove that if A is diagonalizable, then (a) A^T is diagonalizable, and (b) A^k is diagonalizable, where k is a positive integer.

30. Let $\lambda_1, \lambda_2, \ldots, \lambda_k$ be distinct eigenvalues of a matrix A with associated eigenvectors $\mathbf{x}_1, \mathbf{x}_2, \ldots, \mathbf{x}_k$. Prove that $\{\mathbf{x}_1, \mathbf{x}_2, \ldots, \mathbf{x}_k\}$ is linearly independent. (*Hint:* See the proof of Theorem 6.4.)

31. Let A and B be $n \times n$ matrices such that $A\mathbf{x} = \lambda\mathbf{x}$ and $B\mathbf{x} = \mu\mathbf{x}$. Show that:

(a) $(A + B)\mathbf{x} = (\lambda + \mu)\mathbf{x}$.

(b) $(AB)\mathbf{x} = (\lambda\mu)\mathbf{x}$.

32. The **Cayley–Hamilton theorem** states that a matrix satisfies its characteristic equation; that is, if A is an $n \times n$ matrix with characteristic polynomial

$$f(\lambda) = \lambda^n + a_1\lambda^{n-1} + \cdots + a_{n-1}\lambda + a_n,$$

then

$$A^n + a_1A^{n-1} + \cdots + a_{n-1}A + a_nI_n = O.$$

The proof and applications of this result, unfortunately, lie beyond the scope of this book. Verify the Cayley–Hamilton theorem for the following matrices.

(a) $\begin{bmatrix} 1 & 2 & 3 \\ 2 & -1 & 5 \\ 3 & 2 & 1 \end{bmatrix}$. (b) $\begin{bmatrix} 1 & 2 & 3 \\ 0 & 2 & 2 \\ 0 & 0 & -3 \end{bmatrix}$.

(c) $\begin{bmatrix} 3 & 3 \\ 2 & 4 \end{bmatrix}$.

33. Let A be an $n \times n$ matrix whose characteristic polynomial is

$$f(\lambda) = \lambda^n + a_1\lambda^{n-1} + \cdots + a_{n-1}\lambda + a_n.$$

If A is nonsingular, show that

$$A^{-1} = -\frac{1}{a_n}(A^{n-1} + a_1A^{n-2} + \cdots + a_{n-2}A + a_{n-1}I_n).$$

[*Hint:* Use the Cayley–Hamilton theorem (Exercise 32).]

34. Determine if your software has a command for finding the characteristic polynomial of a matrix A. If it does, compare the output from your software with the results in Examples 10 and 13. Software output for a characteristic polynomial often is just the set of coefficients of the polynomial with the powers of λ omitted. Carefully determine the order in which the coefficients are listed. Experiment further with the matrices in Exercises 2 and 3.

35. If your software has a command for finding the characteristic polynomial of a matrix A (see Exercise 34), it probably has another command for finding the roots of polynomials. Investigate the use of these commands in your software. The roots of the characteristic polynomial of A are the eigenvalues of A. (In this book, eigenvalues are defined to be real numbers.)

36. Assuming that your software has the commands discussed in Exercises 34 and 35, apply them to find the eigenvalues of $A = \begin{bmatrix} 0 & 1 \\ -1 & 0 \end{bmatrix}$. If your software is successful, the results should be $\lambda = i, -i$, where $i = \sqrt{-1}$. (See Appendix B and Example 13.) (**Caution:** Some software does not handle complex roots and may not permit complex elements in a matrix. Determine the situation for the software you use.)

37. Most linear algebra software has a command for automatically determining the eigenvalues of a matrix A. Determine the command available in your software. Test its behavior on Examples 11, 13, and 14. Often such a command uses techniques which are different than finding the roots of the characteristic polynomial. Use the documentation accompanying your software to find the method used. (**Warning:** It may involve ideas from Section 6.2 or more sophisticated procedures.)

38. Following the ideas in Exercise 37, determine the command in your software for obtaining the eigenvectors of a matrix. Often it is a variation of the eigenvalue command. Test it on the matrices in Examples 11, 13, 14, and 15. These examples cover the types of cases for eigenvectors that you will encounter in this course.

6.2. *Diagonalization of Symmetric Matrices*

In this section we consider the diagonalization of symmetric matrices ($A = A^T$). We restrict our attention to symmetric matrices, because they are easier to handle than general matrices and because they arise in many applied problems. One of these applications will be discussed in Section 6.3.

Theorem 6.4 assures us that an $n \times n$ matrix A is diagonalizable if it has n distinct eigenvalues; if this is not so, then A may fail to be diagonalizable. However, every symmetric matrix can be diagonalized; that is, if A is symmetric, there exists a nonsingular matrix P such that $P^{-1}AP = D$, where D is a diagonal matrix. Moreover, P has some noteworthy properties that we remark upon. We thus turn to the study of symmetric matrices in this section.

We first prove that all the roots of the characteristic polynomial of a symmetric matrix are real. Section B.2 contains examples of matrices with complex eigenvalues and provides more background and motivation for considering the case of symmetric matrices. A review of complex arithmetic appears in Section B.1.

Theorem 6.6. *All the roots of the characteristic polynomial of a real symmetric matrix are real numbers.*

Proof: Let $\lambda = a + bi$ be any root of the characteristic polynomial of A. We shall prove that $b = 0$, so that λ is a real number. Now

$$|\lambda I_n - A| = 0 = |(a + bi)I_n - A|.$$

This means that the homogeneous system

$$((a + bi)I_n - A)(\mathbf{x} + \mathbf{y}i) = \mathbf{0} = \mathbf{0} + \mathbf{0}i \tag{1}$$

has a nontrivial solution $\mathbf{x} + \mathbf{y}i$, where $\mathbf{x}$ and $\mathbf{y}$ are vectors in R^n that are not both the zero vector. Carrying out the multiplication in (1), we obtain

$$(aI_n\mathbf{x} - A\mathbf{x} - bI_n\mathbf{y}) + i(aI_n\mathbf{y} + bI_n\mathbf{x} - A\mathbf{y}) = \mathbf{0} + \mathbf{0}i. \tag{2}$$

Setting the real and imaginary parts equal to $\mathbf{0}$, we have

$$\begin{aligned} aI_n\mathbf{x} - A\mathbf{x} - bI_n\mathbf{y} &= \mathbf{0} \\ aI_n\mathbf{y} - A\mathbf{y} + bI_n\mathbf{x} &= \mathbf{0}. \end{aligned} \tag{3}$$

Forming the inner products of both sides of the first equation in (3) with $\mathbf{y}$ and of both sides of the second equation of (3) with $\mathbf{x}$, we have

$$\begin{aligned} (\mathbf{y}, aI_n\mathbf{x} - A\mathbf{x} - bI_n\mathbf{y}) &= (\mathbf{y}, \mathbf{0}) = 0 \\ (aI_n\mathbf{y} - A\mathbf{y} + bI_n\mathbf{x}, \mathbf{x}) &= (\mathbf{0}, \mathbf{x}) = 0 \end{aligned}$$

or

$$\begin{aligned} a(\mathbf{y}, I_n\mathbf{x}) - (\mathbf{y}, A\mathbf{x}) - b(\mathbf{y}, I_n\mathbf{y}) &= 0 \\ a(I_n\mathbf{y}, \mathbf{x}) - (A\mathbf{y}, \mathbf{x}) + b(I_n\mathbf{x}, \mathbf{x}) &= 0. \end{aligned} \tag{4}$$

Now, by Equation (2) in Section 3.3, we see that $(I_n\mathbf{y}, \mathbf{x}) = (\mathbf{y}, I_n^T\mathbf{x}) = (\mathbf{y}, I_n\mathbf{x})$ and that $(A\mathbf{y}, \mathbf{x}) = (\mathbf{y}, A^T\mathbf{x}) = (\mathbf{y}, A\mathbf{x})$. Note that we have used the fact that $I_n^T = I_n$ and that since A is symmetric, we have $A^T = A$. Subtracting the two equations in (4), we now get

$$-b(\mathbf{y}, I_n\mathbf{y}) - b(I_n\mathbf{x}, \mathbf{x}) = 0 \tag{5}$$

or

$$-b[(\mathbf{y}, \mathbf{y}) + (\mathbf{x}, \mathbf{x})] = 0. \tag{6}$$

Since $\mathbf{x}$ and $\mathbf{y}$ are not both the zero vector, $(\mathbf{x}, \mathbf{x}) > 0$ or $(\mathbf{y}, \mathbf{y}) > 0$. From (6) we conclude that $b = 0$. Hence every root of the characteristic polynomial of A is a real number. ■

Once we have established this result, we know that complex numbers do not enter into the study of the diagonalization problem for real symmetric matrices. Thus throughout the remainder of this book, except for Appendix B, we again deal only with real numbers.

Corollary 6.1. *If A is a symmetric $n \times n$ matrix and all the eigenvalues of A are distinct, then A is diagonalizable.*

Proof: Since A is symmetric, all the roots of its characteristic polynomial are real. From Theorem 6.4 it now follows that A can be diagonalized. Moreover, if D is the diagonal matrix that is similar to A, then the elements on the main diagonal of D are the eigenvalues of A. ■

Theorem 6.7. *If A is a symmetric $n \times n$ matrix, then eigenvectors that belong to distinct eigenvalues of A are orthogonal.*

Proof: Let $\mathbf{x}_1$ and $\mathbf{x}_2$ be eigenvectors of A that are associated with the distinct eigenvalues λ_1 and λ_2 of A. We then have

$$A\mathbf{x}_1 = \lambda_1\mathbf{x}_1 \quad \text{and} \quad A\mathbf{x}_2 = \lambda_2\mathbf{x}_2.$$

Now

$$\begin{aligned}\lambda_1(\mathbf{x}_1, \mathbf{x}_2) &= (\lambda_1\mathbf{x}_1, \mathbf{x}_2) = (A\mathbf{x}_1, \mathbf{x}_2)\\ &= (\mathbf{x}_1, A^T\mathbf{x}_2) = (\mathbf{x}_1, A\mathbf{x}_2)\\ &= (\mathbf{x}_1, \lambda_2\mathbf{x}_2) = \lambda_2(\mathbf{x}_1, \mathbf{x}_2).\end{aligned}$$

Thus

$$\lambda_1(\mathbf{x}_1, \mathbf{x}_2) = \lambda_2(\mathbf{x}_1, \mathbf{x}_2)$$

and subtracting, we obtain

$$\begin{aligned}0 &= \lambda_1(\mathbf{x}_1, \mathbf{x}_2) - \lambda_2(\mathbf{x}_1, \mathbf{x}_2)\\ &= (\lambda_1 - \lambda_2)(\mathbf{x}_1, \mathbf{x}_2).\end{aligned}$$

Since $\lambda_1 \neq \lambda_2$, we conclude that $(\mathbf{x}_1, \mathbf{x}_2) = 0$. ■

Example 1. Let $A = \begin{bmatrix} 0 & 0 & -2 \\ 0 & -2 & 0 \\ -2 & 0 & 3 \end{bmatrix}$. The characteristic polynomial of A is

$$f(\lambda) = (\lambda + 2)(\lambda - 4)(\lambda + 1)$$

(verify), so the eigenvalues of A are $\lambda_1 = -2$, $\lambda_2 = 4$, $\lambda_3 = -1$. Associated eigenvectors are the nontrivial solutions of the homogeneous system [Equation (6) in Section 6.1]

$$\begin{bmatrix} \lambda & 0 & 2 \\ 0 & \lambda + 2 & 0 \\ 2 & 0 & \lambda - 3 \end{bmatrix}\begin{bmatrix} a_1 \\ a_2 \\ a_3 \end{bmatrix} = \begin{bmatrix} 0 \\ 0 \\ 0 \end{bmatrix}.$$

For $\lambda_1 = -2$, we find that $\mathbf{x}_1$ is any vector of the form $\begin{bmatrix} 0 \\ r \\ 0 \end{bmatrix}$, where r is any nonzero real number (verify). Thus we may take $\mathbf{x}_1 = \begin{bmatrix} 0 \\ 1 \\ 0 \end{bmatrix}$. For $\lambda_2 = 4$, we find that $\mathbf{x}_2$ is any vector of the form $\begin{bmatrix} -\dfrac{r}{2} \\ 0 \\ r \end{bmatrix}$, where r is any nonzero real number (verify). Thus we may take $\mathbf{x}_2 = \begin{bmatrix} -1 \\ 0 \\ 2 \end{bmatrix}$. For $\lambda_3 = -1$, we find that $\mathbf{x}_3$ is any vector of the form $\begin{bmatrix} 2r \\ 0 \\ r \end{bmatrix}$, where r is any nonzero real number (verify). Thus we may take $\mathbf{x}_3 = \begin{bmatrix} 2 \\ 0 \\ 1 \end{bmatrix}$. It is clear that $\{\mathbf{x}_1, \mathbf{x}_2, \mathbf{x}_3\}$ is orthogonal and linearly independent. Thus A is similar to $D = \begin{bmatrix} -2 & 0 & 0 \\ 0 & 4 & 0 \\ 0 & 0 & -1 \end{bmatrix}$. ■

If A can be diagonalized, then there exists a nonsingular matrix P such that $P^{-1}AP$ is diagonal. Moreover, the columns of P are eigenvectors of A. Now, if the eigenvectors of A form an orthogonal set S, as happens when A is symmetric and the eigenvalues of A are distinct, then since any scalar multiple of an eigenvector of A is also an eigenvector of A, we can normalize S to obtain an orthonormal set $T = \{\mathbf{x}_1, \mathbf{x}_2, \ldots, \mathbf{x}_n\}$ of eigenvectors of A. The jth column of P is the eigenvector $\mathbf{x}_j$ and we now see what type of matrix P must be. We can write P as a partitioned matrix in the form $P = [\mathbf{x}_1\ \mathbf{x}_2\ \ldots\ \mathbf{x}_n]$. Then $P^T = \begin{bmatrix} \mathbf{x}_1^T \\ \mathbf{x}_2^T \\ \vdots \\ \mathbf{x}_n^T \end{bmatrix}$, where $\mathbf{x}_i^T$ is the transpose of the $n \times 1$ matrix (or vector) $\mathbf{x}_i$. We find that the (i, j) entry in P^TP is $(\mathbf{x}_i, \mathbf{x}_j)$. Since $(\mathbf{x}_i, \mathbf{x}_j) = 1$ if $i = j$ and $(\mathbf{x}_i, \mathbf{x}_j) = 0$ if $i \neq j$, we have $P^TP = I_n$, which means that $P^T = P^{-1}$. Such matrices are important enough to have a special name.

Definition 6.4. A square matrix A is called **orthogonal** if $A^{-1} = A^T$. Of course, we can also say that A is orthogonal if $A^TA = I_n$. ■

Example 2. Let $A = \begin{bmatrix} \frac{2}{3} & -\frac{2}{3} & \frac{1}{3} \\ \frac{2}{3} & \frac{1}{3} & -\frac{2}{3} \\ \frac{1}{3} & \frac{2}{3} & \frac{2}{3} \end{bmatrix}$. It is easy to check that $A^TA = I_n$. ■

Example 3. Let A be the matrix defined in Example 1. We already know that the set of eigenvectors $\left\{\begin{bmatrix}0\\1\\0\end{bmatrix}, \begin{bmatrix}-1\\0\\2\end{bmatrix}, \begin{bmatrix}2\\0\\1\end{bmatrix}\right\}$ is orthogonal. If we normalize these vectors, we find that $T = \left\{\begin{bmatrix}0\\1\\0\end{bmatrix}, \begin{bmatrix}-\frac{1}{\sqrt{5}}\\0\\\frac{2}{\sqrt{5}}\end{bmatrix}, \begin{bmatrix}\frac{2}{\sqrt{5}}\\0\\\frac{1}{\sqrt{5}}\end{bmatrix}\right\}$ is an orthonormal basis for R^3. A matrix P such that $P^{-1}AP$ is diagonal is the matrix whose columns are the vectors in T. Thus $P = \begin{bmatrix}0 & -\frac{1}{\sqrt{5}} & \frac{2}{\sqrt{5}}\\1 & 0 & 0\\0 & \frac{2}{\sqrt{5}} & \frac{1}{\sqrt{5}}\end{bmatrix}$. We leave it to the reader to verify that P is an orthogonal matrix and that

$$P^{-1}AP = P^TAP = \begin{bmatrix}-2 & 0 & 0\\0 & 4 & 0\\0 & 0 & -1\end{bmatrix}.$$ ■

The following theorem is not difficult to prove.

Theorem 6.8. The $n \times n$ matrix A is orthogonal if and only if the columns (and rows) of A form an orthonormal set.

Proof: Exercise. ■

If A is an orthogonal matrix, then it is easy to show that $\det(A) = \pm 1$ (Exercise 8). We now look at some of the geometrical properties of orthogonal matrices. If A is an orthogonal $n \times n$ matrix, let $L: R^n \to R^n$ be the linear transformation defined by $L(\mathbf{x}) = A\mathbf{x}$, for $\mathbf{x}$ in R^n (recall Chapter 4). If $\det(A) = 1$, it then follows from Exercise 9(b) in this section that L is a counterclockwise rotation. It can also be shown that if $\det(A) = -1$, then A is a reflection about the x-axis followed by a counterclockwise rotation.

Again, let A be an orthogonal $n \times n$ matrix and let $L: R^n \to R^n$ be defined by $L(\mathbf{x}) = A\mathbf{x}$ for $\mathbf{x}$ in R^n. We now compute $(L(\mathbf{x}), L(\mathbf{y}))$ for any vectors $\mathbf{x}$ and $\mathbf{y}$ in R^n, using the standard inner product on R^n. We have

$$(L(\mathbf{x}), L(\mathbf{y})) = (A\mathbf{x}, A\mathbf{y}) = (\mathbf{x}, A^TA\mathbf{y}) = (\mathbf{x}, A^{-1}A\mathbf{y}) = (\mathbf{x}, I_n\mathbf{y}) = (\mathbf{x}, \mathbf{y}), \quad (7)$$

where we have used Equation (2) in Section 3.3. This means that L preserves the inner product of two vectors, and consequently, L preserves length. It is, of course, clear that if θ is the angle between vectors $\mathbf{x}$ and $\mathbf{y}$ in R^n, then the angle between $L(\mathbf{x})$ and $L(\mathbf{y})$ is also θ. A linear transformation satisfying Equation (7) is called an **isometry.** Conversely, let $L: R^n \to R^n$ be an isometry, so that $(L(\mathbf{x}), L(\mathbf{y})) = (\mathbf{x}, \mathbf{y})$ for any $\mathbf{x}$ and $\mathbf{y}$ in R^n. Let A be the matrix representing L with respect to the natural basis for R^n. Then $L(\mathbf{x}) = A\mathbf{x}$. We now have

$$(\mathbf{x}, \mathbf{y}) = (L(\mathbf{x}), L(\mathbf{y})) = (A\mathbf{x}, A\mathbf{y}) = (\mathbf{x}, A^TA\mathbf{y}).$$

Since this holds for all $\mathbf{x}$ in R^n, then, by Exercise 7(e) in Section 3.3, we conclude that $A^TA\mathbf{y} = \mathbf{y}$ for any $\mathbf{y}$ in R^n. It follows that $A^TA = I_n$, so A is an orthogonal matrix. Other properties of orthogonal matrices and isometries are examined in the exercises. (See also Supplementary Exercises 14 and 16 in Chapter 4.)

We now turn to the general situation for a symmetric matrix; even if A has eigenvalues whose multiplicities are greater than one, it turns out that we can still diagonalize A. We omit the proof of the following theorem.

Theorem 6.9. *If A is a symmetric $n \times n$ matrix, then there exists an orthogonal matrix P such that $P^{-1}AP = P^TAP = D$, a diagonal matrix. The eigenvalues of A lie on the main diagonal of D.* ■

It can be shown that if the symmetric matrix A has an eigenvalue λ of multiplicity k, then the solution space of the homogeneous system $(\lambda I_n - A)\mathbf{x} = \mathbf{0}$ [Equation (6) in Section 6.1] has dimension k. This means that there exist k linearly independent eigenvectors of A associated with the eigenvalue λ. By the Gram–Schmidt process we can choose an orthonormal basis for this solution space. Thus we obtain a set of k orthonormal eigenvectors associated with the eigenvalue λ. Since eigenvectors associated with distinct eigenvalues are orthogonal, if we form the set of all eigenvectors we get an orthonormal set. Hence the matrix P whose columns are the eigenvectors is orthogonal.

Example 4. Let

$$A = \begin{bmatrix} 0 & 2 & 2 \\ 2 & 0 & 2 \\ 2 & 2 & 0 \end{bmatrix}.$$

The characteristic polynomial of A is

$$f(\lambda) = (\lambda + 2)^2(\lambda - 4)$$

(verify), so its eigenvalues are

$$\lambda_1 = -2, \qquad \lambda_2 = -2, \qquad \text{and} \qquad \lambda_3 = 4.$$

That is, -2 is an eigenvalue of multiplicity 2. To find the eigenvectors associated with -2, we solve the homogeneous system $(-2I_3 - A)\mathbf{x} = \mathbf{0}$:

$$\begin{bmatrix} -2 & -2 & -2 \\ -2 & -2 & -2 \\ -2 & -2 & -2 \end{bmatrix} \begin{bmatrix} a_1 \\ a_2 \\ a_3 \end{bmatrix} = \begin{bmatrix} 0 \\ 0 \\ 0 \end{bmatrix}. \tag{8}$$

A basis for the solution space of (8) consists of the eigenvectors $\mathbf{x}_1 = \begin{bmatrix} -1 \\ 1 \\ 0 \end{bmatrix}$ and $\mathbf{x}_2 = \begin{bmatrix} -1 \\ 0 \\ 1 \end{bmatrix}$ (verify). Now $\mathbf{x}_1$ and $\mathbf{x}_2$ are not orthogonal, since $(\mathbf{x}_1, \mathbf{x}_2) \neq 0$. We can use the Gram–Schmidt process to obtain an orthonormal basis for the solution space of (8) (the eigenspace associated with -2) as follows. Let $\mathbf{y}_1 = \mathbf{x}_1$ and

$$\mathbf{y}_2 = \mathbf{x}_2 - \frac{(\mathbf{x}_2, \mathbf{y}_1)}{(\mathbf{y}_1, \mathbf{y}_1)}\mathbf{x}_1 = \begin{bmatrix} -\frac{1}{2} \\ -\frac{1}{2} \\ 1 \end{bmatrix}.$$

To eliminate fractions, we let $\mathbf{y}_2^* = 2\mathbf{y}_2 = \begin{bmatrix} -1 \\ -1 \\ 2 \end{bmatrix}$. The set $\{\mathbf{y}_1, \mathbf{y}_2^*\}$ is an orthogonal set of vectors. Normalizing, we obtain

$$\mathbf{z}_1 = \frac{\mathbf{y}_1}{|\mathbf{y}_1|} = \frac{1}{\sqrt{2}}\begin{bmatrix} -1 \\ 1 \\ 0 \end{bmatrix} \qquad \text{and} \qquad \mathbf{z}_2 = \frac{\mathbf{y}_2^*}{|\mathbf{y}_2^*|} = \frac{1}{\sqrt{6}}\begin{bmatrix} -1 \\ -1 \\ 2 \end{bmatrix}.$$

The set $\{\mathbf{z}_1, \mathbf{z}_2\}$ is an orthonormal basis for the eigenspace associated with $\lambda = -2$. Now we find a basis for the eigenspace associated with $\lambda = 4$ by solving the homogeneous system $(4I_3 - A)\mathbf{x} = \mathbf{0}$:

$$\begin{bmatrix} 4 & -2 & -2 \\ -2 & 4 & -2 \\ -2 & -2 & 4 \end{bmatrix} \begin{bmatrix} a_1 \\ a_2 \\ a_3 \end{bmatrix} = \begin{bmatrix} 0 \\ 0 \\ 0 \end{bmatrix}.$$

A basis for this eigenspace consists of the vector $\mathbf{x}_3 = \begin{bmatrix} 1 \\ 1 \\ 1 \end{bmatrix}$ (verify).

Normalizing this vector, we have the eigenvector $\mathbf{z}_3 = \frac{1}{\sqrt{3}}\begin{bmatrix} 1 \\ 1 \\ 1 \end{bmatrix}$, as a basis for the eigenspace associated with $\lambda = 4$. Since eigenvectors associated with distinct eigenvalues are orthogonal, $\mathbf{z}_3$ is orthogonal to both $\mathbf{z}_1$ and $\mathbf{z}_2$. Thus the set $\{\mathbf{z}_1, \mathbf{z}_2, \mathbf{z}_3\}$ is an orthonormal basis for R^3 consisting of eigenvectors of A. The matrix P is the matrix whose jth column is $\mathbf{z}_j$:

$$P = \begin{bmatrix} -\frac{1}{\sqrt{2}} & -\frac{1}{\sqrt{6}} & \frac{1}{\sqrt{3}} \\ \frac{1}{\sqrt{2}} & -\frac{1}{\sqrt{6}} & \frac{1}{\sqrt{3}} \\ 0 & \frac{2}{\sqrt{6}} & \frac{1}{\sqrt{3}} \end{bmatrix}.$$

We leave it to the reader to verify that

$$P^{-1}AP = P^TAP = \begin{bmatrix} -2 & 0 & 0 \\ 0 & -2 & 0 \\ 0 & 0 & 4 \end{bmatrix}. \qquad \blacksquare$$

Example 5. Let $A = \begin{bmatrix} 1 & 2 & 0 & 0 \\ 2 & 1 & 0 & 0 \\ 0 & 0 & 1 & 2 \\ 0 & 0 & 2 & 1 \end{bmatrix}$. Either by straightforward computation, or by Exercise 14 in Section 5.2, we find that the characteristic polynomial of A is

$$f(\lambda) = (\lambda + 1)^2(\lambda - 3)^2,$$

so its eigenvalues are

$$\lambda_1 = -1, \qquad \lambda_2 = -1, \qquad \lambda_3 = 3, \qquad \text{and} \qquad \lambda_4 = 3.$$

We now compute the associated eigenvectors and the orthogonal matrix P. The eigenspace associated with the eigenvalue -1, of multiplicity 2, is the solution space of the homogeneous system $(-1I_4 - A)\mathbf{x} = \mathbf{0}$:

$$\begin{bmatrix} -2 & -2 & 0 & 0 \\ -2 & -2 & 0 & 0 \\ 0 & 0 & -2 & -2 \\ 0 & 0 & -2 & -2 \end{bmatrix} \begin{bmatrix} a_1 \\ a_2 \\ a_3 \\ a_4 \end{bmatrix} = \begin{bmatrix} 0 \\ 0 \\ 0 \\ 0 \end{bmatrix},$$

which is the set of all vectors of the form

$$\begin{bmatrix} r \\ -r \\ s \\ -s \end{bmatrix} = r \begin{bmatrix} 1 \\ -1 \\ 0 \\ 0 \end{bmatrix} + s \begin{bmatrix} 0 \\ 0 \\ 1 \\ -1 \end{bmatrix},$$

where r and s are any real numbers. Thus the eigenvectors

$$\begin{bmatrix} 1 \\ -1 \\ 0 \\ 0 \end{bmatrix} \quad \text{and} \quad \begin{bmatrix} 0 \\ 0 \\ 1 \\ -1 \end{bmatrix}$$

form a basis for the eigenspace associated with -1, and the dimension of this eigenspace is 2. Note that the eigenvectors

$$\begin{bmatrix} 1 \\ -1 \\ 0 \\ 0 \end{bmatrix} \quad \text{and} \quad \begin{bmatrix} 0 \\ 0 \\ 1 \\ -1 \end{bmatrix}$$

happen to be orthogonal. Since we are looking for an orthonormal basis for this eigenspace, we take

$$\mathbf{x}_1 = \begin{bmatrix} \frac{1}{\sqrt{2}} \\ -\frac{1}{\sqrt{2}} \\ 0 \\ 0 \end{bmatrix} \quad \text{and} \quad \mathbf{x}_2 = \begin{bmatrix} 0 \\ 0 \\ \frac{1}{\sqrt{2}} \\ -\frac{1}{\sqrt{2}} \end{bmatrix}$$

as eigenvectors associated with λ_1 and λ_2, respectively. Then $\{\mathbf{x}_1, \mathbf{x}_2\}$ is an orthonormal basis for the eigenspace associated with -1. The eigenspace associated with the eigenvalue 3, of multiplicity 2, is the solution space of the homogeneous system $(3I_4 - A)\mathbf{x} = \mathbf{0}$:

$$\begin{bmatrix} 2 & -2 & 0 & 0 \\ -2 & 2 & 0 & 0 \\ 0 & 0 & 2 & -2 \\ 0 & 0 & -2 & 2 \end{bmatrix} \begin{bmatrix} a_1 \\ a_2 \\ a_3 \\ a_4 \end{bmatrix} = \begin{bmatrix} 0 \\ 0 \\ 0 \\ 0 \end{bmatrix},$$

which is the set of all vectors of the form

$$\begin{bmatrix} r \\ r \\ s \\ s \end{bmatrix} = r \begin{bmatrix} 1 \\ 1 \\ 0 \\ 0 \end{bmatrix} + s \begin{bmatrix} 0 \\ 0 \\ 1 \\ 1 \end{bmatrix},$$

where r and s are any real numbers. Thus the eigenvectors $\begin{bmatrix} 1 \\ 1 \\ 0 \\ 0 \end{bmatrix}$ and $\begin{bmatrix} 0 \\ 0 \\ 1 \\ 1 \end{bmatrix}$ form a basis for the eigenspace associated with 3, and the dimension of this eigenspace is 2. Since these eigenvectors are orthogonal, we normalize them and let

$$\mathbf{x}_3 = \begin{bmatrix} \frac{1}{\sqrt{2}} \\ \frac{1}{\sqrt{2}} \\ 0 \\ 0 \end{bmatrix} \quad \text{and} \quad \mathbf{x}_4 = \begin{bmatrix} 0 \\ 0 \\ \frac{1}{\sqrt{2}} \\ \frac{1}{\sqrt{2}} \end{bmatrix}$$

be eigenvectors associated with λ_3 and λ_4, respectively. Then $\{\mathbf{x}_3, \mathbf{x}_4\}$ is an orthonormal basis for the eigenspace associated with 3. Now eigenvectors associated with distinct eigenvalues are orthogonal, so $\{\mathbf{x}_1, \mathbf{x}_2, \mathbf{x}_3, \mathbf{x}_4\}$ is an orthonormal basis for R^4. The matrix P is the matrix whose jth column is $\mathbf{x}_j$, $j = 1, 2, 3, 4$. Thus

$$P = \begin{bmatrix} \frac{1}{\sqrt{2}} & 0 & \frac{1}{\sqrt{2}} & 0 \\ -\frac{1}{\sqrt{2}} & 0 & \frac{1}{\sqrt{2}} & 0 \\ 0 & \frac{1}{\sqrt{2}} & 0 & \frac{1}{\sqrt{2}} \\ 0 & -\frac{1}{\sqrt{2}} & 0 & \frac{1}{\sqrt{2}} \end{bmatrix}.$$

We leave it to the reader to verify that P is an orthogonal matrix and that

$$P^{-1}AP = P^TAP = \begin{bmatrix} -1 & 0 & 0 & 0 \\ 0 & -1 & 0 & 0 \\ 0 & 0 & 3 & 0 \\ 0 & 0 & 0 & 3 \end{bmatrix}. \quad \blacksquare$$

If A is an $n \times n$ symmetric matrix, we know that we can find an orthogonal matrix P such that $P^{-1}AP$ is diagonal. Conversely, suppose that A is a matrix for which we can find an orthogonal matrix P such that $P^{-1}AP = D$ is a diagonal matrix. What type of matrix is A? Since $P^{-1}AP = D$, $A = PDP^{-1}$. Also, $P^{-1} = P^T$ since P is orthogonal. Then

$$A^T = (PDP^T)^T = (P^T)^TD^TP^T = PDP^T = A,$$

which means that A is symmetric.

Some remarks about nonsymmetric matrices are in order at this point. Theorem 6.4 assures us that A is diagonalizable if all the roots of its characteristic polynomial are real and distinct. We also studied examples, in Section 6.1, of nonsymmetric matrices that had repeated eigenvalues which were diagonalizable and others which were not diagonalizable. There are some striking differences between the symmetric and nonsymmetric cases, which we now summarize. Thus, if A is nonsymmetric, then the roots of its characteristic polynomial need not all be real numbers; if an eigenvalue λ has multiplicity k, then the solution space of $(\lambda I_n - A)\mathbf{x} = \mathbf{0}$ may have dimension less than k; if the roots of the characteristic polynomial of A are all real, it is still possible for the eigenvectors not to form a basis for R^n; eigenvectors associated with distinct eigenvalues need not be orthogonal. Thus, in Example 15 of Section 6.1, the eigenvectors $\mathbf{x}_1$ and $\mathbf{x}_3$ associated with the eigenvalues $\lambda_1 = 0$ and $\lambda_3 = 1$ are not orthogonal. If a matrix A cannot be diagonalized, then we can often find a matrix B similar to A

that is "nearly diagonal." The matrix B is said to be in **Jordan canonical form.** The study of such matrices lies beyond the scope of this book but they are studied in advanced books on linear algebra [e.g., K. Hoffman and R. Kunze, *Linear Algebra,* 2nd ed. (Englewood Cliffs, N.J.: Prentice-Hall, Inc., 1971)]; they play a key role in many applications of linear algebra.

It should also be noted that, in many applications, we need only find a diagonal matrix D that is similar to the given matrix A; that is, we do not explicitly have to know the orthogonal matrix P such that $P^{-1}AP = D$.

Eigenvalue problems arise in all applications involving vibrations; they occur in aerodynamics, elasticity, nuclear physics, mechanics, chemical engineering, biology, differential equations, and so on. Many of the matrices to be diagonalized in applied problems are either symmetric or all the roots of their characteristic polynomial are real. Of course, the methods for finding eigenvalues that have been presented in this chapter are not recommended for matrices of large order because of the need to evaluate determinants.

6.2. EXERCISES

1. Verify that $P = \begin{bmatrix} \frac{2}{3} & -\frac{2}{3} & \frac{1}{3} \\ \frac{2}{3} & \frac{1}{3} & -\frac{2}{3} \\ \frac{1}{3} & \frac{2}{3} & \frac{2}{3} \end{bmatrix}$ is an orthogonal matrix.

2. Find the inverse of each of the following orthogonal matrices.

(a) $A = \begin{bmatrix} 1 & 0 & 0 \\ 0 & \cos\phi & \sin\phi \\ 0 & -\sin\phi & \cos\phi \end{bmatrix}$.

(b) $B = \begin{bmatrix} 1 & 0 & 0 \\ 0 & \frac{1}{\sqrt{2}} & -\frac{1}{\sqrt{2}} \\ 0 & -\frac{1}{\sqrt{2}} & -\frac{1}{\sqrt{2}} \end{bmatrix}$.

3. Show that if A and B are orthogonal matrices, then AB is an orthogonal matrix.

4. Show that if A is an orthogonal matrix, then A^{-1} is orthogonal.

5. Prove Theorem 6.8.

6. Verify Theorem 6.8 for the matrices in Exercise 2.

7. Verify that the matrix P in Example 3 is an orthogonal matrix and that

$$P^{-1}AP = P^TAP = \begin{bmatrix} -2 & 0 & 0 \\ 0 & 4 & 0 \\ 0 & 0 & -1 \end{bmatrix}.$$

8. Show that if A is an orthogonal matrix, then $\det(A) = \pm 1$.

9. (a) Verify that the matrix $\begin{bmatrix} \cos\phi & \sin\phi \\ -\sin\phi & \cos\phi \end{bmatrix}$ is orthogonal.

(b) Prove that if A is an orthogonal 2×2 matrix, then there exists a real number ϕ such that

$$A = \begin{bmatrix} \cos\phi & \sin\phi \\ -\sin\phi & \cos\phi \end{bmatrix}$$

$$\text{or } A = \begin{bmatrix} \cos\phi & \sin\phi \\ \sin\phi & -\cos\phi \end{bmatrix}.$$

10. For the orthogonal matrix

$$A = \begin{bmatrix} \frac{1}{\sqrt{2}} & -\frac{1}{\sqrt{2}} \\ -\frac{1}{\sqrt{2}} & -\frac{1}{\sqrt{2}} \end{bmatrix},$$

verify that $(A\mathbf{x}, A\mathbf{y}) = (\mathbf{x}, \mathbf{y})$ for any vectors $\mathbf{x}$ and $\mathbf{y}$ in R^2.

11. Let A be an $n \times n$ orthogonal matrix, and let $L: R^n \to R^n$ be the linear transformation associated with A; that is, $L(\mathbf{x}) = A\mathbf{x}$ for $\mathbf{x}$ in R^n. Let θ be the angle between vectors $\mathbf{x}$ and $\mathbf{y}$ in R^n. Prove that the angle between $L(\mathbf{x})$ and $L(\mathbf{y})$ is also θ.

12. A linear transformation $L: V \to V$, where V is an n-dimensional Euclidean space is called **orthogonal** if $(L(\mathbf{x}), L(\mathbf{y})) = (\mathbf{x}, \mathbf{y})$. Let S be an orthonormal basis for V and let the matrix A represent the orthogonal linear transformation L with respect to S. Prove that A is an orthogonal matrix.

13. Let $L: R^2 \to R^2$ be the linear transformation performing a counterclockwise rotation through 45°, and let A be the matrix representing L with respect to the natural basis for R^2. Prove that A is orthogonal.

14. Let A be an $n \times n$ matrix and let $B = P^{-1}AP$ be similar to A. Prove that if $\mathbf{x}$ is an eigenvector of A associated with the eigenvalue λ of A, then $P^{-1}\mathbf{x}$ is an eigenvector of B associated with the eigenvalue λ of B.

In Exercises 15 through 20, diagonalize each given matrix and find an orthogonal matrix P such that $P^{-1}AP$ is diagonal.

15. $A = \begin{bmatrix} 2 & 2 \\ 2 & 2 \end{bmatrix}$.

16. $A = \begin{bmatrix} 0 & 0 & 1 \\ 0 & 0 & 0 \\ 1 & 0 & 0 \end{bmatrix}$.

17. $A = \begin{bmatrix} 0 & 0 & 0 \\ 0 & 2 & 2 \\ 0 & 2 & 2 \end{bmatrix}$.

18. $A = \begin{bmatrix} 0 & 0 & 0 & 0 \\ 0 & 0 & 0 & 0 \\ 0 & 0 & 0 & 1 \\ 0 & 0 & 1 & 0 \end{bmatrix}$.

19. $A = \begin{bmatrix} 0 & -1 & -1 \\ -1 & 0 & -1 \\ -1 & -1 & 0 \end{bmatrix}$.

20. $A = \begin{bmatrix} -1 & 2 & 2 \\ 2 & -1 & 2 \\ 2 & 2 & -1 \end{bmatrix}$.

In Exercises 21 through 28, diagonalize each given matrix.

21. $A = \begin{bmatrix} 2 & 1 \\ 1 & 2 \end{bmatrix}$.

22. $A = \begin{bmatrix} 2 & 2 & 0 & 0 \\ 2 & 2 & 0 & 0 \\ 0 & 0 & 2 & 2 \\ 0 & 0 & 2 & 2 \end{bmatrix}$.

23. $A = \begin{bmatrix} 1 & 1 & 0 \\ 1 & 1 & 0 \\ 0 & 0 & 1 \end{bmatrix}$.

24. $A = \begin{bmatrix} 1 & 0 & 0 \\ 0 & 3 & -2 \\ 0 & -2 & 3 \end{bmatrix}$.

25. $A = \begin{bmatrix} 1 & 0 & 0 \\ 0 & 1 & 1 \\ 0 & 1 & 1 \end{bmatrix}$.

26. $A = \begin{bmatrix} 0 & 0 & 0 & 1 \\ 0 & 0 & 0 & 0 \\ 0 & 0 & 0 & 0 \\ 1 & 0 & 0 & 0 \end{bmatrix}$.

27. $A = \begin{bmatrix} 1 & -1 & 2 \\ -1 & 1 & 2 \\ 2 & 2 & 2 \end{bmatrix}$.

28. $A = \begin{bmatrix} -3 & 0 & -1 \\ 0 & -2 & 0 \\ -1 & 0 & -3 \end{bmatrix}$.

29. Prove Theorem 6.9 for the 2×2 case by studying the two possible cases for the roots of the characteristic polynomial of A.

30. Let $L: V \to V$ be an orthogonal linear transformation (see Exercise 12), where V is an n-dimensional Euclidean space. Show that if λ is an eigenvalue of L, then $|\lambda| = 1$.

31. Let $L: R^2 \to R^2$ be defined by

$$L\left(\begin{bmatrix} x \\ y \end{bmatrix}\right) = \begin{bmatrix} \frac{1}{\sqrt{2}} & \frac{1}{\sqrt{2}} \\ \frac{1}{\sqrt{2}} & -\frac{1}{\sqrt{2}} \end{bmatrix} \begin{bmatrix} x \\ y \end{bmatrix}.$$

Show that L is an isometry of R^2.

32. Let $L: R^n \to R^n$ be a linear operator defined by $L(\mathbf{x}) = A\mathbf{x}$. Prove that if L is an isometry, then L^{-1} is an isometry.

33. Let $L: R^n \to R^n$ be a linear operator and $S = \{\mathbf{v}_1, \mathbf{v}_2, \ldots, \mathbf{v}_n\}$ an orthonormal basis for R^n. Prove that L is an isometry if and only if $T = \{L(\mathbf{v}_1), L(\mathbf{v}_2), \ldots, L(\mathbf{v}_n)\}$ is an orthonormal basis for R^n.

34. Assuming that the software you use has a command for eigenvalues and eigenvectors (see Exercises 37 and 38 in Section 6.1), determine if a set of orthonormal eigenvectors is returned when the input matrix A is symmetric (see Theorem 6.9). Experiment with the matrices in Examples 4 and 5.

35. If the answer to Exercise 35 is no, you can use the Gram–Schmidt procedure to obtain an orthonormal set of eigenvectors (see Exercise 31 in Section 3.4). Experiment with the matrices in Examples 4 and 5 if necessary.

6.3. *Real Quadratic Forms*

In your precalculus and calculus courses you have seen that the graph of the equation

$$ax^2 + 2bxy + cy^2 = d, \tag{1}$$

where a, b, c, and d are real numbers, is a **conic section** centered at the origin of a rectangular Cartesian coordinate system in two-dimensional space. Similarly, the graph of the equation

$$ax^2 + 2dxy + 2exz + by^2 + 2fyz + cz^2 = g, \tag{2}$$

where $a, b, c, d, e, f,$ and g are real numbers, is a **quadric surface** centered at the origin of a rectangular Cartesian coordinate system in three-dimensional space. If a conic section or quadric surface is not centered at the origin, its equations are more complicated than those given in (1) and (2).

The identification of the conic section or quadric surface that is the graph of a given equation often requires the rotation and translation of the coordinate axes. These methods can best be understood as an application of eigenvalues and eigenvectors of matrices and will be discussed in Sections 6.4 and 6.5.

The expressions on the left sides of Equations (1) and (2) are examples of quadratic forms. Quadratic forms arise in statistics, mechanics, and in other problems in physics; in quadratic programming; in the study of maxima and minima of functions of several variables; and in other applied problems. In this section we use our results on eigenvalues and eigenvectors of matrices to give a brief treatment of real quadratic forms in n variables. In Section 6.4 we apply these results to the classification of the conic sections, and in Section 6.5 to the classification of the quadric surfaces.

Definition 6.5. If A is a symmetric matrix, then the function $g: R^n \rightarrow R^1$ (a real-valued function on R^n) defined by

$$g(\mathbf{x}) = \mathbf{x}^T A \mathbf{x},$$

where

$$\mathbf{x} = \begin{bmatrix} x_1 \\ x_2 \\ \vdots \\ x_n \end{bmatrix},$$

is called a **real quadratic form in the n variables $x_1, x_2, \ldots, x_n$**. The matrix A is called the **matrix of the quadratic form g**. We shall also denote the quadratic form by $g(\mathbf{x})$. ■

Example 1. Write the left side of (1) as the quadratic form in the variables x and y.

Solution: Let

$$\mathbf{x} = \begin{bmatrix} x \\ y \end{bmatrix} \quad \text{and} \quad A = \begin{bmatrix} a & b \\ b & c \end{bmatrix}.$$

Then the left side of (1) is the quadratic form

$$g(\mathbf{x}) = \mathbf{x}^T A\mathbf{x}. \qquad \blacksquare$$

Example 2. Write the left side of (2) is the quadratic form

Solution: Let

$$\mathbf{x} = \begin{bmatrix} x \\ y \\ z \end{bmatrix} \quad \text{and} \quad A = \begin{bmatrix} a & d & e \\ d & b & f \\ e & f & c \end{bmatrix}.$$

Then the left side of (2) is the quadratic form

$$g(\mathbf{x}) = \mathbf{x}^T A\mathbf{x}. \qquad \blacksquare$$

Example 3. The following expressions are quadratic forms:

(a) $3x^2 - 5xy - 7y^2 = [x \quad y]\begin{bmatrix} 3 & -\frac{5}{2} \\ -\frac{5}{2} & -7 \end{bmatrix}\begin{bmatrix} x \\ y \end{bmatrix}.$

(b) $3x^2 - 7xy + 5xz + 4y^2 - 4yz - 3z^2 = [x \quad y \quad z]\begin{bmatrix} 3 & -\frac{7}{2} & \frac{5}{2} \\ -\frac{7}{2} & 4 & -2 \\ \frac{5}{2} & -2 & -3 \end{bmatrix}\begin{bmatrix} x \\ y \\ z \end{bmatrix}.$ $\blacksquare$

Suppose now that $g(\mathbf{x}) = \mathbf{x}^T A\mathbf{x}$ is a quadratic form. To simplify the quadratic form, we change from the variables $x_1, x_2, \ldots, x_n$ to the variables $y_1, y_2, \ldots, y_n$, where we assume that the old variables are related to the new variables by $\mathbf{x} = P\mathbf{y}$ for some orthogonal matrix P. Then

$$g(\mathbf{x}) = \mathbf{x}^T A\mathbf{x} = (P\mathbf{y})^T A(P\mathbf{y}) = \mathbf{y}^T(P^T AP)\mathbf{y} = \mathbf{y}^T B\mathbf{y},$$

where $B = P^T AP$. We shall let you verify that if A is a symmetric matrix, then $P^T AP$ is also symmetric (Exercise 25). Thus

$$h(\mathbf{y}) = \mathbf{y}^T B\mathbf{y}$$

is another quadratic form and $g(\mathbf{x}) = h(\mathbf{y})$.

This situation is important enough to formulate the following definitions.

Definition 6.6. If A and B are $n \times n$ matrices, we say that B is **congruent** to A if $B = P^TAP$ for a nonsingular matrix P. ■

In light of Exercise 26, "A is congruent to B," and "B is congruent to A" can both be replaced by "A and B are congruent."

Definition 6.7. Two quadratic forms g and h with matrices A and B, respectively, are said to be **equivalent** if A and B are congruent. ■

The congruence of matrices and equivalence of forms are more general concepts since the matrix P is only required to be nonsingular (not necessarily orthogonal). We shall here consider the more restrictive situation with P orthogonal.

Example 4. Consider the quadratic form in the variables x and y defined by

$$g(\mathbf{x}) = 2x^2 + 2xy + 2y^2 = \begin{bmatrix} x & y \end{bmatrix} \begin{bmatrix} 2 & 1 \\ 1 & 2 \end{bmatrix} \begin{bmatrix} x \\ y \end{bmatrix}. \tag{3}$$

We now change from the variables x and y to the variables x' and y'. Suppose that the old variables are related to the new variables by the equations

$$x = \frac{1}{\sqrt{2}}x' - \frac{1}{\sqrt{2}}y' \quad \text{and} \quad y = \frac{1}{\sqrt{2}}x' + \frac{1}{\sqrt{2}}y', \tag{4}$$

which can be written in matrix form as

$$\mathbf{x} = \begin{bmatrix} x \\ y \end{bmatrix} = \begin{bmatrix} \frac{1}{\sqrt{2}} & -\frac{1}{\sqrt{2}} \\ \frac{1}{\sqrt{2}} & \frac{1}{\sqrt{2}} \end{bmatrix} \begin{bmatrix} x' \\ y' \end{bmatrix} = P\mathbf{y},$$

where the orthogonal (hence nonsingular) matrix

$$P = \begin{bmatrix} \frac{1}{\sqrt{2}} & -\frac{1}{\sqrt{2}} \\ \frac{1}{\sqrt{2}} & \frac{1}{\sqrt{2}} \end{bmatrix} \quad \text{and} \quad \mathbf{y} = \begin{bmatrix} x' \\ y' \end{bmatrix}.$$

We shall soon see why and how this particular matrix P was selected. Substituting in (3), we obtain

$$
\begin{aligned}
g(\mathbf{x}) &= \mathbf{x}^T A \mathbf{x} = (P\mathbf{y})^T A (P\mathbf{y}) = \mathbf{y}^T P^T A P \mathbf{y} \\
&= \begin{bmatrix} x' & y' \end{bmatrix} \begin{bmatrix} \frac{1}{\sqrt{2}} & -\frac{1}{\sqrt{2}} \\ \frac{1}{\sqrt{2}} & \frac{1}{\sqrt{2}} \end{bmatrix}^T \begin{bmatrix} 2 & 1 \\ 1 & 2 \end{bmatrix} \begin{bmatrix} \frac{1}{\sqrt{2}} & -\frac{1}{\sqrt{2}} \\ \frac{1}{\sqrt{2}} & \frac{1}{\sqrt{2}} \end{bmatrix} \begin{bmatrix} x' \\ y' \end{bmatrix} \\
&= \begin{bmatrix} x' & y' \end{bmatrix} \begin{bmatrix} 3 & 0 \\ 0 & 1 \end{bmatrix} \begin{bmatrix} x' \\ y' \end{bmatrix} = h(\mathbf{y}) \\
&= 3x'^2 + y'^2
\end{aligned}
$$

Thus the matrices

$$
\begin{bmatrix} 2 & 1 \\ 1 & 2 \end{bmatrix} \quad \text{and} \quad \begin{bmatrix} 3 & 0 \\ 0 & 1 \end{bmatrix}
$$

are congruent and the quadratic forms g and h are equivalent. ■

We now turn to the question of how to select the matrix P.

Theorem 6.10 (Principal Axes Theorem) *Any quadratic form in n variables $g(\mathbf{x}) = \mathbf{x}^T A \mathbf{x}$ is equivalent by means of an orthogonal matrix P to a quadratic form, $h(\mathbf{y}) = \lambda_1 y_1^2 + \lambda_2 y_2^2 + \cdots + \lambda_n y_n^2$, where*

$$
\mathbf{y} = \begin{bmatrix} y_1 \\ y_2 \\ \vdots \\ y_n \end{bmatrix}
$$

and $\lambda_1, \lambda_2, \ldots, \lambda_n$ are the eigenvalues of the matrix A of g.

Proof: If A is the matrix of g, then, since A is symmetric, we know, by Theorem 6.9, that A can be diagonalized by an orthogonal matrix. This means that there exists an orthogonal matrix P such that $D = P^{-1}AP$ is a diagonal matrix. Since P is orthogonal, $P^{-1} = P^T$, so $D = P^T A P$. Moreover, the elements on the main diagonal of D are the eigenvalues, $\lambda_1, \lambda_2, \ldots, \lambda_n$ of A, which are real numbers. The quadratic form h with matrix D is given by

$$h(\mathbf{y}) = \lambda_1 y_1^2 + \lambda_2 y_2^2 + \cdots + \lambda_n y_n^2;$$

g and h are equivalent. ■

Example 5. Consider the quadratic form g in the variables x, y, and z, defined by

$$g(\mathbf{x}) = 2x^2 + 4y^2 + 6yz - 4z^2.$$

Determine a quadratic form h of the form in Theorem 6.10 to which g is equivalent.

Solution: The matrix of g is

$$A = \begin{bmatrix} 2 & 0 & 0 \\ 0 & 4 & 3 \\ 0 & 3 & -4 \end{bmatrix},$$

and the eigenvalues of A are

$$\lambda_1 = 2, \qquad \lambda_2 = 5, \qquad \text{and} \qquad \lambda_3 = -5. \qquad \text{(verify)}$$

Let h be the quadratic form in the variables x', y', and z' defined by

$$h(\mathbf{y}) = 2x'^2 + 5y'^2 - 5z'^2.$$

Then g and h are equivalent by means of some orthogonal matrix. Note that $\hat{h}(\mathbf{y}) = -5x'^2 + 2y'^2 + 5z'^2$ is also equivalent to g. ■

Note that to apply Theorem 6.10 to diagonalize a given quadratic form, as shown in Example 5, we do not need to know the eigenvectors of A (nor the matrix P); we only require the eigenvalues of A.

To understand the significance of Theorem 6.10, we consider quadratic forms in two and three variables. As we have already observed at the beginning of this section, the graph of the equation

$$g(\mathbf{x}) = \mathbf{x}^T A \mathbf{x} = 1,$$

where $\mathbf{x}$ is a vector in R^2 and A is a symmetric 2×2 matrix, is a conic section centered at the origin of the xy-plane. From Theorem 6.10 it follows that there is a Cartesian coordinate system in the xy-plane with respect to which the equation of this conic section is

$$ax'^2 + by'^2 = 1,$$

where a and b are real numbers. Similarly, the graph of the equation

$$g(\mathbf{x}) = \mathbf{x}^T A\mathbf{x} = 1,$$

where $\mathbf{x}$ is a vector in R^3 and A is a symmetric 3×3 matrix, is a quadric surface centered at the origin of the xyz Cartesian coordinate system. From Theorem 6.10 it follows that there is a Cartesian coordinate system in 3-space with respect to which the equation of the quadric surface is

$$ax'^2 + by'^2 + cz'^2 = 1,$$

where a, b, and c are real numbers. The principal axes of the conic or surface lie along the new coordinate axes, and this is the reason for calling Theorem 6.10 the **principal axes theorem.**

Example 6. Consider the conic section whose equation is

$$g(\mathbf{x}) = 2x^2 + 2xy + 2y^2 = 9.$$

From Example 4 it follows that this conic section can also be described by the equation

$$h(\mathbf{y}) = 3x'^2 + y'^2 = 9,$$

which can be rewritten as

$$\frac{x'^2}{3} + \frac{y'^2}{9} = 1.$$

The graph of this equation is an ellipse (Figure 6.3) whose major axis is along the y'-axis. The major axis is of length 6; the minor axis is of length $2\sqrt{3}$. We now note that there is a very close connection between the eigenvectors of the matrix of (3) and the location of the x'- and y'-axes.

Since $\mathbf{x} = P\mathbf{y}$, we have $\mathbf{y} = P^{-1}\mathbf{x} = P^T\mathbf{x} = P\mathbf{x}$ (P is orthogonal and, in this example, also symmetric). Thus

$$x' = \frac{1}{\sqrt{2}}x + \frac{1}{\sqrt{2}}y \qquad \text{and} \qquad y' = -\frac{1}{\sqrt{2}}x + \frac{1}{\sqrt{2}}y.$$

This means that, in terms of the x- and y-axes, the x'-axis lies along the vector

$$\mathbf{x}_1 = \begin{bmatrix} \frac{1}{\sqrt{2}} \\ \frac{1}{\sqrt{2}} \end{bmatrix}$$

and the y'-axis lies along the vector

$$\mathbf{x}_2 = \begin{bmatrix} -\frac{1}{\sqrt{2}} \\ \frac{1}{\sqrt{2}} \end{bmatrix}.$$

Now $\mathbf{x}_1$ and $\mathbf{x}_2$ are the columns of the matrix

$$P = \begin{bmatrix} \frac{1}{\sqrt{2}} & -\frac{1}{\sqrt{2}} \\ \frac{1}{\sqrt{2}} & \frac{1}{\sqrt{2}} \end{bmatrix},$$

which in turn are eigenvectors of the matrix of (3). Thus the x'- and y'-axes lie along the eigenvectors of the matrix of (3) (see Figure 6.3). ■

Figure 6.3

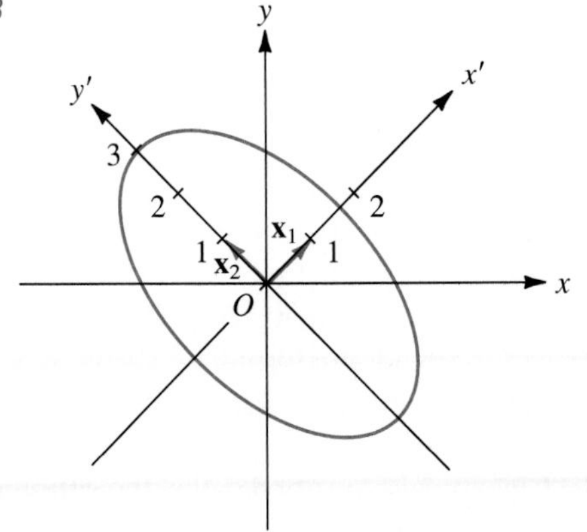

The situation described in Example 6 is true in general. That is, the principal axes of a conic section or quadric surface lie along the eigenvectors of the matrix of the quadratic form.

Let $g(\mathbf{x}) = \mathbf{x}^T A\mathbf{x}$ be a quadratic form in n variables. Then we know that g is equivalent to the quadratic form

$$h(y) = \lambda_1 y_1^2 + \lambda_2 y_2^2 + \cdots + \lambda_n y_n^2,$$

where $\lambda_1, \lambda_2, \ldots, \lambda_n$ are the eigenvalues of the symmetric matrix A of g, and hence are all real. We can label the eigenvalues so that all the positive eigenvalues of A, if any, are listed first, followed by all the negative eigenvalues, if any, followed by the zero eigenvalues, if any. Thus let $\lambda_1, \lambda_2, \ldots, \lambda_p$ be positive, $\lambda_{p+1}, \lambda_{p+2}, \ldots, \lambda_r$ be negative, and $\lambda_{r+1}, \lambda_{r+2}, \ldots, \lambda_n$ be zero. We now define the diagonal matrix H whose entries on the main diagonal are

$$\frac{1}{\sqrt{\lambda_1}}, \frac{1}{\sqrt{\lambda_2}}, \ldots, \frac{1}{\sqrt{\lambda_p}}, \frac{1}{\sqrt{-\lambda_{p+1}}}, \frac{1}{\sqrt{-\lambda_{p+2}}}, \ldots, \frac{1}{\sqrt{-\lambda_r}}, 1, 1, \ldots, 1,$$

with $n - r$ ones. Let D be the diagonal matrix whose entries on the main diagonal are $\lambda_1, \lambda_2, \ldots, \lambda_p, \lambda_{p+1}, \ldots, \lambda_r, \lambda_{r+1}, \ldots, \lambda_n$; A and D are congruent. Let $D_1 = H^T DH$ be the matrix whose diagonal elements are $1, 1, \ldots, 1, -1, \ldots, -1, 0, 0, \ldots, 0$ (p ones, $r - p - 1$ negative ones, and $n - r$ zeros); D and D_1 are then congruent. From Exercise 4 it follows that A and D_1 are congruent. In terms of quadratic forms, we have established Theorem 6.11.

Theorem 6.11. *A quadratic form $g(\mathbf{x}) = \mathbf{x}^T A\mathbf{x}$ in n variables is equivalent to a quadratic form*

$$h(\mathbf{y}) = y_1^2 + y_2^2 + \cdots + y_p^2 - y_{p+1}^2 - y_{p+2}^2 - \cdots - y_r^2. \qquad \blacksquare$$

It is clear that the rank of the matrix D_1 is r, the number of nonzero entries on its main diagonal. Now it can be shown that congruent matrices have equal ranks. Since the rank of D_1 is r, the rank of A is also r. We also refer to r as the **rank** of the quadratic form g whose matrix is A. It can be shown that the number p of positive terms in the quadratic form h of Theorem 6.11 is unique; that is, no matter how we simplify the given quadratic form g to obtain an equivalent quadratic form, the latter will always have p positive terms. Hence the quadratic form h in Theorem 6.11 is unique; it is often called the **canonical form** of a quadratic form in n variables. The difference between the number of positive eigenvalues and the number of negative eigenvalues is $s = p - (r - p) = 2p - r$ and is called the **signature** of the quadratic form. Thus, if g and h are equivalent quadratic forms, then they have equal ranks and signatures. However, it can also be shown that if g and h have equal ranks and signatures, then they are equivalent.

Example 7. Consider the quadratic form in x_1, x_2, x_3, given by

$$g(\mathbf{x}) = 3x_2^2 + 8x_2x_3 - 3x_3^2 = \mathbf{x}^T A\mathbf{x}$$
$$= \begin{bmatrix} x_1 & x_2 & x_3 \end{bmatrix} \begin{bmatrix} 0 & 0 & 0 \\ 0 & 3 & 4 \\ 0 & 4 & -3 \end{bmatrix} \begin{bmatrix} x_1 \\ x_2 \\ x_3 \end{bmatrix}.$$

The eigenvalues of A are (verify)

$$\lambda_1 = 5, \qquad \lambda_2 = -5, \qquad \text{and} \qquad \lambda_3 = 0.$$

In this case A is congruent to

$$D = \begin{bmatrix} 5 & 0 & 0 \\ 0 & -5 & 0 \\ 0 & 0 & 0 \end{bmatrix}.$$

If we let

$$H = \begin{bmatrix} \frac{1}{\sqrt{5}} & 0 & 0 \\ 0 & \frac{1}{\sqrt{5}} & 0 \\ 0 & 0 & 1 \end{bmatrix},$$

then

$$D_1 = H^T D H = \begin{bmatrix} 1 & 0 & 0 \\ 0 & -1 & 0 \\ 0 & 0 & 0 \end{bmatrix}$$

and A are congruent, and the given quadratic form is equivalent to the canonical form

$$h(\mathbf{y}) = y_1^2 - y_2^2.$$

The rank of g is 2, and since $p = 1$, the signature $s = 2p - r = 0$. ■

As a final application of quadratic forms we consider positive definite, symmetric matrices. We recall that, in Section 3.3, a symmetric $n \times n$ matrix A was called positive definite if $\mathbf{x}^T A\mathbf{x} > 0$ for every nonzero vector $\mathbf{x}$ in R^n.

If A is a symmetric matrix, then $\mathbf{x}^T A\mathbf{x}$ is a quadratic form $g(\mathbf{x}) = \mathbf{x}^T A\mathbf{x}$ and, by Theorem 6.10, g is equivalent to h, where

$$h(\mathbf{y}) = \lambda_1 y_1^2 + \lambda_2 y_2^2 + \cdots + \lambda_p y_p^2 + \lambda_{p+1} y_{p+1}^2 + \lambda_{p+2} y_{p+2}^2 + \cdots + \lambda_r y_r^2.$$

Now A is positive definite if and only if $h(\mathbf{y}) > 0$ for each $\mathbf{y} \neq \mathbf{0}$. However, this can happen if and only if all summands in $h(\mathbf{y})$ are positive and $r = n$. These remarks have established the following theorem.

Theorem 6.12. *A symmetric matrix A is positive definite if and only if all the eigenvalues of A are positive.* ■

A quadratic form is then called **positive definite** if its matrix is positive definite.

6.3. EXERCISES

In Exercises 1 and 2, write each quadratic form as $\mathbf{x}^T A\mathbf{x}$, where A is a symmetric matrix.

1. (a) $-3x^2 + 5xy - 2y^2$.
(b) $2x_1^2 + 3x_1x_2 - 5x_1x_3 + 7x_2x_3$.
(c) $3x_1^2 + x_2^2 - 2x_3^2 + x_1x_2 - x_1x_3 - 4x_2x_3$.

2. (a) $x_1^2 - 3x_2^2 + 4x_3^2 - 4x_1x_2 + 6x_2x_3$.
(b) $4x^2 - 6xy + 2y^2$.
(c) $-2x_1x_2 + 4x_1x_3 + 6x_2x_3$.

In Exercises 3 and 4, for each given symmetric matrix A find a diagonal matrix D that is congruent to A.

3. (a) $A = \begin{bmatrix} -1 & 0 & 0 \\ 0 & 1 & 1 \\ 0 & 1 & 1 \end{bmatrix}$.

(b) $A = \begin{bmatrix} 1 & 1 & 1 \\ 1 & 1 & 1 \\ 1 & 1 & 1 \end{bmatrix}$.

(c) $A = \begin{bmatrix} 0 & 2 & 2 \\ 2 & 0 & 2 \\ 2 & 2 & 0 \end{bmatrix}$.

4. (a) $A = \begin{bmatrix} 3 & 4 & 0 \\ 4 & -3 & 0 \\ 0 & 0 & 5 \end{bmatrix}$.

(b) $A = \begin{bmatrix} 2 & 1 & 1 \\ 1 & 2 & 1 \\ 1 & 1 & 2 \end{bmatrix}$.

(c) $A = \begin{bmatrix} 0 & 0 & 1 \\ 0 & 1 & 0 \\ 1 & 0 & 0 \end{bmatrix}$.

In Exercises 5 through 10, find a quadratic form of the type in Theorem 6.10 that is equivalent to the given quadratic form.

5. $2x^2 - 4xy - y^2$.

6. $x_1^2 + x_2^2 + x_3^2 + 2x_2x_3$.

7. $2x_1x_3$.

8. $2x_2^2 + 2x_3^2 + 4x_2x_3$.

9. $-2x_1^2 - 4x_2^2 + 4x_3^2 - 6x_2x_3$.

10. $6x_1x_2 + 8x_2x_3$.

In Exercises 10 through 16, find a quadratic form of the type in Theorem 6.11 that is equivalent to the given quadratic form.

11. $2x^2 + 4xy + 2y^2$.

12. $x_1^2 + x_2^2 + x_3^2 + 2x_1x_2$.

13. $2x_1^2 + 4x_2^2 + 4x_3^2 + 10x_2x_3$.

14. $2x_1^2 + 3x_2^2 + 3x_3^2 + 4x_2x_3$.

15. $-3x_1^2 + 2x_2^2 + 2x_3^2 + 4x_2x_3$.

16. $-3x_1^2 + 5x_2^2 + 3x_3^2 - 8x_1x_3$.

17. Let $g(\mathbf{x}) = 4x_2^2 + 4x_3^2 - 10x_2x_3$ be a quadratic form in three variables. Find a quadratic form of the type in Theorem 6.11 that is equivalent to g. What is the rank of g? What is the signature of g?

18. Let $g(\mathbf{x}) = 3x_1^2 - 3x_2^2 - 3x_3^2 + 4x_2x_3$ be a quadratic form in three variables. Find a quadratic form of the type in Theorem 6.11 that is equivalent to g. What is the rank of g? What is the signature of g?

19. Find all quadratic forms $g(\mathbf{x}) = \mathbf{x}^TA\mathbf{x}$ in two variables of the type described in Theorem 6.11. What conics do the equations $\mathbf{x}^TA\mathbf{x} = 1$ represent?

20. Find all quadratic forms $g(\mathbf{x}) = \mathbf{x}^TA\mathbf{x}$ in two variables of rank 1 of the type described in Theorem 6.11. What conics do the equations $\mathbf{x}^TA\mathbf{x} = 1$ represent?

In Exercises 21 and 22, which of the given quadratic forms in three variables are equivalent?

21. $g_1(\mathbf{x}) = x_1^2 + x_2^2 + x_3^2 + 2x_1x_2$.
$g_2(\mathbf{x}) = 2x_2^2 + 2x_3^2 + 2x_2x_3$.
$g_3(\mathbf{x}) = 3x_2^2 - 3x_3^2 + 8x_2x_3$.
$g_4(\mathbf{x}) = 3x_2^2 + 3x_3^2 - 4x_2x_3$.

22. $g_1(\mathbf{x}) = x_2^2 + 2x_1x_3$.
$g_2(\mathbf{x}) = 2x_1^2 + 2x_2^2 + x_3^2 + 2x_1x_2 + 2x_1x_3 + 2x_2x_3$.
$g_3(\mathbf{x}) = 2x_1x_2 + 2x_1x_3 + 2x_2x_3$.
$g_4(\mathbf{x}) = 4x_1^2 + 3x_2^2 + 4x_3^2 + 10x_1x_3$.

In Exercises 23 and 24, which of the given matrices are positive definite?

23. (a) $\begin{bmatrix} 2 & -1 \\ -1 & 2 \end{bmatrix}$. (b) $\begin{bmatrix} 2 & 1 \\ 1 & 2 \end{bmatrix}$.

(c) $\begin{bmatrix} 3 & 1 & 0 \\ 1 & 3 & 0 \\ 0 & 0 & 3 \end{bmatrix}$. (d) $\begin{bmatrix} 1 & 0 & 0 \\ 0 & 2 & 0 \\ 0 & 0 & -3 \end{bmatrix}$.

(e) $\begin{bmatrix} 2 & 2 \\ 2 & 2 \end{bmatrix}$.

24. (a) $\begin{bmatrix} 0 & -1 \\ -1 & 0 \end{bmatrix}$. (b) $\begin{bmatrix} 1 & 1 \\ 1 & 1 \end{bmatrix}$.

(c) $\begin{bmatrix} 0 & 0 & 0 \\ 0 & 1 & 2 \\ 0 & 2 & 1 \end{bmatrix}$.

(d) $\begin{bmatrix} 7 & 4 & 4 \\ 4 & 7 & 4 \\ 4 & 4 & 7 \end{bmatrix}$.

(e) $\begin{bmatrix} 2 & 0 & 0 & 0 \\ 0 & 1 & 0 & 0 \\ 0 & 0 & 3 & 4 \\ 0 & 0 & 4 & -3 \end{bmatrix}$.

25. Prove that if A is a symmetric matrix, then P^TAP is also symmetric.

26. If A, B, and C are $n \times n$ symmetric matrices, prove the following.

(a) A and A are congruent.
(b) If A and B are congruent, then B and A are congruent.
(c) If A and B are congruent and if B and C are congruent, then A and C are congruent.

27. Prove that if A is symmetric, then A is congruent to a diagonal matrix D.

28. Let $A = \begin{bmatrix} a & b \\ b & d \end{bmatrix}$ be a 2×2 symmetric matrix. Prove that A is positive definite if and only if $\det(A) > 0$ and $a > 0$.

29. Prove that a symmetric matrix A is positive definite if and only if $A = P^T P$ for a nonsingular matrix P.

6.4. Conic Sections

In this section we discuss the classification of the conic sections in the plane. A **quadratic equation** in the variables x and y has the form

$$ax^2 + 2bxy + cy^2 + dx + ey + f = 0, \tag{1}$$

where $a, b, c, d, e,$ and f are real numbers. The graph of Equation (1) is a **conic section**, a curve so named because it is obtained by intersecting a plane with a right circular cone that has two nappes. In Figure 6.4 we show that a plane cuts the cone in an ellipse, circle, parabola, or hyperbola. Degenerate cases of the conic sections are a point, a line, a pair of lines, or the empty set.

The nondegenerate conics are said to be in **standard position** if their graphs and equations are as given in Figure 6.5. The equation is said to be in **standard form**.

Figure 6.4 The nondegenerate conic sections.

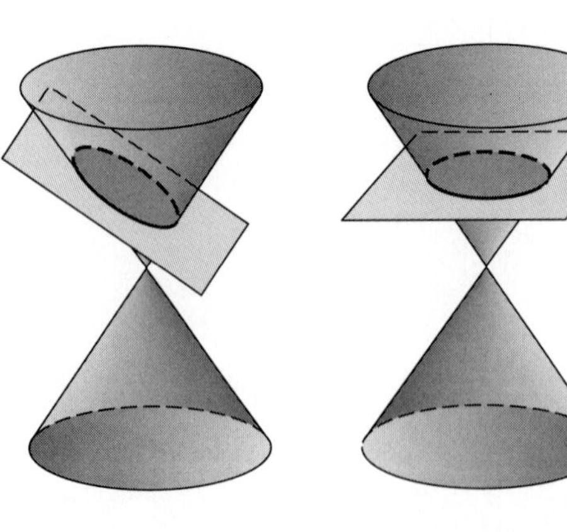

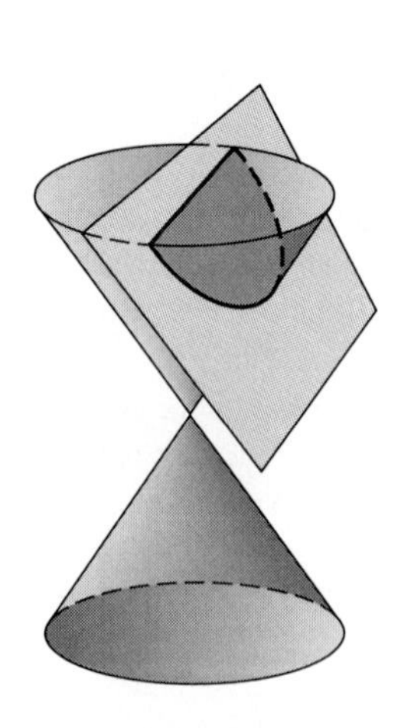

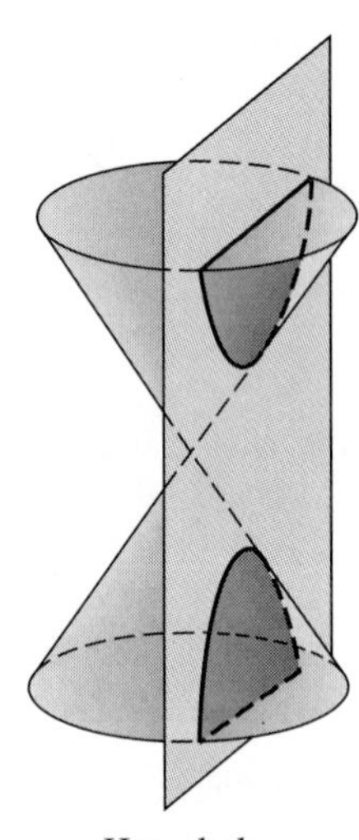

Ellipse. Circle. Parabola. Hyperbola.

Figure 6.5 The conic sections in standard position.

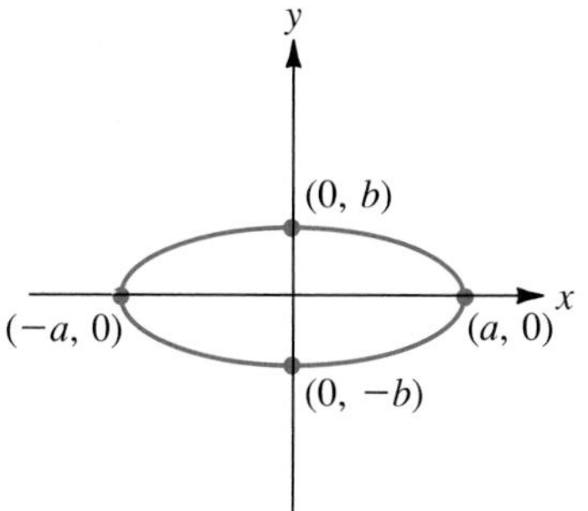

Ellipse.
$$\frac{x^2}{a^2} + \frac{y^2}{b^2} = 1$$
$a > b > 0$

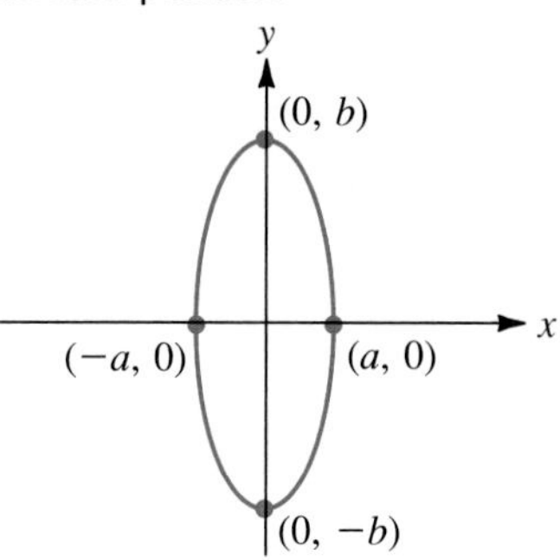

Ellipse.
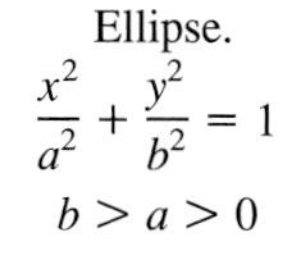
$$\frac{x^2}{a^2} + \frac{y^2}{b^2} = 1$$
$b > a > 0$

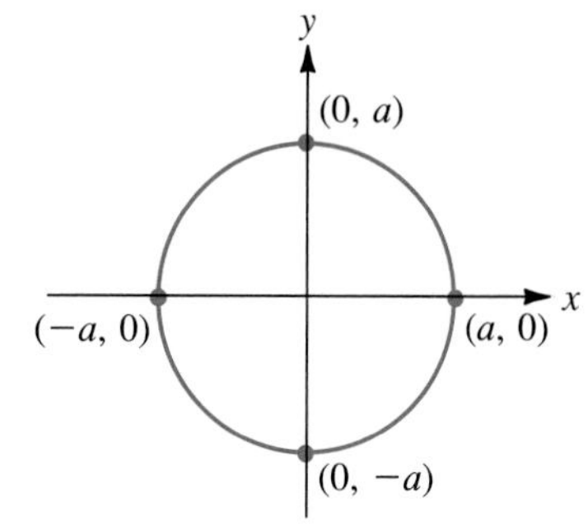

Circle.
$$\frac{x^2}{a^2} + \frac{y^2}{a^2} = 1$$

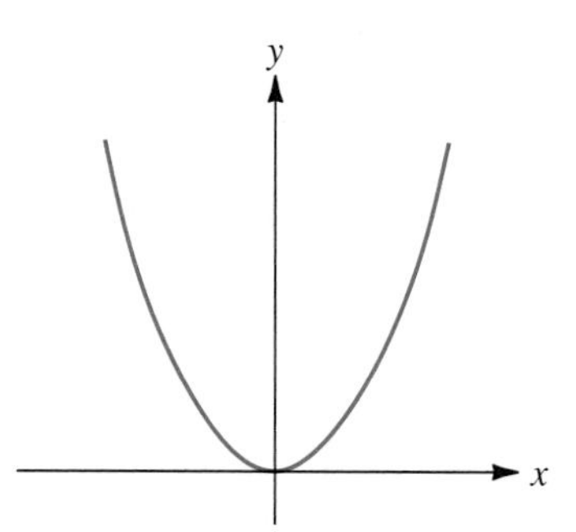

Parabola.
$x^2 = ay$
$a > 0$

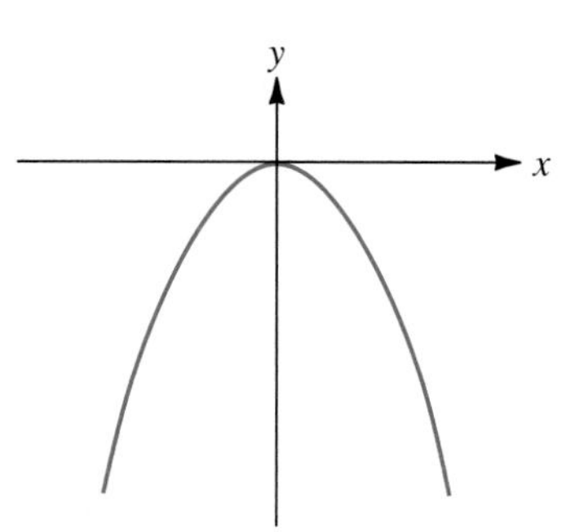

Parabola.
$x^2 = ay$
$a < 0$

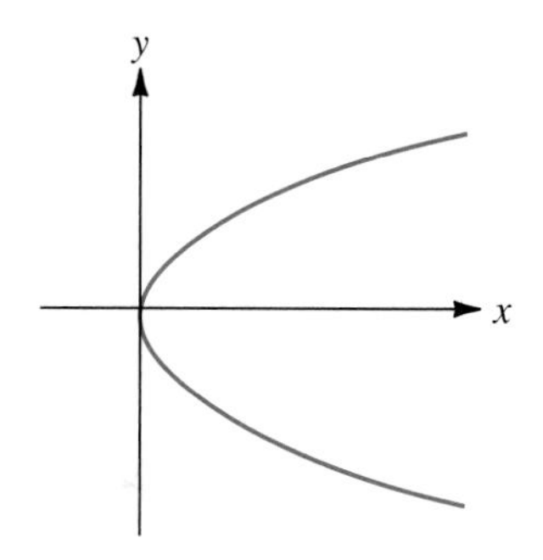

Parabola.
$y^2 = ax$
$a > 0$

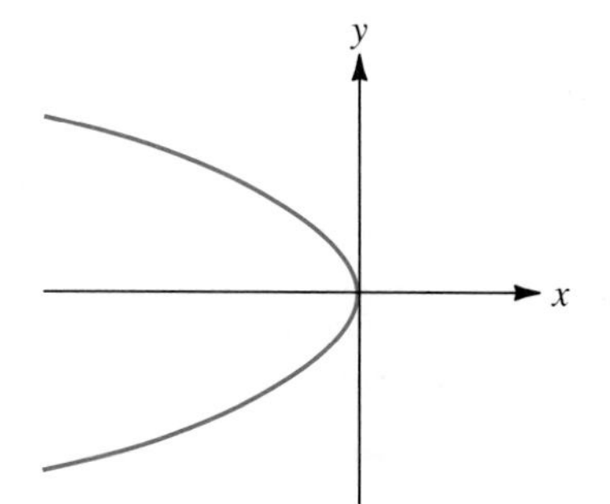

Parabola.
$y^2 = ax$
$a < 0$

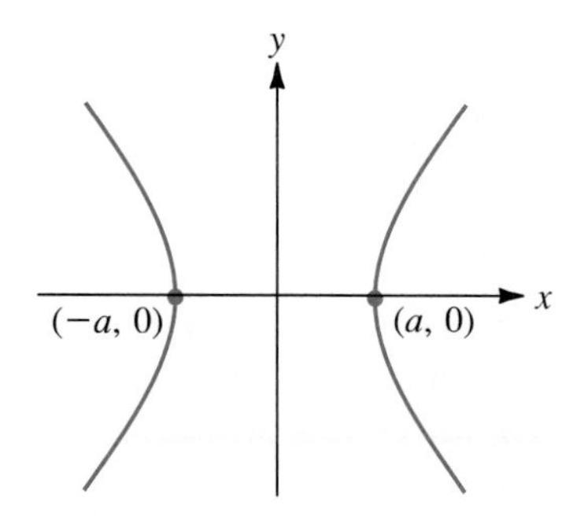

Hyperbola.
$$\frac{x^2}{a^2} - \frac{y^2}{b^2} = 1$$
$a > 0, b > 0$

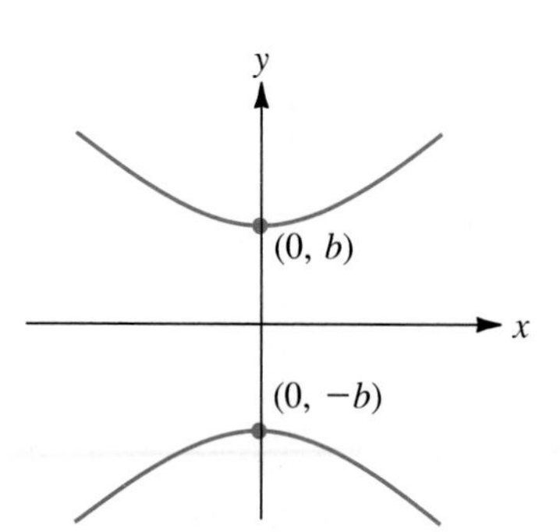

Hyperbola.
$$\frac{y^2}{a^2} - \frac{x^2}{b^2} = 1$$
$a > 0, b > 0$

Example 1. Identify the graph of the given equation.

(a) $4x^2 + 25y^2 - 100 = 0$.
(b) $9y^2 - 4x^2 = -36$.
(c) $x^2 + 4y = 0$.
(d) $y^2 = 0$.
(e) $x^2 + 9y^2 + 9 = 0$.
(f) $x^2 + y^2 = 0$.

Solution: (a) We rewrite the given equation as

$$\frac{4}{100}x^2 + \frac{25}{100}y^2 = \frac{100}{100}$$

or

$$\frac{x^2}{25} + \frac{y^2}{4} = 1,$$

whose graph is an ellipse in standard position with $a = 5$ and $b = 2$. Thus the x-intercepts are $(5, 0)$ and $(-5, 0)$ and the y-intercepts are $(0, 2)$ and $(0, -2)$.

(b) Rewriting the given equation as

$$\frac{x^2}{9} - \frac{y^2}{4} = 1,$$

we see that its graph is a hyperbola in standard position with $a = 3$ and $b = 2$. The x-intercepts are $(3, 0)$ and $(-3, 0)$.

(c) Rewriting the given equation as

$$x^2 = -4y,$$

we see that its graph is a parabola in standard position with $a = -4$, so it opens downward.

(d) Every point satisfying the given equation must have a y-coordinate equal to zero. Thus the graph of this equation consists of all the points on the x-axis.

(e) Rewriting the given equation as

$$x^2 + 9y^2 = -9,$$

we conclude that there are no points in the plane whose coordinates satisfy the given equation.

(f) The only point satisfying the equation is the origin (0, 0), so the graph of this equation is the single point consisting of the origin. ■

We next turn to the study of conic sections whose graphs are not in standard position. First, notice that the equations of the conic sections whose graphs are in standard position do not contain an xy-term (called a **cross-product term**). If a cross-product term appears in the equation, the graph is a conic section that has been rotated from its standard position [see Figure 6.6(a)]. Also notice that none of the equations in Figure 6.5(b) contain an x^2-term and an x-term or a y^2-term and a y-term. If either of these cases occurs and there is no xy-term in the equation, the graph is a conic section that has been translated from its standard position [see Figure 6.6(b)]. On the other hand, if an xy-term is present, the graph is a conic section that has been rotated and possibly also translated [see Figure 6.6(c)].

Figure 6.6

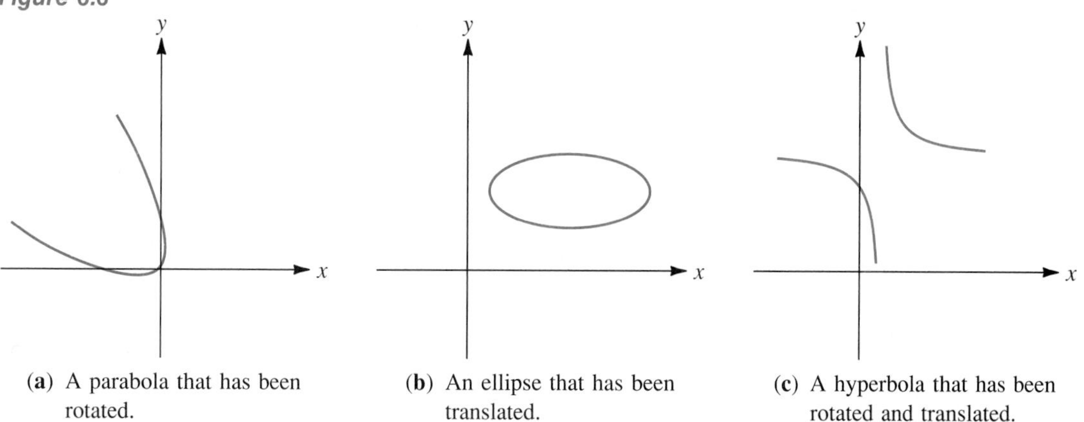

(**a**) A parabola that has been rotated.

(**b**) An ellipse that has been translated.

(**c**) A hyperbola that has been rotated and translated.

> To identify a nondegenerate conic section whose graph is not in standard position, we proceed as follows:
>
> 1. If a cross-product term is present in the given equation, rotate the xy-coordinate axes by means of an orthogonal linear transformation so that in the resulting equation the xy-term no longer appears.
> 2. If an xy-term is not present in the given equation, but an x^2-term and an x-term, or a y^2-term and a y-term appear, translate the xy-coordinate axes by completing the square so that the graph of the resulting equation will be in standard position with respect to the origin of the new coordinate system.

Thus, if an xy-term appears in a given equation, we first rotate the xy-coordinate axes and then, if necessary, translate the rotated axes. In the next example, we deal with the case requiring only a translation of axes.

Example 2. Identify and sketch the graph of the equation

$$x^2 - 4y^2 + 6x + 16y - 23 = 0. \tag{2}$$

Also write its equation in standard form.

Solution: Since there is no cross-product term, we only need to translate axes. Completing the squares in the x- and y-terms, we have

$$\begin{aligned} x^2 + 6x + 9 - 4(y^2 - 4y + 4) - 23 &= 9 - 16 \\ (x + 3)^2 - 4(y - 2)^2 = 23 + 9 - 16 &= 16. \end{aligned} \tag{3}$$

Letting

$$x' = x + 3 \qquad \text{and} \qquad y' = y - 2,$$

we can rewrite Equation (3) as

$$x'^2 - 4y'^2 = 16$$

or in standard form as

$$\frac{x'^2}{16} - \frac{y'^2}{4} = 1. \tag{4}$$

Figure 6.7

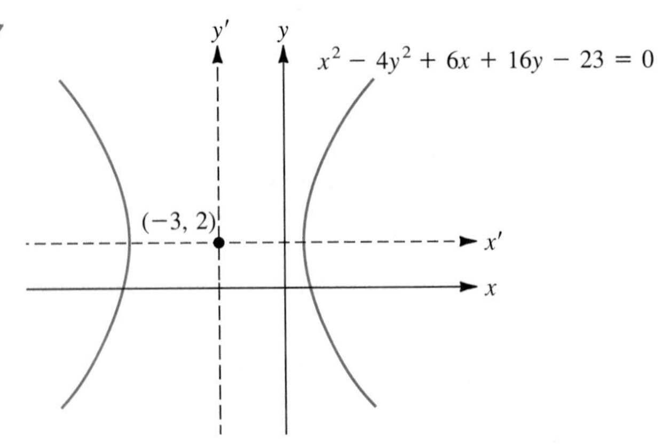

If we translate the xy-coordinate system to the $x'y'$-coordinate system, whose origin is at $(-3, 2)$, then the graph of Equation (4) is a hyperbola in standard position with respect to the $x'y'$-coordinate system (see Figure 6.7). ■

We now turn to the problem of identifying the graph of Equation (1), where we assume that $b \neq 0$, that is, a cross-product term is present. This equation can be written in matrix form as

$$\mathbf{x}^T A\mathbf{x} + B\mathbf{x} + f = 0, \tag{5}$$

where

$$\mathbf{x} = \begin{bmatrix} x \\ y \end{bmatrix}, \qquad A = \begin{bmatrix} a & b \\ b & c \end{bmatrix}, \qquad \text{and} \qquad B = \begin{bmatrix} d & e \end{bmatrix}.$$

Since A is a symmetric matrix, we know from Section 6.2 that it can be diagonalized by an orthogonal matrix P. Thus

$$P^T AP = \begin{bmatrix} \lambda_1 & 0 \\ 0 & \lambda_2 \end{bmatrix},$$

where λ_1 and λ_2 are the eigenvalues of A and the columns of P are $\mathbf{x}_1$ and $\mathbf{x}_2$, orthonormal eigenvectors of A associated with λ_1 and λ_2, respectively.

Letting

$$\mathbf{x} = P\mathbf{y}, \qquad \text{where} \quad \mathbf{y} = \begin{bmatrix} x' \\ y' \end{bmatrix},$$

we can rewrite Equation (5) as

$$\begin{aligned} (P\mathbf{y})^T A(P\mathbf{y}) + B(P\mathbf{y}) + f &= 0 \\ \mathbf{y}^T(P^T AP)\mathbf{y} + BP\mathbf{y} + f &= 0 \end{aligned}$$

or

$$\begin{bmatrix} x' & y' \end{bmatrix} \begin{bmatrix} \lambda_1 & 0 \\ 0 & \lambda_2 \end{bmatrix} \begin{bmatrix} x' \\ y' \end{bmatrix} + BP\mathbf{y} + f = 0 \tag{6}$$

or

$$\lambda_1 x'^2 + \lambda_2 y'^2 + d'x' + e'y' + f = 0. \tag{7}$$

Equation (7) is the resulting equation for the given conic section and it has no cross-product term.

As discussed in Section 6.3, the x' and y' coordinate axes lie along the eigenvectors $\mathbf{x}_1$ and $\mathbf{x}_2$, respectively. Since P is an orthogonal matrix, $|P| = \pm 1$ and, if necessary, we can interchange the columns of P (the eigenvectors $\mathbf{x}_1$ and $\mathbf{x}_2$ of A) or multiply a column of P by -1, so that $|P| = 1$. As noted in Section 6.3, it then follows that P is the matrix of a counterclockwise rotation of R^2 through an angle θ that can be determined as follows. If

$$\mathbf{x}_1 = \begin{bmatrix} x_{11} \\ x_{21} \end{bmatrix},$$

then

$$\theta = \tan^{-1}\left(\frac{x_{21}}{x_{11}}\right),$$

a result that is frequently developed in a calculus course.

Example 3. Identify and sketch the graph of the equation

$$5x^2 - 6xy + 5y^2 - 24\sqrt{2}x + 8\sqrt{2}y + 56 = 0. \tag{8}$$

Write the equation in standard form.

Solution: Rewriting the given equation in matrix form, we obtain

$$[x \quad y]\begin{bmatrix} 5 & -3 \\ -3 & 5 \end{bmatrix}\begin{bmatrix} x \\ y \end{bmatrix} + [-24\sqrt{2} \quad 8\sqrt{2}]\begin{bmatrix} x \\ y \end{bmatrix} + 56 = 0.$$

We now find the eigenvalues of the matrix

$$A = \begin{bmatrix} 5 & -3 \\ -3 & 5 \end{bmatrix}.$$

Thus

$$\begin{aligned} |\lambda I_2 - A| &= \begin{vmatrix} \lambda - 5 & 3 \\ 3 & \lambda - 5 \end{vmatrix} \\ &= (\lambda - 5)(\lambda - 5) - 9 = \lambda^2 - 10\lambda + 16 \\ &= (\lambda - 2)(\lambda - 8), \end{aligned}$$

so the eigenvalues of A are

$$\lambda_1 = 2, \qquad \lambda_2 = 8.$$

Associated eigenvectors are obtained by solving the homogeneous system

$$(\lambda I_2 - A)\mathbf{x} = \mathbf{0}.$$

Thus, for $\lambda_1 = 2$, we have

$$\begin{bmatrix} -3 & 3 \\ 3 & -3 \end{bmatrix}\mathbf{x} = \mathbf{0},$$

so an eigenvector of A associated with $\lambda_1 = 2$ is

$$\begin{bmatrix} 1 \\ 1 \end{bmatrix}.$$

For $\lambda_2 = 8$ we have

$$\begin{bmatrix} 3 & 3 \\ 3 & 3 \end{bmatrix}\mathbf{x} = \mathbf{0},$$

so an eigenvector of A associated with $\lambda_2 = 8$ is

$$\begin{bmatrix} -1 \\ 1 \end{bmatrix}.$$

Normalizing these eigenvectors, we obtain the orthogonal matrix

$$P = \begin{bmatrix} \frac{1}{\sqrt{2}} & -\frac{1}{\sqrt{2}} \\ \frac{1}{\sqrt{2}} & \frac{1}{\sqrt{2}} \end{bmatrix}.$$

Then

$$P^TAP = \begin{bmatrix} 2 & 0 \\ 0 & 8 \end{bmatrix}.$$

Letting $\mathbf{x} = P\mathbf{y}$, we write the transformed equation for the given conic section, Equation (6), as

$$2x'^2 + 8y'^2 - 16x' + 32y' + 56 = 0$$

or

$$x'^2 + 4y'^2 - 8x' + 16y' + 28 = 0.$$

To identify the graph of this equation, we need to translate axes, so we complete the squares, obtaining

$$(x' - 4)^2 + 4(y' + 2)^2 + 28 = 16 + 16$$

$$(x' - 4)^2 + 4(y' + 2)^2 = 4 \tag{9}$$

$$\frac{(x' - 4)^2}{4} + \frac{(y' + 2)^2}{1} = 1.$$

Letting

$$x'' = x' - 4 \quad \text{and} \quad y'' = y' + 2,$$

we find that Equation (9) becomes

$$\frac{x''^2}{4} + \frac{y''^2}{1} = 1, \tag{10}$$

whose graph is an ellipse in standard position with respect to the $x''y''$-coordinate axes, as shown in Figure 6.8, where the origin of the $x''y''$-coordinate system is at $(4, -2)$. Equation (10) is the standard form of the equation of the ellipse. Since

$$\mathbf{x}_1 = \begin{bmatrix} \frac{1}{\sqrt{2}} \\ \frac{1}{\sqrt{2}} \end{bmatrix},$$

the xy-coordinate axes have been rotated through the angle θ, where

$$\theta = \tan^{-1}\left(\frac{\frac{1}{\sqrt{2}}}{\frac{1}{\sqrt{2}}}\right) = \tan^{-1} 1,$$

so $\theta = 45°$. ■

Figure 6.8

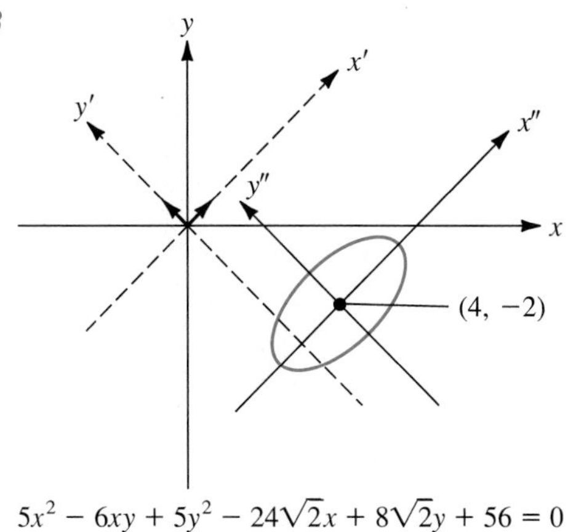

$$5x^2 - 6xy + 5y^2 - 24\sqrt{2}x + 8\sqrt{2}y + 56 = 0$$

The graph of a given quadratic equation in x and y can be identified from the equation that is obtained after rotating axes, that is, from Equation (6) or (7). The identification of the conic section given by these equations is shown in Table 6.1.

Table 6.1 Identification of the Conic Sections

λ_1, λ_2 both nonzero		Exactly one of λ_1, λ_2 is zero
$\lambda_1\lambda_2 > 0$	$\lambda_1\lambda_2 < 0$	
Ellipse	Hyperbola	Parabola

6.4. EXERCISES

In Exercises 1 through 12, identify the graph of the equation.

1. $x^2 + 9y^2 - 9 = 0.$

2. $x^2 = 2y.$

3. $25y^2 - 4x^2 = 100.$

4. $y^2 - 16 = 0.$

5. $3x^2 - y^2 = 0.$

6. $y = 0.$

7. $4x^2 + 4y^2 - 9 = 0.$

8. $-25x^2 + 9y^2 + 225 = 0.$

9. $4x^2 + y^2 = 0.$

10. $9x^2 + 4y^2 + 36 = 0.$

In Exercises 11 through 18, translate axes to identify the graph of the equation and write the equation in standard form.

11. $x^2 + 2y^2 - 4x - 4y + 4 = 0.$

12. $x^2 - y^2 + 4x - 6y - 9 = 0.$

13. $x^2 + y^2 - 8x - 6y = 0.$

14. $x^2 - 4x + 4y + 4 = 0.$

15. $y^2 - 4y = 0.$

16. $4x^2 + 5y^2 - 30y + 25 = 0.$

17. $x^2 + y^2 - 2x - 6y + 10 = 0.$

18. $2x^2 + y^2 - 12x - 4y + 24 = 0.$

In Exercises 19 through 24, rotate axes to identify the graph of the equation and write the equation in standard form.

19. $x^2 + xy + y^2 = 6.$

20. $xy = 1.$

21. $9x^2 + y^2 + 6xy = 4.$

22. $x^2 + y^2 + 4xy = 9.$

23. $4x^2 + 4y^2 - 10xy = 0.$

24. $9x^2 + 6y^2 + 4xy - 5 = 0.$

In Exercises 25 through 30, identify the graph of the equation and write the equation in standard form.

25. $9x^2 + y^2 + 6xy - 10\sqrt{10}x + 10\sqrt{10}y + 90 = 0.$

26. $5x^2 + 5y^2 - 6xy - 30\sqrt{2}x + 18\sqrt{2}y + 82 = 0.$

27. $5x^2 + 12xy - 12\sqrt{13}x = 36.$

28. $6x^2 + 9y^2 - 4xy - 4\sqrt{5}x - 18\sqrt{5}y = 5.$

29. $x^2 - y^2 + 2\sqrt{3}xy + 6x = 0.$

30. $8x^2 + 8y^2 - 16xy + 33\sqrt{2}x - 31\sqrt{2}y + 70 = 0.$

6.5. *Quadric Surfaces*

In Section 6.4 conic sections were used to provide geometric models for quadratic forms in two variables. In this section we investigate quadratic forms in three variables and use particular surfaces called quadric surfaces as geometric models. Quadric surfaces are often studied and sketched in analytic geometry and calculus. Here we use Theorems 6.10 and 6.11 to develop a classification scheme for quadric surfaces.

A **second-degree polynomial equation** in three variables x, y, and z has the form

$$ax^2 + by^2 + cz^2 + 2dxy + 2exz + 2fyz + gx + hy + iz = j, \tag{1}$$

where coefficients a through j are real numbers with $a, b, \ldots, f$ not all zero. Equation (1) can be written in matrix form as

$$\mathbf{x}^T A\mathbf{x} + \mathbf{b}\mathbf{x} = j, \tag{2}$$

where $A = \begin{bmatrix} a & d & e \\ d & b & f \\ e & f & c \end{bmatrix}$, $\mathbf{b} = [g \quad h \quad i]$, and $\mathbf{x} = \begin{bmatrix} x \\ y \\ z \end{bmatrix}$. We call $\mathbf{x}^T A\mathbf{x}$ the **quadratic form (in three variables) associated with the second degree polynomial** in (1). As in Section 6.3, the symmetric matrix A is called the matrix of the quadratic form.

The graph of (1) in R^3 is called a **quadric surface.** As in the case of the classification of conic sections in Section 6.4, the classification of (1) as to the type of surface represented depends on the matrix A. Using the ideas in Section 6.4, we have the following strategies to determine a simpler equation for a quadric surface.

1. If A is not diagonal, then a rotation of axes is used to eliminate any cross-product terms xy, xz, or yz.
2. If $\mathbf{b} = [g \quad h \quad i] \neq \mathbf{0}$, then a translation of axes is used to eliminate any first-degree terms.

The resulting equation will have the standard form

$$\lambda_1 x''^2 + \lambda_2 y''^2 + \lambda_3 z''^2 = k$$

or in matrix form

$$\mathbf{y}^T C\mathbf{y} = k, \tag{3}$$

where $\mathbf{y} = \begin{bmatrix} x'' \\ y'' \\ z'' \end{bmatrix}$, k is some real constant, and C is a diagonal matrix with diagonal entries $\lambda_1, \lambda_2, \lambda_3$ which are the eigenvalues of A.

We now turn to the classification of quadric surfaces.

Definition 6.8. Let A be an $n \times n$ symmetric matrix. The **inertia** of A, denoted In (A), is an ordered triple of numbers

$$(\text{pos, neg, zer}),$$

where pos, neg, and zer are the number of positive, negative, and zero eigenvalues of A, respectively. ■

Example 1. Find the inertia of each of the following matrices:

$$A = \begin{bmatrix} 2 & 2 \\ 2 & 2 \end{bmatrix}, \quad B = \begin{bmatrix} 2 & 1 \\ 1 & 2 \end{bmatrix}, \quad C = \begin{bmatrix} 0 & 2 & 2 \\ 2 & 0 & 2 \\ 2 & 2 & 0 \end{bmatrix}.$$

Solution: We determine the eigenvalues of each of the matrices. It follows that (verify)

$\det(\lambda I_1 - A) = \lambda(\lambda - 4) = 0,$ so $\lambda_1 = 0,\ \lambda_2 = 4$, and $\text{In}(A) = (1, 0, 1)$.

$\det(\lambda I_2 - B) = (\lambda - 1)(\lambda - 3) = 0,$ so $\lambda_1 = 1,\ \lambda_2 = 3$, and $\text{In}(B) = (2, 0, 0)$.

$\det(\lambda I_3 - C) = (\lambda + 2)^2(\lambda - 4) = 0,$ so $\lambda_1 = \lambda_2 = -2,\ \lambda_3 = 4$, and $\text{In}(C) = (1, 2, 0)$. ■

From Section 6.3 the signature of a quadratic form $\mathbf{x}^T A\mathbf{x}$ is the difference between the number of positive eigenvalues and the number of negative eigenvalues of A. In terms of inertia, the signature of $\mathbf{x}^T A\mathbf{x}$ is $s = \text{pos} - \text{neg}$.

In order to use inertia for classification of quadric surfaces (or conic sections), we assume that the eigenvalues of an $n \times n$ symmetric matrix A of a quadratic form in n variables are denoted by

$$\begin{aligned} \lambda_1 \geq \cdots \geq \lambda_{\text{pos}} &> 0 \\ \lambda_{\text{pos}+1} \leq \cdots \leq \lambda_{\text{pos}+\text{neg}} &< 0 \\ \lambda_{\text{pos}+\text{neg}+1} = \cdots = \lambda_n &= 0. \end{aligned}$$

The largest positive eigenvalue is denoted by λ_1 and the smallest one by λ_{pos}. We also assume that $\lambda_1 > 0$ and $j \geq 0$ in (2), which eliminates redundant and impossible cases. For example, if

$$A = \begin{bmatrix} -1 & 0 & 0 \\ 0 & -2 & 0 \\ 0 & 0 & -3 \end{bmatrix}, \qquad \mathbf{b} = \begin{bmatrix} 0 & 0 & 0 \end{bmatrix}, \qquad \text{and} \qquad j = 5,$$

then the second-degree polynomial is $-x^2 - 2y^2 - 3z^2 = 5$, which has an empty solution set. That is, the surface represented has no points. However, if $j = -5$, then the second-degree polynomial is $-x^2 - 2y^2 - 3z^2 = -5$, which is identical to $x^2 + 2y^2 + 3z^2 = 5$. The assumptions $\lambda_1 > 0$ and $j \geq 0$ avoid such a redundant representation.

Example 2. Consider a quadratic form in two variables with matrix A and assume that $\lambda_1 > 0$ and $f \geq 0$ in Equation (1) of Section 6.4. Then there are only three possible cases for the inertia of A, which we summarize as follows.

1. In $(A) = (2, 0, 0)$; then the quadratic form represents an ellipse.
2. In $(A) = (1, 1, 0)$; then the quadratic form represents a hyperbola.
3. In $(A) = (1, 0, 1)$; then the quadratic form represents a parabola.

This classification is identical to that given in Table 6.1, taking the assumptions into account. ■

Note that the classification of the conic sections in Example 2 does not distinguish between special cases within a particular geometric class. For example, both $y = x^2$ and $x = y^2$ have inertia (1, 0, 1).

Before classifying quadric surfaces using inertia, we present the quadric surfaces in the standard forms met in analytic geometry and calculus. (In the following, a, b, and c are positive unless otherwise stated.)

Ellipsoid. (See Figure 6.9.)

$$\frac{x^2}{a^2} + \frac{y^2}{b^2} + \frac{z^2}{c^2} = 1.$$

The special case $a = b = c$ is a sphere.

Figure 6.9 Ellipsoid.

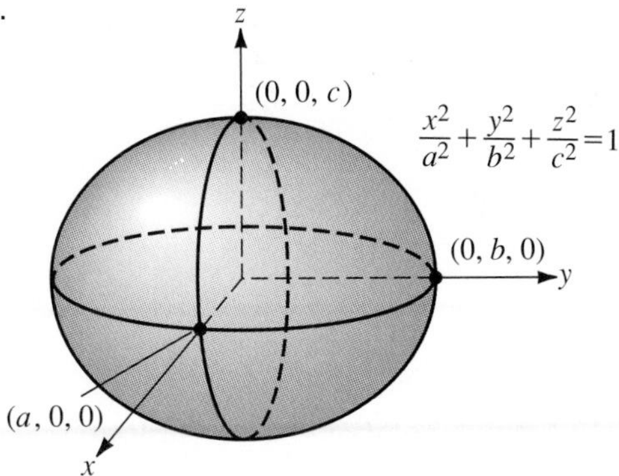

Elliptic Paraboloid. (See Figure 6.10.)

$$z = \frac{x^2}{a^2} + \frac{y^2}{b^2}, \qquad y = \frac{x^2}{a^2} + \frac{z^2}{c^2}, \qquad x = \frac{y^2}{b^2} + \frac{z^2}{c^2}.$$

Figure 6.10 Elliptic paraboloid.

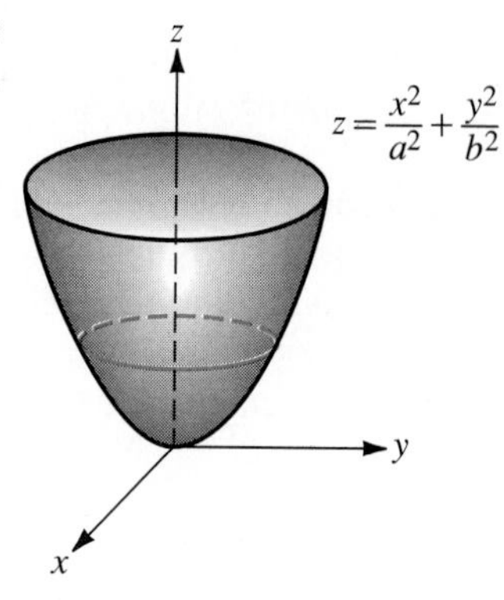

A degenerate case of a parabola is a line, so a degenerate case of an elliptic paraboloid is an **elliptic cylinder** (see Figure 6.11), which is given by

$$\frac{x^2}{a^2} + \frac{y^2}{b^2} = 1, \qquad \frac{x^2}{a^2} + \frac{z^2}{c^2} = 1, \qquad \frac{y^2}{b^2} + \frac{z^2}{c^2} = 1.$$

Figure 6.11 Elliptic cylinder.

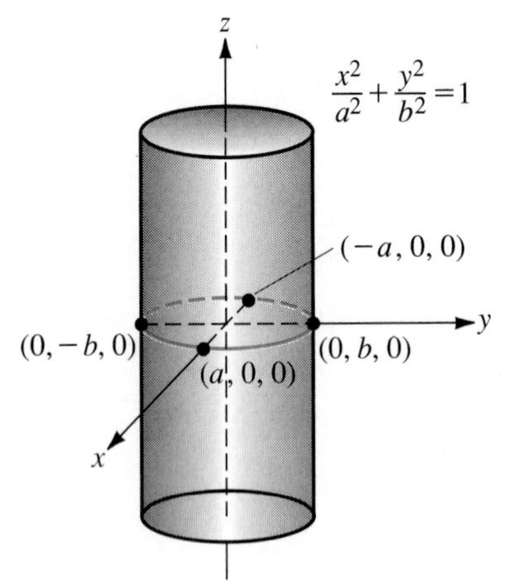

Hyperboloid of One Sheet. (See Figure 6.12.)

$$\frac{x^2}{a^2} + \frac{y^2}{b^2} - \frac{z^2}{c^2} = 1, \qquad \frac{x^2}{a^2} - \frac{y^2}{b^2} + \frac{z^2}{c^2} = 1, \qquad -\frac{x^2}{a^2} + \frac{y^2}{b^2} + \frac{z^2}{c^2} = 1.$$

Figure 6.12 Hyperboloid of one sheet.

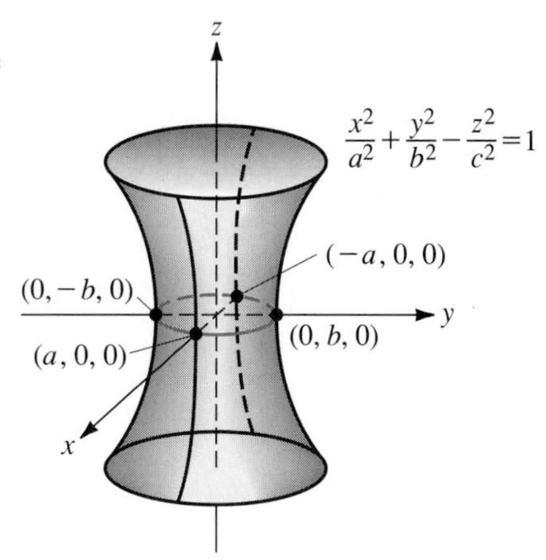

A degenerate case of a hyperbola is a pair of lines through the origin; hence a degenerate case of a hyperboloid of one sheet is a **cone** (Figure 6.13), which is given by

$$\frac{x^2}{a^2}+\frac{y^2}{b^2}-\frac{z^2}{c^2}=0, \qquad \frac{x^2}{a^2}-\frac{y^2}{b^2}+\frac{z^2}{c^2}=0, \qquad \frac{-x^2}{a^2}+\frac{y^2}{b^2}+\frac{z^2}{c^2}=0.$$

Figure 6.13 Cone.

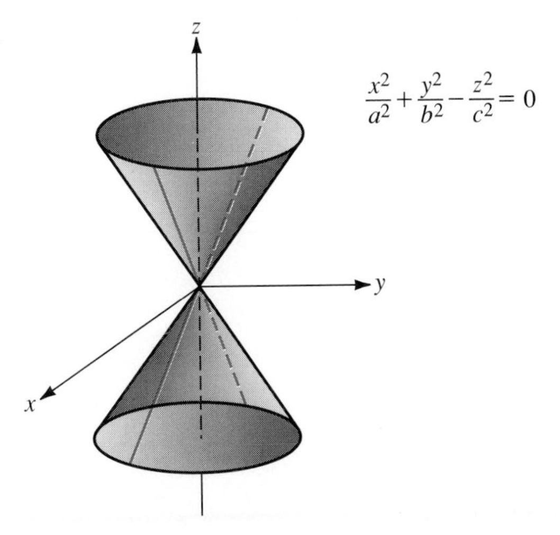

Hyperboloid of Two Sheets. (See Figure 6.14.)

$$\frac{x^2}{a^2}-\frac{y^2}{b^2}-\frac{z^2}{c^2}=1, \qquad -\frac{x^2}{a^2}-\frac{y^2}{b^2}+\frac{z^2}{c^2}=1, \qquad -\frac{x^2}{a^2}+\frac{y^2}{b^2}-\frac{z^2}{c^2}=1.$$

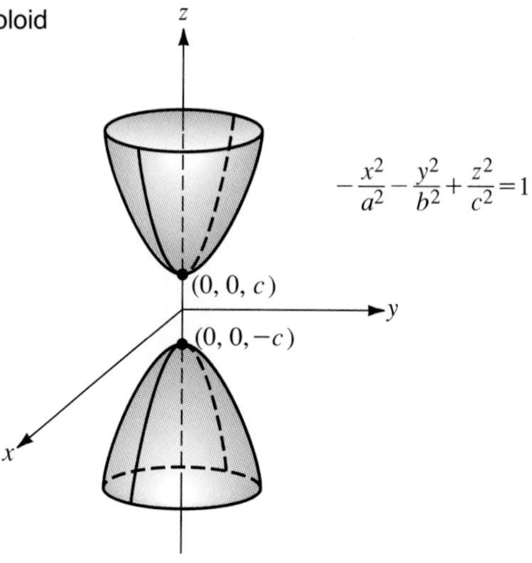

Figure 6.14 Hyperboloid of two sheets.

Hyperbolic Paraboloid. (See Figure 6.15.)

$$\pm z = \frac{x^2}{a^2} - \frac{y^2}{b^2}, \qquad \pm y = \frac{x^2}{a^2} - \frac{z^2}{b^2}, \qquad \pm x = \frac{y^2}{a^2} - \frac{z^2}{b^2}.$$

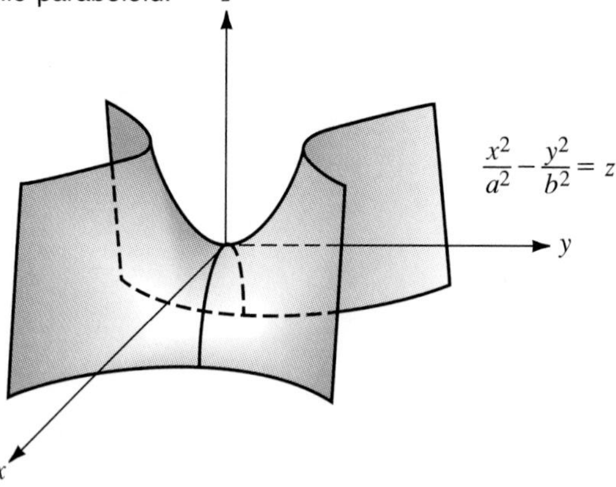

Figure 6.15 Hyperbolic paraboloid.

A degenerate case of a parabola is a line, so a degenerate case of a hyperbolic paraboloid is a hyperbolic cylinder (see Figure 6.16), which is given by

$$\frac{x^2}{a^2} - \frac{y^2}{b^2} = \pm 1, \qquad \frac{x^2}{a^2} - \frac{z^2}{b^2} = \pm 1, \qquad \frac{y^2}{a^2} - \frac{z^2}{b^2} = \pm 1.$$

Figure 6.16 Hyperbolic cylinder.

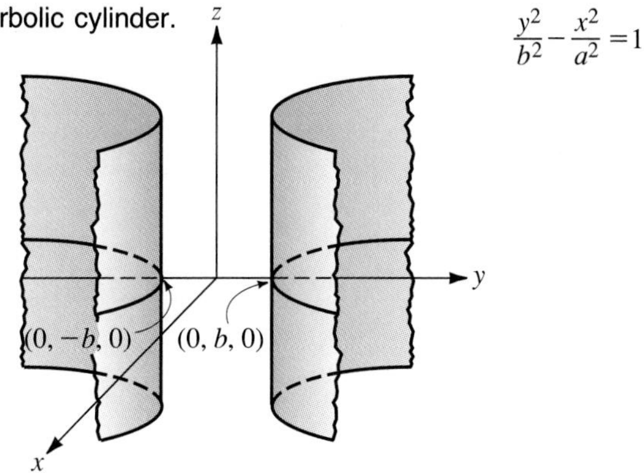

Parabolic Cylinder. (See Figure 6.17.) One of a or b is not zero.

$$x^2 = ay + bz, \qquad y^2 = ax + bz, \qquad z^2 = ax + by.$$

Figure 6.17 Parabolic cylinder.

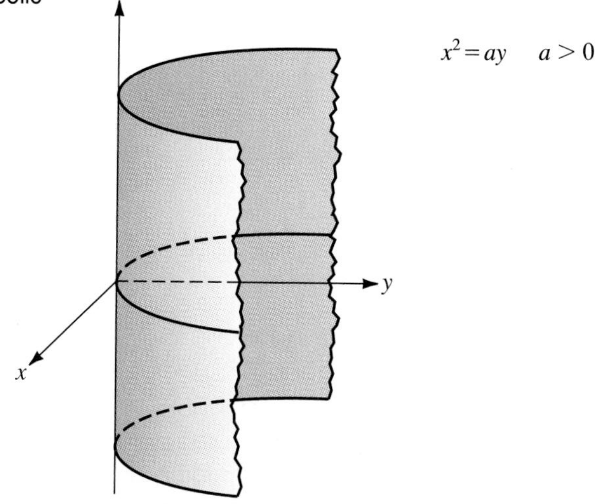

For a quadratic form in three variables with matrix A, under assumptions $\lambda_1 > 0$ and $j \geq 0$ in (2), there are exactly six possibilities for the inertia of A. We present these in Table 6.2. As with the conic section classification in Example 2, the classification of quadric surfaces in Table 6.2 does not distinguish between special cases within a particular geometric class.

Table 6.2 Identification of Quadric Surfaces

In $(A) = (3, 0, 0)$	Ellipsoid
In $(A) = (2, 0, 1)$	Elliptic paraboloid
In $(A) = (2, 1, 0)$	Hyperboloid of one sheet
In $(A) = (1, 2, 0)$	Hyperboloid of two sheets
In $(A) = (1, 1, 1)$	Hyperbolic paraboloid
In $(A) = (1, 0, 2)$	Parabolic cylinder

Example 3. Classify the quadric surface represented by the quadratic form $\mathbf{x}^T A\mathbf{x} = 3$, where

$$A = \begin{bmatrix} 0 & 2 & 2 \\ 2 & 0 & 2 \\ 2 & 2 & 0 \end{bmatrix} \quad \text{and} \quad \mathbf{x} = \begin{bmatrix} x \\ y \\ z \end{bmatrix}.$$

Solution: From Example 1 we have that In $(A) = (1, 2, 0)$ and hence the quadric surface is a hyperboloid of two sheets. ■

Example 4. Classify the quadric surface given by

$$2x^2 + 4y^2 - 4z^2 + 6yz - 5x + 3y = 2.$$

Solution: Rewrite the second-degree polynomial as a quadratic form in three variables to identify the matrix A of the quadratic form. We have $A = \begin{bmatrix} 2 & 0 & 0 \\ 0 & 4 & 3 \\ 0 & 3 & -4 \end{bmatrix}$. Its eigenvalues are $\lambda_1 = 5$, $\lambda_2 = 2$, and $\lambda_3 = -5$ (verify). Thus In $(A) = (2, 1, 0)$ and hence the quadric surface is a hyperboloid of one sheet. ■

The classification of a quadric surface is much easier than the problem of transforming it to the standard forms that are used in analytic geometry and calculus. The algebraic steps to obtain an equation in standard form from a second-degree polynomial equation (1) require, in general, a rotation and translation of axes, as mentioned earlier. The rotation requires both the eigenvalues and eigenvectors of the matrix A of the quadratic form. The eigenvectors of A are used to form an orthogonal matrix P so that $|P| = 1$, and hence the change of variables $\mathbf{x} = P\mathbf{y}$ represents a rotation. The resulting associated quadratic form is that obtained in the principal axes theorem, Theorem 6.10, that is, all cross-product terms are eliminated. We illustrate this with the next example.

Example 5. For the quadric surface in Example 4,

$$\mathbf{x}^T A\mathbf{x} + [-5 \quad 3 \quad 0]\mathbf{x} = 2,$$

determine the rotation so that all cross-product terms are eliminated.

Solution: The eigenvalues and eigenvectors of $A = \begin{bmatrix} 2 & 0 & 0 \\ 0 & 4 & 3 \\ 0 & 3 & -4 \end{bmatrix}$ are, respectively, $\lambda_1 = 5$, $\lambda_2 = 2$, $\lambda_3 = -5$ and $\mathbf{v}_1 = \begin{bmatrix} 0 \\ 3 \\ 1 \end{bmatrix}$, $\mathbf{v}_2 = \begin{bmatrix} 1 \\ 0 \\ 0 \end{bmatrix}$, $\mathbf{v}_3 = \begin{bmatrix} 0 \\ 1 \\ -3 \end{bmatrix}$ (verify). The eigenvectors $\mathbf{v}_i$ are mutually orthogonal, since they correspond to distinct eigenvalues of a symmetric matrix (see Theorem 6.7). We normalize the eigenvectors as

$$\mathbf{u}_1 = \frac{1}{\sqrt{10}}\begin{bmatrix} 0 \\ 3 \\ 1 \end{bmatrix}, \qquad \mathbf{u}_2 = \mathbf{v}_2, \qquad \mathbf{u}_3 = \frac{1}{\sqrt{10}}\begin{bmatrix} 0 \\ 1 \\ -3 \end{bmatrix}$$

and define $P = [\mathbf{u}_1 \quad \mathbf{u}_2 \quad \mathbf{u}_3]$. Then $|P| = 1$ (verify), so we let $\mathbf{x} = P\mathbf{y}$ and obtain the representation

$$(P\mathbf{y})^T A(P\mathbf{y}) + [-5 \quad 3 \quad 0]P\mathbf{y} = 2$$
$$\mathbf{y}^T(P^T AP)\mathbf{y} + [-5 \quad 3 \quad 0]P\mathbf{y} = 2.$$

Since $P^T AP = D$, and letting $\mathbf{y} = \begin{bmatrix} x' \\ y' \\ z' \end{bmatrix}$, we have

$$\mathbf{y}^T D\mathbf{y} + [-5 \quad 3 \quad 0]P\mathbf{y} = 2,$$

$$\mathbf{y}^T\begin{bmatrix} 5 & 0 & 0 \\ 0 & 2 & 0 \\ 0 & 0 & -5 \end{bmatrix}\mathbf{y} + \begin{bmatrix} \frac{9}{\sqrt{10}} & -5 & \frac{3}{\sqrt{10}} \end{bmatrix}\mathbf{y} = 2$$

(if $|P| \neq 1$, we redefine P by reordering its columns until we get its determinant to be 1), or

$$5x'^2 + 2y'^2 - 5z'^2 + \frac{9}{\sqrt{10}}x' - 5y' + \frac{3}{\sqrt{10}}z' = 2. \qquad \blacksquare$$

To complete the transformation to standard form, we introduce a change of variable to perform a translation that eliminates any first-degree terms. Algebraically, we complete the square in each of the three variables.

Example 6. Continue with Example 5 to eliminate the first-degree terms.

Solution: The last expression for the quadric surface in Example 5 can be written as

$$5x'^2 + \frac{9}{\sqrt{10}}x' + 2y'^2 - 5y' - 5z'^2 + \frac{3}{\sqrt{10}}z' = 2.$$

Completing the square in each variable, we have

$$\begin{aligned} 5\left(x'^2 + \frac{9}{5\sqrt{10}}x' + \frac{81}{1000}\right) + 2\left(y'^2 - \frac{5}{2}y' + \frac{25}{16}\right) & \\ - 5\left(z'^2 - \frac{3}{5\sqrt{10}}z' + \frac{9}{1000}\right) & \\ = 5\left(x' + \frac{9}{10\sqrt{10}}\right)^2 + 2\left(y' - \frac{5}{4}\right)^2 - 5\left(z' - \frac{3}{10\sqrt{10}}\right)^2 & \\ = 2 + \frac{405}{1000} + \frac{50}{16} - \frac{45}{1000}. & \end{aligned}$$

Letting

$$x'' = x' + \frac{9}{10\sqrt{10}}, \qquad y'' = y' - \frac{5}{4}, \qquad z'' = z' - \frac{3}{10\sqrt{10}},$$

we can write the equation of the quadric surface as

$$5x''^2 + 2y''^2 - 5z''^2 = \frac{5485}{1000} = 5.485.$$

This can be written in standard form as

$$\frac{x''^2}{5.485/5} + \frac{y''^2}{5.485/2} - \frac{z''^2}{5.485/5} = 1. \qquad \blacksquare$$

6.5. EXERCISES

In Exercises 1 through 14, use inertia to classify the quadric surface in each equation.

1. $x^2 + y^2 + 2z^2 - 2xy - 4xz - 4yz + 4x = 8.$

2. $x^2 + 3y^2 + 2z^2 - 6x - 6y + 4z - 2 = 0.$

3. $z = 4xy.$

4. $x^2 + y^2 + z^2 + 2xy = 4.$

5. $x^2 - y = 0.$

6. $2xy + z = 0.$

7. $5y^2 + 20y + z - 23 = 0.$

8. $x^2 + y^2 + 2z^2 - 2xy + 4xz + 4yz = 16.$

9. $4x^2 + 9y^2 + z^2 + 8x - 18y - 4z - 19 = 0.$

10. $y^2 - z^2 - 9x - 4y + 8z - 12 = 0.$

11. $x^2 + 4y^2 + 4x + 16y - 16z - 4 = 0.$

12. $4x^2 - y^2 + z^2 - 16x + 8y - 6z + 5 = 0.$

13. $x^2 - 4z^2 - 4x + 8z = 0.$

14. $2x^2 + 2y^2 + 4z^2 + 2xy - 2xz - 2yz + 3x - 5y + z = 7.$

In Exercises 15 through 28, classify the quadric surface with each equation and determine its standard form.

15. $x^2 + 2y^2 + 2z^2 + 2yz = 1.$

16. $x^2 + y^2 + 2z^2 - 2xy + 4xz + 4yz = 16.$

17. $2xz - 2z - 4y - 4z + 8 = 0.$

18. $x^2 + 3y^2 + 3z^2 - 4yz = 9.$

19. $x^2 + y^2 + z^2 + 2xy = 8.$

20. $-x^2 - y^2 - z^2 + 4xy + 4xz + 4yz = 3.$

21. $2x^2 + 2y^2 + 4z^2 - 4xy - 8xz - 8yz + 8x = 15.$

22. $4x^2 + 4y^2 + 8z^2 + 4xy - 4xz - 4yz + 6x - 10y + 2z = \frac{9}{4}.$

23. $2y^2 + 2z^2 + 4yz + \frac{16}{\sqrt{2}}x + 4 = 0.$

24. $x^2 + y^2 - 2z^2 + 2xy + 8xz + 8yz + 3x + z = 0.$

25. $-x^2 - y^2 - z^2 + 4xy + 4xz + 4yz + \frac{3}{\sqrt{2}}x - \frac{3}{\sqrt{2}}y = 6.$

26. $2x^2 + 3y^2 + 3z^2 - 2yz + 2x + \frac{1}{\sqrt{2}}y + \frac{1}{\sqrt{2}}z = \frac{3}{8}.$

27. $x^2 + y^2 - z^2 - 2x - 4y - 4z + 1 = 0.$

28. $-8x^2 - 8y^2 + 10z^2 + 32xy - 4xz - 4yz = 24.$

Supplementary Exercises

1. Find the eigenvalues and corresponding eigenvectors for each of the matrices in Supplementary Exercise 2 in Chapter 5. Which of these matrices are similar to a diagonal matrix?

2. Let

$$A = \begin{bmatrix} 1 & 0 & -4 \\ 0 & 5 & 4 \\ -4 & 4 & 3 \end{bmatrix}.$$

(a) Find the eigenvalues and corresponding eigenvectors of A.
(b) Is A similar to a diagonal matrix? If so, find a nonsingular matrix P such that $P^{-1}AP$ is diagonal. Is P unique? Explain.
(c) Find the eigenvalues of A^{-1}.
(d) Find the eigenvalues and corresponding eigenvectors of A^2.

3. Let A be any $n \times n$ real matrix.
(a) Prove that the coefficient of λ^{n-1} in the characteristic polynomial of A is given by $-\text{Tr}(A)$ (see Supplementary Exercise 1 in Chapter 4).
(b) Prove that $\text{Tr}(A)$ is the sum of the eigenvalues of A.

4. Prove or disprove: Every nonsingular matrix is similar to a diagonal matrix.

5. Let $p(x) = a_0 + a_1x + a_2x^2 + \cdots + a_kx^k$ be a polynomial in x. Show that the eigenvalues of matrix $p(A) = a_0I_n + a_1A + a_2A^2 + \cdots + a_kA^k$ are $p(\lambda_i)$, $i = 1, 2, \ldots, n$, where λ_i are the eigenvalues of A.

6. Let $p_1(\lambda)$ be the characteristic polynomial of A_{11} and $p_2(\lambda)$ the characteristic polynomial of A_{22}. What is the characteristic polynomial of each of the following partitioned matrices?

(a) $A = \begin{bmatrix} A_{11} & O \\ O & A_{22} \end{bmatrix}$. (b) $A = \begin{bmatrix} A_{11} & A_{21} \\ O & A_{22} \end{bmatrix}$.

(*Hint:* See Exercises 14 and 17 in Section 5.2.)

7. Let $L: P_1 \rightarrow P_1$ be the linear transformation defined by $L(at + b) = \dfrac{a+b}{2}t$.

Let $S = \{2 - t, 3 + t\}$ be a basis for P_1.
(a) Find $[L(2 - t)]_S$ and $[L(3 + t)]_S$.
(b) Find a matrix A representing L with respect to S.
(c) Find the eigenvalues and eigenvectors of A.
(d) Find the eigenvalues and eigenvectors of L.
(e) Describe the eigenspace for each eigenvalue of L.

8. Let $V = {}_2R_2$ and let $L: V \rightarrow V$ be the linear transformation defined by $L(A) = A^T$, for A in V. Let

$S = \{A_1, A_2, A_3, A_4\}$, where $A_1 = \begin{bmatrix} 1 & 0 \\ 0 & 0 \end{bmatrix}$,

$A_2 = \begin{bmatrix} 0 & 1 \\ 0 & 0 \end{bmatrix}$, $A_3 = \begin{bmatrix} 0 & 0 \\ 1 & 0 \end{bmatrix}$, and

$A_4 = \begin{bmatrix} 0 & 0 \\ 0 & 1 \end{bmatrix}$, be a basis for V.

(a) Find $[L(A_i)]_S$ for $i = 1, 2, 3, 4$.
(b) Find the matrix B representing L with respect to S.
(c) Find the eigenvalues and eigenvectors of B.
(d) Find the eigenvalues and eigenvectors of L.
(e) Show that one of the eigenspaces is the set of all 2×2 symmetric matrices and that the other is the set of all 2×2 skew symmetric matrices.

9. Let V be the *real* vector space of trigonometric polynomials of the form $a + b \sin x + c \cos x$. Let $L: V \rightarrow V$ be the linear transformation defined by $L(\mathbf{v}) = \dfrac{d}{dx}[\mathbf{v}]$. Find the eigenvalues and eigenvectors of L. (*Hint:* Use the basis $S = \{1, \sin x, \cos x\}$ for V.)

10. Let V be the *complex* vector space (see Section B.2) of trigonometric polynomials

$$a + b \sin x + c \cos x.$$

For L as defined in Exercise 9, find the eigenvalues and corresponding eigenvectors.

11. Prove that if the matrix A is similar to a diagonal matrix, then A is similar to A^T.

7 Differential Equations

A **differential equation** is an equation that involves an unknown function and its derivatives. An important, simple example of a differential equation is

$$\frac{dy}{dx} = ry,$$

where r is a constant. The idea here is to find a function f that will satisfy the given differential equation; that is, $f' = rf$. This differential equation is further discussed below. Differential equations occur often in all branches of science and engineering; linear algebra is helpful in the formulation and solution of differential equations. In this section we provide only a brief survey of the approach; books on differential equations deal with the subject in much greater detail and several suggestions for further reading are given at the end of this chapter.

Homogeneous Linear Systems

We consider the **homogeneous linear system** of differential equations

$$\begin{aligned}
x_1'(t) &= a_{11}x_1(t) + a_{12}x_2(t) + \cdots + a_{1n}x_n(t) \\
x_2'(t) &= a_{21}x_1(t) + a_{22}x_2(t) + \cdots + a_{2n}x_n(t) \\
\vdots \quad & \quad\; \vdots \qquad\qquad\quad \vdots \qquad\qquad\qquad\quad \vdots \\
x_n'(t) &= a_{n1}x_1(t) + a_{n2}x_2(t) + \cdots + a_{nn}x_n(t),
\end{aligned} \tag{1}$$

where the a_{ij} are known constants. We seek functions $x_1(t), x_2(t), \ldots, x_n(t)$ defined and differentiable on the real line satisfying (1).

We can write (1) in matrix form by letting

$$\mathbf{x}(t) = \begin{bmatrix} x_1(t) \\ x_2(t) \\ \vdots \\ x_n(t) \end{bmatrix}, \qquad A = \begin{bmatrix} a_{11} & a_{12} & \cdots & a_{1n} \\ a_{21} & a_{22} & \cdots & a_{2n} \\ \vdots & \vdots & & \vdots \\ a_{n1} & a_{n2} & \cdots & a_{nn} \end{bmatrix},$$

and defining

$$\mathbf{x}'(t) = \begin{bmatrix} x_1'(t) \\ x_2'(t) \\ \vdots \\ x_n'(t) \end{bmatrix}.$$

Then (1) can be written as

$$\mathbf{x}'(t) = A\mathbf{x}(t). \tag{2}$$

We shall often write (2) more briefly as

$$\mathbf{x}' = A\mathbf{x}.$$

With this notation, a vector function

$$\mathbf{x}(t) = \begin{bmatrix} x_1(t) \\ x_2(t) \\ \vdots \\ x_n(t) \end{bmatrix}$$

satisfying (2) is called a **solution** to the given system.

We leave it to the reader to verify that if $\mathbf{x}^{(1)}(t), \mathbf{x}^{(2)}(t), \ldots, \mathbf{x}^{(n)}(t)$ are all solutions to (2), then any linear combination

$$\mathbf{x}(t) = b_1\mathbf{x}^{(1)}(t) + b_2\mathbf{x}^{(2)}(t) + \cdots + b_n\mathbf{x}^{(n)}(t) \tag{3}$$

is also a solution to (2).

A set of vector functions $\{\mathbf{x}^{(1)}(t), \mathbf{x}^{(2)}(t), \ldots, \mathbf{x}^{(n)}(t)\}$ is said to be a **fundamental system** for (1) if every solution to (1) can be written in the form (3). In this case, the right side of (3), where $b_1, b_2, \ldots, b_n$ are arbitrary constants, is said to be the **general solution** to (2).

It can be shown that any system of the form (2) has a fundamental system (in fact, infinitely many).

The problem of finding a solution $\mathbf{x}(t)$ to (2) such that $\mathbf{x}(0) = \mathbf{x}_0$, an **initial condition,** where $\mathbf{x}_0$ is a given vector, is called an **initial value problem.** If the general solution (3) is known, then the initial value problem can be solved by setting $t = 0$ in (3) and determining the constants $b_1, b_2, \ldots, b_n$ so that

$$\mathbf{x}_0 = b_1\mathbf{x}^{(1)}(0) + b_2\mathbf{x}^{(2)}(0) + \cdots + b_n\mathbf{x}^{(n)}(0).$$

It is easily seen that this is actually an $n \times n$ linear system with unknowns $b_1, b_2, \ldots, b_n$. This linear system can also be written as

$$C\mathbf{b} = \mathbf{x}_0, \tag{4}$$

where

$$\mathbf{b} = \begin{bmatrix} b_1 \\ b_2 \\ \vdots \\ b_n \end{bmatrix}$$

and C is the $n \times n$ matrix whose columns are $\mathbf{x}^{(1)}(0), \mathbf{x}^{(2)}(0), \ldots, \mathbf{x}^{(n)}(0)$, respectively. It can be shown that if $\mathbf{x}^{(1)}(t), \mathbf{x}^{(2)}(t), \ldots, \mathbf{x}^{(n)}(t)$ form a fundamental system for (1), then C is nonsingular, so (4) always has a unique solution.

Example 1. The simplest system of the form (1) is the single equation

$$\frac{dx}{dt} = ax, \tag{5}$$

where a is a constant. From calculus, the solutions to this equation are of the form

$$x = be^{at}; \tag{6}$$

that is, this is the general solution to (5). To solve the initial value problem

$$\frac{dx}{dt} = ax, \qquad x(0) = x_0,$$

we set $t = 0$ in (6) and obtain $b = x_0$. Thus the solution to the initial value problem is

$$x = x_0 e^{at}. \quad \blacksquare$$

The system (2) is said to be **diagonal** if the matrix A is diagonal. Then (1) can be rewritten as

$$\begin{aligned} x_1'(t) &= a_{11}x_1(t) \\ x_2'(t) &= \qquad a_{22}x_2(t) \\ &\vdots \\ x_n'(t) &= \qquad\qquad a_{nn}x_n(t). \end{aligned} \tag{7}$$

This system is easy to solve, since the equations can be solved separately. Applying the results of Example 1 to each equation in (7), we obtain

$$\begin{aligned} x_1(t) &= b_1 e^{a_{11}t} \\ x_2(t) &= b_2 e^{a_{22}t} \\ \vdots &\qquad \vdots \\ x_n(t) &= b_n e^{a_{nn}t}, \end{aligned} \tag{8}$$

where $b_1, b_2, \ldots, b_n$ are arbitrary constants. Writing (8) in vector form yields

$$\mathbf{x}(t) = \begin{bmatrix} b_1 e^{a_{11}t} \\ b_2 e^{a_{22}t} \\ \\ \vdots \\ b_n e^{a_{nn}t} \end{bmatrix} = b_1 \begin{bmatrix} 1 \\ 0 \\ 0 \\ \vdots \\ 0 \end{bmatrix} e^{a_{11}t} + b_2 \begin{bmatrix} 0 \\ 1 \\ 0 \\ \vdots \\ 0 \end{bmatrix} e^{a_{22}t} + \cdots + b_n \begin{bmatrix} 0 \\ 0 \\ \vdots \\ 0 \\ 1 \end{bmatrix} e^{a_{nn}t}.$$

This implies that the vector functions

$$\mathbf{x}^{(1)}(t) = \begin{bmatrix} 1 \\ 0 \\ 0 \\ \vdots \\ 0 \end{bmatrix} e^{a_{11}t}, \quad \mathbf{x}^{(2)}(t) = \begin{bmatrix} 0 \\ 1 \\ 0 \\ \vdots \\ 0 \end{bmatrix} e^{a_{22}t}, \quad \ldots, \quad \mathbf{x}^{(n)}(t) = \begin{bmatrix} 0 \\ 0 \\ \vdots \\ 0 \\ 1 \end{bmatrix} e^{a_{nn}t}$$

form a fundamental system for the diagonal system (7).

Example 2. The diagonal system

$$\begin{bmatrix} x_1' \\ x_2' \\ x_3' \end{bmatrix} = \begin{bmatrix} 3 & 0 & 0 \\ 0 & -2 & 0 \\ 0 & 0 & 4 \end{bmatrix} \begin{bmatrix} x_1 \\ x_2 \\ x_3 \end{bmatrix} \tag{9}$$

can be written as three equations:

$$\begin{aligned} x_1' &= 3x_1 \\ x_2' &= -2x_2 \\ x_3' &= 4x_3. \end{aligned}$$

Integrating these equations, we obtain

$$x_1 = b_1 e^{3t}, \qquad x_2 = b_2 e^{-2t}, \qquad x_3 = b_3 e^{4t},$$

where b_1, b_2, and b_3 are arbitrary constants. Thus

$$\mathbf{x}(t) = \begin{bmatrix} b_1 e^{3t} \\ b_2 e^{-2t} \\ b_3 e^{4t} \end{bmatrix} = b_1 \begin{bmatrix} 1 \\ 0 \\ 0 \end{bmatrix} e^{3t} + b_2 \begin{bmatrix} 0 \\ 1 \\ 0 \end{bmatrix} e^{-2t} + b_3 \begin{bmatrix} 0 \\ 0 \\ 1 \end{bmatrix} e^{4t}$$

is the general solution to (9) and the functions

$$\mathbf{x}^{(1)}(t) = \begin{bmatrix} 1 \\ 0 \\ 0 \end{bmatrix} e^{3t}, \qquad \mathbf{x}^{(2)}(t) = \begin{bmatrix} 0 \\ 1 \\ 0 \end{bmatrix} e^{-2t}, \qquad \mathbf{x}^{(3)}(t) = \begin{bmatrix} 0 \\ 0 \\ 1 \end{bmatrix} e^{4t}$$

form a fundamental system for (9). ■

If the system (2) is not diagonal, then it cannot be solved as simply as the system in the preceding example. However, there is an extension of this method that yields the general solution in the case where A is diagonalizable. Suppose that A is diagonalizable and P is a nonsingular matrix such that

$$P^{-1}AP = D, \tag{10}$$

where D is diagonal. Then multiplying the given system

$$\mathbf{x}' = A\mathbf{x}$$

on the left by P^{-1}, we obtain

$$P^{-1}\mathbf{x}' = P^{-1}A\mathbf{x}.$$

Since $P^{-1}P = I_n$, we can rewrite the last equation as

$$P^{-1}\mathbf{x}' = (P^{-1}AP)(P^{-1}\mathbf{x}). \tag{11}$$

Temporarily, let

$$\mathbf{u} = P^{-1}\mathbf{x}. \tag{12}$$

Since P^{-1} is a constant matrix,

$$\mathbf{u}' = P^{-1}\mathbf{x}'. \tag{13}$$

Therefore, substituting (10), (12), and (13) into (11), we obtain

$$\mathbf{u}' = D\mathbf{u}. \tag{14}$$

Equation (14) is a diagonal system and can be solved by the methods just discussed. Before proceeding, however, let us recall from Theorem 6.2 that

$$D = \begin{bmatrix} \lambda_1 & 0 & \cdots & 0 \\ 0 & \lambda_2 & \cdots & 0 \\ \vdots & \vdots & & \vdots \\ 0 & 0 & \cdots & \lambda_n \end{bmatrix},$$

where $\lambda_1, \lambda_2, \ldots, \lambda_n$ are the eigenvalues of A, and that the columns of P are linearly independent eigenvectors of A associated, respectively, with λ_1, λ_2,

$\ldots, \lambda_n$. From the discussion just given for diagonal systems, the general solution to (14) is

$$\mathbf{u}(t) = b_1\mathbf{u}^{(1)}(t) + b_2\mathbf{u}^{(2)}(t) + \cdots + b_n\mathbf{u}^{(n)}(t)$$
$$= \begin{bmatrix} b_1e^{\lambda_1 t} \\ b_2e^{\lambda_2 t} \\ \vdots \\ b_ne^{\lambda_n t} \end{bmatrix},$$

where

$$\mathbf{u}^{(1)}(t) = \begin{bmatrix} 1 \\ 0 \\ 0 \\ \vdots \\ 0 \end{bmatrix} e^{\lambda_1 t}, \quad \mathbf{u}^{(2)}(t) = \begin{bmatrix} 0 \\ 1 \\ 0 \\ \vdots \\ 0 \end{bmatrix} e^{\lambda_2 t}, \quad \ldots, \quad \mathbf{u}^{(n)}(t) = \begin{bmatrix} 0 \\ 0 \\ \vdots \\ 0 \\ 1 \end{bmatrix} e^{\lambda_n t} \tag{15}$$

and $b_1, b_2, \ldots, b_n$ are arbitrary constants. From Equation (12), $\mathbf{x} = P\mathbf{u}$, so the general solution to the given system $\mathbf{x}' = A\mathbf{x}$ is

$$\mathbf{x}(t) = P\mathbf{u}(t) = b_1P\mathbf{u}^{(1)}(t) + b_2P\mathbf{u}^{(2)}(t) + \cdots + b_nP\mathbf{u}^{(n)}(t). \tag{16}$$

However, since the constant vectors in (15) are the columns of the identity matrix and $PI_n = P$, (16) can be rewritten as

$$\mathbf{x}(t) = b_1\mathbf{p}_1e^{\lambda_1 t} + b_2\mathbf{p}_2e^{\lambda_2 t} + \cdots + b_n\mathbf{p}_ne^{\lambda_n t}, \tag{17}$$

where $\mathbf{p}_1, \mathbf{p}_2, \ldots, \mathbf{p}_n$ are the columns of P, and therefore eigenvectors of A associated with $\lambda_1, \lambda_2, \ldots, \lambda_n$, respectively.

We summarize the discussion above in the following theorem.

Theorem 7.1. *If the $n \times n$ matrix A has n linearly independent eigenvectors $\mathbf{p}_1, \mathbf{p}_2, \ldots, \mathbf{p}_n$ associated with the eigenvalues $\lambda_1, \lambda_2, \ldots, \lambda_n$, respectively, then the general solution to the system of differential equations*

$$\mathbf{x}' = A\mathbf{x}$$

is given by (17). ■

Example 3. For the system

$$\mathbf{x}' = \begin{bmatrix} 1 & -1 \\ 2 & 4 \end{bmatrix}\mathbf{x},$$

the matrix

$$A = \begin{bmatrix} 1 & -1 \\ 2 & 4 \end{bmatrix}$$

has eigenvalues $\lambda_1 = 2$ and $\lambda_2 = 3$ with associated eigenvectors (verify)

$$\mathbf{p}_1 = \begin{bmatrix} 1 \\ -1 \end{bmatrix} \quad \text{and} \quad \mathbf{p}_2 = \begin{bmatrix} 1 \\ -2 \end{bmatrix}.$$

These eigenvectors are automatically linearly independent, since they are associated with distinct eigenvalues (proof of Theorem 6.4). Hence the general solution to the given system is

$$\mathbf{x}(t) = b_1\begin{bmatrix} 1 \\ -1 \end{bmatrix}e^{2t} + b_2\begin{bmatrix} 1 \\ -2 \end{bmatrix}e^{3t}.$$

In terms of components, this can be written as

$$\begin{aligned} x_1(t) &= b_1e^{2t} + b_2e^{3t} \\ x_2(t) &= -b_1e^{2t} - 2b_2e^{3t}. \end{aligned}$$

■

Example 4. Consider the following linear system of differential equations:

$$\mathbf{x}' = \begin{bmatrix} x_1' \\ x_2' \\ x_3' \end{bmatrix} = \begin{bmatrix} 0 & 1 & 0 \\ 0 & 0 & 1 \\ 8 & -14 & 7 \end{bmatrix}\begin{bmatrix} x_1 \\ x_2 \\ x_3 \end{bmatrix}.$$

The characteristic polynomial of A is (verify)

$$f(\lambda) = \lambda^3 - 7\lambda^2 + 14\lambda - 8$$

or

$$f(\lambda) = (\lambda - 1)(\lambda - 2)(\lambda - 4),$$

so the eigenvalues of A are $\lambda_1 = 1$, $\lambda_2 = 2$, and $\lambda_3 = 4$. Associated eigenvectors are (verify)

$$\begin{bmatrix} 1 \\ 1 \\ 1 \end{bmatrix}, \begin{bmatrix} 1 \\ 2 \\ 4 \end{bmatrix}, \begin{bmatrix} 1 \\ 4 \\ 16 \end{bmatrix},$$

respectively. The general solution is then given by

$$\mathbf{x}(t) = b_1 \begin{bmatrix} 1 \\ 1 \\ 1 \end{bmatrix} e^t + b_2 \begin{bmatrix} 1 \\ 2 \\ 4 \end{bmatrix} e^{2t} + b_3 \begin{bmatrix} 1 \\ 4 \\ 16 \end{bmatrix} e^{4t},$$

where b_1, b_2, and b_3 are arbitrary constants. ■

Example 5. For the linear system of Example 4 solve the initial value problem determined by the **initial conditions** $x_1(0) = 4$, $x_2(0) = 6$, and $x_3(0) = 8$.

Solution: We write our general solution in the form $\mathbf{x} = P\mathbf{u}$ as

$$\mathbf{x}(t) = \begin{bmatrix} 1 & 1 & 1 \\ 1 & 2 & 4 \\ 1 & 4 & 16 \end{bmatrix} \begin{bmatrix} b_1 e^t \\ b_2 e^{2t} \\ b_3 e^{4t} \end{bmatrix}.$$

Now

$$\mathbf{x}(0) = \begin{bmatrix} 4 \\ 6 \\ 8 \end{bmatrix} = \begin{bmatrix} 1 & 1 & 1 \\ 1 & 2 & 4 \\ 1 & 4 & 16 \end{bmatrix} \begin{bmatrix} b_1 e^0 \\ b_2 e^0 \\ b_3 e^0 \end{bmatrix}$$

or

$$\begin{bmatrix} 1 & 1 & 1 \\ 1 & 2 & 4 \\ 1 & 4 & 16 \end{bmatrix} \begin{bmatrix} b_1 \\ b_2 \\ b_3 \end{bmatrix} = \begin{bmatrix} 4 \\ 6 \\ 8 \end{bmatrix}. \tag{18}$$

Solving (18) by Gauss–Jordan reduction, we obtain (verify)

$$b_1 = \tfrac{4}{3}, \qquad b_2 = 3, \qquad b_3 = -\tfrac{1}{3}.$$

Therefore, the solution to the initial value problem is

$$\mathbf{x}(t) = \tfrac{4}{3}\begin{bmatrix}1\\1\\1\end{bmatrix}e^{t} + 3\begin{bmatrix}1\\2\\4\end{bmatrix}e^{2t} - \tfrac{1}{3}\begin{bmatrix}1\\4\\16\end{bmatrix}e^{4t}.$$

■

We now recall several facts from Chapter 6. If A does not have distinct eigenvalues, then we may or may not be able to diagonalize A. Let λ be an eigenvalue of A of multiplicity k. Then A can be diagonalized if and only if the dimension of the eigenspace associated with λ is k, that is, if and only if the rank of the matrix $(\lambda I_n - A)$ is $n - k$ (verify). If the rank of $(\lambda I_n - A)$ is $n - k$, then we can find k linearly independent eigenvectors of A associated with λ.

Example 6. Consider the linear system

$$\mathbf{x}' = A\mathbf{x} = \begin{bmatrix}1 & 0 & 0\\0 & 3 & -2\\0 & -2 & 3\end{bmatrix}\mathbf{x}.$$

The eigenvalues of A are $\lambda_1 = \lambda_2 = 1$ and $\lambda_3 = 5$ (verify). The rank of the matrix $(1I_3 - A) = \begin{bmatrix}0 & 0 & 0\\0 & -2 & 2\\0 & 2 & -2\end{bmatrix}$ is 1 and the linearly independent eigenvectors $\begin{bmatrix}1\\0\\0\end{bmatrix}$ and $\begin{bmatrix}0\\1\\1\end{bmatrix}$ are associated with the eigenvalue 1 (verify). The eigenvector $\begin{bmatrix}0\\1\\-1\end{bmatrix}$ is associated with the eigenvalue 5 (verify). The general solution to the given system is then

$$\mathbf{x}(t) = b_1\begin{bmatrix}1\\0\\0\end{bmatrix}e^{t} + b_2\begin{bmatrix}0\\1\\1\end{bmatrix}e^{t} + b_3\begin{bmatrix}0\\1\\-1\end{bmatrix}e^{5t},$$

where b_1, b_2, and b_3 are arbitrary constants. ■

If we cannot diagonalize A, we are in a considerably more difficult situation. Suppose that the rank of $(\lambda I_n - A)$ is greater than $n - k$. The following example indicates a procedure followed in a specific case. Other cases of this type are not considered here.

Example 7. Consider the following linear system of differential equations:

$$\mathbf{x}' = A\mathbf{x} = \begin{bmatrix} 0 & 1 & 0 \\ 0 & 0 & 1 \\ 12 & -16 & 7 \end{bmatrix} \begin{bmatrix} x_1 \\ x_2 \\ x_3 \end{bmatrix}.$$

The characteristic polynomial of A is (verify)

$$f(\lambda) = \lambda^3 - 7\lambda^2 + 16\lambda - 12 = (\lambda - 2)^2(\lambda - 3).$$

The distinct eigenvalues of A are $\lambda_1 = 2$ and $\lambda_2 = 3$, with respective multiplicities $k_1 = 2$ and $k_2 = 1$.

Observe that the rank of the matrix

$$2I_3 - A = \begin{bmatrix} 2 & -1 & 0 \\ 0 & 2 & -1 \\ -12 & 16 & -5 \end{bmatrix}$$

is 2. For $\lambda_1 = 2$ we find that $\mathbf{p}_1 = \begin{bmatrix} 1 \\ 2 \\ 4 \end{bmatrix}$ is an eigenvector of A associated with λ_1.

Let $\mathbf{p}_2$ satisfy the equation $(A - 2I_3)\mathbf{p}_2 = \mathbf{p}_1$. We are thus solving the system of linear equations

$$\begin{bmatrix} -2 & 1 & 0 \\ 0 & -2 & 1 \\ 12 & -16 & 5 \end{bmatrix} \begin{bmatrix} a_1 \\ a_2 \\ a_3 \end{bmatrix} = \begin{bmatrix} 1 \\ 2 \\ 4 \end{bmatrix}.$$ A solution is $\mathbf{p}_2 = \begin{bmatrix} -1 \\ -1 \\ 0 \end{bmatrix}$ (verify).

Now $\mathbf{p}_3 = \begin{bmatrix} 1 \\ 3 \\ 9 \end{bmatrix}$ is an eigenvector of A associated with $\lambda_2 = 3$ (verify). We can then easily verify that the general solution is

$$\mathbf{x}(t) = b_1\mathbf{p}_1 e^{2t} + b_2(\mathbf{p}_1 te^{2t} + \mathbf{p}_2 e^{2t}) + b_3\mathbf{p}_3 e^{3t}$$

or

$$\mathbf{x}(t) = b_1 \begin{bmatrix} 1 \\ 2 \\ 4 \end{bmatrix} e^{2t} + b_2 \left(\begin{bmatrix} 1 \\ 2 \\ 4 \end{bmatrix} te^{2t} + \begin{bmatrix} -1 \\ -1 \\ 0 \end{bmatrix} e^{2t} \right) + b_3 \begin{bmatrix} 1 \\ 3 \\ 9 \end{bmatrix} e^{3t}$$

$$= \left(b_1 \begin{bmatrix} 1 \\ 2 \\ 4 \end{bmatrix} + b_2 \begin{bmatrix} -1 \\ -1 \\ 0 \end{bmatrix} \right) e^{2t} + b_2 \begin{bmatrix} 1 \\ 2 \\ 4 \end{bmatrix} te^{2t} + b_3 \begin{bmatrix} 1 \\ 3 \\ 9 \end{bmatrix} e^{3t}$$

$$= (b_1\mathbf{p}_1 + b_2\mathbf{p}_2)e^{2t} + b_2\mathbf{p}_1 te^{2t} + b_3\mathbf{p}_3 e^{3t}.$$ ■

*n*th-Order Homogeneous Linear Differential Equations

We now consider an ***n*th-order homogeneous linear differential equation with constant coefficients:**

$$\frac{d^n x(t)}{dt^n} + a_1 \frac{d^{n-1}x(t)}{dt^{n-1}} + \cdots + a_{n-1}\frac{dx(t)}{dt} + a_n x(t) = 0, \tag{19}$$

where $a_1, a_2, \ldots, a_n$ are real numbers. A function $x = x(t)$ that satisfies (19) is called a **solution** to (19).

We first show that every nth-order homogeneous linear differential equation with constant coefficients is equivalent to a system of n first-order homogeneous linear differential equations with constant coefficients that can then be expressed in terms of matrices. If $x(t)$ is the unknown function, we let

$$x_1(t) = x(t),\ x'(t) = \frac{dx(t)}{dt} = x_2(t),$$
$$x''(t) = \frac{d^2x(t)}{dt^2} = x_3(t), \ldots, \frac{d^{n-1}x(t)}{dt^{n-1}} = x_n(t).$$

We next obtain the following system of n first-order homogeneous linear differential equations with constant coefficients:

$$\begin{aligned} x_1' &= \frac{dx_1(t)}{dt} = x_2(t) \\ x_2' &= \frac{dx_2(t)}{dt} = x_3(t) \\ &\vdots \\ x_{n-1}' &= \frac{dx_{n-1}(t)}{dt} = x_n(t) \\ x_n' &= \frac{dx_n(t)}{dt} = -a_n x_1(t) - a_{n-1}x_2(t) - \cdots - a_1 x_n(t). \end{aligned} \tag{20}$$

We see that if $x(t)$ is a solution to (19), then $x_1(t), x_2(t), \ldots, x_n(t)$ is a solution to (20); conversely, any solution to (20) also gives a solution $x_1(t) = x(t)$ to (19). Thus (19) and (20) are equivalent. We can write (20) in matrix form as $\mathbf{x}'(t) = A\mathbf{x}(t)$, where

$$\mathbf{x}(t) = \begin{bmatrix} x_1(t) \\ x_2(t) \\ \vdots \\ x_n(t) \end{bmatrix}, \qquad \mathbf{x}'(t) = \begin{bmatrix} x_1'(t) \\ x_2'(t) \\ \vdots \\ x_n'(t) \end{bmatrix},$$

and

$$A = \begin{bmatrix} 0 & 1 & 0 & \cdots & & 0 \\ 0 & 0 & 1 & 0 & & 0 \\ \vdots & \vdots & \vdots & \vdots & & \vdots \\ 0 & 0 & \cdots & & 0 & 1 \\ -a_n & -a_{n-1} & \cdots & & & -a_1 \end{bmatrix}.$$

The general solution to (20) is called the **general solution** to (19).

Example 8. Consider the differential equation

$$x''' - 7x'' + 14x' - 8x = 0.$$

We let $x_1 = x$, $x_2 = x'$, $x_3 = x''$. Then (19) becomes

$$\begin{aligned} x_1' &= x' = x_2 \\ x_2' &= x'' = x_3 \\ x_3' &= x''' = 8x_1 - 14x_2 + 7x_3. \end{aligned}$$

Writing this in matrix form, we obtain (20) as

$$\mathbf{x}' = \begin{bmatrix} x_1' \\ x_2' \\ x_3' \end{bmatrix} = \begin{bmatrix} 0 & 1 & 0 \\ 0 & 0 & 1 \\ 8 & -14 & 7 \end{bmatrix} \begin{bmatrix} x_1 \\ x_2 \\ x_3 \end{bmatrix}.$$

This is the linear system considered in Example 4. Thus every solution to (20) is of the form

$$x(t) = x_1(t) = b_1e^t + b_2e^{2t} + b_3e^{4t}, \tag{21}$$

where b_1, b_2, and b_3 are arbitrary constants. Equation (21) is the **general solution** to the given differential equation. ■

Example 9. Consider the differential equation $x''' - 7x'' + 16x' - 12x = 0$. We can write this in matrix form as

$$\mathbf{x}' = \begin{bmatrix} x_1' \\ x_2' \\ x_3' \end{bmatrix} = \begin{bmatrix} 0 & 1 & 0 \\ 0 & 0 & 1 \\ 12 & -16 & 7 \end{bmatrix} \begin{bmatrix} x_1 \\ x_2 \\ x_3 \end{bmatrix} = A\mathbf{x}.$$

This is precisely the linear system considered in Example 7. Thus the general solution is

$$x(t) = x_1(t) = (b_1 - b_2)e^{2t} + b_2te^{2t} + b_3e^{3t} = d_1e^{2t} + d_2te^{2t} + d_3e^{3t},$$

where d_1, d_2, and d_3 are arbitrary constants. ■

It can be shown that the nth-order homogeneous linear differential equation with constant coefficients when written in matrix form as $\mathbf{x}' = A\mathbf{x}$ has the property that if λ is an eigenvalue of A of multiplicity $k > 1$, then the general solution will include the terms

$$b_1e^{\lambda t},\ b_2te^{\lambda t},\ b_3t^2e^{\lambda t},\ \ldots,\ b_kt^{k-1}e^{\lambda t},$$

where $b_1, b_2, \ldots, b_k$ are arbitrary constants. Note that to obtain the general solution to (19), we merely need $x(t) = x_1(t)$, which is the first row of P times the solution $\mathbf{x}(t)$ to the resulting system (20). We do not need to know the first row of P, because the general solution involves arbitrary constants. Thus we merely need the eigenvalues of the resulting matrix A and not associated eigenvectors. We may also note that it is quite easy to determine the characteristic polynomial $f(\lambda)$ of the matrix A involved in writing (19) as $\mathbf{x}' = A\mathbf{x}$, since we can show that

$$f(\lambda) = \lambda^n + a_1\lambda^{n-1} + \cdots + a_{n-1}\lambda + a_n.$$

Thus, in Example 9, the characteristic polynomial associated with $x''' - 7x'' + 16x' - 12x = 0$ is $f(\lambda) = \lambda^3 - 7\lambda^2 + 16\lambda - 12$. Also, in Example 8, the characteristic polynomial of $x''' - 7x'' + 14x' - 8x = 0$ is $f(\lambda) = \lambda^3 - 7\lambda^2 + 14\lambda - 8$.

Example 10. Consider the differential equation $x^{(v)} - 15x''' + 10x'' + 60x' - 72x = 0$, which we can write in matrix form as

$$\mathbf{x}' = \begin{bmatrix} x_1' \\ x_2' \\ x_3' \\ x_4' \\ x_5' \end{bmatrix} = \begin{bmatrix} 0 & 1 & 0 & 0 & 0 \\ 0 & 0 & 1 & 0 & 0 \\ 0 & 0 & 0 & 1 & 0 \\ 0 & 0 & 0 & 0 & 1 \\ 72 & -60 & -10 & 15 & 0 \end{bmatrix} \begin{bmatrix} x_1 \\ x_2 \\ x_3 \\ x_4 \\ x_5 \end{bmatrix} = A\mathbf{x}.$$

The characteristic polynomial of A is $f(\lambda) = \lambda^5 - 15\lambda^3 + 10\lambda^2 + 60\lambda - 72 = (\lambda - 2)^3(\lambda + 3)^2$, so the distinct eigenvalues are $\lambda_1 = 2$ and $\lambda_2 = -3$ with multiplicities $k_1 = 3$ and $k_2 = 2$, respectively. The general solution is then

$$x(t) = b_1e^{2t} + b_2te^{2t} + b_3t^2e^{2t} + b_4e^{-3t} + b_5te^{-3t},$$

where b_1, b_2, b_3, b_4, and b_5 are arbitrary constants. ■

Example 11. Find the solution to Example 8 determined by the initial conditions

$$x(0) = 6, \qquad x'(0) = 10, \qquad x''(0) = 30,$$

Solution: From (21) we have

$$x'(t) = b_1e^t + 2b_2e^{2t} + 4b_3e^{4t} \tag{22}$$
$$x''(t) = b_1e^t + 4b_2e^{2t} + 16b_3e^{4t}. \tag{23}$$

Letting $t = 0$ in (21), (22), and (23), we obtain the linear system

$$\begin{aligned} b_1 + b_2 + b_3 &= 6 \\ b_1 + 2b_2 + 4b_3 &= 10 \\ b_1 + 4b_2 + 16b_3 &= 30, \end{aligned}$$

whose solution is (verify)

$$b_1 = 6, \qquad b_2 = -2, \qquad b_3 = 2.$$

Hence the solution to the initial value problem is

$$x(t) = 6e^t - 2e^{2t} + 2e^{4t}.$$ ■

It should be noted that in more advanced courses, methods for solving nth-order homogeneous linear differential equations with constant coefficients are studied, which are more efficient than the method discussed here.

Application—A Diffusion Process

The following example is a modification of an example presented by Derrick and Grossman in *Elementary Differential Equations with Applications* (see Further Reading).

Example 12. Consider two adjoining cells separated by a permeable membrane and suppose that a fluid flows from the first cell to the second one at a rate (in milliliters per minute) that is numerically equal to three times the volume (in milliliters) of the fluid in the first cell. It then flows out of the second cell at a rate (in milliliters per minute) that is numerically equal to twice the volume in the second cell. Let $x_1(t)$ and $x_2(t)$ denote the volumes of the fluid in the first and second cells at time t, respectively. Assume that initially the first cell has 40 milliliters of fluid, while the second one has 5 milliliters of fluid. Find the volume of fluid in each cell at time t.

Solution: The change in volume of the fluid in each cell is the difference between the amount flowing in and the amount flowing out. Since no fluid flows into the first cell, we have

$$\frac{dx_1(t)}{dt} = -3x_1(t),$$

where the minus sign indicates that the fluid is flowing out of the cell. The flow $3x_1(t)$ from the first cell flows into the second cell. The flow out of the second cell is $2x_2(t)$. Thus the change in volume of the fluid in the second cell is given by

$$\frac{dx_2(t)}{dt} = 3x_1(t) - 2x_2(t).$$

We have then obtained the linear system

$$\begin{aligned} \frac{dx_1(t)}{dt} &= -3x_1(t) \\ \frac{dx_2(t)}{dt} &= 3x_1(t) - 2x_2(t), \end{aligned}$$

which can be written in matrix form as

$$\begin{bmatrix} x_1'(t) \\ x_2'(t) \end{bmatrix} = \begin{bmatrix} -3 & 0 \\ 3 & -2 \end{bmatrix} \begin{bmatrix} x_1(t) \\ x_2(t) \end{bmatrix}.$$

The eigenvalues of the matrix

$$A = \begin{bmatrix} -3 & 0 \\ 3 & -2 \end{bmatrix}$$

are (verify)

$$\lambda_1 = -3, \qquad \lambda_2 = -2$$

and corresponding associated eigenvectors are (verify)

$$\begin{bmatrix} 1 \\ -3 \end{bmatrix}, \quad \begin{bmatrix} 0 \\ 1 \end{bmatrix}.$$

Hence the general solution is given by

$$\mathbf{x}(t) = \begin{bmatrix} x_1(t) \\ x_2(t) \end{bmatrix} = b_1 \begin{bmatrix} 1 \\ -3 \end{bmatrix} e^{-3t} + b_2 \begin{bmatrix} 0 \\ 1 \end{bmatrix} e^{-2t}.$$

Using the initial conditions, we find that (verify)

$$b_1 = 40, \qquad b_2 = 125.$$

Thus the volume of fluid in each cell at time t is given by

$$\begin{aligned} x_1(t) &= 40e^{-3t} \\ x_2(t) &= -120e^{-3t} + 125e^{-2t}. \end{aligned}$$

■

FURTHER READING

Bentley, Donald L., and Kenneth L. Cooke. *Linear Algebra with Differential Equations*. New York: Holt, Rinehart and Winston, 1973.

Boyce, William E., and Richard C. DiPrima. *Elementary Differential Equations,* 4th ed. New York: John Wiley & Sons, Inc., 1986.

Brauer, Fred, and John A. Nohel. *Ordinary Differential Equations: A First Course*. Reading, Mass.: Cummings Publishing Company, Inc., 1973.

Cullen, Charles. *Linear Algebra and Differential Equations*. Boston: PWS Publishers, 1979.

Derrick, William R., and Stanley I. Grossman. *Elementary Differential Equations with Applications,* 2nd ed. Reading, Mass.: Addison-Wesley Publishing Company, Inc., 1981.

Dettman, John H. *Introduction to Linear Algebra and Differential Equations*. New York: McGraw-Hill Book Company, 1974.

Rabenstein, Albert L. *Elementary Differential Equations with Linear Algebra,* 3rd ed. Orlando, Fla.: Academic Press, Inc., 1982.

Wolfenstein, Samuel. *Introduction to Linear Algebra and Differential Equations*. San Francisco: Holden-Day, Inc., 1969.

7.1 EXERCISES

1. Consider the linear system of differential equations

$$\begin{bmatrix} x_1' \\ x_2' \\ x_3' \end{bmatrix} = \begin{bmatrix} -3 & 0 & 0 \\ 0 & 4 & 0 \\ 0 & 0 & 2 \end{bmatrix} \begin{bmatrix} x_1 \\ x_2 \\ x_3 \end{bmatrix}.$$

(a) Find the general solution.

(b) Find the solution to the initial value problem determined by the initial conditions $x_1(0) = 3$, $x_2(0) = 4$, and $x_3(0) = 5$.

2. Consider the linear system of differential equations

$$\begin{bmatrix} x_1' \\ x_2' \\ x_3' \end{bmatrix} = \begin{bmatrix} 1 & 0 & 0 \\ 0 & -2 & 1 \\ 0 & 0 & 3 \end{bmatrix} \begin{bmatrix} x_1 \\ x_2 \\ x_3 \end{bmatrix}.$$

(a) Find the general solution.

(b) Find the solution to the initial value problem determined by the initial conditions $x_1(0) = 2$, $x_2(0) = 7$, and $x_3(0) = 20$.

3. Find the general solution to the linear system of differential equations

$$\begin{bmatrix} x_1' \\ x_2' \\ x_3' \end{bmatrix} = \begin{bmatrix} 4 & 0 & 0 \\ 3 & -5 & 0 \\ 2 & 1 & 2 \end{bmatrix} \begin{bmatrix} x_1 \\ x_2 \\ x_3 \end{bmatrix}.$$

4. Prove that the set of all solutions to a linear system of differential equations $\mathbf{x}' = A\mathbf{x}$ is a subspace of the vector space of all real-valued functions. This subspace is called the **solution space** of the given linear system.

5. Find the general solution to the linear system of differential equations

$$\begin{bmatrix} x_1' \\ x_2' \\ x_3' \end{bmatrix} = \begin{bmatrix} 5 & 0 & 0 \\ 0 & -4 & 3 \\ 0 & 3 & 4 \end{bmatrix} \begin{bmatrix} x_1 \\ x_2 \\ x_3 \end{bmatrix}.$$

6. Find the general solution to the linear system of differential equations.

$$\begin{bmatrix} x_1' \\ x_2' \end{bmatrix} = \begin{bmatrix} 3 & -2 \\ -2 & 3 \end{bmatrix} \begin{bmatrix} x_1 \\ x_2 \end{bmatrix}.$$

7. Find the general solution to the linear system of differential equations

$$\begin{bmatrix} x_1' \\ x_2' \\ x_3' \end{bmatrix} = \begin{bmatrix} -2 & -2 & 3 \\ 0 & -2 & 2 \\ 0 & 2 & 1 \end{bmatrix} \begin{bmatrix} x_1 \\ x_2 \\ x_3 \end{bmatrix}.$$

8. Find the general solution to the linear system of differential equations

$$\begin{bmatrix} x_1' \\ x_2' \\ x_3' \end{bmatrix} = \begin{bmatrix} 1 & 1 & 2 \\ 0 & 1 & 0 \\ 0 & 1 & 3 \end{bmatrix} \begin{bmatrix} x_1 \\ x_2 \\ x_3 \end{bmatrix}.$$

9. For each of the following nth-order homogeneous linear differential equations with constant coefficients: (1) find an equivalent system of n first-order homogeneous linear differential equations with constant coefficients, and (2) write the system obtained in (1) in matrix form.
(a) $x'' - 2x' + x = 0$.
(b) $x''' - 2x'' + x' - x = 0$.
(c) $x''' - x = 0$.

10. Repeat Exercise 9 for each of the following.
(a) $x''' + 2x'' + 3x' - 5x = 0$.
(b) $x'' - x = 0$.
(c) $x''' + x'' - x = 0$.

11. Consider the differential equation

$$x'' - 2x' - 3x = 0.$$

(a) Find the general solution.
(b) Find the solution to the initial value problem determined by the initial conditions $x(0) = x'(0) = 2$.

12. Find the general solution to the differential equation $x''' - 5x'' + 7x' - 3x = 0$.

13. Find the general solution to the differential equation $x'' - 2x' + x = 0$.

14. Find the general solution to the differential equation $x^{(iv)} - 5x'' + 4x = 0$.

15. Find the general solution to the differential equation $x^{(iv)} - 10x''' + 37x'' - 60x' + 36x = 0$ (the set of eigenvalues of the associated matrix is $\{2, 2, 3, 3\}$).

16. Find the general solution to the differential equation $x''' - 3x'' - 4x' + 12x = 0$.

17. Find the general solution to the differential equation $x''' - 3x'' - 10x' + 24x = 0$.

18. Find the general solution to the differential equation

$$x^{(vi)} - 9x^{(v)} + 33x^{(iv)} - 63x''' + 66x'' - 36x' + 8x = 0.$$

(The set of eigenvalues is $\{1, 1, 1, 2, 2, 2\}$.)

19. Consider two competing species that live in the same forest and let $x_1(t)$ and $x_2(t)$ denote the respective populations of the species at time t. Suppose that the initial populations are $x_1(0) = 500$ and $x_2(0) = 200$. If the growth rates of the species are given by

$$x_1'(t) = -3x_1(t) + 6x_2(t)$$
$$x_2'(t) = x_1(t) - 2x_2(t),$$

what is the population of each species at time t?

Supplementary Exercises

1. Let $A(t) = [a_{ij}(t)]$ be an $n \times n$ matrix whose entries are all functions of t; $A(t)$ is called a **matrix function.** The derivative and integral of $A(t)$ is defined componentwise; that is,

$$\frac{d}{dt}[A(t)] = \left[\frac{d}{dt}a_{ij}(t)\right]$$

and

$$\int_a^t A(s)\, ds = \left[\int_a^t a_{ij}(s)\, ds\right].$$

For each of the following matrices $A(t)$, compute $\frac{d}{dt}[A(t)]$ and $\int_0^t A(s)\, ds$.

(a) $A(t) = \begin{bmatrix} t^2 & \dfrac{1}{t+1} \\ 4 & e^{-t} \end{bmatrix}$.

(b) $A(t) = \begin{bmatrix} \sin 2t & 0 & 0 \\ 0 & 1 & -t \\ 0 & te^{t^2} & \dfrac{t}{t^2+1} \end{bmatrix}$.

2. The usual rules for differentiation and integration of functions introduced in calculus also apply to matrix functions. Let $A(t)$ and $B(t)$ be $n \times n$ matrix functions whose entries are differentiable and let c_1 and c_2 be real numbers. Prove the following properties.

(a) $\dfrac{d}{dt}[c_1A(t) + c_2B(t)] =$

$$c_1\frac{d}{dt}[A(t)] + c_2\frac{d}{dt}[B(t)].$$

(b) $\displaystyle\int_a^t (c_1A(s) + c_2B(s))\, ds =$

$$c_1\int_a^t A(s)\, ds + c_2\int_a^t B(s)\, ds.$$

(c) $\dfrac{d}{dt}[A(t)B(t)] = \dfrac{d}{dt}[A(t)]B(t) + A(t)\dfrac{d}{dt}[B(t)].$

3. If A is an $n \times n$ matrix, then the matrix function

$$B(t) = I_n + At + A^2\frac{t^2}{2!} + A^3\frac{t^3}{3!} + \cdots$$

is called the **matrix exponential** function and we use the notation $B(t) = e^{At}$.

(a) Prove that $\dfrac{d}{dt}[e^{At}] = Ae^{At}$.

(b) Let

$$A = \begin{bmatrix} 0 & 1 \\ 0 & 0 \end{bmatrix} \quad \text{and} \quad B = \begin{bmatrix} 0 & 0 \\ 1 & 0 \end{bmatrix}.$$

Prove or disprove that $e^Ae^B = e^{A+B}$.

(c) Prove that $e^{iA} = \cos A + i \sin A$, where $i = \sqrt{-1}$ (see Section B.1).

4. Let A and B be $n \times n$ matrices that commute, that is, $AB = BA$. Prove that $e^Ae^B = e^{A+B}$.

5. Let $B(t) = [b_{ij}(t)]$ be a diagonal matrix function with $b_{ii}(t) = e^{t\lambda_{ii}}$, where λ_{ii} is a scalar, $i = 1, 2, \ldots, n$ and $b_{ij}(t) = 0$ if $i \neq j$. Let D be the diagonal matrix with diagonal entries λ_{ii}, $i = 1, 2, \ldots, n$. Prove that $B(t) = e^{Dt}$.

6. Let A be an $n \times n$ matrix that is diagonalizable with eigenvalues λ_i and associated eigenvectors $\mathbf{x}_i$, $i = 1, 2, \ldots, n$. Then we can choose the eigenvectors $\mathbf{x}_i$ so that they form a linearly independent set; the matrix P whose jth column is $\mathbf{x}_j$ is nonsingular, and $P^{-1}AP = D$, where D is the diagonal matrix whose diagonal entries are the eigenvalues of A. Prove that Equation (17) can be written as

$$\mathbf{x}(t) = Pe^{Dt}B,$$

where

$$B = \begin{bmatrix} b_1 \\ b_2 \\ \vdots \\ b_n \end{bmatrix}.$$

7. Let A be an $n \times n$ matrix and

$$\mathbf{x}(t) = \begin{bmatrix} x_1(t) \\ x_2(t) \\ \vdots \\ x_n(t) \end{bmatrix}.$$

Assume that A is diagonalizable as in Exercise 6, and prove that the solution to the initial value problem

$$\mathbf{x}' = A\mathbf{x}$$
$$\mathbf{x}(0) = \mathbf{x}_0$$

can be written as

$$\begin{aligned} \mathbf{x}(t) &= Pe^{Dt}P^{-1}\mathbf{x}_0 \\ &= e^{At}\mathbf{x}_0. \end{aligned}$$

8. For $\mathbf{x}_0 = \begin{bmatrix} 1 \\ 1 \end{bmatrix}$ and each of the following matrices A, solve the initial value problem of Exercise 7.

(a) $A = \begin{bmatrix} 2 & -1 \\ -1 & 2 \end{bmatrix}$. (b) $A = \begin{bmatrix} -1 & 1 \\ 1 & -1 \end{bmatrix}$.

9. For $\mathbf{x}_0 = \begin{bmatrix} 1 \\ 0 \\ 1 \end{bmatrix}$ and each of the following matrices A, solve the initial value problem of Exercise 7.

(a) $A = \begin{bmatrix} -1 & 1 & 0 \\ 0 & 3 & -12 \\ 1 & -1 & 0 \end{bmatrix}$.

(b) $A = \begin{bmatrix} 0 & 1 & 0 \\ 0 & 0 & 1 \\ 0 & 8 & -2 \end{bmatrix}$.

A Preliminaries

In this appendix, which can be consulted as the need arises, we present the basic ideas of sets and functions that are used in Chapters 2, 3, 4, and 6.

A.1. *Sets*

A **set** is a collection, class, aggregate, or family of objects, which are called **elements** or **members** of the set. A set will be denoted by a capital letter, and an element of a set by a lowercase letter. A set S is specified either by describing all the elements of S, or by stating a property that determines, unequivocally, whether an element is or is not an element of S. Let $S = \{1, 2, 3\}$ be the set of all positive integers < 4. Then a real number belongs to S if it is a positive integer < 4. Thus S has been described in both ways. Sets A and B are said to be **equal** if each element of A belongs to B and if each element of B belongs to A. We write $A = B$. Thus $\{1, 2, 3\} = \{3, 2, 1\} = \{2, 1, 3\}$, and so on. If A and B are sets such that every element of A belongs to B, then A is said to be a **subset** of B. The set of all rational numbers is a subset of the set of all real numbers; the set $\{1, 2, 3\}$ is a subset of $\{1, 2, 3\}$; the set of all isosceles triangles is a subset of the set of all triangles. We can see that every set is a subset of itself. The **empty set** is the set that has no elements in it. The set of all real numbers whose squares equal -1 is empty because the square of a real number is never negative.

A.2. Functions

A **function** f from a set S into a set T is a rule that assigns to each element s of S a unique element t of T. We denote the function f by $f\colon S \to T$ and write $t = f(s)$. Functions constitute the basic ingredient of the calculus and other branches of mathematics, and the reader has dealt extensively with them. The set S is called the **domain** of f; the subset $f(S)$ of T consisting of all the elements $f(s)$, for s in S, is called the **range** of f or the **image** of S under f. As examples of functions we consider the following:

1. Let $S = T =$ the set of all real numbers. Let $f\colon S \to T$ be defined by the rule $f(s) = s^2$, for s in S.
2. Let $S =$ the set of all real numbers and let $T =$ the set of all nonnegative real numbers. Let $f\colon S \to T$ be defined by the rule $f(s) = s^2$, for s in S.
3. Let $S =$ three-dimensional space, where each point is described by x-, y-, and z-coordinates (x_1, x_2, x_3). Let $T =$ the (x, y)-plane as a subset of S. Let $f\colon S \to T$ be defined by the rule $f((x_1, x_2, x_3)) = (x_1, x_2, 0)$. To see what f does, we take a point (x_1, x_2, x_3) in S, draw a line from (x_1, x_2, x_3) perpendicular to T, the (x, y)-plane, and find the point of intersection $(x_1, x_2, 0)$ of this line with the (x, y)-plane. This point is the image of (x_1, x_2, x_3) under f; f is called a **projection function** (Figure A.1).
4. Let $S = T =$ the set of all real numbers. Let $f\colon S \to T$ be defined by the rule $f(s) = 2s + 1$, for s in S.

Figure A.1

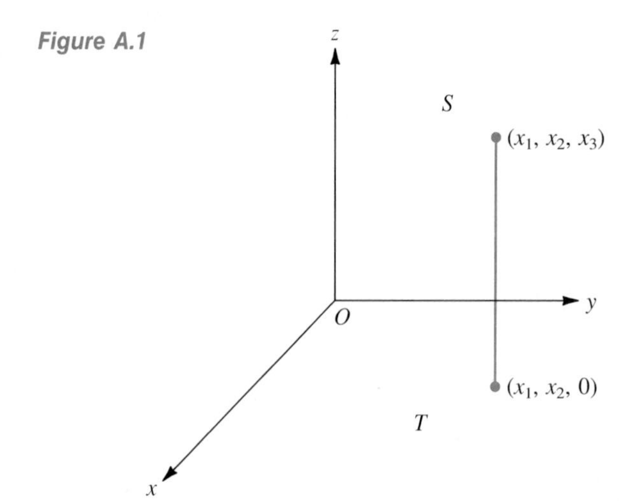

5. Let $S =$ the x-axis in the (x, y)-plane and let $T =$ the (x, y)-plane. Let $f\colon S \to T$ be defined by the rule $f((s, 0)) = (s, 1)$, for s in S.
6. Let $S =$ the set of all real numbers. Let $T =$ the set of all positive real numbers. Let $f\colon S \to T$ be defined by the rule $f(s) = e^s$, for s in S.

There are two properties of functions that we need to distinguish. A function $f\colon S \to T$ is called **one-to-one** if $f(s_1) \neq f(s_2)$ whenever s_1 and s_2 are distinct elements of S. An equivalent statement is that if $f(s_1) = f(s_2)$, then we must have $s_1 = s_2$ (see Figure A.2). A function $f\colon S \to T$ is called **onto** if the range of f is all of T, that is, if for any given t in T there is at least one s in S such that $f(s) = t$ (see Figure A.3).

Figure A.2

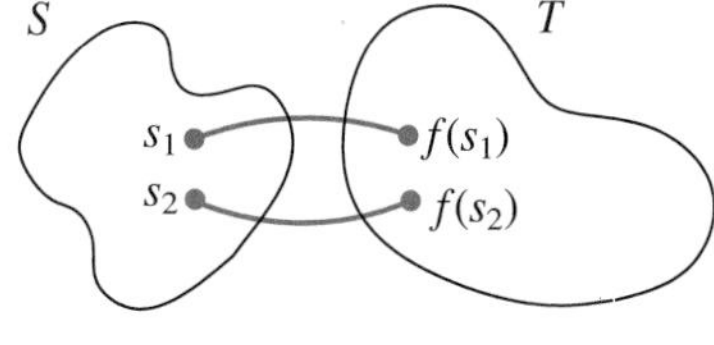

(**a**) f is one-to-one.

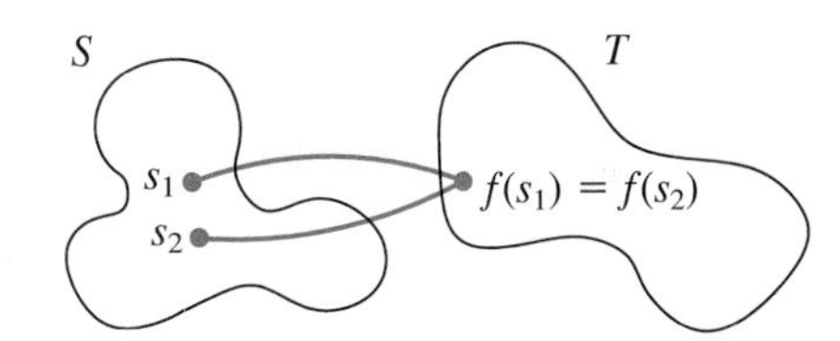

(**b**) f is not one-to-one.

Figure A.3 $S = \{2, 3, 4\}$; $T = \{3, 4, 5\}$

2 ——— $3 = f(2)$
3 ——— $4 = f(3)$
4 ——— $5 = f(4)$

(**a**) f is onto.

$S = \{2, 3, 4\}$; $T = \{3, 4, 5, 6\}$

2 ——— $3 = f(2)$
3 ——— $4 = f(3)$
4 ——— $5 = f(4)$

(**b**) f is not onto.

We now examine the listed functions:

1. f is not, one-to-one, for if $f(s_1) = f(s_2)$, it need not follow that

$$s_1 = s_2 \qquad [f(2) = f(-2) = 4].$$

Since the range of f is the set of nonnegative real numbers, f is not onto. Thus if $t = -4$, then there is no s such that $f(s) = -4$.
2. f is not one-to-one, but is onto. For if t is a given nonnegative real number, then $s = \sqrt{t}$ is in S and $f(s) = t$.
3. f is not one-to-one, for if $f((a_1, a_2, a_3)) = f((b_1, b_2, b_3))$, then $(a_1, a_2, 0) = (b_1, b_2, 0)$ so $a_1 = b_1$ and $a_2 = b_2$. However, b_3 need not equal a_3. The range of f is T; that is, f is onto. For let $(x_1, x_2, 0)$ be any element of T. Can we find

an element (a_1, a_2, a_3) of S such that $f((a_1, a_2, a_3)) = (x_1, x_2, 0)$? We merely let $a_1 = x_1$, $a_2 = x_2$, and let $a_3 =$ any real number we wish, say, $a_3 = 5$.

4. f is one-to-one, for if $f(s_1) = f(s_2)$, then $2s_1 + 1 = 2s_2 + 1$, which means that $s_1 = s_2$. Also, f is onto, for given a real number t we seek a real number s so that $f(s) = t$; that is, we need to solve $2s + 1 = t$ for s, which we can always do, obtaining $s = \frac{1}{2}(t - 1)$.
5. f is one-to-one but f is not onto because not every element in T has 1 for its y-coordinate.
6. f is one-to-one and onto because $e^{s_1} \neq e^{s_2}$ if $s_1 \neq s_2$, and for any positive t we can always solve $t = e^s$, obtaining $s = \ln t$.

If $f: S \to T$ and $g: T \to U$ are functions, then we can define a new function, $g \circ f$, by $(g \circ f)(s) = g(f(s))$, for s in S. The function $g \circ f: S \to U$ is called the **composite** of f and g. Thus, if f and g are the functions 4 and 6 above, then $g \circ f$ is defined by $(g \circ f)(s) = g(f(s)) = e^{2s+1}$, and $f \circ g$ is defined by $(f \circ g)(s) = f(g(s)) = 2e^s + 1$. The function $i: S \to S$ defined by $i(s) = s$, for s in S, is called the **identity function** on S. A function $f: S \to T$ for which there is a function $g: T \to S$ such that $g \circ f = i_s =$ identity function on S and $f \circ g = i_T =$ identity function on T is called an **invertible function** and g is called the **inverse** of f. It is not difficult to show, and we do so in Chapter 4 for a special case, that a function $f: S \to T$ is invertible if and only if it is one-to-one and onto. The inverse of f is written as f^{-1}. If f is invertible, then f^{-1} is also one-to-one and onto. We recall that "if and only if" means that both the statement and its converse are true. That is, if $f: S \to T$ is invertible, then f is one-to-one and onto; if $f: S \to T$ is one-to-one and onto, then f is invertible. Functions 4 and 6 are invertible; the inverse of function 4 is $g: T \to S$ defined by $g(t) = (t - 1)/2$ for t in T; the inverse of function 6 is $g: T \to S$, defined by $g(t) = \ln t$.

B Complex Numbers

B.1. Complex Numbers

Complex numbers are usually introduced in an algebra course to "complete" the solution of the quadratic equation

$$ax^2 + bx + c = 0.$$

In using the quadratic formula

$$x = \frac{-b \pm \sqrt{b^2 - 4ac}}{2a},$$

the case in which $b^2 - 4ac < 0$ is not resolved unless we can cope with the square roots of negative numbers. In the sixteenth century mathematicians and scientists justified this "completion" of the solution of quadratic equations by intuition. Naturally, a controversy arose, with some mathematicians denying their existence and others using them along with real numbers. The use of complex numbers did not lead to any contradictions, and the idea proved to be an important milestone in the development of mathematics.

A **complex number** c is of the form $c = a + bi$, where a and b are real numbers and where $i = \sqrt{-1}$; a is called the **real part** of c and b is called the **imaginary part** of c. The term "imaginary part" arose from the mysticism surrounding the beginnings of complex numbers; however, these numbers are as "real" as the real numbers.

Example 1. (a) $5 - 3i$ has real part 5 and imaginary part -3; (b) $-6 + \sqrt{2}i$ has real part -6 and imaginary part $\sqrt{2}$. ■

The symbol $i = \sqrt{-1}$ has the property that $i^2 = -1$ and we can deduce the following relationships:

$$i^3 = -i,\ i^4 = 1,\ i^5 = i,\ i^6 = -1,\ i^7 = -i,\ \ldots .$$

These results will be handy for simplifying operations involving complex numbers.

We say that two complex numbers $c_1 = a_1 + b_1 i$ and $c_2 = a_2 + b_2 i$ are **equal** if their real and imaginary parts are equal, that is, if $a_1 = a_2$ and $b_1 = b_2$. Of course, every real number a is a complex number with its imaginary part zero: $a = a + 0i$.

Operations on Complex Numbers

If $c_1 = a_1 + b_1 i$ and $c_2 = a_2 + b_2 i$ are complex numbers, then their **sum** is

$$c_1 + c_2 = (a_1 + a_2) + (b_1 + b_2)i,$$

and their **difference** is

$$c_1 - c_2 = (a_1 - a_2) + (b_1 - b_2)i.$$

In words, to form the sum of two complex numbers, add the real parts and add the imaginary parts. The **product** of c_1 and c_2 is

$$\begin{aligned} c_1 c_2 &= (a_1 + b_1 i) \cdot (a_2 + b_2 i) = a_1 a_2 + (a_1 b_2 + b_1 a_2)i + b_1 b_2 i^2 \\ &= (a_1 a_2 - b_1 b_2) + (a_1 b_2 + b_1 a_2)i. \end{aligned}$$

A special case of multiplication of complex numbers occurs when c_1 is real. In this case we obtain the simple result

$$c_1 c_2 = c_1 \cdot (a_2 + b_2 i) = c_1 a_2 + c_1 b_2 i.$$

If $c = a + bi$ is a complex number, then the **conjugate** of c is the complex number $\overline{c} = a - bi$. It is easy to show that if c and d are complex numbers, then the following basic properties of complex arithmetic hold:

1. $\overline{\overline{c}} = c$.
2. $\overline{c + d} = \overline{c} + \overline{d}$.
3. $\overline{cd} = \overline{c}\overline{d}$.
4. c is a real number if and only if $c = \overline{c}$.
5. $c\overline{c}$ is a nonnegative real number and $c\overline{c} = 0$ if and only if $c = 0$.

We prove property 4 here and leave the others as exercises. Let $c = a + bi$ so that $\overline{c} = a - bi$. If $c = \overline{c}$, then $a + bi = a - bi$, so $b = 0$ and c is real. On the other hand, if c is real, then $c = a$ and $\overline{c} = a$, so $c = \overline{c}$.

Example 2. Let $c_1 = 5 - 3i$, $c_2 = 4 + 2i$, and $c_3 = -3 + i$.

(a) $c_1 + c_2 = (5 - 3i) + (4 + 2i) = 9 - i$.

(b) $c_2 - c_3 = (4 + 2i) - (-3 + i) = (4 - (-3)) + (2 - 1)i = 7 + i$.

(c) $c_1c_2 = (5 - 3i) \cdot (4 + 2i) = 20 + 10i - 12i - 6i^2 = 26 - 2i$.

(d) $c_1\overline{c_3} = (5 - 3i) \cdot \overline{(-3 + i)} = (5 - 3i) \cdot (-3 - i) = -15 - 5i + 9i + 3i^2$
$= -18 + 4i$.

(e) $3c_1 + 2\overline{c_2} = 3(5 - 3i) + 2\overline{(4 + 2i)} = (15 - 9i) + 2(4 - 2i)$
$= (15 - 9i) + (8 - 4i) = 23 - 13i$.

(f) $c_1\overline{c_1} = (5 - 3i)\overline{(5 - 3i)} = (5 - 3i)(5 + 3i) = 34$. ■

When we consider systems of linear equations with complex coefficients, we will need to divide complex numbers to complete the solution process and obtain a reasonable form for the solution. Let $c_1 = a_1 + b_1i$ and $c_2 = a_2 + b_2i$. If $c_2 \neq 0$, that is, if $a_2 \neq 0$ or $b_2 \neq 0$, then we can **divide** c_1 by c_2:

$$\frac{c_1}{c_2} = \frac{a_1 + b_1i}{a_2 + b_2i}.$$

To conform to our practice of expressing a complex number in the form real part + imaginary part · i, we must simplify the foregoing expression for c_1/c_2. To simplify this complex fraction, we multiply the numerator and the denominator by the conjugate of the denominator. Thus, dividing c_1 by c_2 gives the complex number

$$\frac{c_1}{c_2} = \frac{a_1 + b_1i}{a_2 + b_2i} = \frac{(a_1 + b_1i)(a_2 - b_2i)}{(a_2 + b_2i)(a_2 - b_2i)} = \frac{a_1a_2 + b_1b_2}{a_2^2 + b_2^2} - \frac{a_1b_2 + a_2b_1}{a_2^2 + b_2^2}i.$$

Example 3. Let $c_1 = 2 - 5i$ and $c_2 = -3 + 4i$. Then

$$\frac{c_1}{c_2} = \frac{2-5i}{-3+4i} = \frac{(2-5i)(-3-4i)}{(-3+4i)(-3-4i)} = \frac{-26+7i}{(-3)^2+(4)^2} = -\frac{26}{25} + \frac{7}{25}i. \quad \blacksquare$$

Finding the reciprocal of a complex number is a special case of division of complex numbers. If $c = a + bi$, $c \neq 0$, then

$$\begin{aligned}\frac{1}{c} = \frac{1}{a+bi} &= \frac{a-bi}{(a+bi)(a-bi)} = \frac{a-bi}{a^2+b^2} \\ &= \frac{a}{a^2+b^2} - \frac{b}{a^2+b^2}i.\end{aligned}$$

Example 4. (a) $\dfrac{1}{2+3i} = \dfrac{2-3i}{(2+3i)(2-3i)} = \dfrac{2-3i}{2^2+3^2} = \dfrac{2}{13} - \dfrac{3}{13}i.$

(b) $\dfrac{1}{i} = \dfrac{-i}{i(-i)} = \dfrac{-i}{-i^2} = \dfrac{-i}{-(-1)} = -i.$ ■

Summarizing, we can say that complex numbers are mathematical objects for which addition, subtraction, multiplication, and division are defined in such a way that these operations on real numbers can be derived as special cases. In fact, it is easy to show that complex numbers form a mathematical system that is called a field.

Geometric Representation of Complex Numbers

A complex number $c = a + bi$ may be regarded as an ordered pair (a, b) of real numbers. This ordered pair of real numbers corresponds to a point in the plane. Such a correspondence naturally suggests that we represent $a + bi$ as a point in the **complex plane,** where the horizontal axis is used to represent the real part of c and the vertical axis is used to represent the imaginary part of c. To simplify matters, we call these the **real axis** and **imaginary axis,** respectively (see Figure B.1).

Example 5. Plot the complex numbers $c = 2 - 3i$, $d = 1 + 4i$, $e = -3$, and $f = 2i$ in the complex plane.

Solution: See Figure B.2. ■

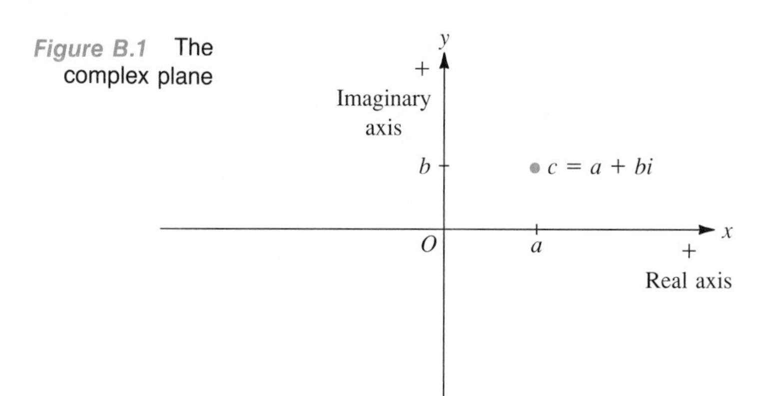

Figure B.1 The complex plane

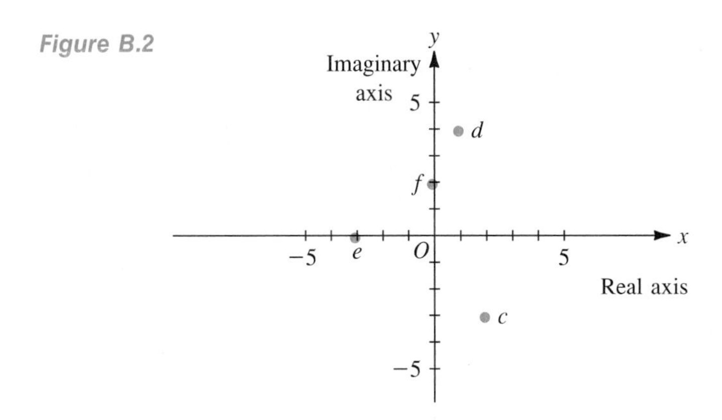

Figure B.2

The rules concerning inequality of real numbers, such as less than and greater than, *do not apply to complex numbers*. There is no way to arrange the complex numbers according to size. However, using the geometric representation from the complex plane, we can attach a notion of size to a complex number by measuring its distance from the origin. The distance from the origin to $c = a + bi$ is called the **absolute value** or **modulus** of the complex number, and is denoted by $|c| = |a + bi|$. Using the formula for the distance between ordered pairs of real numbers, we obtain

$$|c| = |a + bi| = \sqrt{a^2 + b^2}.$$

It follows that $c\bar{c} = |c|^2$ (verify).

Example 6. Referring to Example 5: $|c| = \sqrt{13}$; $|d| = \sqrt{17}$; $|e| = 3$; $|f| = 2$. ■

A different, but related, interpretation of a complex number is obtained if we associate with $c = a + bi$ the vector OP, where O is the origin $(0, 0)$ and P is the point (a, b). There is an obvious correspondence between this representation and vectors in the plane discussed in calculus, which we reviewed in Section 2.1. Using a vector representation, addition and subtraction of complex numbers can be viewed as the corresponding vector operations. These are represented in Figures 2.5, 2.6, and 2.9. We will not pursue the manipulation of complex numbers by vector operations here, but such a point of view is important for the development and study of complex variables.

Matrices with Complex Entries

If the entries of a matrix are complex numbers, we can perform the matrix operations of addition, subtraction, multiplication, and scalar multiplication in a manner completely analogous to that for real matrices. The validity of these operations can be verified using properties of complex arithmetic, just imitating the proofs for real matrices presented in the text. We illustrate these concepts in the following example.

Example 7. Let

$$A = \begin{bmatrix} 4+i & -2+3i \\ 6+4i & -3i \end{bmatrix}, \quad B = \begin{bmatrix} 2-i & 3-4i \\ 5+2i & -7+5i \end{bmatrix}, \quad C = \begin{bmatrix} 1+2i & i \\ 3-i & 8 \\ 4+2i & 1-i \end{bmatrix}.$$

(a) $$A + B = \begin{bmatrix} (4+i)+(2-i) & (-2+3i)+(3-4i) \\ (6+4i)+(5+2i) & (-3i)+(-7+5i) \end{bmatrix} = \begin{bmatrix} 6 & 1-i \\ 11+6i & -7+2i \end{bmatrix}.$$

(b) $$B - A = \begin{bmatrix} (2-i)-(4+i) & (3-4i)-(-2+3i) \\ (5+2i)-(6+4i) & (-7+5i)-(-3i) \end{bmatrix} = \begin{bmatrix} -2-2i & 5-7i \\ -1-2i & -7+8i \end{bmatrix}.$$

(c) $$CA = \begin{bmatrix} 1+2i & i \\ 3-i & 8 \\ 4+2i & 1-i \end{bmatrix} \begin{bmatrix} 4+i & -2+3i \\ 6+4i & -3i \end{bmatrix}$$

$$= \begin{bmatrix} (1+2i)(4+i)+(i)(6+4i) & (1+2i)(-2+3i)+(i)(-3i) \\ (3-i)(4+i)+(8)(6+4i) & (3-i)(-2+3i)+(8)(-3i) \\ (4+2i)(4+i)+(1-i)(6+4i) & (4+2i)(-2+3i)+(1-i)(-3i) \end{bmatrix}$$

$$= \begin{bmatrix} -2+15i & -5-i \\ 61+31i & -3-13i \\ 24+10i & -17+5i \end{bmatrix}.$$

$$\text{(d)}\ (2+i)B = \begin{bmatrix} (2+i)(2-i) & (2+i)(3-4i) \\ (2+i)(5+2i) & (2+i)(-7+5i) \end{bmatrix} = \begin{bmatrix} 5 & 10-5i \\ 8+9i & -19+3i \end{bmatrix}. \quad \blacksquare$$

Just as we can compute the conjugate of a complex number, we can compute the **conjugate of a matrix** by computing the conjugate of each entry of the matrix. We denote the conjugate of matrix A by $\overline{A}$, and write

$$\overline{A} = [\overline{a_{ij}}].$$

Example 8. Referring to Example 7, we find that

$$\overline{A} = \begin{bmatrix} 4-i & -2-3i \\ 6-4i & 3i \end{bmatrix} \quad \text{and} \quad \overline{B} = \begin{bmatrix} 2+i & 3+4i \\ 5-2i & -7-5i \end{bmatrix}. \quad \blacksquare$$

The following properties of the conjugate of a matrix hold:

1. $\overline{\overline{A}} = A$.
2. $\overline{A+B} = \overline{A} + \overline{B}$.
3. $\overline{AB} = \overline{A}\,\overline{B}$.
4. For any real number k, $\overline{kA} = k\overline{A}$.
5. For any complex number c, $\overline{cA} = \overline{c}\,\overline{A}$.
6. $(\overline{A})^T = \overline{A^T}$.
7. If A is nonsingular, then $(\overline{A})^{-1} = \overline{A^{-1}}$.

We prove properties 5 and 6 here and leave the others as exercises. First property 5: If c is complex, the (i, j) entry of $\overline{cA}$ is

$$\overline{ca_{ij}} = \overline{c}\,\overline{a}_{ij},$$

which is the (i, j) entry of $\overline{c}\overline{A}$. Next property 6: The (i, j) entry of $(\overline{A})^T$ is $\overline{a}_{ji}$, which is the (i, j) entry of $\overline{A^T}$.

Complex Numbers and Roots of Polynomials

A polynomial of degree n with real coefficients has n complex roots, some, all, or none of which may be real numbers. Thus the polynomial $f_1(x) = x^4 - 1$ has the roots i, $-i$, 1, and -1; the polynomial $f_2(x) = x^2 - 1$ has the roots 1 and -1; and the polynomial $f_3(x) = x^2 + 1$ has the roots i and $-i$. We shall make use of this result later, when we extend the definition of an eigenvalue (Section 6.1) to allow it to be a complex number.

B.1. EXERCISES

1. Let $c_1 = 3 + 4i$, $c_2 = 1 - 2i$, and $c_3 = -1 + i$. Compute each of the following and simplify as much as possible.

(a) $c_1 + c_2$. (b) $c_3 - c_1$.
(c) c_1c_2. (d) $c_2\bar{c}_3$.
(e) $4c_3 + \bar{c}_2$. (f) $(-i) \cdot c_2$.
(g) $\overline{3c_1 - ic_2}$. (h) $c_1c_2c_3$.

2. Write in the form $a + bi$.

(a) $\dfrac{1+2i}{3-4i}$. (b) $\dfrac{2-3i}{3-i}$.
(c) $\dfrac{(2+i)^2}{i}$. (d) $\dfrac{1}{(3+2i)(1+i)}$.

3. Represent each complex number as a point and as a vector in the complex plane.

(a) $4 + 2i$. (b) $-3 + i$. (c) $3 - 2i$.
(d) $i(4 + i)$.

4. Find the modulus of each complex number in Exercise 3.

5. If $c = a + bi$, then we can denote the real part of c by $\text{Re}(c)$ and the imaginary part of c by $\text{Im}(c)$.

(a) For any complex numbers $c_1 = a_1 + b_1i$, $c_2 = a_2 + b_2i$, prove that $\text{Re}(c_1 + c_2) = \text{Re}(c_1) + \text{Re}(c_2)$ and $\text{Im}(c_1 + c_2) = \text{Im}(c_1) + \text{Im}(c_2)$.
(b) For any real number k, prove that $\text{Re}(kc) = k\,\text{Re}(c)$ and $\text{Im}(kc) = k\,\text{Im}(c)$.
(c) Is part (b) true if k is a complex number?
(d) Prove or disprove:

$$\text{Re}(c_1c_2) = \text{Re}(c_1) \cdot \text{Re}(c_2).$$

6. In the complex plane sketch the vectors corresponding to c and $\bar{c}$ for $c = 2 + 3i$ and $c = -1 + 4i$. Geometrically, we can say that $\bar{c}$ is the reflection of c with respect to the real axis. (See also Example 7 in Section 4.1.)

7. Let

$$A = \begin{bmatrix} 2+2i & -1+3i \\ -2 & 1-i \end{bmatrix},$$

$$B = \begin{bmatrix} 2i & 1+2i \\ 0 & 3-i \end{bmatrix}, \qquad C = \begin{bmatrix} 2+i \\ -i \end{bmatrix}.$$

Compute each of the following and simplify each entry as $a + bi$.

(a) $A + B$. (b) $(1 - 2i)C$. (c) AB.
(d) BC. (e) $A - 2I_2$. (f) $\bar{B}$.
(g) $A\bar{C}$. (h) $\overline{(A + B)C}$.

8. Let A and B be $m \times n$ complex matrices, and let C be an $n \times n$ nonsingular matrix.

(a) Prove that $\overline{A + B} = \bar{A} + \bar{B}$.
(b) Prove that for any real number k, $\overline{kA} = k\bar{A}$.
(c) Prove that $(\bar{C})^{-1} = \overline{C^{-1}}$.

9. If $A = \begin{bmatrix} 0 & i \\ i & 0 \end{bmatrix}$, compute A^2, A^3, and A^4. Give a general rule for A^n, n a positive integer.

10. Find all the roots.

(a) $x^2 + x + 1 = 0$.
(b) $x^3 + 2x^2 + x + 2 = 0$.
(c) $x^5 + x^4 - x - 1 = 0$.

11. Let $p(x)$ denote a polynomial and let A be a square matrix. Then $p(A)$ is called a **matrix polynomial** or a **polynomial in the matrix** A. For $p(x) = 2x^2 + 5x - 3$, compute $p(A) = 2A^2 + 5A - 3I_n$ for each of the following.

(a) $A = \begin{bmatrix} -3 & 0 \\ 0 & -3 \end{bmatrix}$. (b) $A = \begin{bmatrix} 1 & 2 \\ 0 & 1 \end{bmatrix}$.

(c) $A = \begin{bmatrix} 0 & i \\ i & 0 \end{bmatrix}$. (d) $A = \begin{bmatrix} 1 & i \\ 0 & 0 \end{bmatrix}$.

12. Let $p(x) = x^2 + 1$.

(a) Determine two different 2×2 matrices A of the form kI_2 that satisfy $p(A) = O_2$.
(b) Verify that $p(A) = O_2$, for $A = \begin{bmatrix} 1 & 2 \\ -1 & -1 \end{bmatrix}$.

13. Find all the 2×2 matrices A of the form kI_2 that satisfy $p(A) = O_2$ for $p(x) = x^2 - x - 2$.

14. In Supplementary Exercise 4 in Chapter 1, we introduced the concept of a square root of a matrix with real entries. We can generalize the notion of a square root of a matrix if we permit complex entries.

(a) Compute a complex square root of

$$A = \begin{bmatrix} -1 & 0 \\ 0 & 0 \end{bmatrix}.$$

(b) Compute a complex square root of

$$A = \begin{bmatrix} -2 & 2 \\ 2 & -2 \end{bmatrix}.$$

B.2. *Complex Numbers in Linear Algebra*

Almost everything in the first five chapters of this book remains true if we cross out the word *real* and replace it by the word *complex*. In Section B.1 we introduced matrices with complex entries. Here we consider linear systems with complex entries, determinants of complex matrices, complex vector spaces, and eigenvalues and eigenvectors of complex matrices. The theory we have developed for these topics using real numbers is immediately applicable to the case with complex numbers. We could have developed Chapters 1 through 5 with complex numbers, but we limited ourselves to the real numbers for the following reasons:

1. The results would not look markedly different.
2. For a good many applications the real number situation is adequate; the transition to complex numbers is, as we have just remarked, not too difficult.
3. The computational effort of complex arithmetic is, in most cases, more than double that resulting with only real numbers.

As a matter of fact, we can do most of the first five chapters not only over the real numbers and the complex numbers, but also over any *field* (the real numbers and the complex numbers are familiar examples of a field).

The primary goal of this appendix is to provide an easy transition to complex numbers in linear algebra. This is of particular importance in Chapter 6, where complex eigenvalues and eigenvectors arise naturally for matrices with real entries. Hence we do not restate our theorems in the complex case, but provide a discussion and examples of the major ideas needed to accomplish this transition. It will soon be evident that the increased computational effort of complex arithmetic becomes quite tedious if done by hand.

Solving Linear Systems with Complex Entries

The results and techniques dealing with the solution of linear systems that we developed in Chapter 1 carry over directly to linear systems with complex coefficients. We shall illustrate row operations and echelon forms for such systems with Gauss–Jordan reduction using complex arithmetic.

Example 1. Solve the linear system

$$\begin{aligned}(1+i)x_1 + (2+i)x_2 &= 5\\ (2-2i)x_1 + \qquad ix_2 &= 1+2i.\end{aligned}$$

Solution: We form the augmented matrix and use elementary row operations to transform it to reduced row echelon form. For the augmented matrix $[A \mid B]$,

$$\left[\begin{array}{cc|c} 1+i & 2+i & 5 \\ 2-2i & i & 1+2i \end{array}\right],$$

multiply the first row by $1/(1+i)$ to obtain

$$\left[\begin{array}{cc|c} 1 & \frac{3}{2}-\frac{1}{2}i & \frac{5}{2}-\frac{5}{2}i \\ 2-2i & i & 1+2i \end{array}\right].$$

We now add $-(2-2i)$ times the first row to the second row to get

$$\left[\begin{array}{cc|c} 1 & \frac{3}{2}-\frac{1}{2}i & \frac{5}{2}-\frac{5}{2}i \\ 0 & -2+5i & 1+12i \end{array}\right].$$

Multiply the second row by $1/(-2+5i)$ to obtain

$$\left[\begin{array}{cc|c} 1 & \frac{3}{2}-\frac{1}{2}i & \frac{5}{2}-\frac{5}{2}i \\ 0 & 1 & 2-i \end{array}\right],$$

which is in row echelon form. To get to reduced row echelon form, we add $-(\frac{3}{2}-\frac{1}{2}i)$ times the second row to the first row to get

$$\left[\begin{array}{cc|c} 1 & 0 & 0 \\ 0 & 1 & 2-i \end{array}\right].$$

Hence the solution is $x_1 = 0$ and $x_2 = 2 - i$. ■

If you carry out the arithmetic for the row operations in the preceding example, you will feel the burden of the complex arithmetic even though there were just two equations in two unknowns. Gaussian elimination with back substitution can also be used on linear systems with complex coefficients.

Example 2. Suppose that the augmented matrix of a linear system has been transformed to the following matrix in row echelon form:

$$\left[\begin{array}{ccc|c} 1 & 0 & 1+i & -1 \\ 0 & 1 & 3i & 2+i \\ 0 & 0 & 1 & 2i \end{array}\right].$$

The back-substitution procedure gives us

$$\begin{aligned} x_3 &= 2i \\ x_2 &= 2 + i - 3i(2i) = 2 + i + 6 = 8 + i \\ x_1 &= -1 - (1 + i)(2i) = -1 - 2i + 2 = 3 - 2i. \end{aligned}$$ ■

We can alleviate the tediousness of complex arithmetic by using a computer code for linear systems with complex entries. However, we must still pay a high price because the execution time will be approximately twice as long as that for the same size linear system with all real entries. We can illustrate this by showing how to transform an $n \times n$ linear system with complex coefficients to a $2n \times 2n$ linear system with only real coefficients.

Example 3. Consider the linear system

$$\begin{aligned} (2 + i)x_1 + \ (1 + i)x_2 &= 3 + 6i \\ (3 - i)x_1 + (2 - 2i)x_2 &= 7 - i. \end{aligned}$$

If we let $x_1 = a_1 + b_1 i$ and $x_2 = a_2 + b_2 i$, with a_1, b_1, a_2, and b_2 real numbers, then we can write this system in matrix form as

$$\begin{bmatrix} 2+i & 1+i \\ 3-i & 2-2i \end{bmatrix} \begin{bmatrix} a_1 + b_1 i \\ a_2 + b_2 i \end{bmatrix} = \begin{bmatrix} 3+6i \\ 7-i \end{bmatrix}.$$

We first rewrite the given linear system as

$$\left(\begin{bmatrix} 2 & 1 \\ 3 & 2 \end{bmatrix} + i\begin{bmatrix} 1 & 1 \\ -1 & -2 \end{bmatrix}\right)\left(\begin{bmatrix} a_1 \\ a_2 \end{bmatrix} + i\begin{bmatrix} b_1 \\ b_2 \end{bmatrix}\right) = \begin{bmatrix} 3 \\ 7 \end{bmatrix} + i\begin{bmatrix} 6 \\ -1 \end{bmatrix}.$$

Multiplying, we have

$$\left(\begin{bmatrix} 2 & 1 \\ 3 & 2 \end{bmatrix}\begin{bmatrix} a_1 \\ a_2 \end{bmatrix} - \begin{bmatrix} 1 & 1 \\ -1 & -2 \end{bmatrix}\begin{bmatrix} b_1 \\ b_2 \end{bmatrix}\right) + i\left(\begin{bmatrix} 2 & 1 \\ 3 & 2 \end{bmatrix}\begin{bmatrix} b_1 \\ b_2 \end{bmatrix} + \begin{bmatrix} 1 & 1 \\ -1 & -2 \end{bmatrix}\begin{bmatrix} a_1 \\ a_2 \end{bmatrix}\right) = \begin{bmatrix} 3 \\ 7 \end{bmatrix} + i\begin{bmatrix} 6 \\ -1 \end{bmatrix}.$$

The real and imaginary parts on both sides of the equation must agree, respectively, and so we have

$$\begin{bmatrix} 2 & 1 \\ 3 & 2 \end{bmatrix}\begin{bmatrix} a_1 \\ a_2 \end{bmatrix} - \begin{bmatrix} 1 & 1 \\ -1 & -2 \end{bmatrix}\begin{bmatrix} b_1 \\ b_2 \end{bmatrix} = \begin{bmatrix} 3 \\ 7 \end{bmatrix}$$

and

$$\begin{bmatrix} 2 & 1 \\ 3 & 2 \end{bmatrix}\begin{bmatrix} b_1 \\ b_2 \end{bmatrix} + \begin{bmatrix} 1 & 1 \\ -1 & -2 \end{bmatrix}\begin{bmatrix} a_1 \\ a_2 \end{bmatrix} = \begin{bmatrix} 6 \\ -1 \end{bmatrix}.$$

This leads to the linear system

$$\begin{aligned} 2a_1 + a_2 - b_1 - b_2 &= 3 \\ 3a_1 + 2a_2 + b_1 + 2b_2 &= 7 \\ a_1 + a_2 + 2b_1 + b_2 &= 6 \\ -a_1 - 2a_2 + 3b_1 + 2b_2 &= -1, \end{aligned}$$

which can be written as

$$\begin{bmatrix} 2 & 1 & -1 & -1 \\ 3 & 2 & 1 & 2 \\ 1 & 1 & 2 & 1 \\ -1 & -2 & 3 & 2 \end{bmatrix}\begin{bmatrix} a_1 \\ a_2 \\ b_1 \\ b_2 \end{bmatrix} = \begin{bmatrix} 3 \\ 7 \\ 6 \\ -1 \end{bmatrix}.$$

This linear system of four equations in four unknowns is now solved as in Chapter 1. The solution is (verify) $a_1 = 1$, $a_2 = 2$, $b_1 = 2$, and $b_2 = -1$. Thus $x_1 = 1 + 2i$ and $x_2 = 2 - i$ is the solution to the given linear system. ■

Determinants of Complex Matrices

The definition of a determinant and all the properties derived in Chapter 5 apply to matrices with complex entries. The following example is an illustration.

Example 4. Compute $|A|$ for the coefficient matrix A in Example 3.

Solution:

$$\begin{aligned}\begin{vmatrix} 2+i & 1+i \\ 3-i & 2-2i \end{vmatrix} &= (2+i)(2-2i) - (3-i)(1+i) \\ &= (6-2i) - (4+2i) \\ &= 2-4i. \end{aligned}$$

■

Complex Vector Spaces

A **complex vector space** is defined exactly as was a real vector space in Definition 2.4, except that the scalars in properties 5 through 8 are permitted to be complex numbers. The terms *complex* vector space and *real* vector space emphasize the field from which the scalars are chosen. It happens that in order to satisfy the closure property of scalar multiplication [Definition 2.4(b)] in a complex vector space, we must, in most examples, consider vectors that involve complex numbers.

Most of the real vector spaces of Chapter 2 have complex vector space analogues.

Example 5. (a) Consider C^n, the set of all $n \times 1$ matrices $\begin{bmatrix} a_1 \\ a_2 \\ \vdots \\ a_n \end{bmatrix}$ with complex entries. Let the operation $\oplus$ be matrix addition and let the operation $\odot$ be multiplication of a matrix by a complex number. We can verify that C^n is a complex vector space by using the properties of matrices established in Section 1.3 and the properties of complex arithmetic established in Section B.1. (Note that if the operation $\odot$ is taken as multiplication of a matrix by a real number, then C^n is a real vector space whose vectors have complex components.)

(b) The set of all $m \times n$ matrices with complex entries with matrix addition as $\oplus$ and multiplication of a matrix by a complex number as $\odot$ is a complex vector space (verify). We denote this vector space by ${}_mC_n$.

(c) The set of polynomials with complex coefficients with polynomial addition as $\oplus$ and multiplication of a polynomial by a complex constant as $\odot$ forms a complex vector space. Verification follows the pattern of Example 5 in Section 2.2.

(d) The set of complex-valued continuous functions defined on an interval

$[a, b]$ (i.e., all functions of the form $f(t) = f_1(t) + if_2(t)$, where f_1 and f_2 are real-valued continuous functions on $[a, b]$), with $\oplus$ defined by $(f \oplus g)(t) = f(t) + g(t)$ and $\odot$ defined by $(c \odot f)(t) = c \cdot f(t)$ for a complex scalar c, forms a complex vector space. The corresponding real vector space is given in Example 6 in Section 2.2 for the interval $(-\infty, \infty)$. ■

A **complex vector subspace** W of a complex vector space V is defined as in Definition 2.5, but with real scalars replaced by complex ones. The analogue of Theorem 2.3 can be proved to show that a nonempty subset W of a complex vector space V is a complex vector subspace if and only if the following conditions hold:

(a) If $\mathbf{u}$ and $\mathbf{v}$ are any vectors in W, then $\mathbf{u} \oplus \mathbf{v}$ is in W.
(b) If c is any complex number and $\mathbf{u}$ is any vector in W, then $c \odot \mathbf{u}$ is in W.

Example 6. (a) Let W be the set of all vectors in ${}_3C_1$ of the form $\begin{bmatrix} a \\ 0 \\ b \end{bmatrix}$, where a and b are complex numbers. It easily follows that

$$\begin{bmatrix} a \\ 0 \\ b \end{bmatrix} \oplus \begin{bmatrix} d \\ 0 \\ e \end{bmatrix} = \begin{bmatrix} a + d \\ 0 \\ b + e \end{bmatrix}$$

belongs to W and for any complex scalar c,

$$c \odot \begin{bmatrix} a \\ 0 \\ b \end{bmatrix} = \begin{bmatrix} ca \\ 0 \\ cb \end{bmatrix}$$

belongs to W. Hence W is a complex vector subspace of ${}_3C_1$.

(b) Let W be the set of all vectors in ${}_mC_n$ having only real entries. If $A = [a_{ij}]$ and $B = [b_{ij}]$ belong to W, then so will $A \oplus B$, because if a_{ij} and b_{ij} are real, then so is their sum. However, if c is any complex scalar and A belongs to W, then $c \odot A = cA$ can have entries ca_{ij} that need not be real numbers. It follows that $c \odot A$ need not belong to W, so W is not a complex vector subspace. ■

Linear Independence and Basis in Complex Vector Spaces

The notions of linear combinations, spanning sets, linear dependence, linear independence, and basis are unchanged for complex vector spaces, except that we use complex scalars (see Sections 2.3 and 2.4).

Example 7. Let V be the complex vector space C^3. Let

$$\mathbf{v}_1 = \begin{bmatrix} 1 \\ i \\ 0 \end{bmatrix}, \quad \mathbf{v}_2 = \begin{bmatrix} i \\ 0 \\ 1+i \end{bmatrix}, \quad \text{and} \quad \mathbf{v}_3 = \begin{bmatrix} 1 \\ 1 \\ 1 \end{bmatrix}.$$

(a) Determine whether $\mathbf{v} = \begin{bmatrix} -1 \\ -3+3i \\ -4+i \end{bmatrix}$ is a linear combination of $\mathbf{v}_1$, $\mathbf{v}_2$, $\mathbf{v}_3$.
(b) Determine whether $\{\mathbf{v}_1, \mathbf{v}_2, \mathbf{v}_3\}$ spans C^3.
(c) Determine whether $\{\mathbf{v}_1, \mathbf{v}_2, \mathbf{v}_3\}$ is a linearly independent subset of C^3.
(d) Is $\{\mathbf{v}_1, \mathbf{v}_2, \mathbf{v}_3\}$ a basis for C^3?

Solution: (a) We proceed as in Example 1 of Section 2.3. We form a linear combination of $\mathbf{v}_1$, $\mathbf{v}_2$, and $\mathbf{v}_3$ with unknown coefficients a_1, a_2, and a_3, respectively, and set it equal to $\mathbf{v}$:

$$a_1\mathbf{v}_1 + a_2\mathbf{v}_2 + a_3\mathbf{v}_3 = \mathbf{v}.$$

If we substitute the vectors $\mathbf{v}_1$, $\mathbf{v}_2$, $\mathbf{v}_3$, and $\mathbf{v}$ into this expression, we obtain the linear system

$$\begin{aligned} a_1 + \quad ia_2 + a_3 &= -1 \\ ia_1 \qquad\qquad + a_3 &= -3 + 3i \\ (1+i)a_2 + a_3 &= -4 + i. \end{aligned}$$

We next investigate the consistency of this system by using elementary row operations to transform its augmented matrix to either row echelon or reduced row echelon form. The row echelon form is (verify)

$$\left[\begin{array}{ccc|c} 1 & i & 1 & -1 \\ 0 & 1 & 1-i & -3+4i \\ 0 & 0 & 1 & -3 \end{array}\right],$$

which implies that the system is consistent, hence $\mathbf{v}$ is a linear combination of $\mathbf{v}_1$, $\mathbf{v}_2$, and $\mathbf{v}_3$. In fact, back substitution gives $a_1 = 3$, $a_2 = i$, and $a_3 = -3$.

(b) Let $\mathbf{v} = \begin{bmatrix} c_1 \\ c_2 \\ c_3 \end{bmatrix}$ be an arbitrary vector of C^3. We form the linear combination

$$a_1\mathbf{v}_1 + a_2\mathbf{v}_2 + a_3\mathbf{v}_3 = \mathbf{v}$$

and solve for a_1, a_2, and a_3. The resulting linear system is

$$\begin{aligned} a_1 + \quad\quad ia_2 + a_3 &= c_1 \\ ia_1 \quad\quad\quad\quad + a_3 &= c_2 \\ (1+i)a_2 + a_3 &= c_3. \end{aligned}$$

Transforming the augmented matrix to row echelon form, we obtain (verify)

$$\left[\begin{array}{ccc|c} 1 & i & 1 & c_1 \\ 0 & 1 & 1-i & c_2 - ic_1 \\ 0 & 0 & 1 & -c_3 + (1+i)(c_2 - ic_1) \end{array}\right].$$

Hence we can solve for a_1, a_2, a_3 for any choice of complex numbers c_1, c_2, c_3, which implies that $\{\mathbf{v}_1, \mathbf{v}_2, \mathbf{v}_3\}$ spans C^3.

(c) Proceeding as in Example 8 of Section 2.3, we form the equation

$$a_1\mathbf{v}_1 + a_2\mathbf{v}_2 + a_3\mathbf{v}_3 = \mathbf{0}$$

and solve for a_1, a_2, and a_3. The resulting homogeneous system is

$$\begin{aligned} a_1 + \quad\quad ia_2 + a_3 &= 0 \\ ia_1 \quad\quad\quad\quad + a_3 &= 0 \\ (1+i)a_2 + a_3 &= 0. \end{aligned}$$

Transforming the augmented matrix to row echelon form, we obtain (verify)

$$\left[\begin{array}{ccc|c} 1 & i & 1 & 0 \\ 0 & 1 & 1-i & 0 \\ 0 & 0 & 1 & 0 \end{array}\right],$$

and hence the only solution is $a_1 = a_2 = a_3 = 0$, showing that $\{\mathbf{v}_1, \mathbf{v}_2, \mathbf{v}_3\}$ is linearly independent.

(d) Yes, because $\mathbf{v}_1$, $\mathbf{v}_2$, and $\mathbf{v}_3$ span C^3 [part (b)] and they are linearly independent [part (c)]. ■

Just as in real vector spaces, the questions of spanning sets, linearly independent or linearly dependent sets, and basis in a complex vector space are resolved by using an appropriate linear system. The definition of the dimension of a complex vector space is the same as that given in Definition 2.10. In discussing the dimension of a complex vector space like C^n, we must adjust our intuitive picture. For example, C^1 consists of all complex multiples of a single nonzero vector. This collection can be put into a one-to-one correspondence with the complex numbers themselves, that is, all the points in the complex plane (see Figure B.1). Since the elements of a two-dimensional real vector space can be put into a one-to-one correspondence with the points of R^2 (see Section 2.1), we see that a complex vector space of dimension one has a geometric model that is in one-to-one correspondence with a geometric model of a two-dimensional real vector space. Similarly, a complex vector space of dimension two is the same, geometrically, as a four-dimensional real vector space.

Complex Eigenvalues and Eigenvectors

Let V be an n-dimensional complex vector space and let $L: V \to V$ be a linear transformation. If $S = \{\mathbf{x}_1, \mathbf{x}_2, \ldots, \mathbf{x}_n\}$ is a basis for V such that L is represented by the $n \times n$ complex matrix A, then for $\mathbf{x}$ in V

$$L(\mathbf{x}) = A\mathbf{x}. \tag{1}$$

Imitating Definition 6.2, we say that the complex number λ is an **eigenvalue** of L if there exists a nonzero vector $\mathbf{x}$ in V such that

$$L(\mathbf{x}) = \lambda\mathbf{x}. \tag{2}$$

Every nonzero complex vector $\mathbf{x}$ satisfying this equation is called an **eigenvector** of L **associated with the eigenvalue $\boldsymbol{\lambda}$.** Substituting Equation (1) into Equation (2), we have the relationship

$$A\mathbf{x} = \lambda\mathbf{x}. \tag{3}$$

If λ is a complex scalar and $\mathbf{x} \neq \mathbf{0}_V$ is a vector in the complex vector space V such that Equation (3) is satisfied, then we say that λ is an **eigenvalue** of the complex matrix A and $\mathbf{x}$ is an **eigenvector** of A **associated with $\boldsymbol{\lambda}$.** Equation (3) can be rewritten as the homogeneous system

$$(\lambda I_n - A)\mathbf{x} = \mathbf{0}. \tag{4}$$

This homogeneous system has a nonzero solution $\mathbf{x}$ if and only if

$$\det(\lambda I_n - A) = 0$$

has a solution. Following the development in Section 6.1, we shall call $\det(\lambda I_n - A)$ the **characteristic polynomial** of the matrix A, which is again a complex polynomial of degree n in λ. As in Section 6.1, the eigenvalues of the complex matrix A are the complex roots of the characteristic polynomial. According to the Fundamental Theorem of Algebra, an nth-degree polynomial has exactly n roots, if multiple roots are counted. Even if $\det(\lambda I_n - A)$ has only one root of multiplicity n, we then know that the characteristic polynomial of any complex matrix A, or equivalently of any linear transformation L mapping a complex vector space into itself always has at least one root (possibly complex). This is quite different from the case of real vector spaces discussed in Section 6.1, because there are real matrices that do not have real eigenvalues (see Example 13 in Section 6.1).

Example 8. Let

$$A = \begin{bmatrix} 0 & 1 \\ -1 & 0 \end{bmatrix}.$$

Then the characteristic polynomial of A is

$$\det(\lambda I_2 - A) = \lambda^2 + 1.$$

If we interpret A as representing a linear transformation between real vector spaces, then A has no (real) eigenvalues because the roots of $f(\lambda) = \lambda^2 + 1$ are i and $-i$. However, if we interpret A as representing a linear transformation between complex vector spaces, then we know that there is always at least one (complex) eigenvalue, and in this case the eigenvalues are i and $-i$. The corresponding (complex) eigenvectors are obtained by finding a nontrivial solution of the homogeneous systems

$$(iI_2 - A)\mathbf{x} = \mathbf{0} \quad \text{and} \quad (-iI_2 - A)\mathbf{x} = \mathbf{0},$$

respectively. An eigenvector associated with eigenvalue i is $\begin{bmatrix} 1 \\ i \end{bmatrix}$ and an eigenvector associated with eigenvalue $-i$ is $\begin{bmatrix} -1 \\ i \end{bmatrix}$ (verify). ■

Every real $n \times n$ matrix can be considered as an $n \times n$ complex matrix. Hence the eigenvalue and eigenvector computations for real matrices carried out in Chapter 6 can "be fixed up" by permitting complex roots of the characteristic polynomial to be called eigenvalues. Of course, this means that we would need to use complex arithmetic to solve for the associated eigenvector. However, this "fix" does not apply when we consider a linear transformation $L: V \to V$ where V is a real vector space, because a complex number with nonzero imaginary part does not belong to the field of scalars (the real numbers) in this case. Thus, if $\lambda = a + bi$, $b \neq 0$, is a root of the characteristic polynomial of L, the equation

$$L(\mathbf{x}) = \lambda\mathbf{x}$$

cannot hold. However, the information derived from such a procedure can reveal fundamental properties of L.

So if, as in Chapter 6, we restrict ourselves to real vector spaces, we must permit only real numbers to be eigenvalues. The eigen-problem for symmetric matrices, Section 6.2, needs no "fix" because the eigenvalues of any real symmetric matrix are all real numbers (see Theorem 6.6).

B.2. EXERCISES

1. Solve using Gauss–Jordan reduction.

(a) $\begin{aligned} (1 + 2i)x_1 + (-2 + i)x_2 &= 1 - 3i \\ (2 + i)x_1 + (-1 + 2i)x_2 &= -1 - i. \end{aligned}$

(b) $\begin{aligned} 2ix_1 - (1 - i)x_2 &= 1 + i \\ (1 - i)x_1 + x_2 &= 1 - i. \end{aligned}$

(c) $\begin{aligned} (1 + i)x_1 - x_2 &= -2 + i \\ 2ix_1 + (1 - i)x_2 &= i. \end{aligned}$

2. Transform the given augmented matrix of a linear system to row echelon form and solve by back substitution.

(a) $\left[\begin{array}{ccc|c} 2 & i & 0 & 1 - i \\ 0 & 3i & -2 + i & 4 \\ 0 & 0 & 2 + i & 2 - i \end{array}\right].$

(b) $\left[\begin{array}{ccc|c} i & 2 & 1 + i & 3i \\ 0 & 1 - i & 0 & 2 + i \\ 0 & 0 & 3 & 6 - 3i \end{array}\right].$

3. Solve by Gaussian elimination with back substitution.

(a)
$$\begin{aligned} ix_1 + (1+i)x_2 \qquad &= i \\ (1-i)x_1 + \qquad x_2 - ix_3 &= 1 \\ ix_2 + x_3 &= 1. \end{aligned}$$

(b)
$$\begin{aligned} x_1 + ix_2 + (1-i)x_3 &= 2+i \\ ix_1 \qquad + (1+i)x_3 &= -1+i \\ 2ix_2 - \qquad x_3 &= 2-i. \end{aligned}$$

4. Compute the determinant and simplify as much as possible.

(a) $\begin{vmatrix} 1+i & -1 \\ 2i & 1+i \end{vmatrix}$. (b) $\begin{vmatrix} 2-i & 1+i \\ 1+2i & -(1-i) \end{vmatrix}$.

(c) $\begin{vmatrix} 1+i & 2 & 2-i \\ i & 0 & 3+i \\ -2 & 1 & 1+2i \end{vmatrix}$. (d) $\begin{vmatrix} 2 & 1-i & 0 \\ 1+i & -1 & i \\ 0 & -i & 2 \end{vmatrix}$.

5. Find the inverse of each of the following matrices if possible.

(a) $\begin{bmatrix} i & 2 \\ 1+i & -i \end{bmatrix}$. (b) $\begin{bmatrix} 2 & i & 3 \\ 1+i & 0 & 1-i \\ 2 & 1 & 2+i \end{bmatrix}$.

6. Determine whether or not the following subsets W of ${}_2C_2$ are complex vector subspaces.

(a) W is the set of all 2×2 complex matrices with zeros on the main diagonal.

(b) W is the set of all 2×2 complex matrices that have diagonal entries with real part equal to zero.

(c) W is the set of all symmetric 2×2 complex matrices.

7. An $n \times n$ complex matrix A is called **Hermitian** if $\overline{A^T} = A$. This is equivalent to saying that $a_{ij} = \overline{a_{ji}}$ for all i and j.

(a) Prove that the diagonal entries of a Hermitian matrix must be real.

(b) Prove that every Hermitian matrix A can be written as $A = B + iC$, where B is real and symmetric and C is real and skew symmetric (see Definition 1.8). [*Hint:* Consider $B = (A + \overline{A})/2$ and $C = (A - \overline{A})/2i$.]

(c) Prove or disprove: The set W of all $n \times n$ Hermitian matrices is a complex vector subspace of ${}_nC_n$.

(d) Prove or disprove: The set W of all $n \times n$ Hermitian matrices is a real vector subspace of the real vector space of all $n \times n$ complex matrices.

(e) Prove that every real symmetric matrix is Hermitian.

8. An $n \times n$ complex matrix A is called **unitary** if $(\overline{A^T})A = A(\overline{A^T}) = I_n$. This is equivalent to saying that $\overline{A^T} = A^{-1}$.

(a) If A is unitary, show that A^T is unitary.

(b) If A is unitary, show that A^{-1} is unitary.

(c) Prove or disprove: The set W of all $n \times n$ unitary matrices is a complex vector subspace of ${}_nC_n$.

9. Let $W = \text{span }\{\mathbf{v}_1, \mathbf{v}_2, \mathbf{v}_3\}$, where

$$\mathbf{v}_1 = \begin{bmatrix} -1+i \\ 2 \\ 1 \end{bmatrix}, \quad \mathbf{v}_2 = \begin{bmatrix} 1 \\ 1+i \\ i \end{bmatrix}, \quad \mathbf{v}_3 = \begin{bmatrix} -5+2i \\ -1-3i \\ 2-3i \end{bmatrix}.$$

(a) Does $\mathbf{v} = \begin{bmatrix} i \\ 0 \\ 0 \end{bmatrix}$ belong to W?

(b) Is the set $\{\mathbf{v}_1, \mathbf{v}_2, \mathbf{v}_3\}$ linearly independent or linearly dependent?

10. Let $\{\mathbf{v}_1, \mathbf{v}_2, \mathbf{v}_3\}$ be a basis for a complex vector space V. Determine whether or not $\mathbf{w}$ is in span $\{\mathbf{w}_1, \mathbf{w}_2\}$.

(a)
$$\begin{aligned} \mathbf{w}_1 &= i\mathbf{v}_1 + (1-i)\mathbf{v}_2 + 2\mathbf{v}_3 \\ \mathbf{w}_2 &= (2+i)\mathbf{v}_1 + 2i\mathbf{v}_2 + (3-i)\mathbf{v}_3 \\ \mathbf{w} &= (-2-3i)\mathbf{v}_1 + (3-i)\mathbf{v}_2 + (-2-2i)\mathbf{v}_3. \end{aligned}$$

(b)
$$\begin{aligned} \mathbf{w}_1 &= 2i\mathbf{v}_1 + \mathbf{v}_2 + (1-i)\mathbf{v}_3 \\ \mathbf{w}_2 &= 3i\mathbf{v}_1 + (1+i)\mathbf{v}_2 + 3\mathbf{v}_3 \\ \mathbf{w} &= (2+3i)\mathbf{v}_1 + (2+i)\mathbf{v}_2 + (4-2i)\mathbf{v}_3. \end{aligned}$$

11. Find the eigenvalues and associated eigenvectors of the following complex matrices.

(a) $A = \begin{bmatrix} 1 & 1 \\ -1 & 1 \end{bmatrix}$. (b) $A = \begin{bmatrix} 1 & i \\ -i & 1 \end{bmatrix}$.

(c) $A = \begin{bmatrix} 2 & 0 & 0 \\ 0 & 2 & i \\ 0 & -i & 2 \end{bmatrix}$.

12. For each of the parts in Exercise 11, find a matrix P such that $P^{-1}AP = D$, a diagonal matrix. For part (c), find three different matrices P that diagonalize A.

13. (a) Prove that if A is Hermitian (see Exercise 7), then the eigenvalues of A are real.
(b) Verify that A in Exercise 11(c) is Hermitian.
(c) Are the eigenvectors associated with an eigenvalue of a Hermitian matrix guaranteed to be real vectors? Explain.

14. Prove that an $n \times n$ complex matrix A is unitary (see Exercise 8) if and only if the columns (and rows) of A form an orthonormal set with respect to the complex inner product. $(\mathbf{u}, \mathbf{v}) = \overline{\mathbf{u}^T}\mathbf{v}$ for $\mathbf{u}$ and $\mathbf{v}$ in C^n. (*Hint:* See Theorem 6.8.)

15. Let A be an $n \times n$ complex matrix.
(a) Show that A can be written as $B + iC$, where B and C are Hermitian.
(b) A is called **normal** if

$$\overline{A^T}A = A\overline{A^T}.$$

Show that A is normal if and only if

$$BC = CB.$$

[*Hint:* Consider $B = (A + \overline{A^T})/2$ and $C = (A - \overline{A^T})/2i$.]

16. (a) Prove that any Hermitian matrix is normal.
(b) Prove that any unitary matrix is normal.
(c) Find a 2×2 normal matrix that is neither symmetric, Hermitian, nor unitary.

Answers to Odd-Numbered Exercises

Chapter 1

Section 1.1, p. 8

1. $x_1 = 4$, $x_2 = 2$.

3. $x_1 = -4$, $x_2 = 2$, $x_3 = 10$.

5. $x_1 = 2$, $x_2 = -1$, $x_3 = -2$.

7. $x_1 = -20$, $x_2 = (\frac{1}{4})x_3 + 8$, $x_3 =$ any real number.

9. This linear system has no solution. It is inconsistent.

11. $x_1 = 5$, $x_2 = 1$.

13. This linear system has no solution. It is inconsistent.

15. (a) $t = 10$. (b) One value is $t = 3$.
(c) The choice $t = 3$ in part (b) was arbitrary. Any choice for t, other than $t = 10$ makes the system inconsistent. Hence there are infinitely many ways to choose a value for t in part (b).

19. $x_1 = 2$, $x_2 = 1$, $x_3 = 0$.

21. There is no such value of r.

Section 1.2, p. 18

1. (a) $\begin{bmatrix} 5 & -5 & 8 \\ 4 & 2 & 9 \\ 5 & 3 & 4 \end{bmatrix}$.

(b) $AB = \begin{bmatrix} 14 & 8 \\ 16 & 9 \end{bmatrix}$, $BA = \begin{bmatrix} 1 & 2 & 3 \\ 4 & 5 & 10 \\ 7 & 8 & 17 \end{bmatrix}$.

(c) $\begin{bmatrix} 0 & 10 & -9 \\ 8 & -1 & -2 \\ -5 & -4 & 3 \end{bmatrix}$.

(d) Impossible.

(e) $\begin{bmatrix} 19 & -8 \\ 32 & 30 \end{bmatrix}$.

3. (a) $\begin{bmatrix} -2 & 12 \\ 2 & 17 \\ 10 & 13 \end{bmatrix}$. (b) $\begin{bmatrix} 8 & 12 \\ 12 & -1 \end{bmatrix}$.

(c) $3(2A) = \begin{bmatrix} 6 & 12 & 18 \\ 12 & 6 & 24 \end{bmatrix} = 6A$.

(d) Impossible. (e) Impossible.

5. (a) $\begin{bmatrix} 1 & 2 \\ 2 & 1 \\ 3 & 4 \end{bmatrix}$. (b) $(A^T)^T = A$.

(c) $\begin{bmatrix} 14 & 16 \\ 8 & 9 \end{bmatrix}$. (d) Same as (c).

(e) $(C + E)^T = C^T + E^T = \begin{bmatrix} 5 & 4 & 5 \\ -5 & 2 & 3 \\ 8 & 9 & 4 \end{bmatrix}$.

(f) $A(2B) = 2(AB) = \begin{bmatrix} 28 & 16 \\ 32 & 18 \end{bmatrix}$.

9. There are infinitely many choices. For example, $r = 1$, $s = 0$; or $r = 0$, $s = 2$; or $r = 10$, $s = -18$.

15.
$$\begin{aligned} -2x_1 - x_2 + 4x_4 &= 5 \\ -3x_1 + 2x_2 + 7x_3 + 8x_4 &= 3 \\ x_1 + 2x_4 &= 4 \\ 3x_1 + x_3 + 3x_4 &= 6. \end{aligned}$$

17. $\begin{bmatrix} 2 & 3 & 0 \\ 0 & 3 & 1 \\ 2 & 0 & -1 \end{bmatrix} \begin{bmatrix} x_1 \\ x_2 \\ x_3 \end{bmatrix} = \begin{bmatrix} 0 \\ 0 \\ 0 \end{bmatrix}$.

19. The linear systems are equivalent. That is, they have the same solutions.

21. (a)
$$\begin{aligned} x_1 + 3x_2 &= 3 \\ 2x_1 &= 1 \\ x_1 - x_2 &= 4. \end{aligned}$$

(b)
$$\begin{aligned} 2x_1 + 3x_2 &= 0 \\ x_1 - x_2 + x_3 &= 0 \\ 2x_2 - x_3 &= 0 \\ x_1 + 2x_2 + 3x_3 &= 0. \end{aligned}$$

Section 1.3, p. 28

9. One such pair is $A = \begin{bmatrix} 1 & 1 \\ 2 & 2 \end{bmatrix}$ and $B = \begin{bmatrix} 1 & 1 \\ -1 & -1 \end{bmatrix}$. Another pair is $A = \begin{bmatrix} 2 & 6 \\ 4 & 12 \end{bmatrix}$ and $B = \begin{bmatrix} -3 & -3 \\ 1 & 1 \end{bmatrix}$. Yet another pair is $A = O_2$ and $B =$ any 2×2 matrix with at least one nonzero element. There are infinitely many such pairs.

11. There are many such pairs of matrices. For example, $A = \begin{bmatrix} 1 & 1 \\ 0 & 1 \end{bmatrix}$ and $B = \begin{bmatrix} 1 & -1 \\ 0 & 1 \end{bmatrix}$ or $A = \begin{bmatrix} 1 & 1 \\ 1 & 2 \end{bmatrix}$ and $B = \begin{bmatrix} 2 & -1 \\ -1 & 1 \end{bmatrix}$. Note also that for $A = k\begin{bmatrix} 1 & 0 \\ 0 & 1 \end{bmatrix}$ and $B = \left(\frac{1}{k}\right)\begin{bmatrix} 1 & 0 \\ 0 & 1 \end{bmatrix}$, $k \neq 0, \pm 1$, we have $A \neq B$ and $AB = \begin{bmatrix} 1 & 0 \\ 0 & 1 \end{bmatrix}$.

21. $r = 2$.

23. $s = r^2$.

29. $k = \pm\sqrt{\frac{1}{6}}$.

Section 1.4, p. 40

9. $B = \begin{bmatrix} 1 & 3 \\ 3 & 1 \end{bmatrix}$ is such that $AB = BA$. There are infinitely many such matrices B.

21. (a) $A^{-1} = \begin{bmatrix} -\frac{2}{13} & \frac{3}{13} \\ \frac{5}{13} & -\frac{1}{13} \end{bmatrix}$. (b) $A^{-1} = \begin{bmatrix} -\frac{1}{3} & \frac{2}{3} \\ \frac{2}{3} & -\frac{1}{3} \end{bmatrix}$.

23. $\begin{bmatrix} 11 & 19 \\ 7 & 0 \end{bmatrix}$.

25. (a) $\begin{bmatrix} \frac{6}{13} \\ \frac{11}{13} \end{bmatrix}$. (b) $\begin{bmatrix} \frac{8}{13} \\ \frac{19}{13} \end{bmatrix}$.

Section 1.5, p. 60

1. (a) Possible answer: $B = \begin{bmatrix} 1 & 2 & -3 & 1 \\ 0 & 1 & 2 & -1 \\ 0 & 0 & 1 & -\frac{7}{4} \\ 0 & 0 & 0 & 1 \end{bmatrix}$.

(b) $C = I_4$.

5. (a) $x_1 = 1,\ x_2 = 2,\ x_3 = -2$.
(b) $x_1 = 1,\ x_2 = 2,\ x_3 = -2$.

7. (a) $x_1 = -1,\ x_2 = 4,\ x_3 = -3$.
(b) $x_1 = x_2 = x_3 = 0$.
(c) $x_1 = r,\ x_2 = -2r,\ x_3 = r$, where r = any real number.
(d) $x_1 = -2r,\ x_2 = r,\ x_3 = 0$, where r = any real number.

9. (a) $x_1 = 1,\ x_2 = 2,\ x_3 = 2$.
(b) $x_1 = x_2 = x_3 = 0$.

11. $X = \begin{bmatrix} r \\ r \end{bmatrix}$, $r \neq 0$.

13. $X = \begin{bmatrix} -\frac{1}{2}r \\ \frac{1}{2}r \\ r \end{bmatrix}$, $r \neq 0$.

15. (a) $a = \pm\sqrt{3}$. (b) $a \neq \pm\sqrt{3}$.
(c) There is no value of a such that this system has infinitely many solutions.

17. (a) $a = -3$. (b) $a \neq \pm 3$. (c) $a = 3$.

23. (a) $\begin{bmatrix} 1 & 0 & 0 & 0 & 0 \\ 2 & 1 & 0 & 0 & 0 \\ 3 & \frac{5}{3} & 1 & 0 & 0 \end{bmatrix}$. (b) $\begin{bmatrix} 1 & 0 & 0 & 0 & 0 \\ 0 & 1 & 0 & 0 & 0 \\ 0 & 0 & 1 & 0 & 0 \end{bmatrix}$.

Section 1.6, p. 69

3. (a) $\begin{bmatrix} 1 & -4 & 0 & 0 \\ 0 & 1 & 0 & 0 \\ 0 & 0 & 1 & 0 \\ 0 & 0 & 0 & 1 \end{bmatrix}$. (b) $\begin{bmatrix} 1 & 0 & 0 & 0 \\ 0 & 0 & 1 & 0 \\ 0 & 1 & 0 & 0 \\ 0 & 0 & 0 & 1 \end{bmatrix}$.

(c) $\begin{bmatrix} 1 & 0 & 0 & 0 \\ 0 & 1 & 0 & 0 \\ 0 & 0 & 4 & 0 \\ 0 & 0 & 0 & 1 \end{bmatrix}$.

5. $A^{-1} = \begin{bmatrix} -2 & \frac{3}{2} \\ 1 & -\frac{1}{2} \end{bmatrix}$.

7. (a) Singular. (b) $A^{-1} = \begin{bmatrix} \frac{1}{2} & -\frac{1}{4} \\ \frac{1}{6} & \frac{1}{12} \end{bmatrix}$.

(c) $A^{-1} = \begin{bmatrix} 0 & 1 & -1 \\ 2 & -2 & -1 \\ -1 & 1 & 1 \end{bmatrix}$.

(d) Singular.

9. (a) $A^{-1} = \begin{bmatrix} 1 & 0 & -1 \\ 1 & -1 & 2 \\ -1 & 1 & -1 \end{bmatrix}$.

(b) $A^{-1} = \begin{bmatrix} \frac{7}{3} & -\frac{1}{3} & -\frac{1}{3} & -\frac{2}{3} \\ \frac{4}{9} & -\frac{1}{9} & -\frac{4}{9} & \frac{1}{9} \\ -\frac{1}{9} & -\frac{2}{9} & \frac{1}{9} & \frac{2}{9} \\ -\frac{5}{3} & \frac{2}{3} & \frac{2}{3} & \frac{1}{3} \end{bmatrix}$.

(c) Singular.

(d) $A^{-1} = \begin{bmatrix} \frac{3}{2} & -1 & \frac{1}{2} \\ \frac{1}{2} & 0 & -\frac{1}{2} \\ -\frac{3}{2} & 1 & \frac{1}{2} \end{bmatrix}$.

(e) Singular.

11. $A = \begin{bmatrix} 1 & 0 \\ 3 & 1 \end{bmatrix}\begin{bmatrix} 1 & 0 \\ 0 & -2 \end{bmatrix}\begin{bmatrix} 1 & 2 \\ 0 & 1 \end{bmatrix}$.

13. $A = \begin{bmatrix} \frac{1}{2} & -1 \\ -\frac{1}{2} & 2 \end{bmatrix}$.

15. (a) and (b).

17. A^{-1} exists for $a \neq 0$. Then

$$A^{-1} = \begin{bmatrix} 0 & 1 & 0 \\ 1 & -1 & 0 \\ -\frac{2}{a} & \frac{1}{a} & \frac{1}{a} \end{bmatrix}.$$

Section 1.7, p. 75

3. (a) $\begin{bmatrix} 1 & 0 & 0 & 0 \\ 0 & 1 & 0 & 0 \\ 0 & 0 & 0 & 0 \end{bmatrix}$. (b) $\begin{bmatrix} 1 & 0 & 0 \\ 0 & 1 & 0 \\ 0 & 0 & 1 \\ 0 & 0 & 0 \end{bmatrix}$.

(c) $\begin{bmatrix} 1 & 0 & 0 & 0 \\ 0 & 1 & 0 & 0 \\ 0 & 0 & 1 & 0 \\ 0 & 0 & 0 & 0 \\ 0 & 0 & 0 & 0 \end{bmatrix}$. (d) I_3.

7. $B = \begin{bmatrix} 1 & 0 & 0 & 0 \\ 0 & 1 & 0 & 0 \\ 0 & 0 & 0 & 0 \\ 0 & 0 & 0 & 0 \end{bmatrix}$, $P = \begin{bmatrix} -1 & 1 & 0 & 0 \\ 2 & -1 & 0 & 0 \\ -4 & 0 & 1 & 0 \\ 0 & 0 & 0 & 1 \end{bmatrix}$,

$Q = \begin{bmatrix} 1 & 0 & -1 & 2 \\ 0 & 1 & 1 & 5 \\ 0 & 0 & 1 & 0 \\ 0 & 0 & 0 & 1 \end{bmatrix}$.

Section 1.8, p. 85

1. $X = \begin{bmatrix} 1 \\ 2 \\ 1 \end{bmatrix}$.

3. $X = \begin{bmatrix} 1 \\ 0 \\ 2 \\ -4 \end{bmatrix}$.

5. $L = \begin{bmatrix} 1 & 0 & 0 \\ 2 & 1 & 0 \\ 2 & -2 & 1 \end{bmatrix}$, $U = \begin{bmatrix} 2 & 3 & 4 \\ 0 & -1 & 2 \\ 0 & 0 & -2 \end{bmatrix}$,

$X = \begin{bmatrix} 4 \\ -2 \\ 1 \end{bmatrix}$.

7. $L = \begin{bmatrix} 1 & 0 & 0 \\ 0.5 & 1 & 0 \\ 0.25 & -1.5 & 1 \end{bmatrix}$, $U = \begin{bmatrix} 4 & 2 & 3 \\ 0 & -1 & 3.5 \\ 0 & 0 & 5.5 \end{bmatrix}$,

$X = \begin{bmatrix} 2 \\ -2 \\ -1 \end{bmatrix}$.

9. $L = \begin{bmatrix} 1 & 0 & 0 & 0 \\ 0.5 & 1 & 0 & 0 \\ -1 & 0.2 & 1 & 0 \\ 2 & -0.4 & 2 & 1 \end{bmatrix}$,

$U = \begin{bmatrix} 2 & 1 & 0 & -4 \\ 0 & -0.5 & 0.25 & 1 \\ 0 & 0 & 0.2 & 2 \\ 0 & 0 & 0 & 2 \end{bmatrix}$, $X = \begin{bmatrix} 0.5 \\ 2 \\ -2 \\ 1.5 \end{bmatrix}$.

Supplementary Exercises, p. 86

1. (a) 3. (b) 6. (c) 10. (d) $\frac{n}{2}(n+1)$.

3. $\begin{bmatrix} 1 & 0 \\ 0 & 1 \end{bmatrix}$, $\begin{bmatrix} -1 & 0 \\ 0 & -1 \end{bmatrix}$, $\begin{bmatrix} 1 & b \\ 0 & -1 \end{bmatrix}$, $\begin{bmatrix} -1 & b \\ 0 & 1 \end{bmatrix}$,

where b is any real number.

13. (a) $a = -4$ or $a = 2$.
(b) Any real number.

Chapter 2

Section 2.1, p. 101

1.

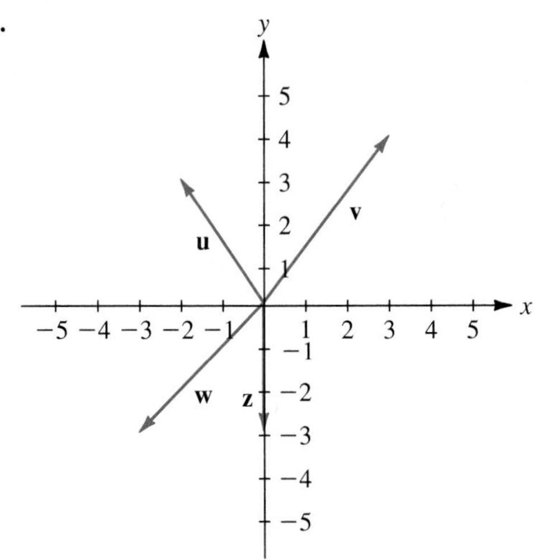

3. Tail at $(-1, -4)$.

5. $a = 3$, $b = -1$.

7. (a) $\begin{bmatrix} 2 \\ 3 \end{bmatrix}$. (b) $\begin{bmatrix} -1 \\ 3 \\ -1 \end{bmatrix}$.

9. (a) $\begin{bmatrix} 4 \\ 3 \end{bmatrix}$. (b) $\begin{bmatrix} 2 \\ 3 \\ 7 \end{bmatrix}$.

11. (a) $\mathbf{u} + \mathbf{v} = \begin{bmatrix} 0 \\ 8 \end{bmatrix}$, $\mathbf{u} - \mathbf{v} = \begin{bmatrix} 4 \\ -2 \end{bmatrix}$,

$2\mathbf{u} = \begin{bmatrix} 4 \\ 6 \end{bmatrix}$, $3\mathbf{u} - 2\mathbf{v} = \begin{bmatrix} 10 \\ -1 \end{bmatrix}$.

(b) $\mathbf{u} + \mathbf{v} = \begin{bmatrix} 3 \\ 5 \end{bmatrix}$, $\mathbf{u} - \mathbf{v} = \begin{bmatrix} -3 \\ 1 \end{bmatrix}$,

$2\mathbf{u} = \begin{bmatrix} 0 \\ 6 \end{bmatrix}$, $3\mathbf{u} - 2\mathbf{v} = \begin{bmatrix} -6 \\ 5 \end{bmatrix}$.

(c) $\mathbf{u} + \mathbf{v} = \begin{bmatrix} 5 \\ 8 \end{bmatrix}$, $\mathbf{u} - \mathbf{v} = \begin{bmatrix} -1 \\ 4 \end{bmatrix}$,

$2\mathbf{u} = \begin{bmatrix} 4 \\ 12 \end{bmatrix}$, $3\mathbf{u} - 2\mathbf{v} = \begin{bmatrix} 0 \\ 14 \end{bmatrix}$.

13. (a) $\begin{bmatrix} 1 \\ 5 \\ 3 \end{bmatrix}$. (b) $\begin{bmatrix} -4 \\ -3 \\ -1 \end{bmatrix}$.

(c) $\begin{bmatrix} 1 \\ 6 \\ 2 \end{bmatrix}$. (d) $\begin{bmatrix} -7 \\ 1 \\ 13 \end{bmatrix}$.

15. (a) $r = \frac{1}{2}$, $s = \frac{3}{2}$.
(b) $r = -2$, $s = 1$, $t = -1$.
(c) $r = -2$, $s = 6$.

17. Impossible.

19. Possible answer: $c_1 = -2$, $c_2 = -1$, $c_3 = 1$.

Section 2.2, p. 114

3. Properties (3), (4), (b), (5), (6), and (7).

5. Properties (5), (6), and (8).

7. Property (8).

13. No.

17. (a) and (c).

19. (b) and (c).

21. (a).

23. (a).

35. (a) and (c).

37. (a) Possible answer: $x = 2 + 2t$, $y = -3 + 5t$, $z = 1 + 4t$.
(b) Possible answer: $x = -3 + 8t$, $y = -2 + 7t$, $z = -2 + 6t$.

Section 2.3, p. 129

1. (b) and (c).

3. (a) Yes. (b) No. (c) Yes. (d) No.

5. (a), (c), and (d).

7. (a) and (d).

9. No.

11. Possible answer: $\left\{\begin{bmatrix} -1 \\ -1 \\ 1 \\ 0 \end{bmatrix}\right\}$.

13. No.

15. (a) and (c) are linearly dependent; (b) is linearly independent.
(a) $[3\ \ 6\ \ 6] = 2[1\ \ 1\ \ 0] + 1[0\ \ 2\ \ 3] + 1[1\ \ 2\ \ 3]$.
(c) $[0\ \ 0\ \ 0] = 0[1\ \ 1\ \ 0] + 0[0\ \ 2\ \ 3] + 0[1\ \ 2\ \ 3]$.

17. (a) and (b) are linearly independent; (c) is linearly dependent: $t + 13 = 3(2t^2 + t + 1) - 2(3t^2 + t - 5)$.

19. (b) is linearly dependent:

$$\begin{bmatrix} 1 \\ 2 \\ -2 \end{bmatrix} = 2\begin{bmatrix} 1 \\ 1 \\ -1 \end{bmatrix} + \begin{bmatrix} 0 \\ 1 \\ 1 \end{bmatrix} - \begin{bmatrix} 1 \\ 1 \\ 1 \end{bmatrix}.$$

Section 2.4, p. 146

1. (a) and (d).

3. (a) and (d).

5. (c).

7. (a) is a basis for R^3 and

$$\begin{bmatrix} 2 \\ 1 \\ 3 \end{bmatrix} = \frac{3}{2}\begin{bmatrix} 1 \\ 1 \\ 1 \end{bmatrix} + \frac{1}{2}\begin{bmatrix} 1 \\ 2 \\ 3 \end{bmatrix} - \frac{3}{2}\begin{bmatrix} 0 \\ 1 \\ 0 \end{bmatrix}.$$

9. (a) forms a basis.
$5t^2 - 3t + 8 = 5(t^2 + t) - 8(t - 1)$.

11. $\left\{\begin{bmatrix} 1 \\ 2 \\ 2 \end{bmatrix}, \begin{bmatrix} 3 \\ 2 \\ 1 \end{bmatrix}\right\}$; $\dim W = 2$.

13. $\{t^3 + t^2 - 2t + 1, t^2 + 1\}$.

15. (a) $\left\{\begin{bmatrix} 1 \\ 1 \\ 0 \end{bmatrix}, \begin{bmatrix} 0 \\ 1 \\ 1 \end{bmatrix}\right\}$. (b) $\left\{\begin{bmatrix} 1 \\ 1 \\ 0 \end{bmatrix}, \begin{bmatrix} 0 \\ 0 \\ 1 \end{bmatrix}\right\}$.

17. (a) 3. (b) 2.

19. (a) 2. (b) 1. (c) 2. (d) 2.

21. (a) 4. (b) 3. (c) 3. (d) 4.

23. $\{t^3 + t, t^2 - t, t^3, 1\}$.

25. $\left\{\begin{bmatrix} -2 \\ 1 \\ 0 \\ 0 \\ 0 \end{bmatrix}, \begin{bmatrix} -3 \\ 0 \\ 1 \\ 1 \\ 0 \end{bmatrix}\right\}$; dim $V = 2$.

27. $\left\{\begin{bmatrix} -1 \\ 1 \\ 0 \\ 1 \\ 0 \end{bmatrix}, \begin{bmatrix} -2 \\ \frac{1}{2} \\ 0 \\ 0 \\ 1 \end{bmatrix}\right\}$; dim $V = 2$.

Section 2.5, p. 164

1. (a) $[\mathbf{v}]_T = \begin{bmatrix} -7 \\ 4 \end{bmatrix}$; $[\mathbf{w}]_T = \begin{bmatrix} 7 \\ -1 \end{bmatrix}$.

(b) $P = \begin{bmatrix} 1 & 2 \\ -1 & -1 \end{bmatrix}$.

(c) $[\mathbf{v}]_S = \begin{bmatrix} 1 \\ 3 \end{bmatrix}$; $[\mathbf{w}]_S = \begin{bmatrix} 5 \\ -6 \end{bmatrix}$.

(d) Same as (c).

(e) $Q = \begin{bmatrix} -1 & -2 \\ 1 & 2 \end{bmatrix}$.

(f) Same as (a).

3. (a) $[\mathbf{v}]_T = \begin{bmatrix} 3 \\ 2 \\ -7 \end{bmatrix}$; $[\mathbf{w}]_T = \begin{bmatrix} 2 \\ 3 \\ -3 \end{bmatrix}$.

(b) $P = \begin{bmatrix} 2 & 1 & 0 \\ 1 & -\frac{2}{5} & \frac{3}{5} \\ 0 & \frac{2}{5} & \frac{2}{5} \end{bmatrix}$.

(c) $[\mathbf{v}]_S = \begin{bmatrix} 8 \\ -2 \\ -2 \end{bmatrix}$; $[\mathbf{w}]_S = \begin{bmatrix} 7 \\ -1 \\ 0 \end{bmatrix}$.

(d) Same as (c).

(e) $Q = \begin{bmatrix} \frac{1}{3} & \frac{1}{3} & -\frac{1}{2} \\ \frac{1}{3} & -\frac{2}{3} & 1 \\ -\frac{1}{3} & \frac{2}{3} & \frac{3}{2} \end{bmatrix}$.

(f) $[\mathbf{v}]_T = Q[\mathbf{v}]_S = \begin{bmatrix} 3 \\ 2 \\ -7 \end{bmatrix}$.

$[\mathbf{w}]_T = Q[\mathbf{w}]_S = \begin{bmatrix} 2 \\ 3 \\ -3 \end{bmatrix}$; same as (a).

5. (a) $[\mathbf{v}]_T = \begin{bmatrix} 1 \\ 1 \\ 1 \\ 0 \end{bmatrix}$; $[\mathbf{w}]_T = \begin{bmatrix} 2 \\ -2 \\ 1 \\ -1 \end{bmatrix}$.

(b) $P = \begin{bmatrix} 1 & 0 & 0 & 1 \\ \frac{1}{3} & \frac{2}{3} & -\frac{2}{3} & 0 \\ \frac{1}{3} & -\frac{1}{3} & \frac{1}{3} & 0 \\ -\frac{1}{3} & \frac{1}{3} & \frac{2}{3} & 0 \end{bmatrix}$.

(c) $[\mathbf{v}]_S = \begin{bmatrix} 1 \\ \frac{1}{3} \\ \frac{1}{3} \\ \frac{2}{3} \end{bmatrix}$; $[\mathbf{w}]_S = \begin{bmatrix} 1 \\ -\frac{4}{3} \\ \frac{5}{3} \\ -\frac{2}{3} \end{bmatrix}$.

(d) Same as (c).

(e) $Q = \begin{bmatrix} 0 & 1 & 2 & 0 \\ 0 & 1 & 0 & 1 \\ 0 & 0 & 1 & 1 \\ 1 & -1 & -2 & 0 \end{bmatrix}$.

(f) Same as (a).

9. Let $L\colon R_n \to R^n$ be defined by

$L([a_1\ a_2\ \ldots\ a_n]) = \begin{bmatrix} a_1 \\ a_2 \\ \vdots \\ a_n \end{bmatrix}$. Verify that L is an isomorphism.

11. (b) 4.

15. (a) $t^2 - 3t + 2$. (b) $t^2 + 1$. (c) $t^2 - t + 2$.

Section 2.6, p. 175

1. A possible basis is $\{\mathbf{v}_1, \mathbf{v}_2, \mathbf{v}_3\}$, where

$$\mathbf{v}_1 = \begin{bmatrix} 1 \\ 0 \\ 0 \end{bmatrix}, \mathbf{v}_2 = \begin{bmatrix} 0 \\ 1 \\ 0 \end{bmatrix}, \text{ and } \mathbf{v}_3 = \begin{bmatrix} 0 \\ 0 \\ 1 \end{bmatrix}.$$

(a) $\begin{bmatrix} 3 \\ 4 \\ 12 \end{bmatrix} = 3\mathbf{v}_1 + 4\mathbf{v}_2 + 12\mathbf{v}_3$.

(b) $\begin{bmatrix} 3 \\ 2 \\ 2 \end{bmatrix} = 3\mathbf{v}_1 + 2\mathbf{v}_2 + 2\mathbf{v}_3$.

(c) $\begin{bmatrix} 1 \\ 2 \\ 6 \end{bmatrix} = \mathbf{v}_1 + 2\mathbf{v}_2 + 6\mathbf{v}_3$.

3. A possible answer is $\left\{\begin{bmatrix} 1 & 0 \\ 0 & 0 \end{bmatrix}, \begin{bmatrix} 0 & 1 \\ 0 & 0 \end{bmatrix}, \begin{bmatrix} 0 & 0 \\ 1 & 0 \end{bmatrix}, \begin{bmatrix} 0 & 0 \\ 0 & 1 \end{bmatrix}\right\}$.

5. (a) 3. (b) 5.

9. (a) 2. (b) 3.

11. B and C are equivalent; A, D, and E are equivalent.

13. Neither.

15. (b).

17. (b).

19. (a) 2. (b) 3.

Supplementary Exercises, p. 178

1. (b) $k = 0$.

3. (a) No. (b) Yes. (c) Yes.

9. $a = 1$ or $a = 2$.

11. $k \neq 1, -1$.

15. (a) $\mathbf{b} = \begin{bmatrix} a \\ b \\ c \end{bmatrix}$, where $b + c - 3a = 0$.

(b) Any $\mathbf{b}$.

23. (a) $T = \left\{\begin{bmatrix} 2 \\ 2 \\ 1 \end{bmatrix}, \begin{bmatrix} 3 \\ 1 \\ 2 \end{bmatrix}, \begin{bmatrix} 5 \\ 4 \\ 4 \end{bmatrix}\right\}$.

(b) $S = \left\{\begin{bmatrix} \frac{6}{5} \\ -\frac{2}{5} \\ -\frac{4}{5} \end{bmatrix}, \begin{bmatrix} \frac{2}{5} \\ \frac{1}{5} \\ -\frac{4}{5} \end{bmatrix}, \begin{bmatrix} -\frac{1}{5} \\ \frac{2}{5} \\ -\frac{3}{5} \end{bmatrix}\right\}$.

Chapter 3

Section 3.1, p. 190

1. (a) 1. (b) 0. (c) $\sqrt{5}$.

3. (a) 1. (b) $\sqrt{2}$.

5. (a) $\sqrt{74}$. (b) $\sqrt{58}$

7. (a) 0. (b) 3. (c) 8.

9. (a) $\dfrac{-14}{\sqrt{5}\sqrt{41}}$. (b) $\dfrac{-6}{\sqrt{5}\sqrt{41}}$.

11. (a) 1, 0, 0.

(b) $\dfrac{1}{\sqrt{14}}, \dfrac{3}{\sqrt{14}}, \dfrac{2}{\sqrt{14}}$.

(c) $\dfrac{-1}{\sqrt{14}}, \dfrac{-2}{\sqrt{14}}, \dfrac{-3}{\sqrt{14}}$.

(d) $\frac{4}{\sqrt{29}}, \frac{-3}{\sqrt{29}}, \frac{2}{\sqrt{29}}$.

17. (a) $\mathbf{v}_1$ and $\mathbf{v}_4$, $\mathbf{v}_1$ and $\mathbf{v}_6$, $\mathbf{v}_3$ and $\mathbf{v}_4$, $\mathbf{v}_3$ and $\mathbf{v}_6$, $\mathbf{v}_4$ and $\mathbf{v}_5$, $\mathbf{v}_5$ and $\mathbf{v}_6$.
(b) $\mathbf{v}_1$ and $\mathbf{v}_5$, $\mathbf{v}_4$ and $\mathbf{v}_6$.
(c) $\mathbf{v}_1$ and $\mathbf{v}_3$, $\mathbf{v}_3$ and $\mathbf{v}_5$.

19. (b).

21.

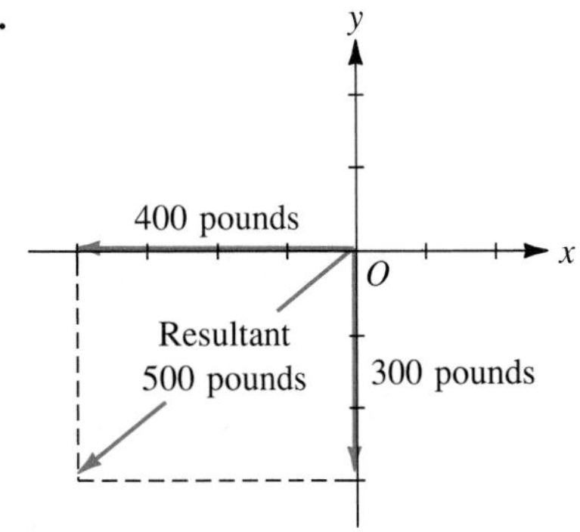

25. $\frac{5}{2}$.

27. $a = -2$, $b = 2$.

Section 3.2, p. 202

1. (a) $-15\mathbf{i} - 2\mathbf{j} + 9\mathbf{k}$.
(b) $-3\mathbf{i} + 3\mathbf{j} + 3\mathbf{k}$.
(c) $7\mathbf{i} + 5\mathbf{j} - \mathbf{k}$.
(d) $0\mathbf{i} + 0\mathbf{j} + 0\mathbf{k}$.

5. (a) $(\mathbf{u} \times \mathbf{v}) \cdot \mathbf{w} = 24$.
(b) $\mathbf{u} \times (\mathbf{v} \times \mathbf{w}) = 27\mathbf{i} + 9\mathbf{j} - 15\mathbf{k}$.

13. $\frac{3}{2}\sqrt{10}$.

15. 1.

17. (a).

19. (a) $x - z + 2 = 0$.
(b) $3x + y - 14z + 47 = 0$.

21. $4x - 4y + z + 16 = 0$.

23. $x = -2 + 2t$, $y = 5 - 3t$, $z = -3 + 4t$.

Section 3.3, p. 216

9. (a) -8. (b) 0. (c) 1.

11. (a) $\frac{3}{2}$. (b) 1. (c) $\frac{1}{2}\sin^2 1$.

13. (a) 2. (b) $2\sqrt{5}$. (c) $2\sqrt{2}$.

15. (a) 1. (b) 0. (c) $\frac{2\sin^2 1}{\sqrt{4 - \sin^2 2}}$.

23. If $\mathbf{u} = [u_1 \quad u_2]$ and $\mathbf{v} = [v_1 \quad v_2]$, then $(\mathbf{u}, \mathbf{v}) = 3u_1v_1 - 2u_1v_2 - 2u_2v_1 + 3u_2v_2$.

25. (a) $\sqrt{1 - \sin^2 1}$. (b) $\sqrt{\frac{1}{30}}$.

27. (a) Orthonormal. (b) Neither. (c) Neither.

29. $a = 0$.

31. $a = 5$.

33. $B = \begin{bmatrix} b_{11} & b_{12} \\ b_{21} & b_{22} \end{bmatrix}$ with $b_{11} + 3b_{21} + 2b_{12} + 4b_{22} = 0$.

Section 3.4, p. 232

1. (a) $\left\{\begin{bmatrix} 1 \\ 2 \end{bmatrix}, \begin{bmatrix} -4 \\ 2 \end{bmatrix}\right\}$.
(b) $\left\{\frac{1}{\sqrt{5}}\begin{bmatrix} 1 \\ 2 \end{bmatrix}, \frac{1}{\sqrt{5}}\begin{bmatrix} -2 \\ 1 \end{bmatrix}\right\}$.

3. $\left\{\frac{1}{\sqrt{3}}[1 \quad 1 \quad -1 \quad 0], \frac{1}{\sqrt{33}}[-2 \quad 4 \quad 2 \quad 3]\right\}$.

5. $\{\sqrt{3}t, 2 - 3t\}$.

7. $\left\{\sqrt{3}t, \dfrac{\sin 2\pi t + (3/2\pi)t}{\sqrt{\frac{1}{2} - 3/4\pi^2}}\right\}$.

9. $\left\{\begin{bmatrix} \frac{2}{3} \\ -\frac{2}{3} \\ \frac{1}{3} \end{bmatrix}, \begin{bmatrix} \frac{2}{3} \\ \frac{1}{3} \\ -\frac{2}{3} \end{bmatrix}, \begin{bmatrix} \frac{1}{3} \\ \frac{2}{3} \\ \frac{2}{3} \end{bmatrix}\right\}$.

11. Possible answer: $\left\{\dfrac{1}{\sqrt{3}}\begin{bmatrix} 1 \\ 1 \\ 1 \end{bmatrix}, \dfrac{1}{\sqrt{6}}\begin{bmatrix} -1 \\ -1 \\ 2 \end{bmatrix}, \dfrac{1}{\sqrt{2}}\begin{bmatrix} -1 \\ 1 \\ 0 \end{bmatrix}\right\}$.

17. (b) $\begin{bmatrix} 9 \\ 9 \\ 9 \end{bmatrix}$.

(c) $\|\mathbf{v}\| = 9\sqrt{3}$.

23. (a) $2 \sin t$.

(b) $\dfrac{\pi^2}{3} - 4 \cos t$.

(c) $\left(\dfrac{e^{\pi} - e^{-\pi}}{2\pi}\right) + \left(\dfrac{e^{-\pi} - e^{\pi}}{2\pi}\right) \cos t + \left(\dfrac{e^{\pi} - e^{-\pi}}{2\pi}\right) \sin t$.

25. 2.

Section 3.5, p. 240

3. $\hat{x} \approx \begin{bmatrix} -1.5333 \\ -1.8667 \\ 4.2667 \end{bmatrix}$.

7. $\hat{x} = \begin{bmatrix} -\frac{5}{11} \\ \frac{4}{11} \\ 0 \end{bmatrix}$.

Supplementary Exercises, p. 241

1. $\left\{\begin{bmatrix} 2 \\ 1 \\ 0 \end{bmatrix}, \begin{bmatrix} -\frac{1}{5} \\ \frac{2}{5} \\ 1 \end{bmatrix}\right\}$.

3. $\mathbf{v} = -\sqrt{2}\begin{bmatrix} \frac{1}{\sqrt{2}} \\ 0 \\ -\frac{1}{\sqrt{2}} \end{bmatrix} + 2\begin{bmatrix} 0 \\ 1 \\ 0 \end{bmatrix} + 2\sqrt{2}\begin{bmatrix} \frac{1}{\sqrt{2}} \\ 0 \\ \frac{1}{\sqrt{2}} \end{bmatrix}$.

5. Vector in P closest to $\mathbf{u}$ is $\dfrac{1}{122}\begin{bmatrix} 171 \\ 223 \\ 66 \end{bmatrix}$; distance is $\dfrac{7}{\sqrt{122}}$.

9. (a) Possible answer: $\left\{\dfrac{1}{\sqrt{30}}\begin{bmatrix} -5 \\ 2 \\ 1 \\ 0 \end{bmatrix}, \dfrac{1}{\sqrt{30}}\begin{bmatrix} 2 \\ 5 \\ 0 \\ 1 \end{bmatrix}\right\}$.

(b) Possible answer: $\left\{\dfrac{1}{\sqrt{30}}\begin{bmatrix} -5 \\ 2 \\ 1 \\ 0 \end{bmatrix}, \dfrac{1}{\sqrt{255}}\begin{bmatrix} -5 \\ -14 \\ 3 \\ 5 \end{bmatrix}\right\}$.

11. (a) $\begin{bmatrix} -1 \\ 0 \\ 1 \end{bmatrix}$.

(c) (i) $\mathbf{v} = \dfrac{1}{2}\begin{bmatrix} 1 \\ 0 \\ 1 \end{bmatrix}$, $\mathbf{w} = -\dfrac{1}{2}\begin{bmatrix} -1 \\ 0 \\ 1 \end{bmatrix}$;

(ii) $\mathbf{v} = \begin{bmatrix} 2 \\ 2 \\ 2 \end{bmatrix}$, $\mathbf{w} = \begin{bmatrix} -1 \\ 0 \\ 1 \end{bmatrix}$.

13. $\left\{\dfrac{1}{\sqrt{2}}, \sqrt{\dfrac{3}{2}}t, \sqrt{\dfrac{5}{8}}(3t^2 - 1)\right\}$.

Chapter 4

Section 4.1, p. 253

1. (b).

3. (a) and (b).

7. (a) $\begin{bmatrix} 15 & 5 & 4 & 8 \\ -5 & -1 & 10 & 2 \end{bmatrix}$.

9. (a) $\begin{bmatrix} 2 \\ 15 \end{bmatrix}$. (b) $\begin{bmatrix} 2a_1 + 3a_2 + 2a_3 \\ -4a_1 - 5a_2 + 3a_3 \end{bmatrix}$.

11. (a) $2t^3 - 5t^2 + 2t + 3$.
(b) $at^3 + bt^2 + at + c$.

13. a = any real number, $b = 0$.

15. (a) Reflection about the y-axis.

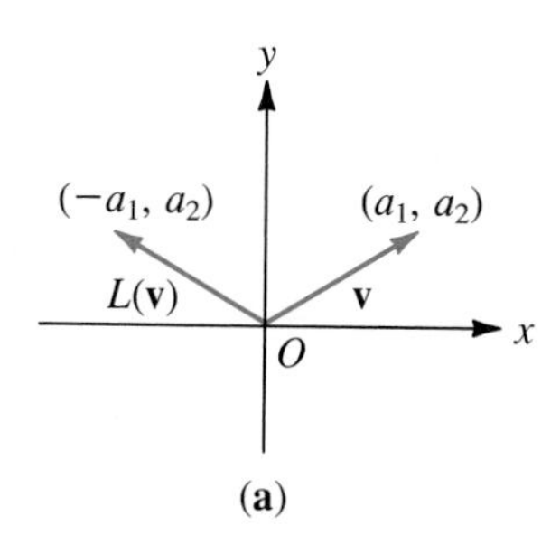

(a)

(b) Reflection about the origin.

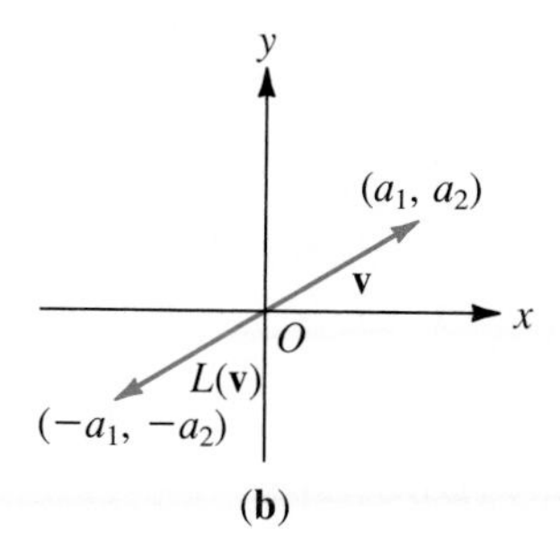

(b)

(c) Counterclockwise rotation through $\pi/2$ radians.

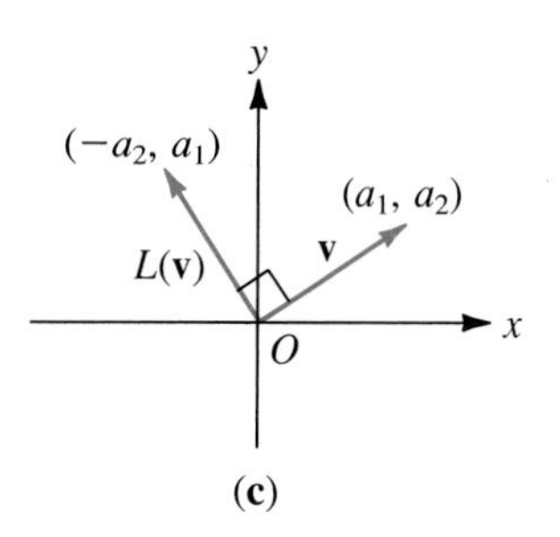

(c)

Section 4.2, p. 267

1. (a) Yes. (b) No. (c) Yes. (d) No.

(e) All vectors $\begin{bmatrix} 0 \\ a \end{bmatrix}$, where a is any real number; that is, the y-axis.

(f) All vectors $\begin{bmatrix} a \\ 0 \end{bmatrix}$, where a is any real number; that is, the x-axis.

3. (a) No. (b) Yes. (c) Yes. (d) Yes.
(e) All vectors of the form $[-r \quad -s \quad r \quad s]$, where r and s are real numbers.
(f) $\{[1 \quad 0], [0 \quad 1]\}$.

5. (a) $\{[1 \quad -1 \quad -1 \quad 1]\}$. (b) 1.
(c) $\{[1 \quad 0 \quad 1], [1 \quad 0 \quad 0], [0 \quad 1 \quad 1]\}$.
(d) 3.

7. (a) 0. (b) 6.

9. (a) dim ker $L = 1$, dim range $L = 2$.
(b) dim ker $L = 1$, dim range $L = 2$.
(c) dim ker $L = 0$, dim range $L = 4$.

13. No.

15. (b) $L^{-1}\left(\begin{bmatrix} 2 \\ 3 \\ 4 \end{bmatrix}\right) = \begin{bmatrix} \frac{3}{2} \\ -\frac{1}{2} \\ \frac{1}{2} \end{bmatrix}$.

17. 1.

19. (b) $\begin{bmatrix} 2a_1 - a_3 \\ -2a_1 - a_2 + 2a_3 \\ a_1 + a_2 - a_3 \end{bmatrix}$.

21. (a) 2. (b) 1.

Section 4.3, p. 279

1. (a) $\begin{bmatrix} 1 & 2 \\ 2 & -1 \end{bmatrix}$. (b) $\begin{bmatrix} 1 & -\frac{1}{2} \\ 1 & \frac{3}{4} \end{bmatrix}$.

(c) $\begin{bmatrix} 3 & 2 \\ -4 & 4 \end{bmatrix}$. (d) $\begin{bmatrix} -2 & 2 \\ \frac{1}{2} & 2 \end{bmatrix}$.

(e) $\begin{bmatrix} 5 \\ 0 \end{bmatrix}$.

3. (a) $\begin{bmatrix} 1 & 0 & 1 & 1 \\ 0 & 1 & 2 & 1 \\ -1 & -2 & 1 & 0 \end{bmatrix}$.

(b) $\begin{bmatrix} 1 & 0 & 2 & 1 \\ 1 & 1 & 3 & 3 \\ -5 & -3 & -4 & -5 \end{bmatrix}$.

5. (a) $\begin{bmatrix} 1 & 2 & 1 \\ 1 & 0 & 0 \\ 0 & 1 & 1 \end{bmatrix}$. (b) $\begin{bmatrix} 8 \\ 1 \\ 5 \end{bmatrix}$.

7. (a) $\begin{bmatrix} 10 \\ 5 \\ 5 \end{bmatrix}$. (b) $\begin{bmatrix} 4 \\ 2 \\ 2 \end{bmatrix}$.

9. $\begin{bmatrix} 0 & 1 & 0 & 0 \\ 0 & 0 & 0 & 0 \\ 0 & 0 & 1 & 1 \\ 0 & 0 & 0 & 1 \end{bmatrix}$.

11. (a) $\begin{bmatrix} 0 & -3 & 2 & 0 \\ -2 & -3 & 0 & 2 \\ 3 & 0 & 3 & -3 \\ 0 & 3 & -2 & 0 \end{bmatrix}$.

(b) $\begin{bmatrix} 0 & 3 & -2 & 3 \\ 0 & -9 & -2 & -6 \\ 0 & 3 & 6 & 0 \\ 0 & 4 & 0 & 3 \end{bmatrix}$.

(c) $\begin{bmatrix} 0 & 3 & 2 & 0 \\ -3 & -6 & 1 & 3 \\ 3 & 0 & 3 & -3 \\ 1 & 3 & -1 & -1 \end{bmatrix}$.

(d) $\begin{bmatrix} 0 & -3 & 2 & -3 \\ 0 & -5 & -2 & -3 \\ 0 & 3 & 6 & 0 \\ 0 & 3 & -2 & 3 \end{bmatrix}$.

13. (a) $\begin{bmatrix} 1 & 0 \\ 0 & -1 \end{bmatrix}$. (b) $\begin{bmatrix} 0 & -1 \\ -1 & 0 \end{bmatrix}$.

(c) $\begin{bmatrix} \frac{1}{2} & -\frac{1}{2} \\ -\frac{1}{2} & -\frac{1}{2} \end{bmatrix}$. (d) $\begin{bmatrix} 1 & -1 \\ -1 & -1 \end{bmatrix}$.

17. (a) $\begin{bmatrix} 1 & 0 \\ 0 & 1 \end{bmatrix}$. (b) $\begin{bmatrix} 1 & 0 \\ 0 & 1 \end{bmatrix}$.

(c) $\begin{bmatrix} \frac{3}{5} & -\frac{2}{5} \\ \frac{1}{5} & \frac{1}{5} \end{bmatrix}$. (d) $\begin{bmatrix} 1 & 2 \\ -1 & 3 \end{bmatrix}$.

19. $\begin{bmatrix} 0 & -1 \\ 1 & 0 \end{bmatrix}$.

Section 4.4, p. 289

3. (a) $[-a_1 + 4a_2 - a_3 \quad 3a_1 - a_2 + 3a_3 \quad 4a_1 + 3a_2 + 5a_3]$.

(b) $[5 \quad -4 \quad -4]$.

(c) $\begin{bmatrix} -1 & 4 & -1 \\ 3 & -1 & 3 \\ 4 & 3 & 5 \end{bmatrix}$.

(d) $[2a_1 + 2a_3 \quad -4a_1 - 2a_2 - 2a_3 \quad -2a_1 - 4a_2 - 6a_3]$.

(e) $[14 \quad -28 \quad 14]$.

(f) $\begin{bmatrix} 2 & 16 & 2 \\ 0 & -10 & 6 \\ 10 & 0 & 2 \end{bmatrix}$.

5. (a) $\begin{bmatrix} 5 \\ 10 \\ 0 \end{bmatrix}$.

(b) $\ker L_1$ = all vectors of the form $\begin{bmatrix} -r - s \\ r \\ s \end{bmatrix}$,

$\ker L_2$ = all vectors of the form $\begin{bmatrix} -r - 2s \\ r \\ s \end{bmatrix}$,

$\ker L_1 \cap \ker L_2$ = all vectors of the form $\begin{bmatrix} -r \\ r \\ 0 \end{bmatrix}$.

(c) All vectors of the form $\begin{bmatrix} -r \\ r \\ 0 \end{bmatrix}$.

(d) They are the same.

7. (a) $C = \begin{bmatrix} 2 & 2 \\ 1 & -1 \\ -1 & 1 \\ 1 & -1 \end{bmatrix}$.

(b) $A = \begin{bmatrix} 1 & 1 \\ \frac{2}{3} & -\frac{2}{3} \\ \frac{1}{3} & -\frac{1}{3} \end{bmatrix}$, $B = \begin{bmatrix} 2 & 0 & 0 \\ 0 & 1 & 1 \\ 0 & -1 & -1 \\ 0 & 1 & 1 \end{bmatrix}$.

9. (a) $\begin{bmatrix} 2 & 8 & -2 \\ 4 & 2 & 6 \\ 2 & -2 & 4 \end{bmatrix}$. (b) $\begin{bmatrix} 10 & 17 & 17 \\ 11 & 8 & 13 \\ 3 & -1 & 4 \end{bmatrix}$.

11. (a) 6. (b) 6. (c) 12. (d) 12.

15. (a) $L(t^2) = t + 3$, $L(t) = 2t + 4$, $L(1) = -2t - 1$.
(b) $(a + 2b - 2c)t + (3a + 4b - c)$.
(c) $-16t - 18$.

17. Possible answers: $L\left(\begin{bmatrix} a_1 \\ a_2 \end{bmatrix}\right) = \begin{bmatrix} a_2 \\ a_1 \end{bmatrix}$;

$L\left(\begin{bmatrix} a_1 \\ a_2 \end{bmatrix}\right) = \begin{bmatrix} -a_1 \\ -a_2 \end{bmatrix}$.

19. Possible answers: $L\left(\begin{bmatrix} a_1 \\ a_2 \end{bmatrix}\right) = \begin{bmatrix} a_1 \\ 0 \end{bmatrix}$;

$L\left(\begin{bmatrix} a_1 \\ a_2 \end{bmatrix}\right) = \begin{bmatrix} \dfrac{a_1 + a_2}{2} \\ \dfrac{a_1 + a_2}{2} \end{bmatrix}$.

21. $\begin{bmatrix} 2 & 0 & -1 \\ -2 & -1 & 2 \\ 1 & 1 & -1 \end{bmatrix}$.

Section 4.5, p. 298

3. $\begin{bmatrix} -2 & 2 \\ \frac{1}{2} & 2 \end{bmatrix}$.

5. (a) $A = \begin{bmatrix} 1 & 0 \\ 0 & -1 \end{bmatrix}$. (b) $B = \begin{bmatrix} \frac{1}{3} & \frac{4}{3} \\ \frac{2}{3} & -\frac{1}{3} \end{bmatrix}$.

(c) $P = \begin{bmatrix} 1 & 1 \\ -1 & 2 \end{bmatrix}$.

9. $\begin{bmatrix} \frac{13}{2} & \frac{1}{2} & -\frac{3}{2} \\ -\frac{5}{2} & \frac{1}{2} & -\frac{1}{2} \end{bmatrix}$.

13. $\begin{bmatrix} 1 & 0 \\ 0 & 1 \end{bmatrix}$.

15. $\begin{bmatrix} 0 & -1 \\ 1 & 0 \end{bmatrix}$.

Supplementary Exercises, p. 299

3. No.

5. (a) $8t + 7$. (b) $\frac{1}{2}(3a + b)t + \frac{1}{2}(3a - b)$.

7. (a) No. (b) No. (c) Yes. (d) No.
(e) $\{t^3 + t^2, t + 1\}$.
(f) $\{t^3, t\}$.

9. (b) ker L = the set of all continuous functions f on $[0, 1]$ such that $f(0) = 0$.
(c) Yes.

11. (a) $\begin{bmatrix} 2 & 0 & 1 \\ 0 & 1 & 2 \\ -1 & 0 & -2 \end{bmatrix}$. (b) $4t^2 - 4t + 1$.

13. $\begin{bmatrix} 0 & 0 & 0 & 0 \\ 3 & 0 & 0 & 0 \\ 0 & 2 & 0 & 0 \\ 0 & 0 & 1 & 0 \end{bmatrix}$.

Chapter 5

Section 5.1, p. 305

1. (a) 5. (b) 7. (c) 4.

3. (a) Even. (b) Odd. (c) Even.

5. (a) $-$. (b) $+$. (c) $-$.

7. (a) 9.
(b) Number of inversions in 416235 is 6; number of inversions in 436215 is 9.

9. (a) 0. (b) 0.

11. (a) 9. (b) 0. (c) 144.

13. (a) $t^2 - 3t - 4$. (b) $t^3 - 4t^2 + 3t$.

15. (a) $t = 4$, $t = -1$. (b) $t = 1$, $t = 0$, $t = 3$.

Section 5.2, p. 314

1. (a) 3. (b) 2. (c) 24. (d) 29.
(e) 4. (f) -30.

3. 3.

5. (a) 2. (b) -120. (c) $(t - 1)(t - 2)(t - 3) = t^3 - 6t^2 + 11t - 6$. (d) $t^2 - 2t - 11$.

21. 32.

23. (b) is nonsingular.

25. The system has a nontrivial solution.

Section 5.3, p. 320

1. (a) 1. (b) 7. (c) 2. (d) 10.

3. (a) -2. (b) 9. (c) -2. (d) -1.

5. (a) 3. (d) 29. (e) 4.

7. (a) 4. (c) -30. (f) 0.

11. (a) $t = 0$, $t = 5$. (b) $t = 1$, $t = 4$.

Section 5.4, p. 326

3. (a) $\begin{bmatrix} 24 & -42 & -30 \\ 19 & -2 & -30 \\ -4 & 32 & 30 \end{bmatrix}$. (b) 150.

5. (a) $A^{-1} = \begin{bmatrix} 1 & 0 & -1 \\ 1 & -1 & 2 \\ -1 & 1 & -1 \end{bmatrix}$.

(b) $A^{-1} = \begin{bmatrix} \frac{7}{3} & -\frac{1}{3} & -\frac{1}{3} & -\frac{2}{3} \\ \frac{4}{9} & -\frac{1}{9} & -\frac{4}{9} & \frac{1}{9} \\ -\frac{1}{9} & -\frac{2}{9} & \frac{1}{9} & \frac{2}{9} \\ -\frac{5}{3} & \frac{2}{3} & \frac{2}{3} & \frac{1}{3} \end{bmatrix}$.

(c) Singular. (d) $A^{-1} = \begin{bmatrix} \frac{3}{2} & -1 & \frac{1}{2} \\ \frac{1}{2} & 0 & -\frac{1}{2} \\ -\frac{3}{2} & 1 & \frac{1}{2} \end{bmatrix}$.

(e) Singular.

7. (a) $-\frac{1}{28}\begin{bmatrix} -30 & -5 & 9 & 46 \\ -32 & 4 & 4 & 36 \\ -12 & -2 & -2 & 24 \\ 16 & -2 & -2 & -32 \end{bmatrix}$.

(b) $\begin{bmatrix} \frac{3}{14} & -\frac{3}{7} & \frac{1}{7} \\ \frac{1}{7} & \frac{5}{7} & -\frac{4}{7} \\ -\frac{1}{14} & \frac{1}{7} & \frac{2}{7} \end{bmatrix}$. (c) $\begin{bmatrix} \frac{2}{9} & -\frac{1}{9} \\ \frac{1}{6} & \frac{1}{6} \end{bmatrix}$.

9. $\begin{bmatrix} \dfrac{d}{ad-bc} & \dfrac{-b}{ad-bc} \\ \dfrac{-c}{ad-bc} & \dfrac{a}{ad-bc} \end{bmatrix}$.

11. $\begin{bmatrix} \frac{1}{4} & 0 & 0 \\ 0 & -\frac{1}{3} & 0 \\ 0 & 0 & \frac{1}{2} \end{bmatrix}$.

Section 5.5, p. 332

1. $x_1 = -2$, $x_2 = 0$, $x_3 = 1$.

3. Yes.

5. $x_1 = 1$, $x_2 = 2$, $x_3 = -2$.

7. $x_1 = x_2 = x_3 = 0$.

9. $x_1 = 1$, $x_2 = \frac{2}{3}$, $x_3 = -\frac{2}{3}$.

11. Yes.

13. For $c \neq \pm 2$.

15. (a) $-15\mathbf{i} - 2\mathbf{j} + 9\mathbf{k}$. (b) $-3\mathbf{i} + 3\mathbf{j} + 3\mathbf{k}$.
(c) $7\mathbf{i} + 5\mathbf{j} - \mathbf{k}$. (d) $3\mathbf{i} - 8\mathbf{j} - \mathbf{k}$.

Supplementary Exercises, p. 334

1. (a) 5. (b) 4. (c) 36. (d) 5.

Chapter 6

Section 6.1, p. 357

1. The only eigenvalue of L is $\lambda = -1$. Every nonzero vector in R^2 is an eigenvector of L associated with λ.

3. (a) $f(\lambda) = \lambda^3$. The eigenvalues of A are $\lambda_1 = \lambda_2 = \lambda_3 = 0$. Associated eigenvectors are

$$\mathbf{x}_1 = \mathbf{x}_2 = \mathbf{x}_3 = r\begin{bmatrix} 1 \\ 0 \\ 0 \end{bmatrix},$$

where r = any nonzero real number.

(b) $f(\lambda) = (\lambda - 1)(\lambda - 3)(\lambda + 2)$. The eigenvalues are $\lambda_1 = 1$, $\lambda_2 = 3$, and $\lambda_3 = -2$. Associated eigenvectors are

$$\mathbf{x}_1 = \begin{bmatrix} 6 \\ 3 \\ 8 \end{bmatrix}, \quad \mathbf{x}_2 = \begin{bmatrix} 0 \\ 5 \\ 2 \end{bmatrix}, \quad \text{and}$$

$$\mathbf{x}_3 = \begin{bmatrix} 0 \\ 0 \\ 1 \end{bmatrix}.$$

(c) $f(\lambda) = \lambda^2 - 2\lambda$. The eigenvalues are $\lambda_1 = 0$ and $\lambda_2 = 2$. Associated eigenvectors are

$$\mathbf{x}_1 = \begin{bmatrix} 1 \\ -1 \end{bmatrix} \quad \text{and} \quad \mathbf{x}_2 = \begin{bmatrix} 1 \\ 1 \end{bmatrix}.$$

5. (a) $f(\lambda) = \lambda^2 - 5\lambda + 6$. The eigenvalues are $\lambda_1 = 2$ and $\lambda_2 = 3$. Associated eigenvectors are

$$\mathbf{x}_1 = \begin{bmatrix} 1 \\ -1 \end{bmatrix} \quad \text{and} \quad \mathbf{x}_2 = \begin{bmatrix} 1 \\ -2 \end{bmatrix}.$$

(b) $f(\lambda) = \lambda^3 - 7\lambda^2 + 14\lambda - 8$. The eigenvalues are $\lambda_1 = 1$, $\lambda_2 = 2$, and $\lambda_3 = 4$. Associated eigenvectors are

$$\mathbf{x}_1 = \begin{bmatrix} -1 \\ 1 \\ 1 \end{bmatrix}, \quad \mathbf{x}_2 = \begin{bmatrix} 1 \\ 0 \\ 0 \end{bmatrix}, \quad \text{and} \quad \mathbf{x}_3 = \begin{bmatrix} 7 \\ -4 \\ 2 \end{bmatrix}.$$

(c) $f(\lambda) = \lambda^3 - 5\lambda^2 + 2\lambda + 8$. The eigenvalues are $\lambda_1 = -1$, $\lambda_2 = 2$, and $\lambda_3 = 4$. Associated eigenvectors are

$$\mathbf{x}_1 = \begin{bmatrix} 1 \\ 0 \\ -1 \end{bmatrix}, \quad \mathbf{x}_2 = \begin{bmatrix} -2 \\ -3 \\ 2 \end{bmatrix}, \quad \text{and} \quad \mathbf{x}_3 = \begin{bmatrix} 8 \\ 5 \\ 2 \end{bmatrix}.$$

7. L is not diagonalizable. The eigenvalues of L are $\lambda_1 = \lambda_2 = \lambda_3 = 0$. The set of associated eigenvectors do not form a basis for P_2.

11. The eigenvalues of A^2 are $\lambda_1 = 9$ and $\lambda_2 = 4$. Associated eigenvectors are

$$\mathbf{x}_1 = \begin{bmatrix} 1 \\ -1 \end{bmatrix} \quad \text{and} \quad \mathbf{x}_2 = \begin{bmatrix} 4 \\ 1 \end{bmatrix}.$$

13. No.

19. The eigenvalues of L are $\lambda_1 = 2$, $\lambda_2 = -3$, and $\lambda_3 = 4$. Associated eigenvectors are $\mathbf{x}_1 = t^2$, $\mathbf{x}_2 = t^2 - 5t$, and $\mathbf{x}_3 = 9t^2 + 4t + 14$.

21. Necessary and sufficient conditions are $(a - d)^2 + 4bc > 0$ or $b = c = 0$.

25. (a) and (c).

27. $\begin{bmatrix} 768 & -1280 \\ 256 & -768 \end{bmatrix}$.

Section 6.2, p. 372

7. $P^TP = I_3$.

9. (a) If B is the given matrix, verify that $B^TB = I_2$.

15. A is similar to $D = \begin{bmatrix} 0 & 0 \\ 0 & 4 \end{bmatrix}$ and

$$P = \begin{bmatrix} \frac{1}{\sqrt{2}} & \frac{1}{\sqrt{2}} \\ -\frac{1}{\sqrt{2}} & \frac{1}{\sqrt{2}} \end{bmatrix}.$$

17. A is similar to $D = \begin{bmatrix} 0 & 0 & 0 \\ 0 & 0 & 0 \\ 0 & 0 & 4 \end{bmatrix}$ and

$$P = \begin{bmatrix} 1 & 0 & 0 \\ 0 & -\frac{1}{\sqrt{2}} & \frac{1}{\sqrt{2}} \\ 0 & \frac{1}{\sqrt{2}} & \frac{1}{\sqrt{2}} \end{bmatrix}.$$

19. A is similar to

$$D = \begin{bmatrix} -2 & 0 & 0 \\ 0 & 1 & 0 \\ 0 & 0 & 1 \end{bmatrix} \quad \text{and}$$

$$P = \begin{bmatrix} \frac{1}{\sqrt{3}} & -\frac{1}{\sqrt{2}} & -\frac{1}{\sqrt{6}} \\ \frac{1}{\sqrt{3}} & \frac{1}{\sqrt{2}} & -\frac{1}{\sqrt{6}} \\ \frac{1}{\sqrt{3}} & 0 & \frac{2}{\sqrt{6}} \end{bmatrix}.$$

21. A is similar to $D = \begin{bmatrix} 3 & 0 \\ 0 & 1 \end{bmatrix}$.

23. A is similar to $D = \begin{bmatrix} 1 & 0 & 0 \\ 0 & 2 & 0 \\ 0 & 0 & 0 \end{bmatrix}$.

25. A is similar to $D = \begin{bmatrix} 1 & 0 & 0 \\ 0 & 0 & 0 \\ 0 & 0 & 2 \end{bmatrix}$.

27. A is similar to $D = \begin{bmatrix} 2 & 0 & 0 \\ 0 & -2 & 0 \\ 0 & 0 & 4 \end{bmatrix}$.

Section 6.3, p. 384

1. (a) $\begin{bmatrix} x & y \end{bmatrix} \begin{bmatrix} -3 & \frac{5}{2} \\ \frac{5}{2} & -2 \end{bmatrix} \begin{bmatrix} x \\ y \end{bmatrix}$.

(b) $\begin{bmatrix} x_1 & x_2 & x_3 \end{bmatrix} \begin{bmatrix} 2 & \frac{3}{2} & -\frac{5}{2} \\ \frac{3}{2} & 0 & \frac{7}{2} \\ -\frac{5}{2} & \frac{7}{2} & 0 \end{bmatrix} \begin{bmatrix} x_1 \\ x_2 \\ x_3 \end{bmatrix}$.

(c) $\begin{bmatrix} x_1 & x_2 & x_3 \end{bmatrix} \begin{bmatrix} 3 & \frac{1}{2} & -\frac{1}{2} \\ \frac{1}{2} & 1 & -2 \\ -\frac{1}{2} & -2 & -2 \end{bmatrix} \begin{bmatrix} x_1 \\ x_2 \\ x_3 \end{bmatrix}$.

3. (a) $\begin{bmatrix} -1 & 0 & 0 \\ 0 & 2 & 0 \\ 0 & 0 & 0 \end{bmatrix}$. (b) $\begin{bmatrix} 3 & 0 & 0 \\ 0 & 0 & 0 \\ 0 & 0 & 0 \end{bmatrix}$.

5. $3x'^2 - 2y'^2$.

7. $y_1^2 - y_3^2$.

9. $-2y_1^2 + 5y_2^2 - 5y_3^2$.

11. y'^2.

13. $y_1^2 + y_2^2 - y_3^2$.

15. $y_1^2 - y_2^2$.

17. $y_1^2 - y_2^2$, rank $= 2$, signature $= 0$.

21. g_1, g_2, and g_4.

23. (a), (b), and (c).

Section 6.4, p. 395

1. Ellipse.

3. Hyperbola.

5. Two intersecting lines.

7. Circle.

9. Point.

11. Ellipse; $\dfrac{x'^2}{2} + y'^2 = 1$.

13. Hyperbola; $\dfrac{x'^2}{5^2} - \dfrac{y'^2}{5^2} = 1$.

15. Pair of parallel lines; $y' = 2$, $y' = -2$; $y'^2 = 4$.

17. Point (1, 3); $x'^2 + y'^2 = 0$.

19. Possible answer: ellipse; $\dfrac{x'^2}{12} + \dfrac{y'^2}{4} = 1$.

21. Possible answer: pair of parallel lines $y' = \dfrac{2}{5}$ and $y' = -\dfrac{2}{5}$; $y'^2 = \dfrac{4}{10}$.

23. Possible answer: two intersecting lines $y' = 3x'$ and $y' = -3x'$; $9x'^2 - y'^2 = 0$.

25. Possible answer: parabola; $y''^2 = -4x''$.

27. Possible answer: hyperbola; $\dfrac{x''^2}{4} - \dfrac{y''^2}{9} = 1$.

29. Possible answer: hyperbola; $\dfrac{x''^2}{\frac{9}{8}} - \dfrac{y''^2}{\frac{9}{8}} = 1$.

Exercise 6.5, p. 407

1. Hyperboloid of one sheet.

3. Hyperbolic paraboloid.

5. Parabolic cylinder.

7. Parabolic cylinder.

9. Ellipsoid.

11. Elliptic paraboloid.

13. Hyperbolic paraboloid.

15. Ellipsoid; $x'^2 + y'^2 + \dfrac{z'^2}{\frac{1}{3}} = 1$.

17. Hyperbolic paraboloid; $\dfrac{x''^2}{4} - \dfrac{y''^2}{4} = z''$.

19. Elliptic paraboloid; $\dfrac{x'^2}{4} + \dfrac{y'^2}{8} = 1$.

21. Hyperboloid of one sheet; $\dfrac{x''^2}{2} + \dfrac{y''^2}{4} - \dfrac{z''}{4} = 1$.

23. Parabolic cylinder; $x''^2 = \dfrac{4}{\sqrt{2}}y''$.

25. Hyperboloid of two sheets;
$\dfrac{x''^2}{\frac{21}{4}} - \dfrac{y''^2}{\frac{21}{4}} - \dfrac{z''^2}{\frac{21}{4}} = 1$.

27. Cone; $x''^2 + y''^2 - z''^2 = 0$.

Supplementary Exercises, p. 407

1. (a) $\lambda_1 = 1$, $\lambda_2 = 1$, $\lambda_3 = 4$; corresponding

eigenvectors $\begin{bmatrix}1\\0\\0\end{bmatrix}, \begin{bmatrix}0\\0\\1\end{bmatrix}, \begin{bmatrix}4\\3\\0\end{bmatrix}$.

(b) $\lambda_1 = 3$, $\lambda_2 = 4$, $\lambda_3 = -1$; corresponding

eigenvectors $\begin{bmatrix}0\\0\\1\end{bmatrix}, \begin{bmatrix}1\\-1\\0\end{bmatrix}, \begin{bmatrix}2\\3\\0\end{bmatrix}$.

(c) $\lambda_1 = 1$, $\lambda_2 = 2$, $\lambda_3 = 3$; corresponding

eigenvectors $\begin{bmatrix}1\\1\\1\end{bmatrix}, \begin{bmatrix}1\\2\\4\end{bmatrix}, \begin{bmatrix}1\\3\\9\end{bmatrix}$.

(d) $\lambda_1 = -3$, $\lambda_2 = 1$, $\lambda_3 = -1$; corresponding

eigenvectors $\begin{bmatrix}1\\-3\\9\end{bmatrix}, \begin{bmatrix}1\\1\\1\end{bmatrix}, \begin{bmatrix}-1\\1\\-1\end{bmatrix}$.

7. (a) $\begin{bmatrix}-\frac{3}{10}\\ \frac{1}{5}\end{bmatrix}, \begin{bmatrix}-\frac{6}{5}\\ \frac{4}{5}\end{bmatrix}$.

(b) $A = \dfrac{1}{10}\begin{bmatrix}-3 & -12\\ 2 & 8\end{bmatrix}$.

(c) The eigenvalues are $\lambda_1 = 0$, $\lambda_2 = \frac{1}{2}$. Associated eigenvectors are

$$\mathbf{x}_1 = \begin{bmatrix}-4\\1\end{bmatrix} \quad \text{and} \quad \mathbf{x}_2 = \begin{bmatrix}-3\\2\end{bmatrix}.$$

(d) The eigenvalues are $\lambda_1 = 0$ and $\lambda_2 = \frac{1}{2}$. Associated eigenvectors are $x_1 = 5t - 5$ and $x_2 = 5t$.

(e) The eigenspace for $\lambda_1 = 0$ is the subspace of P_1 with basis $\{5t - 5\}$. The eigenspace for $\lambda_2 = \frac{1}{2}$ is the subspace of P_1 with basis $\{5t\}$.

9. The only eigenvalue is $\lambda_1 = 0$ and an associated eigenvector is $p_1(x) = 1$.

Chapter 7

Chapter 7, p. 426

1. (a) $\mathbf{x}(t) = \begin{bmatrix}x_1(t)\\x_2(t)\\x_3(t)\end{bmatrix} = \begin{bmatrix}b_1e^{-3t}\\b_2e^{4t}\\b_3e^{2t}\end{bmatrix}$

$$= b_1 \begin{bmatrix} 1 \\ 0 \\ 0 \end{bmatrix} e^{-3t} + b_2 \begin{bmatrix} 0 \\ 1 \\ 0 \end{bmatrix} e^{4t} + b_3 \begin{bmatrix} 0 \\ 0 \\ 1 \end{bmatrix} e^{2t}.$$

(b) $$\begin{bmatrix} 3e^{-3t} \\ 4e^{4t} \\ 5e^{2t} \end{bmatrix} = 3 \begin{bmatrix} 1 \\ 0 \\ 0 \end{bmatrix} e^{-3t} + 4 \begin{bmatrix} 0 \\ 1 \\ 0 \end{bmatrix} e^{4t} + 5 \begin{bmatrix} 0 \\ 0 \\ 1 \end{bmatrix} e^{2t}.$$

3. $$\mathbf{x}(t) = b_1 \begin{bmatrix} 6 \\ 2 \\ 7 \end{bmatrix} e^{4t} + b_2 \begin{bmatrix} 0 \\ 7 \\ -1 \end{bmatrix} e^{-5t} + b_3 \begin{bmatrix} 0 \\ 0 \\ 1 \end{bmatrix} e^{2t}.$$

5. $$\mathbf{x}(t) = b_1 \begin{bmatrix} 1 \\ 0 \\ 0 \end{bmatrix} e^{5t} + b_2 \begin{bmatrix} 0 \\ 1 \\ 3 \end{bmatrix} e^{5t} + b_3 \begin{bmatrix} 0 \\ -3 \\ 1 \end{bmatrix} e^{-5t}.$$

7. $$\mathbf{x}(t) = b_1 \begin{bmatrix} 1 \\ 0 \\ 0 \end{bmatrix} e^{-2t} + b_2 \begin{bmatrix} 1 \\ 1 \\ 2 \end{bmatrix} e^{2t} + b_3 \begin{bmatrix} -7 \\ -2 \\ 1 \end{bmatrix} e^{-3t}.$$

9. (a) $$\begin{bmatrix} x_1' \\ x_2' \end{bmatrix} = \begin{bmatrix} 0 & 1 \\ -1 & 2 \end{bmatrix} \begin{bmatrix} x_1 \\ x_2 \end{bmatrix}.$$

(b) $$\begin{bmatrix} x_1' \\ x_2' \\ x_3' \end{bmatrix} = \begin{bmatrix} 0 & 1 & 0 \\ 0 & 0 & 1 \\ 1 & -1 & 2 \end{bmatrix} \begin{bmatrix} x_1 \\ x_2 \\ x_3 \end{bmatrix}.$$

(c) $$\begin{bmatrix} x_1' \\ x_2' \\ x_3' \end{bmatrix} = \begin{bmatrix} 0 & 1 & 0 \\ 0 & 0 & 1 \\ 1 & 0 & 0 \end{bmatrix} \begin{bmatrix} x_1 \\ x_2 \\ x_3 \end{bmatrix}.$$

11. (a) $x(t) = b_1 e^{-t} + b_2 e^{3t}$.
(b) $x(t) = e^{-t} + e^{3t}$.

13. $x(t) = b_1 e^t + b_2 t e^t$.

15. $x(t) = b_1 e^{2t} + b_2 t e^{2t} + b_3 e^{3t} + b_4 t e^{3t}$.

17. $x(t) = b_1 e^{2t} + b_2 e^{-3t} + b_3 e^{4t}$.

19. $$\mathbf{x}(t) = \begin{bmatrix} 440 + 60e^{-5t} \\ 220 - 20e^{-5t} \end{bmatrix}.$$

Supplementary Exercises, p. 427

1. (a) $$\frac{d}{dt}[A(t)] = \begin{bmatrix} 2t & \dfrac{-1}{(t+1)^2} \\ 0 & -e^{-t} \end{bmatrix}$$

$$\int_0^t A(s)\, ds = \begin{bmatrix} \dfrac{t^3}{3} & \ln(1+t) \\ 4t & -e^{-t}+1 \end{bmatrix}.$$

(b) $$\frac{d}{dt}[A(t)] = \begin{bmatrix} 2\cos 2t & 0 & 0 \\ 0 & 0 & -1 \\ 0 & e^{t^2} + 2t^2 e^{t^2} & \dfrac{1-t^2}{(t^2+1)^2} \end{bmatrix}.$$

$$\int_0^t A(s)\, ds = \begin{bmatrix} -\dfrac{\cos 2t}{2} + \dfrac{1}{2} & 0 & 0 \\ 0 & t & -\dfrac{t^2}{2} \\ 0 & \dfrac{e^{t^2}}{2} - \dfrac{1}{2} & \ln(t^2+1) \end{bmatrix}.$$

9. (a) $$\mathbf{x}(t) = \tfrac{2}{5} \begin{bmatrix} 4 \\ 4 \\ 1 \end{bmatrix} + \tfrac{7}{20} \begin{bmatrix} -1 \\ -6 \\ 1 \end{bmatrix} e^{5t} + \tfrac{1}{4} \begin{bmatrix} -1 \\ 2 \\ 1 \end{bmatrix} e^{-3t}.$$

(b) $$\mathbf{x}(t) = \tfrac{7}{8} \begin{bmatrix} 1 \\ 0 \\ 0 \end{bmatrix} + \tfrac{1}{12} \begin{bmatrix} 1 \\ 2 \\ 4 \end{bmatrix} e^{2t} + \tfrac{1}{24} \begin{bmatrix} 1 \\ -4 \\ 16 \end{bmatrix} e^{-4t}.$$

Appendix B

Section B.1, p. 442

1. (a) $4 + 2i$. (b) $-4 - 3i$. (c) $11 - 2i$. (d) $-3 + i$. (e) $-3 + 6i$. (f) $-2 - i$. (g) $7 - 11i$. (h) $-9 + 13i$.

3. (a)

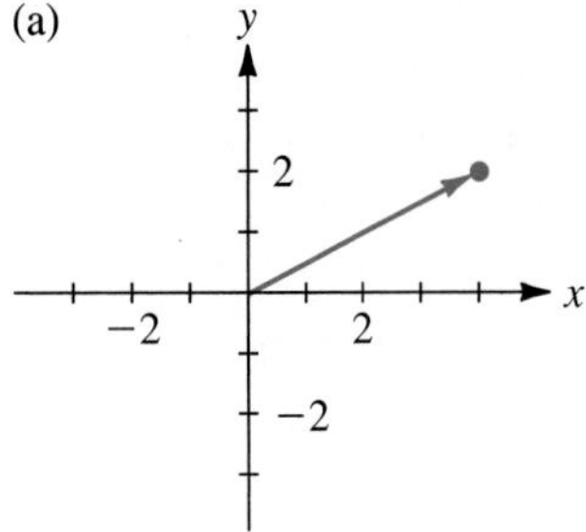

(b)

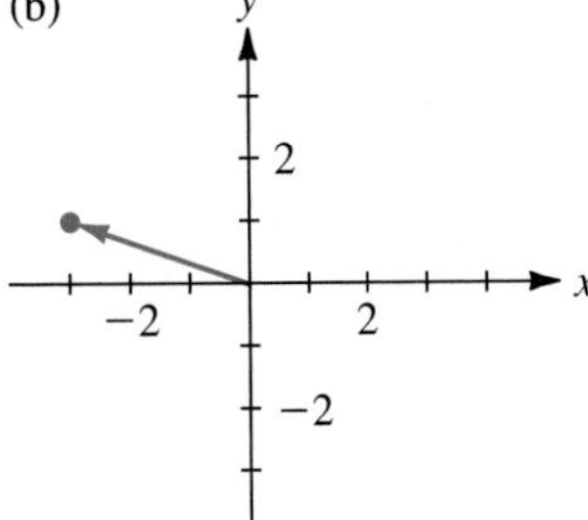

(c)

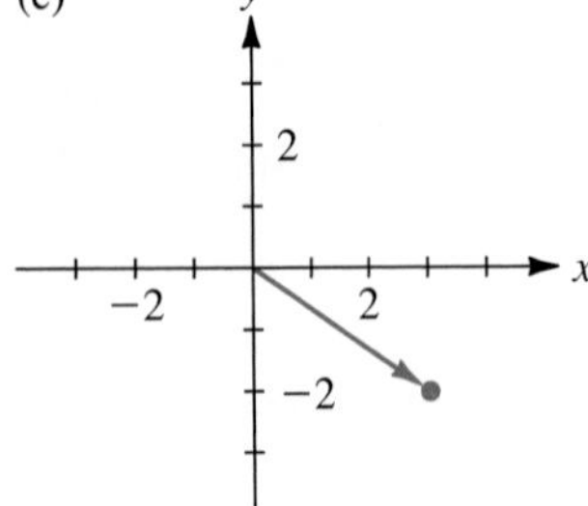

(d)

7. (a) $\begin{bmatrix} 2+4i & 5i \\ -2 & 4-2i \end{bmatrix}$. (b) $\begin{bmatrix} 4-3i \\ -2-i \end{bmatrix}$.

(c) $\begin{bmatrix} -4+4i & -2+16i \\ -4i & -8i \end{bmatrix}$. (d) $\begin{bmatrix} 3i \\ -1-3i \end{bmatrix}$.

(e) $\begin{bmatrix} 2i & -1+3i \\ -2 & -1-i \end{bmatrix}$. (f) $\begin{bmatrix} -2i & 1-2i \\ 0 & 3+i \end{bmatrix}$.

(g) $\begin{bmatrix} 3+i \\ -3+3i \end{bmatrix}$. (h) $\begin{bmatrix} 3-6i \\ -2-6i \end{bmatrix}$.

9. $A^2 = \begin{bmatrix} -1 & 0 \\ 0 & -1 \end{bmatrix}$, $A^3 = \begin{bmatrix} 0 & -i \\ -i & 0 \end{bmatrix}$,

$A^4 = \begin{bmatrix} 1 & 0 \\ 0 & 1 \end{bmatrix}$, $A^{4n} = I_2$, $A^{4n+1} = A$, $A^{4n+2} = A^2 = -I_2$,

$A^{4n+3} = A^3 = -A$.

11. (a) $\begin{bmatrix} 0 & 0 \\ 0 & 0 \end{bmatrix}$. (b) $\begin{bmatrix} 4 & 18 \\ 0 & 4 \end{bmatrix}$.

(c) $\begin{bmatrix} -5 & 5i \\ 5i & -5 \end{bmatrix}$. (d) $\begin{bmatrix} 4 & 7i \\ 0 & -3 \end{bmatrix}$.

13. (a) Possible answers: $\begin{bmatrix} 2 & 0 \\ 0 & 2 \end{bmatrix}$, $\begin{bmatrix} -1 & 0 \\ 0 & -1 \end{bmatrix}$.

Section B.2, p. 453

1. (a) No solution. (b) No solution.

(c) $x_1 = \frac{3}{4} + \frac{5}{4}i$, $x_2 = \frac{3}{2} + i$.

3. (a) $x_1 = i,\ x_2 = 1,\ x_3 = 1 - i$.
(b) $x_1 = 0,\ x_2 = -i,\ x_3 = i$.

5. (a) $\dfrac{1}{5}\begin{bmatrix} 2+i & 2-4i \\ 3-i & -2-i \end{bmatrix}$.

(b) $\dfrac{1}{6}\begin{bmatrix} i & 1-3i & 1 \\ -2-3i & 2i & 3+2i \\ 1 & 2i & -i \end{bmatrix}$.

9. (a) Yes. (b) Linearly independent.

11. (a) The eigenvalues are $\lambda_1 = 1 + i,\ \lambda_2 = 1 - i$. Associated eigenvectors are

$$\mathbf{x}_1 = \begin{bmatrix} -i \\ 1 \end{bmatrix} \quad \text{and} \quad \mathbf{x}_2 = \begin{bmatrix} i \\ 1 \end{bmatrix}.$$

(b) The eigenvalues are $\lambda_1 = 0,\ \lambda_2 = 2$. Associated eigenvectors are

$$\mathbf{x}_1 = \begin{bmatrix} -i \\ 1 \end{bmatrix} \quad \text{and} \quad \mathbf{x}_2 = \begin{bmatrix} i \\ 1 \end{bmatrix}.$$

(c) The eigenvalues are $\lambda_1 = 1,\ \lambda_2 = 2,\ \lambda_3 = 3$. Associated eigenvectors are

$$\mathbf{x}_1 = \begin{bmatrix} 0 \\ -i \\ 1 \end{bmatrix},\ \mathbf{x}_2 = \begin{bmatrix} 1 \\ 0 \\ 0 \end{bmatrix}, \quad \text{and} \quad \mathbf{x}_3 = \begin{bmatrix} 0 \\ i \\ 1 \end{bmatrix}.$$

Index

D

E

F

G

H

I

J

K

L

M

N

O

P

T

U

V

X

Y

Z

Page Index to Lemmas, Theorems, and Corollaries